INSECT–PLANT BIOLOGY

JOIN US ON THE INTERNET VIA WWW, GOPHER, FTP OR EMAIL:

WWW: http://www.thomson.com
GOPHER: gopher.thomson.com
FTP: ftp.thomson.com
EMAIL: findit@kiosk.thomson.com

A service of I(T)P

INSECT–PLANT BIOLOGY

From physiology to evolution

L. M. Schoonhoven
Department of Entomology, Agricultural University, Wageninger, The Netherlands

T. Jermy
Plant Protection Institute, Hungarian Academy of Sciences, Budapest, Hungary

J. J. A. van Loon
Department of Entomology, Agricultural University, Wageningen, The Netherlands

CHAPMAN & HALL
London · Glasgow · New York · Tokyo · Melbourne · Madras

Published by Chapman & Hall, 2–6 Boundary Row, London SE1 8HN, UK

Chapman & Hall, 2–6 Boundary Row, London SE1 8HN, UK

Chapman & Hall GmbH, Pappelallee 3, 69469 Weinheim, Germany

Chapman & Hall USA, 115 Fifth Avenue, New York, NY 10003, USA

Chapman & Hall Japan, ITP-Japan, Kyowa Building, 3F, 2-2-1 Hirakawacho, Chiyoda-ku, Tokyo 102, Japan

Chapman & Hall Australia, 102 Dodds Street, South Melbourne, Victoria 3205, Australia

Chapman & Hall India, R. Seshadri, 32 Second Main Road, CIT East, Madras 600 035, India

First edition 1998

© 1998 L.M. Schoonhoven, T. Jermy and J.J.A. van Loon

Typeset in 10/12pt Palatino by Best-set Typesetter Ltd, Hong Kong
Printed in Great Britain at the University Press, Cambridge

ISBN 0 412 80480 8 (Hb)   0 412 58700 9 (Pb)

Apart from any fair dealing for the purposes of research or private study, or criticism or review, as permitted under the UK Copyright Designs and Patents Act, 1988, this publication may not be reproduced, stored, or transmitted, in any form or by any means, without the prior permission in writing of the publishers, or in the case of reprographic reproduction only in accordance with the terms of the licences issued by the Copyright Licensing Agency in the UK, or in accordance with the terms of licences issued by the appropriate Reproduction Rights Organization outside the UK. Enquiries concerning reproduction outside the terms stated here should be sent to the publishers at the London address printed on this page.
   The publisher makes no representation, express or implied, with regard to the accuracy of the information contained in this book and cannot accept any legal responsibility or liability for any errors or omissions that may be made.

A Catalogue record for this book is available from the British Library

Library of Congress Catalog Card Number: 97-75321

DEDICATION

This book is dedicated to J. de Wilde, V. G. Dethier and J. S. Kennedy, whose pioneering contributions paved the way of modern research in insect-plant interactions.

CONTENTS

Foreword by Professor Sir Richard Southwood   xi

Preface   xiii

1: Introduction   1
    1.1   Increased attention: why?   1
    1.2   Relationships between insects and plants   1
    1.3   Relevance for agriculture   2
    1.4   Insect–plant research involves many biological subdisciplines   3
    1.5   References   4

2: Herbivorous insects: something for everyone   5
    2.1   Host-plant specialization   5
    2.2   Food-plant range and host-plant range   9
    2.3   Specialization on plant parts   11
    2.4   Number of insect species per plant species   13
    2.5   Herbivorous insects: are they plant taxonomists?   15
    2.6   Host plant is more than food plant   16
    2.7   Microclimates around plants   18
    2.8   Extent of insect damage in natural and agricultural ecosystems   18
    2.9   Compensation for herbivore damage   23
    2.10   Conclusions   25
    2.11   References   25

3: Plant chemistry: endless variety   31
    3.1   Plant biochemistry   32
    3.2   Alkaloids   36
    3.3   Terpenoids and steroids   37
    3.4   Phenolics   38
    3.5   Glucosinolates   41
    3.6   Cyanogenics   41
    3.7   Leaf surface chemistry   42
    3.8   Plant volatiles   45
    3.9   Concentrations of secondary plant substances   51
    3.10   Production costs   53
    3.11   Compartmentation   55
    3.12   Temporal variability   57
    3.13   Effects of location and fertilizers   60
    3.14   Induced resistance   64

	3.15	Genotypic variation	70
	3.16	Conclusions	73
	3.17	Literature	73
	3.18	References	74

4: Plants as insect food: not the ideal 83
	4.1	Insect feeding systems	83
	4.2	Plants are suboptimal food	89
	4.3	Artificial diets	93
	4.4	Consumption and utilization	95
	4.5	Symbionts	108
	4.6	Host-plant effects on herbivore susceptibility to pathogens and insecticides	109
	4.7	Plant-mediated effects of air pollution on insects	112
	4.8	Conclusions	113
	4.9	References	113

5: Host-plant selection: how to find a host plant 121
	5.1	Terminology	121
	5.2	Host-plant selection: a catenary process	122
	5.3	Searching mechanisms	126
	5.4	Orientation to host plants	129
	5.5	Chemosensory basis of host-plant odour detection	138
	5.6	Host-plant searching in nature	146
	5.7	Conclusions	148
	5.8	References	148

6: Host-plant selection: when to accept a plant 155
	6.1	Contact phase: plant evaluation	155
	6.2	Physical plant features acting during contact	156
	6.3	Plant chemistry: contact-chemosensory evaluation	159
	6.4	The importance of plant chemistry for host-plant selection: a historical intermezzo	160
	6.5	Stimulation of feeding and oviposition	162
	6.6	Inhibition of feeding and oviposition	169
	6.7	Plant acceptability: a balance between stimulation and deterrence	172
	6.8	Contact-chemosensory basis of host-plant selection behaviour	172
	6.9	Conclusions	186
	6.10	References	187

7: Host-plant selection: why insects do not behave normally 195
	7.1	Geographical variation	195
	7.2	Differences between populations in the same region	198
	7.3	Differences between individuals	199
	7.4	Environmental factors causing changes in host-plant preference	200
	7.5	Internal factors causing changes in host-plant preference	203
	7.6	Experience-induced changes in host-plant preference	204
	7.7	Pre- and early-adult experience	214

	7.8	Adaptive significance of experience-induced changes in host preference	215
	7.9	Genetic variation in host-plant preference	216
	7.10	Conclusions	219
	7.11	References	220

8: The endocrine system of herbivores listens to host-plant signals — 227
- 8.1 Development — 227
- 8.2 Reproduction — 231
- 8.3 Conclusions — 235
- 8.4 References — 236

9: Ecology: living apart together — 239
- 9.1 Host plants affecting herbivorous insect demography — 240
- 9.2 Herbivorous insects affecting plant demography — 244
- 9.3 The composition of insect–plant communities — 247
- 9.4 Interactions in insect–plant communities — 256
- 9.5 Herbivorous insects affecting plant communities — 268
- 9.6 Energy and mineral flow in communities — 270
- 9.7 Conclusions — 272
- 9.8 References — 272

10: Evolution: who drives whom? — 279
- 10.1 Paleontological aspects — 279
- 10.2 Speciation in herbivorous insects — 284
- 10.3 Why does host-plant specificity prevail? — 289
- 10.4 Evolutionary driving forces — 295
- 10.5 Theories — 302
- 10.6 Conclusions — 307
- 10.7 References — 308

11: Insects and flowers: the beauty of mutualism — 315
- 11.1 Mutualism — 317
- 11.2 Flower constancy — 319
- 11.3 Pollination energetics — 324
- 11.4 Pollinator movement within multiple-flower inflorescences — 331
- 11.5 Competition — 331
- 11.6 Evolution — 333
- 11.7 Nature conservation — 337
- 11.8 Economy — 339
- 11.9 Conclusions — 339
- 11.10 References — 339

12: Insects and plants: how to apply our knowledge — 343
- 12.1 Which herbivorous insect species become pests and why? — 343
- 12.2 Host-plant resistance — 346
- 12.3 Polycultures: why fewer pests? — 350
- 12.4 Plant-derived insecticides and antifeedants — 355

x *Contents*

12.5	Weed control by herbivorous insects	359
12.6	Conclusion: diversification holds the clue to the control of pestiferous insects	362
12.7	References	362

Appendix A: Further reading   367
   Books wholly or to a large extent focused on insect–plant interactions   367
   Proceedings of international symposia on insect–plant relationships   368

Appendix B: Structural formulae of selected secondary plant substances   369

Appendix C: Methodology   375
   C.1 Choice of plants and insects   375
   C.2 Behaviour   376
   C.3 Sensory physiology   379
   C.4 Plant chemistry   380
   C.5 References   380

Taxonomic index   385

Author index   391

Subject index   403

FOREWORD

We live in a green world; the organisms that dominate our view are plants. Yet virtually everywhere that there is a plant there will be insects. Though largely unnoticed, they consume on average about 10% of plants' resources. The shift of energy from plants to insects rivals in scale mankind's own demands on the photosynthesizing world. Hence insects most generally come to human notice by competing with us, by eating our crops. Thus insect–plant relationships are fundamental to agricultural science and much research in this area sought, quite simply, to reduce pest damage. However, in the last half century many biologists have recognized that this subject also provides excellent model systems for investigating fundamental aspects of sensory physiology, behaviour, community ecology and evolution. The concept of coevolution was born from studies on the food plants of butterflies. This is not the end of the streams of scientific endeavour that have come together in this book to provide such a stimulating story. With its roots in nineteenth century perfumery and pharmacognosy, the chemistry of secondary plant substances was an arcane field largely closed to entomologists. Biologists will find the clear exposition given here a great help in mastering this aspect. This book provides a unique overview of a subject that has such significance for fundamental biology and for its more applied aspects, whether agriculture, conservation or apiculture. It covers a field in which great strides are being made; the excitement of this progress is captured without losing the sense of wonder that must fortify the investigative drive of the scientist. The authors have been lifelong students of insect–plant relationships and their deep knowledge shines through every page.

Although they disclaim a complete review, which would be well beyond the scope of anything other than a series of volumes, they have sampled the full diversity of existing knowledge – a reflection of their own familiarity with the subject. In successive chapters they lead logically from the composition of plants, through the coming together of insects and plants to the ecology and evolution of these links. In each section the generality of the pattern is revealed, but none of the intricate variation is hidden.

The composition of plants is superficially so similar, but in fact there is endless variation. Both features have contributed to the patterns we see in the physiology and ecology of this relationship. The different theories for host-plant specificity, community structure and evolution are explained. As they are fully justified to do, with their lifetimes of experience, the authors bravely argue their personal opinions on these rival ideas. Although so much is known and so expertly described, one is not left with a sense of a worked-out field. On the contrary, unanswered questions and neglected subjects are drawn to our attention.

In the penultimate chapter the authors turn the coin, so to speak, and review the partnership between flowers and their insect pollinators. In one of the many delightful vignettes that enliven the text, they record how Christian Sprengel, the founder of pollination studies, was encouraged to study nature to alleviate his depression. Research today is not always so relaxing, but with this text as their guide future students will be able to work from a firm knowledge base.

T.R.E. Southwood
Oxford, April 1997

PREFACE

Green plants cover most of the terra firma on planet Earth. Insects are dominant among plant consumers. The interactions between plants trying to avoid consumption, and insects trying to optimize food exploitation, are the subject of this book. It is a rich subject: the primary literature has grown during the past 25 years at an exponential rate. It is also an intellectually challenging subject since, in spite of the wealth of facts, the principles underlying insect–plant interactions are still largely unknown. This book aims to categorize the multitude of facts derived from studies in natural surroundings as well as agricultural environments, and attempts to indicate emerging lines of understanding. Hopefully it will serve as an introduction to students of this area of biology and will highlight to general biologists the complexity of interactions between organisms.

The need for increased agricultural production, together with the necessity to reduce the use of insecticides, forces agricultural entomologists to study how plants in nature have survived insect attack over the aeons, and whether these defence systems can be adopted in agricultural settings. Therefore this book may also be helpful to applied entomologists, who are in search of new ways to protect our daily food production.

The information abounding in the recent literature is too extensive to attempt any complete review. Therefore we have selected studies which are especially appealing to us. In this process we must have missed other equally (or more) important reports and opinions, for which we apologize. In addition to trying to offer an objective representation of facts and thoughts as found in the existing literature we have unavoidably, but also deliberately, given some personal views as well.

We wish to dedicate this book to the memory of three great men who have deeply influenced our thoughts on this subject and who can be considered as founding fathers of the field: Jan de Wilde, Vincent G. Dethier, and John S. Kennedy. Without their foresight, their stimulating enthusiasm, and their perceptiveness of basic mechanisms operative in nature, the field of insect–plant relationships would not have reached its present prominence.

Many people have provided generous assistance in a variety of ways – stimulating discussions, frank criticism, the provision of material for illustrations, and permission to use published diagrams and information. We should especially like to mention those who have read parts of the manuscript and made useful suggestions for improvement: T. A. van Beek, J. Beetsma, M. Dicke, P. Harrewijn, M. van Helden, J. C. van Lenteren S. B. J. Menken, L. Messchendorp, C. Mollema, P. Roessingh, E. Städler, Á. Szentesi, W. F. Tjallingii, and H. H. J. Velthuis. Last but not least our thanks also go to the staff of Chapman & Hall for seeing the book efficiently through production.

Wageningen, Budapest
Autumn 1996

L. M. S.
T. J.
J. J. A. v. L.

INTRODUCTION 1

1.1 Increased attention: why? 1
1.2 Relationships between insects and plants 1
1.3 Relevance for agriculture 2
1.4 Insect–plant research involves many biological subdisciplines 3
1.5 References 4

W. Kirby and W. Spence described more than a century ago in their much read book on entomology [4] the flight of a large white butterfly in pursuit of a suitable plant to lay eggs on: 'she is in search of some plant of the cabbage tribe. Led by an instinct far more unerring than the practised eye of the botanist, she recognizes the desired plant the moment she approaches it; and upon this she places her precious burden.' The authors then pose one of the basic questions, which has occupied scholars of insect–plant relationships till the present day: 'But how is she to distinguish the cabbage plant from the surrounding vegetables?' The answer given: 'She is taught by God!' [4] shows that the precise relationship between herbivorous insects and their host plants has for a long time defied causal analysis. Scientific enquiry into the mechanisms of host-plant selection by herbivorous insects started around the turn of the century [2, 9], but for a long time roused curiosity among only a few biologists. Fairly recently, zoologists have begun the causal analysis of insect behaviour such as host-plant discrimination and gradually some insight has been gained into the underlying mechanisms.

1.1 INCREASED ATTENTION: WHY?

There are several reasons why insect–plant interactions are receiving increasing attention from biologists as well as agronomists. It is now recognized that, from the perspective of fundamental knowledge of earth's biosphere, the relationships between insects and plants are of crucial importance. First there is the quantitative factor: the Plant Kingdom and the class of insects represent two very extensive taxa of living organisms, both in abundance of species and in amount of biomass. Green plants form by far the most voluminous sector of living matter (Fig. 1.1), whereas insects are the leaders in number of species.

As ecologist Robert May [6] puts it: 'To a rough approximation, and setting aside vertebrate chauvinism, it can be said that essentially all organisms are insects'. Certainly not only their variety but also their total volume is colossal, in spite of their small body sizes. For instance, the biomass of all insects in the Brazilian Amazon outweighs that of the total land vertebrate population by about nine to one [3] (Fig. 1.2).

1.2 RELATIONSHIPS BETWEEN INSECTS AND PLANTS

The two empires, herbivorous insects and plants, are united by intricate relationships. Animal life, including that of insects, cannot exist in the absence of green plants, which serve as the primary source of energy-rich compounds for heterotrophic organisms. On the other hand, long-standing exposure to animals has supposedly been a major cause in developing great diversity in the plant world. Insects, with their overwhelming variation in form and life history, may have been one of the forces in shaping the plant world. Such a role has been postulated by Ehrlich and Raven [1]

2 Introduction

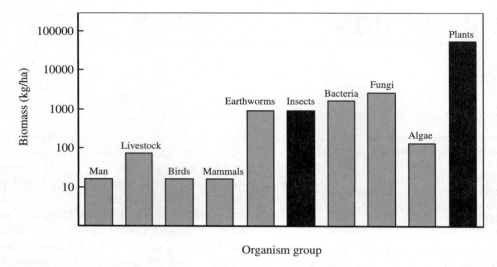

Figure 1.1 The average biomass of man and his livestock, and the estimated biomass of natural biota of some other major groups of organisms per hectare in the USA. Insects include also non-insect arthropods. Note logarithmic scale. (Source: data from Pimentel and Andow, 1984.)

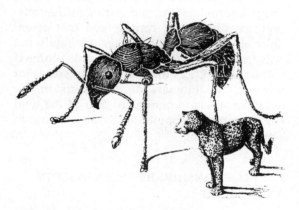

Figure 1.2 In this drawing the ant represents the biomass of all ant populations and the leopard the biomass of all land vertebrates in the Brazilian Amazon. The ants alone outweigh the vertebrates by about four to one. If all other social and non-social insects were included this ratio would be nine to one. (Source: reproduced by courtesy of Ms K. Brown-Wing, Boston, MA.)

who in a seminal paper attributed the plant–herbivore interface as the major zone of interaction for generating the present diversity of terrestrial life forms. The terrestrial flowering plants are the *sine qua non* of the insect tribe, for it is among the insects that feed upon these that herbivory reaches its highest degree of specialization. Such species present a series of complex relationships that are more easily understood if we first consider separately several of their peculiarities. Probably no other interactions between two groups of organisms comparable in type and extent can be found elsewhere in the living world, thus rendering insect–plant interactions an unique and scientifically very fruitful area of biological research.

1.3 RELEVANCE FOR AGRICULTURE

Obviously insect–plant interactions are also of crucial importance from an applied point of view. Insects remain, and may even have increased significance as, the chief pests of crops and stored products, despite expensive and environmentally hazardous control measures (Fig. 1.3).

There is an irrefutable need to better understand the factors governing the relationships between insects and plants, which may help to

1.4 INSECT–PLANT RESEARCH INVOLVES MANY BIOLOGICAL SUBDISCIPLINES

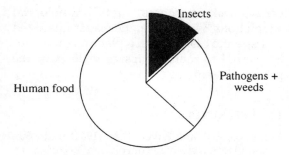

Figure 1.3 Estimated potential crop losses during preharvest to insects (13%), diseases and weeds (24%) in the USA. (Source: data from Pimentel, 1991.)

unravel the causes of insect plague development. Such knowledge is fundamental when attempting to create biologically safe control strategies intended to prevent insect pest outbreaks. The study of insect–plant relationships therefore constitutes, as Lipke and Fraenkel [5] aptly put it, 'the very heart of agricultural entomology'.

Insect–plant interactions include problems at different levels of biological analysis. Questions such as: 'Why do cabbage worms devour cabbage leaves but refuse to eat potato plants?' lie at the level of the organism, whereas the question: 'Why are some forests more prone to insect outbreaks than others?' requires an ecological approach. The focus in this book is upon the mechanistic analysis at the level of the organism. Ecological aspects, however, will not be neglected, because insights derived from studies at the organismal level are often useful elements in ecological models. Another reason for including a discussion of some ecological aspects is that the biological relevance of many facts at the behavioural/physiological level only becomes obvious when put in an ecological perspective.

As in other biological subdisciplines students of insect–plant interactions may be interested in proximate puzzles (how?) or in

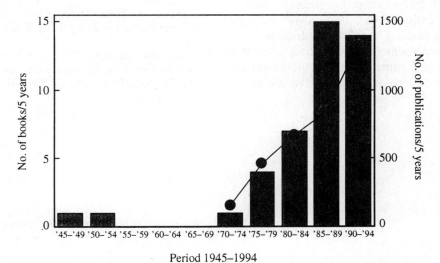

Figure 1.4 The number of books wholly or to a large extent dealing with insect – plant interactions (Appendix A) published during 5 year period ('45–'49 = 1945–1949, etc.). The total period spans 50 years. Dots = primary references (1970–1994) drawn from the BIOSIS database. The figures show all publications that use 'herbivore', 'phytophage' or 'host plant resistance' in the title, abstract or key words and that are cross-referenced in the biosystematic index under Insects. (Source: reproduced by courtesy of Dr J. M. Scriber, East Lansing, MI.)

ultimate factors (why?). Questions like: 'How does an insect recognize its host plant?' and 'How does a herbivore avoid being poisoned by toxic compounds in its food plant?' belong to the first category. Questions like: 'Why do desert plants contain more terpenoids than species occurring in pastures?' and 'To which extent have insects stimulated the evolution of flowering plants?' enquire into ultimate causes. Physiologists are mainly concerned with proximate factors, whereas students of evolution concentrate on finding ultimate causes. Both approaches have their merits, as in fact they are complementary to each other. Both approaches will therefore be employed in this book, but in many instances without explicitly referring to either type.

The topic of insect–plant interactions is too extensive to be covered comprehensively in a book of this size. The rapidly growing interest in this field is evidenced by the real flood of scientific papers, including many extensive reviews and several books, published during the last 5–10 years (Fig. 1.4; Appendix A). The amount of information becoming available cannot be collected, let alone absorbed by a single individual. The following text therefore attempts to present generalizations, illustrated with a limited number of specific examples. Since the species-to-species variation in behavioural responses and physiological adaptations is very great, the reader interested in a specific case of insect–plant relationships is referred to more extensive reviews or the primary literature.

1.5 REFERENCES

1. Ehrlich, P. R. and Raven, P. H. (1964) Butterflies and plants, a study in coevolution. *Evolution*, **18**, 586–608.
2. Errera, L. (1886) Un ordre de recherches trop négligé. L'efficacité des structures défensives des plantes. *Compt. Rend. Soc. Roy. Bot. Belg.*, **25**, 80–99.
3. Holden, C. (1989) Entomologists wane as insects wax. *Science*, **246**, 734–736.
4. Kirby, W. and Spence, W. (1863) *An Introduction to Entomology*, 7th edn, Longman, Green, Longman, Roberts & Green, London.
5. Lipke, H. and Fraenkel, G. S. (1956) Insect nutrition. *Annu. Rev. Entomol.*, **1**, 17–44.
6. May, R. M. (1988) How many species are there on earth? *Science*, **241**, 1441–1449.
7. Pimentel, D. and Andow, D. A. (1984) Pest management and pesticide impacts. *Insect Sci. Appl.*, **5**, 141–149.
8. Pimentel, D. (1991) Diversification of biological control strategies in agriculture. *Crop Prot.*, **10**, 243–253.
9. Verschaffelt, E. (1910) The cause determining the selection of food in some herbivorous insects. *Proc. K. Ned. Akad. Wet.*, **13**, 536–542.

HERBIVOROUS INSECTS: SOMETHING FOR EVERYONE 2

2.1 Host-plant specialization 5
2.2 Food-plant range and host-plant range 9
2.3 Specialization on plant parts 11
2.4 Number of insect species per plant species 13
2.5 Herbivorous insects: are they plant taxonomists? 15
2.6 Host plant is more than food plant 16
2.7 Microclimates around plants 18
2.8 Extent of insect damage in natural and agricultural ecosystems 18
2.9 Compensation for herbivore damage 23
2.10 Conclusions 25
2.11 References 25

Insects have the most species of any class of organisms on earth. Green plants make up the greatest part of all biomass on land. Nearly half of all existing insect species feed on living plants. Thus more than 400000 herbivorous (synonymous with phytophagous) insect species live on roughly 300000 vascular plant species (Fig. 2.1).

According to some recent estimates the total number of insect species is considerably larger than has previously been thought and may reach a figure of 10 million or even more [65, 78]. If this reflects reality the number of vegetarian species probably needs to be adjusted proportionally. Herbivory does not occur in all insect groups to the same extent. The members of some orders of insects are almost exclusively herbivorous, whereas in other orders herbivory occurs less frequently or is even absent. Conspicuous among the herbivores are the Lepidoptera (butterflies and moths), Hemiptera (bugs, leaf-hoppers, aphids, etc.), Orthoptera (grasshoppers and locusts) and some small orders like the Thysanoptera (thrips) and Phasmida (walking sticks). Another great part of the vast horde of herbivorous insects belongs to the large orders Coleoptera, Hymenoptera and Diptera, all of which also include numerous species with predatory and parasitic habits (Table 2.1).

Given the innumerable plant-infesting insect species it is not surprising that indeed all terrestrial tracheophytes (vascular plants) harbour some members of the herbivore tribe. Although at some time it was assumed that evolutionarily ancient plants, like the maidenhair tree (*Ginkgo biloba*), a 'living fossil' (Fig. 2.2), and ferns were devoid of insect consumers, it is now known that this tree [104], as well as ferns [45, 82, 83], mosses [58, 91], lichens [9, 58] and mushrooms [19] serve as food to at least some insect species.

2.1 HOST-PLANT SPECIALIZATION

One of the most striking features of insect–plant relationships is the fact that insects are specialist feeders. This phenomenon forms the heart of these relationships, and all discussions in the following chapters are pervaded with this notion. It is therefore useful to consider the degree of dietary specialization or generalization shown by herbivores. Insects that in nature occur only on one or a few closely related plant species are called **monophagous**. Many lepidopterous larvae, hemipterans and coleopterans fit into this category. **Oligophagous** insects, such as the cabbage white butterfly (*Pieris brassicae*) and

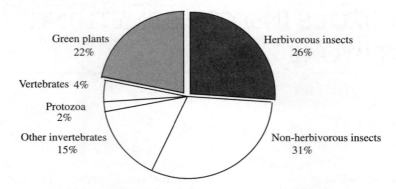

Figure 2.1 The proportions of plant and animal species in major taxa, excluding fungi, algae and microbes. (Source: redrawn from Strong et al., 1984.)

Table 2.1 Numbers of herbivorous species in different insect orders. (Data from various sources)

Insect order	Total no. of species	Herbivorous species No.	%
Coleoptera	349 000	122 000	35
Lepidoptera	119 000	119 000	100
Diptera	119 000	35 700	30
Hymenoptera	95 000	10 500	11
Hemiptera	59 000	53 000	91
Orthoptera	20 000	19 900	100
Thysanoptera	5 000	4 500	90
Phasmida	2 000	2 000	100

Figure 2.2 *Ginkgo biloba*. Shoot with young leaves and male inflorescence. (Source: reproduced from Fitting et al., 1936, with permission.)

the Colorado potato beetle (*Leptinotarsa decemlineata*), feed on a number of plant species, though all belonging to the same plant family. **Polyphagous** insect species seem to exercise little choice and accept many plants belonging to different plant families. The green peach aphid (*Myzus persicae*), for instance, has been recorded to feed on members of more than 50 plant families.

This classification into three categories, however, is fairly arbitrary, because precise definitions of monophagy and oligophagy are difficult to sustain. As a first problem there exists a completely graded spectrum between species that will eat only a single kind of plant and those that regularly consume many very diverse ones. Second, individuals of the same insect species may show different host-plant preferences in different areas of its distribution, and even individuals belonging to the same population may be much more restricted in their choices than the population as a whole [36, 49, 50, 76]. In view of these observations it is often more convenient to distinguish only **specialists** (broadly, monophagous and oligophagous species) from **generalists**

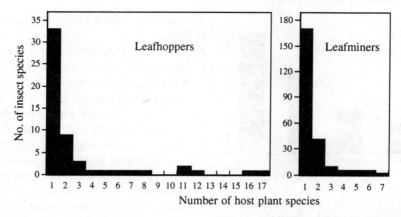

Figure 2.3 Most leafhoppers and leafminers on British trees are strict monophages. (Source: redrawn from Crawley, 1985.)

(polyphagous species). Occasionally the term **stenophagous** is used to denote specialist feeding habits, while the term **euryphagous** is used for generalists.

Host-plant specialization is the rule rather than the exception. It has been calculated [7] that less than 10% of herbivorous species feed on plants in more than three different plant families. Monophagy, the other extreme, is a common feature, and in certain insect groups it is even the dominant habit. An inventory of about 5000 British herbivorous insects shows that more than 80% of them should be regarded as specialists. Different insect groups, however, may show quite different degrees of specialization. Of the 25 British orthopteran species, 51% are polyphagous whereas 41% are restricted feeders on grasses and sedges. Conversely, 76% of all British aphids are strictly monophagous, 18% are oligophagous and only 6% are polyphagous. Monophagy is also a common habit among leaf miners and leafhoppers (Fig. 2.3).

In discussions on host-plant specialization and its terminology [55] it has been argued that some oligophagous or even polyphagous insects should more appropriately be considered as monophagous when their host-plant selection is based upon a specific type of plant chemical. Larvae of the cabbage white butterfly, which are restricted to cruciferous plants, are occasionally also found on nasturtium or *Reseda* species. Both plants belong to different families but, in common with the normal host plants of this insect, they contain glucosinolates, chemicals that typically occur in the Cruciferae. One could say that the cabbage white butterfly is monophagous on glucosinolate-containing plants, but the usage of the term in this narrow sense ignores the fact that additional plant characteristics usually play a role in host-plant selection. The same reasoning has been put forward to characterize the polyphagous larva of the browntail (*Euproctis chrysorrhoea*) as a specialist, since it feeds predominantly on tree species with tannins in their leaves (Fig. 2.4). For practical as well as principal reasons we prefer, however, to relate the classification of host specialization to the range of an insect's natural host plants.

The breadth of the host-plant range shown by a particular insect species is probably one of its major biological characteristics, and is constrained by several morphological, physiological, and ecological factors. In order to uncover these constraints it may be helpful to look for correlations between diet breadth and plant or herbivore characteristics. Several

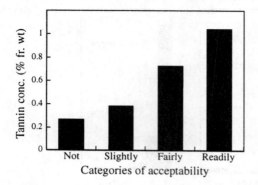

Figure 2.4 Polyphagous larvae of *Euproctis chrysorrhoea* favour plants with high tannin levels. Sixty-one plant species were categorized into four acceptability classes, which appear to correlate with average tannin contents. (Source: data from Grevillius, 1905.)

interesting relationships have been found. For instance, insects living on herbaceous plants often show a higher degree of host specialization than insects on shrubs and trees [10, 38] (Table 2.2).

This may be explained by the fact that herbaceous plant species show a greater diversity, e.g. in life cycles and chemical composition, than woody plants. These features mean that insects adapted to these variables are in a better position to exploit such food plants than generalists. A second noteworthy aspect of host-plant specialization is the relationship between the breadth of an insect's diet and its body size: smaller species are generally more specialized than larger species [61, 71] (Table 2.3).

Perhaps larger species run a greater risk of food depletion and are therefore less choosy. A third observation to be made here is that herbivores with narrow host-plant ranges usually show a preference for young growing leaves while, overall, larvae of polyphagous species prefer mature leaves of their various host plants. Young leaves are generally more nutritious [89] but at the same time often also more toxic than mature leaves [10]. These three trends are most probably not fortuitous correlations but reflect some biological principles. Perhaps the observed patterns are related to completely different biological properties, because the frequency of strong specializations is much higher in some insect taxa than in others (Table 2.3). Probably the evolutionary 'choice' between becoming a specialist or a generalist depends on a large number of heterogenous factors. It is still a long way before we can understand why, for instance, lycaenid butterflies in the tropics are significantly more often generalists than specialists, in contrast to confamilial species in temperate climates [32], and why the opposite holds for the Papilionidae [87].

Polyphagous insect species may feed on a great diversity of plant species but certainly do not indiscriminately accept all green plants. Even notoriously catholic feeders are restricted to a few hundred plant species (Table 2.4), while plants outside this range are

Table 2.2 Host plant specialization of Lepidoptera on herbaceous and woody plant species. (Source: modified from Futuyma, 1976)

	No. of species	% specialists
Moths and butterflies in Great Britain		
on herbaceous plants	143	67
on woody plants	219	54
Butterflies in north America		
on herbaceous plants	110	88
on woody plants	53	68

Table 2.3 Percentage of insect species within taxonomic groups that feed on plants within a single plant genus, or within a single plant family, or on more than one family of plants. Note that the first five groups of insects comprise small insects compared to the other groups. The correlation between size and host-plant range, however, is low. Many examples exist of closely related similarly-sized insects that show large differences in extent of host-plant ranges. NA=Data not available. (Source: modified from Mattson et al., 1988)

Insect group	% of species feeding on			No. of species
	One plant genus only	One plant family only	More than one plant family	
Psyllidae	94	3	0	78
Aphidinae	91	7	2	445
Scolytidae	59	38	3	NA
Diaspididae	58	8	34	64
Thysanoptera	56	15	29	88
Nymphalidae	56	11	33	88
Lycaenidae	55	14	31	89
Pieridae	33	53	14	43
Papilionidae	25	21	54	89
Other Macrolepidoptera	17	23	60	430

Table 2.4 Number of plant species infested by some polyphagous insect species; note that these examples of extreme polyphagy belong to four different orders

Insect species	No. of plant species infested	Reference
Bemisia tabaci (cotton whitefly)	506	14
Lymantria dispar (gypsy moth)	>500	4, 57
Schistocerca gregaria (desert locust)	>400	98
Lygus lineolaris (tarnished plantbug)	385	109
Popilia japonica (Japanese beetle)	295	57

hardly fed upon or are totally rejected, even in the absence of any alternative food source.

Nor must it be imagined that polyphagous and oligophagous species are indiscriminate in what they choose from their acceptable host-plant range. On the contrary, some degree of preference is almost always apparent. Even archetypal polyphages like the desert locust, Schistocerca gregaria, which feeds on a wide range of plants belonging to many different families, exhibits pronounced preferences for particular plants, eating some species in small amounts and others in large amounts [11]. Another insect with a wide spectrum of host plants, the gypsy moth larva (Lymantria dispar), shows still more discrimination. In a choice situation it shows a predilection for leaves grown on the sunny side of a tree over those collected on the north side of the same tree (Fig. 2.5). And, although it will feed greedily on young expanding leaves, when offered a choice it clearly prefers full-grown leaves.

2.2 FOOD-PLANT RANGE AND HOST-PLANT RANGE

In many herbivorous insect species the ovipositing female selects the plant on which its offspring will feed and the question arises whether or not host-plant choice by the ovipositing adult is identical with food plant

10 Herbivorous insects: something for everyone

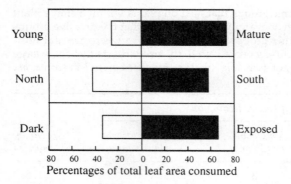

Figure 2.5 When in a choice test leaves of alder (*Alnus glutinosa*) were offered to gypsy moth larvae (*Lymantria dispar*), the caterpillars ate more from mature leaves than from young leaves. Likewise they preferred leaves picked from the south side of the tree over those facing north, and leaves that were exposed to normal light over leaves that were kept in the dark for 24 hours. (Source: data from Schoonhoven, 1977.)

range of the larval stage. Although, as was to be expected, the two host ranges show a fairly close similarity, they are often not identical. This observation indicates that host selection behaviour in the ovipositing female is governed by different parts of the genome from those coding for food selection behaviour in the larva [106]. Interestingly, the breadth of diet of the larvae is often wider than the range of plants acceptable as oviposition substrate to the adult female (Fig. 2.6).

Obviously, natural selection will prevent the development of too great a discrepancy between the preferences of ovipositing females and their offspring. Several studies have addressed the question of whether or not the oviposition preferences of herbivorous insects fully match the performance of their offspring on these food plants in terms of survival, growth and reproduction. In a number

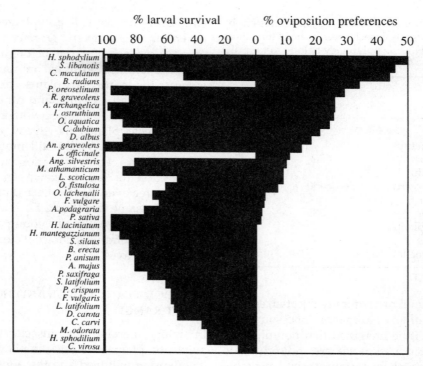

Figure 2.6 Female swallowtail butterflies (*Papilio machaon*) show a hierarchy of oviposition preferences (right side). Most plants are suitable food plants for larvae (left side), though females lay eggs on *Bifora radians*, which does not support larval growth. Larvae show also high survival rates on some plant species that are not selected for oviposition. (Source: redrawn from Wiklund, 1975.)

Specialization on plant parts

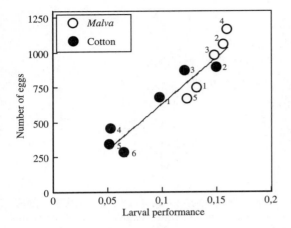

Figure 2.7 Correlation between number of eggs laid by tipworm (*Crocidosema plebejana*) females on two host plants, cotton (*Gossypium hirsutum*) and *Malva parfiflora*, and larval performance. Host plants were offered in no-choice situations and at different developmental stages, varying between the seedling stage (1) and senescing (*Malva*, 5) or open bolls (cotton, 6). There is a strong correlation between acceptability level of plants as an oviposition substrate and its developmental stage, reflecting its nutritional adequacy for larval performance. (Source: redrawn from Hamilton and Zalucki, 1993.)

of cases there is a good association (i.e. females preferentially oviposit on plants where their offspring perform best; Fig. 2.7), but in several others the association is poor [48, 68, 96, 101] (Fig. 2.6). These cases of dissimilarity represent imperfect adaptations, which are perhaps due to a lack of adequate genetic variation in the alleles that determine oviposition preference.

2.3 SPECIALIZATION ON PLANT PARTS

Insects may consume every anatomical part of plants but, in addition to host-plant specialization, also show specialization with regard to the feeding sites they occupy on their hosts. Insects rarely thrive on all parts of their host plant equally well. Many caterpillars, beetles and grasshoppers are leaf foragers, ingesting relatively large chunks of leaf material. Other insects show more specific needs. Thus plant-bugs often penetrate epidermal cells and ingest cell contents, whereas aphids mainly suck from the sap flow of phloem sieve elements. Whiteflies often tap the xylem [69]. Leaf-mining insects live and feed during their larval stage between the upper and lower epidermis of a leaf-blade and devour parenchymal tissues (Fig. 2.8).

Different species may excavate different layers of the leaf parenchyma. Leaves of birch, for example, are attacked by two hymenopterous leafminers, one of which, *Fenusa pumila*, feeds on the entire mesophyll while the larvae of *Messa nana* feed only on palisade parenchyma [24, 54]. Furthermore, leaf-mining species often show a predilection for particular parts of a leaf. Some tunnel near the mid-rib of the leaf whereas others are usually found near the periphery of the lamina (Fig. 2.9).

Thus, different leaf parts taste different, not only to leafminers but also to insects ingesting leaf pieces. Larvae of several moth species, e.g. *Catocala* spp. and *Lymantria dispar*, can discriminate between the basal, lateral and terminal leaflets of their compound-leaved foodplants, and show a dislike of basal leaflets

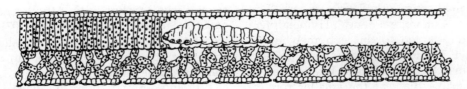

Figure 2.8 Cross-section of a leaf with a beetle larva mining in palisade parenchyma. (Source: from *Insect Biology* by Evans. Copyright (©) 1984 by Addison Wesley Publishing Company. Reprinted by permission.)

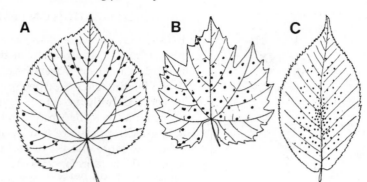

Figure 2.9 Distribution of mines on three hosts. **(A)** 50 mines of *Brachys* on lime (*Tilia* sp.). **(B)** 50 mines of *Antispila viticordifoliella* on grape (*Vitis vinifera*). **(C)** 100 mines of *Lithocolletis ostryarella* on hophornbeam (*Ostrya* sp.). (Source: reproduced from Frost, 1942, with permission.)

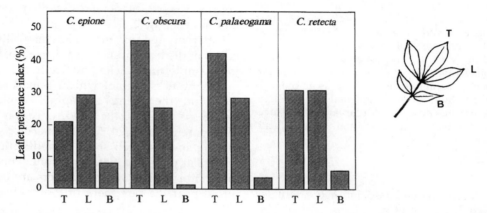

Figure 2.10 Caterpillars of four *Catocala* spp. eat less basal (B) leaf material than material from terminal (T) and lateral (L) parts of hickory (*Carya*) leaves. (Source: redrawn from Gall, 1987.) Right: compound leaf of *C. ovata*.

[39] (Fig. 2.10). Plant stems may harbour stemborers, mainly lepidopterous, dipterous and coleopterous larvae (Fig. 2.11), and the bark of woody plants is often infested by bark beetles (Scolytidae and others).

Wood may contain the larvae of some Lepidoptera, Coleoptera and Hymenoptera, which are adapted to this extremely unbalanced diet. The roots of plants support many sorts of insect. Some of these live in the soil, such as grubs that eat the smaller rootlets. Others bore directly in the roots (for example larvae of onion flies, carrot flies, cabbage root flies), while certain cicadas and some aphid species pierce the roots and drink their liquid food. Other insects are specialist feeders on flowers, fruits or seeds and members of several insect orders induce the formation of galls in various plant parts [88, 107]. Taken together these examples show how all parts of the plant are 'shared out' and can support some insect or other.

The endless variation in adaptations to certain plant tissues is, at least to some extent,

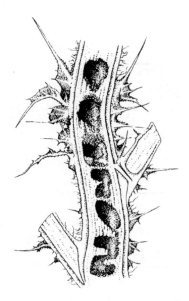

Figure 2.11 Damage caused by weevil larva (*Apion* sp.) in the stem of thistle (*Cirsium vulgare*). (Source: reproduced from Redfern, 1983, with permission.)

due to nutritional factors. The dietary value of different plant parts and even different tissues is so different that it is not surprising to find that most small insects are specialists. The smaller the herbivore's body size, the finer the scale of heterogeneity of the plant tissues it meets. For instance, the larvae of a polyphagous pest insect, *Mamestra configurata*, when feeding on the pods of rape, one of their host plants, remain smaller and show a 30% increase in mortality compared to conspecific larvae feeding on foliage [8]. The larvae of *Dasineura brassicae*, on the other hand, are specialized feeders on the pods of rape and survive only on these plant parts [1]. Nutritional factors are not of course the only determinants of feeding site specialization, which is evinced by almost all herbivorous insect species. Several other physiological and ecological factors must also be involved.

Host specialization thus appears to have two dimensions: host-plant species and host-plant part. Only through the combination of these two features could insects evolve an abundance of species unsurpassed by other animal groups. It is interesting to note that the mechanisms underlying host-plant specialization have been studied in much greater detail than the factors that restrict insects to certain plant parts only.

2.4 NUMBER OF INSECT SPECIES PER PLANT SPECIES

The number of herbivorous insect species, even at a conservative estimate, exceeds the number of vascular plant species (Fig. 2.1). Since insects, except for strictly monophagous species, occur on more than one plant species, each plant may be expected to harbour several different insects at least, as can easily be observed to be the case in nature. Different insect species living on the same plant are not necessarily direct competitors. In addition to spatial separation, as discussed above, they are often also temporally separated because of differences in phenology between insects. Stinging nettle (*Urtica dioica*), for instance, is the host plant of eight insect species, which, because of different life cycle patterns, show seasonal differences in population build-up. As a result there is only limited overlap of population peaks between different species [23] (Fig. 2.12).

Some plants house a larger insect fauna than others. At least 110 different species, including a number of entomophagous species, are associated with stinging nettle; 31 of them are specialized feeders on this plant [23]. Some 423 insect species are found to feed on two species of oaks. By contrast, yew (*Taxus baccata*) supports only six insect species [52]. Of course all plants are continuously visited by a multitude of herbivorous insect species, but only a small fraction of these visitors establish a permanent relationship with the plants. For example, Kogan [56] has recorded the presence of over 400 different herbivorous insects in soybean fields in Illinois (USA), but actual records of colonization are limited to not more than about 40 species.

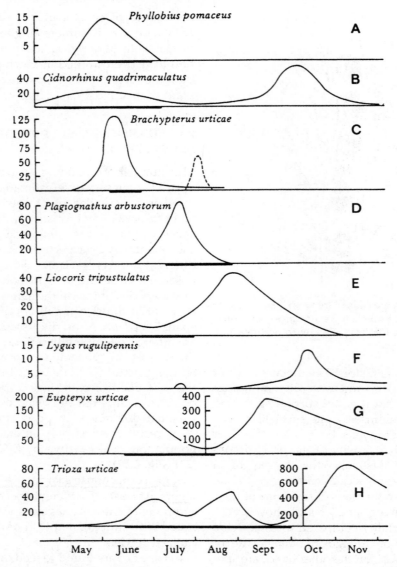

Figure 2.12 Life-cycle patterns of three Coleoptera **(A)–(C)**, three Heteroptera **(D)–(F)**, and two Homoptera **(G)–(H)** that feed on stinging nettles, as determined from weekly samples of adult insects. Thickened lines indicate the presence of adults with eggs. (Source: reproduced from Davis, 1983, with permission.)

Curious differences between the number of insect species associated with different plant phyla appear when ferns are compared with angiosperms. Ferns, although evolutionarily much older than flowering plants, have on average a 30-fold lower ratio of insect to plant species than angiosperms [18]. Conceivably, the data on the insect fauna of ferns has been undercollected in comparison to that of angiosperms. More probably, however, the underpinning of the dramatic difference in insect species richness of both plant groups

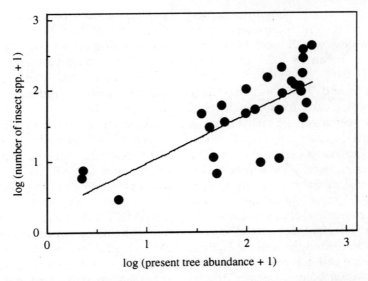

Figure 2.13 Relationship between tree abundance in Great Britain and the number of insect species inhabiting different tree species. (Source: redrawn from Kennedy and Southwood, 1984.)

must be sought at the physiological and/or ecological level.

The differences in the numbers of insects associated with particular plants have been attributed to differences in plant life history, plant abundance (Fig. 2.13), evolutionary duration of cohabitation, plant size and architecture, efficacy of defence mechanisms and still other factors [52]. The relationships between these factors and the numbers of plant denizens will be discussed in Chapters 9 and 10.

2.5 HERBIVOROUS INSECTS: ARE THEY PLANT TAXONOMISTS?

The phenomenon of host-plant specialization requires that an insect must be able to search for and recognize its specific host even when this plant is growing in the middle of a species-rich vegetation. J. H. Fabre [30], in one of his famous books on insect behaviour, concluded that ovipositing females possess a 'botanical instinct' that helps them to recognize their host plants. This term has also been used in a slightly different connotation to indicate that an oligophagous insect is in some way or other able to recognize the taxonomic relationship of plants, enabling it to accept only related plant species [86]. When the leafhopper *Aphrophora alni* was tested for its feeding preferences when exposed to eight of its normal host plants, it appeared that the insect classified the plant species in quite the same order as regards genus and family as plant taxonomists [74]. The 'botanical instinct' of some specialized feeders has in some cases helped botanists to track down mistakes in earlier plant classifications. Thus the larvae of *Thyridia* sp. were found to feed on *Brunsfelsia* spp. (Scrophulariaceae). When taxonomists realized that all known *Thyridia* species live on solanaceous plants, the taxonomic position of *Brunsfelsia* was reinvestigated, leading to the transfer of the genus to the Solanaceae [46]. Several other examples have been reported of the feeding habits of specialized insects providing clues as to taxonomic relationships between various plant taxa [94]. Thus aphids and psyllids have been successfully utilized to solve problems in systematic botany or to distinguish closely related plant species, e.g. in the *Populus* complex, which have been confused by human botanists [25, 47]. In an examination of two cottonwood species (*Populus fremontii* and *P. angustifolia*) their hybrids and complex backcrosses, the level of concordance between a genetic analysis and a classification based upon associated herbivores was 98%. This result exemplifies that the use of insect bioassays may be a more rigorous method of distinguishing closely related plant taxa than reliance solely on, for instance, morphological or chemical characteristics [35].

The foregoing observations might lead to the conclusion that monophagous and oligophagous insects are brilliant botanists, which, aided by a mysterious 'botanic instinct', unerringly recognize taxonomic relationships in the plant world. Our present knowledge of phytochemistry, however, can to a large extent explain the insect's capacity to recognize related plants, because taxonomic relationships are often synonymous with biochemical relatedness. Insects do not search for plants that have been classified by us into a particular taxon, whether it be species, genus or family, but hunt for plants with a chemical profile fitting their search image. This profile may be rather narrow and specific and restricted to plants belonging to a single species, or somewhat broader and more variable and so characteristic of a plant genus or even family. With this explanation we have touched upon a central theme in the study of insect–plant relationships: the chemical constitution of a plant is the prime factor in its interaction with the insect world. Obviously this aspect will be discussed in much more detail in the following chapters.

2.6 HOST PLANT IS MORE THAN FOOD PLANT

The host plant is not merely something fed on, it is something lived on. This statement by J. S. Kennedy [53] recognizes the importance of housing facilities provided by the host-plant: biotic and abiotic factors other than food. Insects living on a plant are confronted with many kinds of cohabitants, including competitors and natural enemies, a specific microclimate, effects induced by host-plant pathogens, etc.

For instance, larvae of *Platyprepia virginalis* collected from hemlock, one of the host plants of this generalist species, appeared to be parasitized by a tachinid fly in 83% of all cases, whereas only 50% of the caterpillars collected from lupin in the same habitat were parasitized [27]. If the heavy toll taken by the parasite on hemlock is not compensated by some physiological or ecological advantage, the insect may be expected to develop an avoidance reaction to this host, which, though nutritionally equivalent to lupin, is suboptimal in terms of host suitability. Insects may even prefer host plants that are nutritionally suboptimal but are not visited by some of their natural enemies and thus provide an 'enemy-free space'. Such plants present better overall survival rates than more nutritious hosts where the herbivore is more vulnerable to parasitization [75].

Another instance of an insect dwelling affecting its mortality rate concerns whiteflies on cucumber in relation to the presence or absence of trichomes, or plant hairs. The minute parasitic wasp *Encarsia formosa* is considerably more efficient in finding its host, whitefly larvae, on glabrous cultivars than on hairy leaves (Fig. 2.14). On glabrous leaves, because it can move faster and manoeuvre better, this parasite achieves parasitization levels of whiteflies 20% higher than on hairy cultivars [99].

Getting a good grip on plants with cuticles covered with slippery wax layers presents a serious problem for many insects [26, 90], but they have evolved many structures to solve it. Many chrysomelid beetles, for instance, have minute setae on the tarsal pulvilli excreting an adhesive material [102]. Some *Empoasca* species can use their tarsal pulvilli as suction

Host plant is more than food plant 17

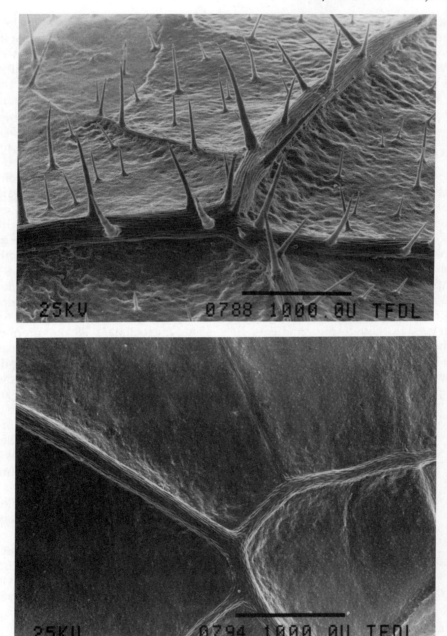

Figure 2.14 SEM of the undersurfaces of a hairy and a glabrous leaf of two cucumber cultivars. The longest trichomes on the hairy leaf are about the size of *Encarsia formosa* adults. (Source: reproduced by courtesy of J. C. van Lenteren, Wageningen, The Netherlands.)

cups [60], and many lepidopteran larvae glue a silk thread 'rope-ladder' to the plant surface to serve as a 'foothold'. The aphid *Myzocallis schreiberi*, which lives on the densely pubescent leaves of *Quercus ilex*, has a pair of claws and a pair of flexible empodia that help it to get a good grip on the leaves [51].

2.7 MICROCLIMATES AROUND PLANTS

A plant provides a unique microclimate for its commensals. These microclimates can vary considerably from standard meteorological measurements to which the vegetation as a whole is exposed. Plant surfaces have boundary layers of relatively still air where, because of frictionary drag, turbulence does not occur. Here temperature and relative humidity, partly as a result of photosynthetic and transpiration processes, can differ markedly from ambient levels. Although the gradients in these boundary layers span only millimetres, or at best a few centimetres (Fig. 2.15), depending on wind velocity and leaf size, they may be very important for any insect living in these zones.

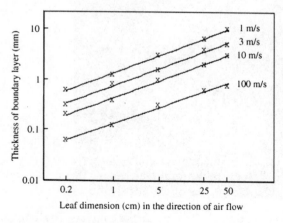

Figure 2.15 Thickness of the boundary layer over a leaf as a function of wind velocity and leaf size. Note that both axis scales are logarithmic. (Source: redrawn from Fitter and Hay, 1987; data from P. S. Nobel, 1974.)

The upper surface of a leaf may be warm or cool depending on the rate of transpiration, its size, shape, reflectance and height above the ground. The undersurface of a leaf is usually cooler and more humid than the upper surface. Temperatures at the leaf surface may be up to 10°C and even more above or below air temperatures (Fig. 2.16; Table 2.5) and, likewise, the relative humidity close to the leaf surface may considerably exceed that of the surrounding air. As a consequence small insects such as aphids and early instars of folivorous insects inevitably experience microclimatic conditions that will significantly influence their temperature and water balance, two basic factors of their physiology [108].

Microclimates may also be studied at the level of whole plants or within natural vegetations (Fig. 2.17) or field crops.

The microclimate at the ground surface under vegetation differs greatly from that at a bare soil surface. When it is realized that the total surface of vegetation growing in a meadow is some 20–40 times the area of the ground on which it grows [40], one can easily appreciate the effect of vegetation in reducing the amount of radiation that reaches ground level. The vegetation also produces gradients of windspeed, temperature and humidity (Fig. 2.18).

Thus an insect living on the inflorescences of tall grasses, such as the aphid *Rhopalosiphum padi*, is exposed to environmental conditions that are totally different from the microclimate experienced by the aphid *Therioaphis trifolii* situated at the underside of clover leaves, though the two insects may live only a few tens of centimetres from each other.

2.8 EXTENT OF INSECT DAMAGE IN NATURAL AND AGRICULTURAL ECOSYSTEMS

Students of insect–plant interactions are confronted with the paradoxical observation that most plants in natural ecosystems show only

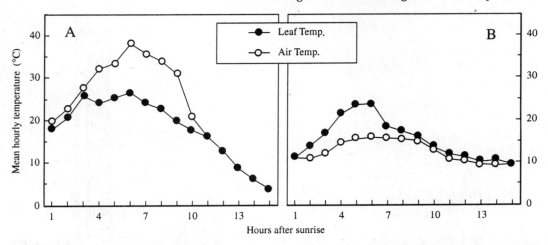

Figure 2.16 Undersurface temperatures of apple leaves and ambient temperatures measured on **(A)** a hot, cloudy summer day and **(B)** a cool, clear day. (Source: redrawn from Ferro et al., 1979.)

Table 2.5 Leaf temperatures (Δt) above or below air temperatures of plants from some tropical, desert and temperate regions. (Source: modified from Stoutjesdijk and Barkman, 1992)

Species	Locality	Air temperature (°C)	Δt
Musa acuminata	Java, Indonesia (1500 m)	22.5	13.1
Rhododendron javanicum	Java, Indonesia (1500 m)	21.8	9.1
Polypodium feei	Java, Indonesia (1500 m)	22.7	10.8
Saccharum officinarum	Java, Indonesia (lowland)	31.5	3.1
Calotropis gigantea	Java, Indonesia (lowland)	32.4	0.6
Citrullus colocynthis	Sahara desert	50.0	−13.0
Rhamnus cathartica	Netherlands	20.4	8.0
Salix cinera	Netherlands	20.4	5.7
Cynoglossum officinale	Netherlands	21.0	5.7
Aster tripolium	Netherlands	18.0	10.1
Phragmites australis	Netherlands	23.0	1.0
Ligustrum vulgare	Netherlands	24.3	9.5
Crataegus monogyna	Netherlands	24.3	8.5
Convolvulus arvensis	Netherlands	18.1	14.2

little or even no obvious damage despite the existence of an innumerable number of herbivorous insect species. Complete defoliation of vegetation happens only sporadically. It is estimated that insects consume in the order of 10% of all annually produced plant biomass [3, 17, 22]. This figure, of course, varies considerably with vegetation type and time. Thus it has been calculated that in tropical forests leaf-cutting ants alone remove approximately 17% of the total annual leaf production [12]. Estimates of losses to sap-feeding insects are harder to obtain, but supposedly range from 1–6% of primary production [103]. When two *Eucalyptus* tree species were insecticide-protected from two sap-feeding coreid bugs, the two tree species showed during a 12-month sampling period an 8.5% and 39%

20 Herbivorous insects: something for everyone

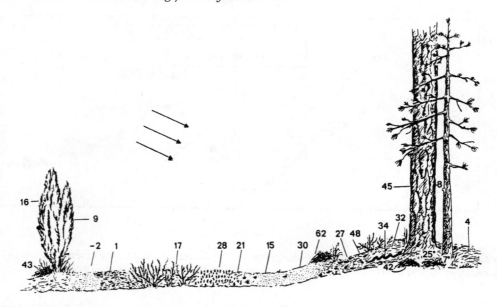

Figure 2.17 Microclimatic variations in relation to sun position and type of vegetation. Surface temperatures on a cloudless day (The Netherlands, 3 March 1976, noon) along a transect perpendicular to the fringe of a forest facing south. Air temperature at a height of 1 m was 11.8°C. The ground was still frozen in the shadow of a *Juniperus* bush, whereas a few meters to the right surface temperatures up to 62°C were recorded. Arrows = direction of sun rays. (Source: reproduced from Stoutjesdijk and Barkman, 1992, with permission.)

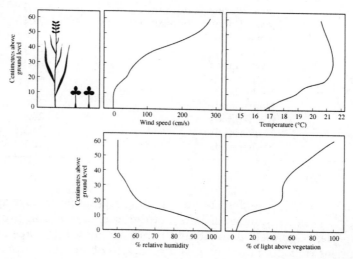

Figure 2.18 Microclimate variables in a grassland vegetation. (Source: redrawn from Cox *et al.*, 1973.)

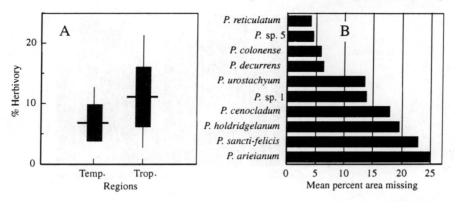

Figure 2.19 Rates of herbivory in forests in different climatic zones and for different species of *Piper*, a genus of climbing shrubs. **(A)** Leaf areas eaten annually in temperate and tropical broad-leaved forests. Plots indicate mean ±SD and range. (Source: redrawn from Coley and Aide, 1991.) **(B)** Mean percentage area missing per *Piper* species for a minimum of 50 freshly abscissed leaves per species. (Source: redrawn from Marquis, 1991.)

height advantage respectively compared to unprotected trees [5]. Whereas Australian *Eucalyptus* trees suffer from chronic levels of insect damage accounting for 10–50% of foliage production [70], other plants, e.g. neem trees (*Azadirachta indica*), *Juniperus* spp. and *Rhododendron* spp. hardly show any losses to insects. Even related plant species may show a considerable interspecific variation of losses to herbivory (Fig. 2.19).

More severe impacts do occur occasionally, such as widespread defoliations and death of birch forest in Fennoscandinavia caused by moth species belonging to the genera *Oporinia* and *Operophtera* [95]. Interestingly, trees in urban and ornamental plantings do not sustain more insect damage than trees in natural forests [73].

The question arises of whether the 10% damage level represents a negligible loss of energy to a plant and, consequently, whether it significantly affects the plant's fitness. Several indications point to marked effects even at low insect damage levels. For instance, it has been calculated that the annual net assimilate devoted to reproduction ranges from 1–15% in herbaceous perennials and from 15–30% in herbaceous annual plant species [43]. Hence, as a rough generalization, a figure of 10% may cover the proportion of biomass plants allocate to reproduction [67]. Thus, as a very general approximation, losses to insects are of the same magnitude as the biomass plants devote to reproduction. In view of the magnitude of these figures it seems unlikely that insect damage is negligible. Of course the 10% loss to herbivory is not the 10% spent on reproduction, since the losses are presumably shared more or less evenly by all functions.

Figures on losses of leaf surface possibly underestimate the real damage inflicted by insects, because many small wounds may have a much greater effect than the complete removal of some leaves. It is known that physiological effects of wounding are systemically transmitted to other plant parts (Chapter 3). It is therefore quite likely that the number of damaged sites is more important than the total size of the damaged area. When in 12 plant species all leaves with some signs of insect damage were scored, it appeared that on average 87% of the leaves were affected, with some plant species even having all their leaves damaged to some extent [22] (Fig. 2.20).

This figure is too different from the 10%

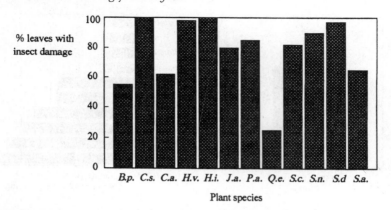

Figure 2.20 Percentages of leaves showing insect damage in 12 dicotyledonous plant species. B. p. = *Betula pubescens*; C. s. = *Calystegia sepium*; C. a. = *Corylus avellana*; H. v. = *Hamamelis vernalis*; H. i. = *Heliconia imbricata*; J. a. = *Juglans arizonica*; P. a. = *Prunus avium*; Q. e. = *Quercus emoryi*; S. c. = *Salix caprea*; S. n. = *Sambucus nigra*; S. d. = *Solanum dulcamara*; S. a. = *Spartina altiniflora*. (It should be noted that damage assessment in many of the given species is based on only one or two plant specimens, thus allowing doubt about the representativeness of the values given.) (Source: data from Damman, 1993.)

damage level to be ignored! Other studies, however, have indicated that there are plant communities that sustain only sporadic insect damage. Price and coworkers [79] have reported that, although tropical savannah is very rich in caterpillar species, the numbers per species are very low. On average, only one lepidopterous larva of all species combined was found per ten plant individuals (1–2 m tall trees of four species). Clearly, the intensity of insect attack may vary tremendously among plant communities and plant species and our limited knowledge prevents us from making any sound generalization at the moment.

An interesting study on oak trees showed that even moderate insect attack may markedly depress seed production. Experimental trees were regularly treated with insecticides, thereby suppressing defoliation below 5%, while water-sprayed control trees suffered twice that amount. Tree growth, as determined from tree rings, was not affected, but the number of acorns produced per shoot was up to four times higher in insecticide-treated trees than in untreated control trees [21]. Whether or not reduced acorn production under

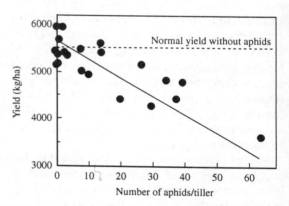

Figure 2.21 Yield of wheat in relation to peak number of grain aphids (*Sitobion avenae*) per tiller. (Source: data from Vereijken, 1979.)

natural circumstances negatively affects the population density of oak trees remains, however, an open question.

Sucking insects, when present in sufficient numbers, may also affect seed production negatively (Fig. 2.21). Thus aphid-infested wood groundsel (*Senecio sylvaticus*) yields 50% less seeds than aphid-free plants [28]. Spittlebugs also caused a severe reduction in plant

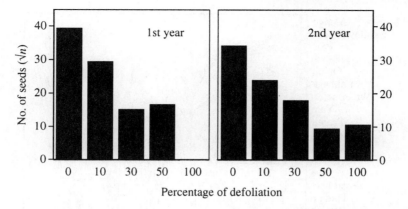

Figure 2.22 Effects of pruning on seed production by *Piper arieianum* shrubs. When ⩾ 30% of the leaf area is removed seed numbers are significantly reduced in the year after the treatment as well as the following year. (Source: redrawn from Marquis, 1984.)

growth rates and seed production in goldenrod (*Solidago altissima*) [66].

Another experimental approach to assess effects of insect attack on seed production involves artificial defoliation. Such an experiment was done on *Piper arieianum* shrubs occurring in the neotropical rainforest, which often suffer from heavy attacks by several weevil species. The plants produce fewer seeds after pruning, an effect that is carried over to the next year as a result of reduced storage allocation (Fig. 2.22). (Some plants, for instance most fruit trees, produce more fruit after pruning.) Thus low to moderate herbivory levels often have potent effects on seed production.

Insects often inflict much more damage in agroecosystems than in natural settings. Despite intensive use of insecticides, crop losses to insect feeding in the USA amount to 13% (see Fig. 1.3), whereas this figure worldwide reaches 15% or more. The phenomenon of host specialization as discussed earlier in this chapter has fortunate consequences for the number of pest species. About 1000 insect species attack agricultural crops in the USA. On a world scale this figure runs to about 9000 species, though less than 5% are considered to be serious pests [77], a relatively small number in view of the many insect species present. True, our agriculture relies on a very small subset of the world's flora, with just four major and 26 minor crop species contributing to 95% of human nutrition [84] (Fig. 2.23), but many of these cultivated plants have covered large areas of land for millennia and thus have offered a plethora of food to numerous insects with a notoriously high degree of adaptability.

Insect pest species are also predominantly specialist feeders, i.e. 75% of temperate and 80% of tropical lepidopterous pests are monophagous or oligophagous [2]. These ratios tally strikingly with the figures presented above for the effects herbivores inflict on natural ecosystems.

2.9 COMPENSATION FOR HERBIVORE DAMAGE

Few plants escape herbivore damage, but they are equipped with mechanisms to reduce the deleterious effects of herbivory. Most plants, as long as insects do not attack their meristemic tissues or apical tips, have remarkable power of regeneration. From a morphological and developmental point of view plants are

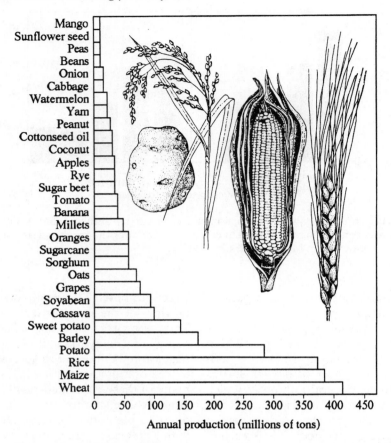

Figure 2.23 Annual production of 30 out of the Earth's 250 000 species of higher plants, which together account for 95% of human nutrition. More than 50% of our food is obtained from only four crop species, potato, rice, maize and wheat. (Source: redrawn from Sattaur, 1989.)

basically different from animals. Structurally, plants are modular organisms, i.e. they consist of repetitive multicellular units each with its own meristem. None of these units is vital for the plant as a whole. This property of modularity reduces the adverse effects of herbivory considerably and allows for easy recovery from tissue removal, in contrast to unitary organisms, such as insects, which are killed or at least seriously disabled when body parts are removed. The surprising ease of recovery from herbivory results from the presence of (often dormant) meristems and the ability to redirect resources, i.e. nutrients and photosynthetic products, to regrowing tissues [44]. Under moderate or good resource conditions, plants can partially or wholly compensate, and sometimes even overcompensate, for losses to insect feeding (overcompensation being defined as production of more biomass than has been lost to herbivory [6]). However, an unequivocal demonstration of the phenomenon of compensation appears to be more difficult than one might expect, because it requires that insect-damaged plants under natural conditions exhibit a significant increase in fitness as compared to undamaged controls. Many factors, including anatomical

Table 2.6 Plant traits and environmental factors which may determine a plant's compensatory response to herbivory. (Source: modified from Whitham et al., 1991)

Undercompensation	Equal or overcompensation
Herbivory late in season	Herbivory early in season
Low water, nutrients and/or light	Abundant water, nutrients and light
High competition	Low competition
Meristem limitation	No meristem limitation
Slow growth	Fast growth
Non-integrated plant modules	Integrated plant modules
Woody perennials	Annuals and biannuals

characteristics (presence of reserve meristems, vascular integration of different parts), mode of reproduction, timing of herbivory, stored reserves and availability of water, nutrients and light, contribute to a plant's capacity to (partially) make up for tissue losses [105]. As a result it is premature in the present state of our knowledge to make generalizations regarding the ecological or agricultural importance of compensatory responses (Table 2.6).

One can only say that the phenomenon of compensatory growth is reported in many studies [97, 105], but hard evidence for the existence of overcompensation is scarce and seems to occur only under growth-chamber conditions and in agricultural crops [6, 80]. Assessing the significance of compensatory reactions is, last but not least, hampered by the fact that our knowledge of the impact of insect herbivory is mainly derived from extreme cases, whereas relatively little is known of the impact of herbivores at low densities and its relationships with effects caused by other herbivores.

2.10 CONCLUSIONS

The key point of this chapter is the high degree of food specialization generally shown by herbivorous insects. This concept will appear to be the core of all further explorations of insect–plant interactions in the remaining chapters of this book. One of the central questions raised in this area of research relates to the observation that herbivorous insects cause relatively little visible damage to plants in natural ecosystems, despite their large number of species and astounding reproduction capacity. At the same time, the omnipresence of plant-feeding insects must have a profound influence on plant structure and function.

Apparently, plants are generally well protected against insect attack. Elucidation of the nature of the protection mechanisms may be useful in developing methods for reducing insect damage in cultivated crop plants.

2.11 REFERENCES

1. Åhman, I. (1985) Larval feeding period and growth of *Dasineura brassicae* (Diptera) on *Brassica* host plants. *Oikos*, **44**, 191–194.
2. Barbosa, P. (1993) Lepidopteran foraging on plants in agroecosystems, constraints and consequences, in *Caterpillars. Ecological and Evolutionary Constraints on Foraging*, (eds N. E. Stamp and T. M. Casey), Chapman & Hall, New York, pp. 523–566.
3. Barbosa, P. and Schultz, J. C. (1987) *Insect Outbreaks*, Academic Press, San Diego, CA.
4. Barbosa, P., Martinat, J. and Waldvogel, M. (1986) Development, fecundity and survival of the herbivore *Lymantria dispar* and the number of plant species in its diet. *Ecol. Entomol.*, **11**, 1–6.
5. Bashford, R. (1992) Observations on the life history of two sap-feeding coreid bugs and their impact on growth of plantation eucalypts. *Tasforests*, **4**, 87–93.
6. Belsky, A. J. (1986) Does herbivory benefit plants? A review of the evidence. *Am. Nat.*, **127**, 870–892.
7. Bernays, E. A. and Graham, M. (1988) On the evolution of host specificity in phytophagous arthropods. *Ecology*, **69**, 886–892.
8. Bracken, G. K. (1984) Within plant preferences of larvae of *Mamestra configurata* (Lepidoptera, Noctuidae) feeding on oilseed rape. *Can. Entomol.*, **116**, 45–49.

9. Callaghan, C. J. (1992) Biology of epiphyll feeding butterflies in a Nigerian cola forest (Lycaenidae, Liptenınae). *J. Lepidop. Soc.*, **42**, 203–214.
10. Cates, R. G. (1980) Feeding patterns of monophagous, oligophagous, and polyphagous insect herbivores, the effect of resource abundance and plant chemistry. *Oecologia*, **46**, 22–31.
11. Chapman, R. F. (1990) Food selection, in *Biology of Grasshoppers*, (eds R. F. Chapman and A. Joern), John Wiley, New York, pp. 39–72.
12. Cherrett, J. M., Powell, R. J. and Stradling, D. J. (1989) The mutualism between leaf-cutting ants and their fungus, in *Insect–Fungus Interactions*, (eds N. Wilding, N. M. Collins, P. M. Hammond and F. J. Weber), Academic Press, New York, pp. 93–120.
13. Claridge, M. F. and Wilson, M. R. (1981) Host plant associations, diversity and species-area relationships of mesophyll-feeding leafhoppers of trees and shrubs in Britain. *Ecol. Entomol.*, **6**, 217–238.
14. Cock, M. J. W. (1986) *Bemisia tabaci*, a literature survey on cotton whitefly with an annotated bibliography. FAO, United Nations. *C. A. B. Inst. Biol. Contr.*, pp. 1–121.
15. Cockburn, A. (1991) *An Introduction to Evolutionary Ecology*, Blackwell, Oxford.
16. Coley, P. D. and Aide, T. M. (1991) Comparison of herbivory and plant defenses in temperate and tropical broad-leaved forests, in *Plant–Animal Interactions. Evolutionary Ecology in Tropical and Temperate Regions*, (eds P. W. Price, T. M. Lewinsohn, G. W. Fernandes and W. W. Benson), John Wiley, New York, pp. 25–49.
17. Coley, P. D., Bryant, J. P. and Chapin, F. S. (1985) Resource availability and plant antiherbivore defense. *Science*, **230**, 895–899.
18. Cooper-Driver, G. A. N. (1978) Insect-fern associations. *Entomol. Exp. Appl.*, **24**, 110–116.
19. Courtney, S. P., Kibota, T. T. and Singleton, T. A. (1990) Ecology of mushroom-feeding Drosophilidae. *Adv. Ecol. Res.*, **20**, 225–274.
20. Cox, C. B. and Moore, P. D. (1980) *Biogeography, an Ecological and Evolutionary Approach*, 3rd edn, Blackwell, Oxford.
21. Crawley, M. J. (1985) Reduction of oak fecundity by low-density herbivore population. *Nature*, **314**, 163–164.
22. Damman, H. (1993) Patterns of interaction among herbivore species, in *Caterpillars. Ecological and Evolutionary Constraints on Foraging*, (eds N. E. Stamp and T. M. Casey), Chapman & Hall, New York, pp. 131–169.
23. Davis, B. N. K. (1983) *Insects on Nettles*, Cambridge University Press, Cambridge.
24. DeClerck, R. A. and Shorthouse, J. D. (1985) Tissue preference and damage by *Fenusa pusilla* and *Mesa nana* (Hymenoptera, Tenthredinidae), leaf mining sawflies on white birch (*Betula papyrifera*). *Can. Entomol.*, **117**, 351–362.
25. Eastop, V. (1979) Sternorrhyncha as angiosperm taxonomists. *Symb. Bot. Upsal.*, **22**(4), 120–134.
26. Eigenbrode, S. D. and Espelie, K. E. (1995) Effects of plant epicuticular lipids on insect herbivores. *Annu. Rev. Entomol.*, **40**, 171–194.
27. English-Loeb, G. M., Brody, A. K. and Karban, R. (1993) Host-plant-mediated interactions between a generalist folivore and its tachinid parasitoid. *J. Anim. Ecol.*, **62**, 465–471.
28. Ernst, W. H. O. (1987) Impact of the aphid *Aulocorthum solani* Kltb, on growth and reproduction of winter and summer annual life forms of *Senecio sylvaticus* L. *Acta Oecol./Oecol. Gener.*, **8**, 537–547.
29. Evans, H. E. (1984) *Insect Biology*, Addison-Wesley, Reading, MA.
30. Fabre, J. H. (1886) *Souvenirs entomologiques*, vol. 3, Delagrave, Paris.
31. Ferro, D. N., Chapman, R. B. and Penman, D. R. (1979) Observations on insect microclimate and insect pest management. *Environ. Entomol.*, **8**, 1000–1003.
32. Fiedler, K. (1996) Host-plant relationships of lycaenid butterflies, large-scale patterns, interactions with plant chemistry, and mutualism with ants. *Entomol. Exp. Appl.*, **80**, 259–267.
33. Fitter, A. H. and Hay, R. K. M. (1987) *Environmental Physiology of Plants*, 2nd edn., Academic Press, London.
34. Fitting, H., Sierp, H., Harder, R. and Karsten, G. (1936) *Lehrbuch der Botanik*, 19th edn., G. Fischer, Jena.
35. Floate, K. D. and Whitham, T. G. (1995) Insects as traits in plant systematics, their use in discriminating between hybrid cottonwoods. *Can. J. Bot.*, **73**, 1–13.
36. Fox, L. R. and Morrow, P. A. (1981) Specialization, species property or local phenomenon? *Science*, **211**, 887–893.
37. Frost, S. W. (1942) *General Entomology*, McGraw-Hill, New York.
38. Futuyma, D. J. (1976) Food plant specialization

and environmental predictability in Lepidoptera. *Am. Nat.*, **110**, 285–292.
39. Gall, L. F. (1987) Leaflet position influences caterpillar feeding and development. *Oikos*, **49**, 172–176.
40. Geiger, R. (1965) *The Climate Near the Ground*, Harvard University Press, Cambridge, MA.
41. Grevillius, A. Y. (1905) Zur Kenntnis der Biologie des Goldafters *Euproctis chrysorrhoea* L. *Bot. Cbl. Beiheft*, **18**, 222–322.
42. Hamilton, J. G. and Zalucki, M. P. (1993) Interactions between a specialist herbivore, *Crocidosema plebejana*, and its host plants *Malva parviflora* and cotton, *Gossypium hirsutum*, oviposition preference. *Entomol. Exp. Appl.*, **66**, 207–212.
43. Harper, J. L. (1977) *Population Biology of Plants*, Academic Press, London.
44. Haukioja, E. (1991) The influence of grazing on the evolution, morphology and physiology of plants as modular organisms. *Phil. Trans. Roy. Soc. Lond. B*, **333**, 241–247.
45. Hendrix, S. D. (1980) An evolutionary and ecological perspective of the insect fauna of ferns. *Am. Nat.*, **115**, 171–196.
46. Hering, M. (1926) *Biologie der Schmetterlinge*, Springer-Verlag, Berlin.
47. Hille Ris Lambers, D. (1979) Aphids as botanists? *Symb. Bot. Upsal.*, **22**(4), 114–119.
48. Horner, J. D. and Abrahamson, W. G. (1992) Influence of plant genotype and environment on oviposition and offspring survival in gall-making herbivores. *Oecologia*, **90**, 323–332.
49. Howard, J. J., Raubenheimer, D. and Bernays, E. A. (1994) Population and individual polyphagy in the grasshopper *Taeniopoda eques* during natural foraging. *Entomol. Exp. Appl.*, **71**, 167–176.
50. Karban, R. and Myers, J. H. (1989) Induced plant responses to herbivory. *Annu. Rev. Ecol. Syst.*, **20**, 331–348.
51. Kennedy, C. E. J. (1986) Attachment may be a basis for specialization in oak aphids. *Ecol. Entomol.*, **11**, 291–300.
52. Kennedy, C. E. J. and Southwood, T. R. E. (1984) The number of species of insects associated with British trees, a reanalysis. *J. Anim. Ecol.*, **53**, 455–478.
53. Kennedy, J. S. (1953) Host plant selection in Aphididae. *Trans. 9th Int. Congr. Entomol. (Amsterdam)*, **2**, 106–113.
54. Kimmerer, T. W. and Potter, D. A. (1987) Nutritional quality of specific leaf tissues and selective feeding by a specialist leafminer. *Oecologia*, **71**, 548–551.
55. Klausnitzer, B. (1983) Bemerkungen über die Ursachen und die Entstehung der Monophagie bei Insekten, in *Verhandlungen des SIEEC X, Budapest*, pp. 5–12.
56. Kogan, M. (1986) Plant defense strategies and host-plant resistance, in *Ecological Theory and Integrated Pest Management Practice*, (ed. M. Kogan), John Wiley, New York, pp. 83–134.
57. Lance, D. R. (1983) Host-seeking behavior of the gypsy moth, the influence of polyphagy and highly apparent host plants, in *Herbivorous Insects, Host Seeking Behaviour and Mechanisms*, (ed. S. Ahmad), Academic Press, New York, pp. 201–224.
58. Lawrey, J. D. (1987) Nutritional ecology of lichen/moss arthropods, in *Nutritional Ecology of Insects, Mites, Spiders and Related Invertebrates*, (eds F. Slansky and J. G. Rodriguez), John Wiley, New York, pp. 209–233.
59. Lawton, J. H. and Price, P. W. (1979) Species richness of parasites on hosts, Agromyzid flies on the British Umbelliferae. *J. Anim. Ecol.*, **48**, 619–637.
60. Lee, Y. L., Kogan, M. and Larsen, J. R. (1986) Attachment of the potato leafhopper to soybean plant surfaces as affected by morphology of the pretarsus. *Entomol. Exp. Appl.*, **42**, 101–108.
61. Lindström, J., Kaila, L. and Niemelä, P. (1994) Polyphagy and adult body size in geometrid moths. *Oecologia*, **78**, 130–132.
62. Marquis, R. J. (1984) Leaf herbivores decrease fitness of a tropical plant. *Science*, **226**, 537–539.
63. Marquis, R. J. (1991) Herbivore fauna of *Piper* (Piperaceae) in a Costa Rican wet forest, diversity, specificity, and impact, in *Plant–Animal Interactions. Evolutionary Ecology in Tropical and Temperate Regions*, (eds P. W. Price, T. M. Lewinsohn, G. W. Fernandes and W. W. Benson) John Wiley, New York, pp. 179–199.
64. Mattson, W. J., Lawrence, R. K., Haack, R. A., Herms, D. A. and Charles, P. J. (1988) Defensive strategies of woody plants against different insect-feeding guilds in relation to plant ecological strategies and intimacy of association with insects, in *Mechanisms of Woody Plant Defenses against Insects*, (eds W. J. Mattson, J. Levieux and C. Bernard-Dagan) Springer-Verlag, Berlin, pp. 3–38.
65. May, R. M. (1992) How many species inhabit the earth? *Sci. Am.*, **267**(4), 18–24.

66. Meyer, G. A. and Root, R. B. (1993) Effects of herbivorous insects and soil fertility on reproduction in goldenrod. *Ecology*, **74**, 1117–1128.
67. Mooney, H. A. (1972) The carbon balance of plants. *Annu. Rev. Ecol. Syst.*, **3**, 315–346.
68. Morgan, F. D. and Cobinah, J. R. (1977) Oviposition and establishment of *Uraba lugens* (Walker) the gum leaf skeletoniser. *Aust. For.*, **40**, 44–55.
69. Neal, J. J. (1993) Xylem transport interruption by *Anasa tristis* feeding causes *Cucurbita pepo* to wilt. *Entomol. Exp. Appl.*, **69**, 195–200.
70. New, T. R. (c. 1988) *Associations Between Insects and Plants*, New South Wales University Press, Kensington, New South Wales.
71. Niemelä, P., Hanhimäki, S. and Mannila, R. (1981) The relationship of adult size in noctuid moths (Lepidoptera, Noctuidae) to breadth of diet and growth form of host plant. *Ann. Entomol. Fenn.*, **47**, 17–20.
72. Nobel, P. S. (1974) *Introduction to Biophysical Plant Physiology*, W. H. Freeman, San Francisco, CA.
73. Nuckols, M. S. and Connor, E. F. (1995) Do trees in urban or ornamental plantings receive more damage by insects than trees in natural forests? *Ecol. Entomol.*, **20**, 253–260.
74. Nuorteva, P. (1952) Die Nahrungspflanzenwahl der Insekten im Lichte von Untersuchungen an Zikaden. *Ann. Acad. Sci. Fenn. Ser. A, IV Biol.*, **19**, 1–90.
75. Ohsaki, N. and Sato, Y. (1994) Food plant choice of *Pieris* butterflies as a trade-off between parasitoid avoidance and quality of plants. *Ecology*, **75**, 59–68.
76. Pashley, D. P. (1986) Host-associated genetic differentiation in fall armyworm (Lepidoptera, Noctuidae), a sibling species complex? *Ann. Entomol. Soc. Am.*, **79**, 898–904.
77. Pimentel, D. (1991) Diversification of biological control strategies in agriculture. *Crop Prot.*, **10**, 243–253.
78. Pimm, S. L., Russell, G. J., Gittleman, J. L. and Brooks, T. M. (1995) The future of biodiversity. *Science*, **269**, 347–350.
79. Price, P. W., Diniz, I. R., Morais, H. C. and Marques, E. S. A. (1995) The abundance of insect herbivore species in the tropics, the high local richness of rare species. *Biotropica*, **27**, 468–478.
80. Prins, A. H. and Verkaar, H. J. (1992) Defoliation, do physiological and morphological responses lead to (over)compensation?, in *Pests and Pathogens. Plant Responses to Foliar Attack*, (ed. P. G. Ayres), Bios Scientific Publishers, Oxford, pp. 13–31.
81. Redfern, M. (1983) *Insects on Thistles*, Cambridge University Press, Cambridge.
82. Robinson, A. G. (1985) *Macrosiphum* (*Sitobia*) species on ferns (revised edn), in *Proceedings of an International Aphidology Symposium, Jablonna (1981)*, Ossolineum, Warsaw, pp. 471–474.
83. Ruehlmann, T. E., Matthews, R. W. and Matthews, J. R. (1988) Roles for structural and temporal shelter-changing by fern-feeding lepidopteran larvae. *Oecologia*, **75**, 228–232.
84. Sattaur, O. (1989) The shrinking gene pool. *N. Sci.* 1675, 37–41.
85. Schoonhoven, L. M. (1977) Feeding behaviour in phytophagous insects, on the complexity of the stimulus situation. *Coll. Int. CNRS*, **265**, 391–398.
86. Schoonhoven, L. M. (1991) Insects and host plants, 100 years of 'botanical instinct'. *Symp. Biol. Hung.*, **39**, 3–14.
87. Scriber, J. M. (1995) Overview of swallowtail butterflies, taxonomic and distributional latitude, in *Swallowtail Butterflies. Their Ecology and Evolutionary Biology*, (eds J. M. Scriber, Y. Tsubaki and R. C. Lederhouse), Scientific Publications, Gainesville, FL, pp. 3–8.
88. Shorthouse, J. D. and Rohfritsch, O. (1992) *Biology of Insect-Induced Galls*, Oxford University Press, Oxford.
89. Slansky, F. and Scriber, J. M. (1985) Food consumption and utilization, in *Comprehensive Insect Physiology, Biochemistry and Pharmacology*, vol. 4, (eds G. A. Kerkut and L. I. Gilbert), Pergamon Press, Oxford, pp. 87–163.
90. Southwood, T. R. E. (1973) The insect/plant relationship – an evolutionary perspective. *Symp. Roy. Entomol. Soc. Lond.*, **6**, 3–30.
91. Spencer, K. C., Hoffman, L. R. and Seigler, D. S. (1984) Host shift of *Ecpantheria defrorata* (Arctiidae) from an Angiosperm to a liverwort. *J. Lepidop. Soc.*, **38**, 192–193.
92. Stoutjesdijk, P. and Barkman, J. J. (1992) *Microclimate, Vegetation, and Fauna*, Opulus, Knivsta.
93. Strong, D. R., Lawton, J. H. and Southwood, T. R. E. (1984) *Insects on Plants. Community Patterns and Mechanisms*, Blackwell, Oxford.
94. Tempère, G. (1969) Un critère méconnu des systématiciens phanérogamistes, l'instinct des insectes phytophages. *Botaniste*, **50**, 473–482.
95. Tenow, O. (1972) The outbreaks of *Oporinia autumnata* Bkh. and *Operophtera* spp. (Lep.,

Geometridae) in the Scandinavian mountain chain and northern Finland 1862–1968. *Zool. Bidrag Uppsala (Suppl.)*, **2**, 1–107.
96. Thompson, J. N. (1988) Evolutionary ecology of the relationship between preference and performance of offspring in phytophagous insects. *Entomol. Exp. Appl.*, **47**, 3–14.
97. Trumble, J. T., Kolodny-Hirsch, D. M. and Ting, I. P. (1993) Plant compensation for arthropod herbivory. *Annu. Rev. Entomol.*, **38**, 93–119.
98. Uvarov, B. (1977) *Grasshoppers and Locusts*, vol. 2, Centre for Overseas Pest Research, London.
99. Van Lenteren, J. C., Li Zhao Hua, Kamerman, J. W. and Xu Rumei (1995) The parasite–host relationship between *Encarsia formosa* (Hym., Aphelinidae) and *Trialeurodes vaporariorum* (Hom., Aleyrodidae) XXVI. Leaf hairs reduce the capacity of *Encarsia* to control greenhouse whitefly on cucumber. *J. Appl. Entomol.*, **139**, 553–559.
100. Vereijken, B. H. (1979) Feeding and multiplication of three cereal aphid species and their effect on yield of winter wheat. *Agricultural Research Report 888*, Centre for Agricultural Publishing and Documentation, Wageningen.
101. Via, S. (1990) Ecological genetics and host adaptation in herbivorous insects, the experimental study of evolution in natural and agricultural systems. *Annu. Rev. Entomol.*, **35**, 421–446.
102. Waterhouse, D. F. (1970) *The Insects of Australia*, Melbourne University Press, Melbourne, Victoria.
103. Weiss, A. E. and Berenbaum, M. R. (1989) Herbivorous insects and green plants, in *Plant–Animal Interactions*, (ed. W. G. Abrahamson), McGraw-Hill, New York, pp. 123–162.
104. Wheeler, A. G. (1975) Insect associates of *Ginkgo biloba*. *Entomol. News*, **86**, 37–44.
105. Whitham, T. G., Maschinski, J., Larson, K. C. and Paige, K. N. (1991) Plant responses to herbivory, the continuum from negative to positive and underlying physiological mechanisms, in *Plant–Animal Interactions. Evolutionary Ecology in Tropical and Temperate Regions*, (eds P. W. Price, T. M. Lewinsohn, G. W. Fernandes and W. W. Benson), John Wiley, New York, pp. 227–256.
106. Wiklund, C. (1975) The evolutionary relationship between adult oviposition preferences and larval host plant range in *Papilio machaon* L. *Oecologia*, **18**, 186–197.
107. Williams, M. A. J. (1994) *Plant Galls, Organisms, Interactions, Populations*, Clarendon Press, Oxford.
108. Willmer, P. (1986) Microclimatic effects on insects at the plant surface, in *Insects and the Plant Surface*, (eds B. Juniper and T. R. E. Southwood), Edward Arnold, London, pp. 65–80.
109. Young, O. P. (1986) Host plants of the tarnished plant bug, *Lygus lineolaris* (Heteroptera, Miridae). *Ann. Entomol. Soc. Am.*, **79**, 747–762.

PLANT CHEMISTRY: ENDLESS VARIETY 3

3.1	Plant biochemistry	32
3.1.1	Primary plant metabolism	32
3.1.2	Secondary plant substances	33
3.2	Alkaloids	36
3.3	Terpenoids and steroids	37
3.4	Phenolics	38
3.5	Glucosinolates	41
3.6	Cyanogenics	41
3.7	Leaf surface chemistry	42
3.8	Plant volatiles	45
3.9	Concentrations of secondary plant substances	51
3.10	Production costs	53
3.11	Compartmentation	55
3.12	Temporal variability	57
3.12.1	Seasonal effects	58
3.12.2	Day/night effects	59
3.12.3	Interyear variation	60
3.13	Effects of location and fertilizers	60
3.13.1	Sun and shade	60
3.13.2	Soil factors	62
3.14	Induced resistance	64
3.15	Genotypic variation	70
3.16	Conclusions	73
3.17	Literature	73
3.18	References	74

On the face of it, plants, which cannot fight or flee and which often have long generation spans and low recombination rates, appear to be at a disadvantage when compared to the herbivores consuming them. Insects especially can often adapt rapidly to changing conditions, because of their small sizes and concomitant relatively short generation spans, combined with a high reproductive capacity. Moreover, insects profit from being winged, which permits them to disperse and invade potential food sources even at considerable distances from their place of birth and larval domicile. Despite the apparent vulnerability of plants to herbivore attack the earth's flora has evolved to a green and highly diverse blanket. Plants clearly possess an effective resistance system, based on a combination of physical, chemical and developmental features. The term **resistance** is (in the context of insect–plant interactions) used to describe a plant's capacity to avoid or reduce damage to herbivory. It is not synonymous with **defence**, because the latter term implies something about the evolutionary *raison d'être* for the trait, and indicates that the resistance trait has evolved or is maintained in the plant population because of selection exerted by herbivores or other natural enemies. The term resistance is more neutral and therefore preferable, since it does not imply assumptions that in many cases are (still) unproved [154].

Chemical characteristics of plants especially have attracted the interests of many students of insect–plant relationships, resulting in a large and flourishing literature on the subject. It is now recognized that the plant world is characterized by a bewildering proliferation of secondary metabolites. Over 80% of the presently known natural compounds have a botanical origin [82]. Since chemicals produced by plants may be more important than any other single factor controlling insect behaviour in nature [169], much of this chapter deals with a description of the nature and dynamics of secondary plant substances. An elementary knowledge of phytochemistry is essential to fully comprehend insect–plant interactions.

Entomologists, though well aware of variations in the morphology and behaviour of insects, may envision plants as a homogeneous resource for herbivores and consider, by and large, that the chemical composition of a specific plant part, for instance a leaf, within each plant and between individual plants is similar. This is a misconception. In this chapter the view is developed that plants are highly heterogeneous hosts in space and time.

There is increasing evidence to support the idea that heterogeneity in chemical and structural composition, together with interplant variation, crucially prevents herbivorous insects from fully exploiting their host plants. Insects, often highly specialized and adapted to certain diets only, face decreased fitness *via* both direct and indirect pathways on resources of varying composition [44].

3.1 PLANT BIOCHEMISTRY

Plants share with all other living organisms a number of biochemical reactions that maintain their basic or primary metabolism, which is involved in formation and breakdown of a limited set of chemicals. These include nucleic acids and proteins with their precursors, particular carbohydrates, carboxylic acids, etc. Based on this primary metabolism, plants have evolved a corona of secondary metabolic pathways producing an extraordinary array of secondary plant substances. The large variety of secondary constituents is produced *via* only three main biogenetic routes, each leading to one or a few key metabolites, from which numerous derivatives are formed, usually by simple enzymatic transformation [87]. To date few biosynthetic routes of secondary compounds have been fully elucidated. Often, they are very complex, as in the case of the synthesis of macapine, an alkaloid derived from two tyrosine molecules. Its manufacture involves 20 enzymatic conversions [114].

It should be emphasized that, although the adjectives 'primary' and 'secondary' might suggest a sharp distinction between both metabolic systems, this is not true. The sugar alcohol sorbitol (**64**), for instance, which is rarely found outside the Rosaceae, functions in hawthorn (*Crataegus monogyna*) as the major soluble carbohydrate. (Bold numbers refer to the structures in Appendix B.) At leaf concentrations of up to 11% dry weight it serves as the primary energy carrier [68] and in this case it seems difficult to attach the label 'secondary' to this compound. Moreover, primary and secondary metabolism are strongly intertwined and the division that has been made between primary and secondary plant substances is therefore arbitrary and for convenience only. In nature the two systems operate as a unit [12].

3.1.1 PRIMARY PLANT METABOLISM

The greater part of a plant's biomass consists of primary plant substances. Some of them occur in great quantities: lignocellulose, for instance, is considered to be the most abundant organic material on earth. Several primary plant metabolites, compounds involved in fundamental plant physiological processes such as proteins, carbohydrates and lipids, form essential nutrients for herbivores. Variation in primary plant compounds can have profound effects on insect preference and performance [11], as will be discussed in Chapter 4.

Working from differences in photosynthetic pathways, plants are classified as C_3 plants, which include the majority of temperate species, and C_4 plants. Species of the latter type are almost exclusively grasses, favouring hot, dry growing seasons. They evolved from plants with the C_3 pathway, possibly in response to a number of environmental changes during the late Miocene/Pliocene epoch (25–2 million years ago) [136]. The differences in carbon fixation processes have important physiological as well as morphological consequences. C_4 plants show increased efficiency in carbon dioxide assimilation and their water requirements are approximately half as much as of C_3 plants [166], which explains the fact that they occur

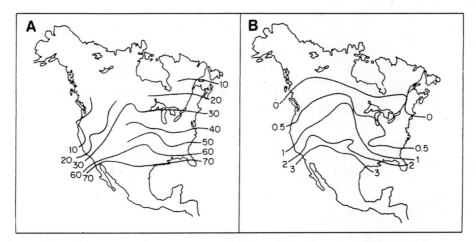

Figure 3.1 'Contour' lines indicating **(A)** the percentage of grass taxa and **(B)** the percentage of dicotyledon taxa that use the C_4 pathway in photosynthesis in North America. (Source: reproduced from Fitter and Hay, 1987, with permission.)

predominantly in (sub)tropical and dry habitats (Fig. 3.1).

The C_4 metabolism is accompanied by some anatomical modifications that have been found to affect insect herbivory. Grasshoppers prefer to feed on C_3 plants rather than on C_4 species because in the latter the veins are surrounded by a layer of large, thick-walled, vascular bundle sheath cells. The edible sheath cells are reinforced with hemicellulose, which apparently cannot be digested [94].

3.1.2 SECONDARY PLANT SUBSTANCES

Secondary plant substances can be defined as 'plant compounds that are not universally found in higher plants, but are restricted to certain plant taxa, or occur in certain plant taxa at much higher concentrations than in others, and are of no nutritional significance to insects' [167]. In the past secondary plant substances were generally considered to represent metabolic waste products. In retrospect this designation is less plausible, because organisms with a primarily anabolic biochemistry should be able to avoid producing wastes.

Some botanists favoured the waste product explanation up till the second half of this century [137], although Justus von Liebig suggested as early as 1858 that secondary metabolites have a function in plant resistance. In a seminal paper Fraenkel [65] stressed the role of secondary plant substances as a defence system against insects and other natural enemies. Though undoubtedly there is much compelling evidence for that supposition, critics of this one-sided concept have emphasized that many secondary plant substances appear to have other (additional) functions within the plant. They argue that the defensive role of these compounds may simply be pleiotropic effects of genes controlling resistance factors that were selected in response to other environmental stresses. Thus competing con- and heterospecific plants (allelopathy), nutrient deficiency (e.g. alkaloids as nitrogen reserves), drought and ultraviolet radiation have been suggested as environmental factors that have stimulated the evolution of the vast biochemical machinery serving the production of secondary plant substances. This point will be discussed in more detail in Chapter 10.

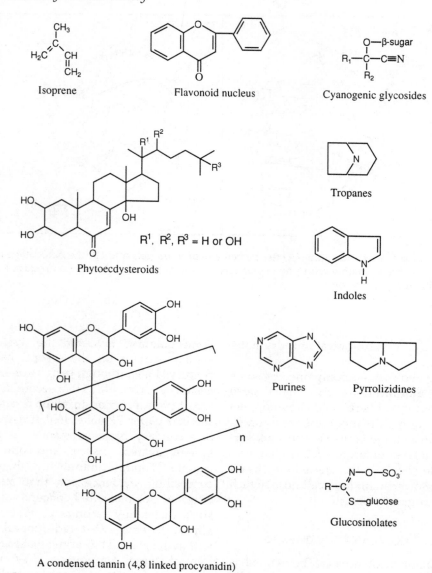

Figure 3.2 General structure of major groups of secondary plant substances.

Because of their ecological role, secondary plant substances can be classified as 'allelochemics', a term coined by C. Whittaker. An allelochemic is defined as a 'non-nutritional chemical produced by an individual of one species that affects the growth, health, behaviour, or population biology of another species' [215].

In contrast to the relative monotony of their primary metabolic profiles, plants produce a bewildering array of secondary metabolites (Fig. 3.2).

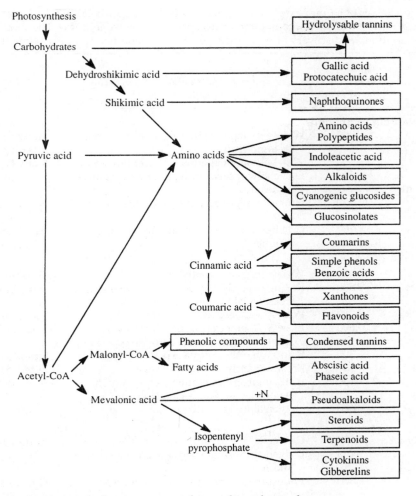

Figure 3.3 Biosynthetic routes of major groups of secondary plant substances.

It has been estimated that the plant kingdom synthesizes several hundreds of thousands of different secondary plant substances. The number of identified compounds may now exceed 100 000 [23] and new structures are reported daily in the scientific literature. It is difficult to construct a satisfactory classification of secondary plant substances, for instance based on molecular structures. Since secondary metabolites are produced from universally present precursors, most often acetyl-CoA, amino acids or shikimate, a classification derived from their biosynthetic pathways [122, 206] (Fig. 3.3) appears to be suitable for most cases.

Starting from these few basic chemicals the synthesis of secondary compounds often involves 'polydimensional networks', i.e. different pathways interconnected in several places. A simplified classification distinguishes (1) N-containing compounds, (2) terpenoids, (3) phenolic compounds and (4) acetylenic compounds (Table 3.1).

Table 3.1 Major classes of secondary plant compounds with significant roles in insect–plant interactions; the approximate numbers of known compounds reflect present knowledge, but will become outdated soon because of the continuous discovery of new structures (Source: modified from Harborne, 1993)

Class	No. of known compounds	Distribution in vascular plants	Physiological activity
Nitrogen-containing compounds			
Alkaloids	10 000	Widely in angiosperms, especially in root, leaf and fruit	Many toxic and bitter-tasting
Amines	100	Widely in angiosperms, often in flowers	Many repellent
Amino acids (non-protein)	400	Especially in seeds of legumes, but relatively widespread	Many toxic
Cyanogenic glycosides	50	Sporadic, especially in fruit and leaf	Poisonous (as HCN)
Glucosinolates	100	Cruciferae and occasionally in ten other families	Acrid and bitter (as isothiocyanates)
Terpenoids			
Monoterpenes	1 000	Widespread in essential oils	Pleasant smells
Sesquiterpenes	3 500	In Angiospermae, especially in Asteraceae, in essential oils and resins	Some bitter and toxic
Diterpenes	3 000	Widespread, especially in latex and resins	Some toxic
Saponins	600	In over 70 plant families, especially Liliflorae, Solanaceae, Scrophulariaceae	Toxic (haemolytic)
Limonoids	300	Predominantly in Rutaceae, Meliaceae	Bitter-tasting
Cucurbitacins	50	Predominantly in Cucurbitaceae	Bitter-tasting and toxic
Cardenolides	150	In 12 angiosperm families, especially in Apocynaceae and Asclepiadaceae	Toxic and bitter
Carotenoids	600	Universal in leaf, often in flower and fruit	Pigments
Phenolics			
Simple phenols	200	Universal in leaf, often also in other tissues	Antimicrobial
Flavonoids (incl. tannins)	8 000	Universal in Angiospermae, Gymnospermae and ferns	Often pigments
Quinones	800	Widespread, especially in Rhamnaceae	Pigments
Polyacetates			
Polyacetylenes	750	Mainly in Compositae and Umbelliferae	Some toxic

3.2 ALKALOIDS

Alkaloids are cyclic nitrogen containing compounds with a limited distribution among living organisms. They include a vast array of chemicals which are often structurally unrelated. Alkaloids are often distinguished on the basis of their precursor molecules. Most of them seem to be derived from a quite restricted range of common amino acids such as lysine, tyrosine, tryptophan, histidine and ornithine. Nicotine (**41**), for instance, is produced from ornithine and nicotinic acid. Among the best known representatives of the **benzyl isoquinoline alkaloids** are papaverine (**42**), berberine (**7**) and morphine. Most of the curare alkaloids also fall into this group, including tubocurarine (**70**). Many of the alkaloids particularly characteristic of the Solanaceae belong to the **tropane alkaloids**. Atropine (**5**), found in deadly nightshade (*Atropa belladona*), and scopolamine (**57**) serve

Table 3.2 Major classes of plant terpenoids

Terpenoid category and general formula	Plant product	Principal types
Hemiterpenoids (C_5H_8)	Essential oils	Tuliposides
Monoterpenoids ($C_{10}H_{16}$)	Essential oils	Iridoids
Sesquiterpenoids ($C_{15}H_{24}$)	Essential oils, resins	Sesquiterpene lactones
Diterpenoids ($C_{20}H_{32}$)	Resins, bitter extracts	Clerodanes, tiglianes, gibberellins
Triterpenoids ($C_{30}H_{48}$)	Resins, latex, corks, cutins	Sterols, cardiac glycosides (cardenolides), phytoecdysteroids, cucurbitacins, saponins
Tetraterpenoids ($C_{40}H_{64}$)	Pigments	Carotenes, xanthophylls
Polyterpenoids [$(C_5H_8)_n$]	Latex	Rubber, balata, gutta

as examples. Cocaine (13) and related alkaloids from the coca plant (*Erythroxylon coca*) are of the same type, but do not occur in the Solanaceae. The so-called **indole alkaloids** owe their name to the presence of an indole nucleus. Two well-known compounds, strychnine (65) and quinine (50), which have a bitter taste to us and are deterrent to many insects, belong to this group of alkaloids. **Pyrrolizidine alkaloids** (PAs) are ester alkaloids. Their biosynthesis has been studied most extensively in *Senecio* species. Senecionine (59) is a noxious macrocyclic diester. The **quinolizidine alkaloids**, which are derived from lysine, are frequently called lupin alkaloids, because of their general abundance in the genus *Lupinus*. Polyhydroxy alkaloids have recently been recognized as compounds that stereochemically mimic sugars, thereby interfering with glycosidases. They act as feeding deterrents against various insects [61]. Some other alkaloids are derived from nicotinic acid, purines, anthranilic acid, polyacetates and terpenes. They include the **purine alkaloids**, e.g. caffeine (9). Most alkaloids have been isolated from angiosperms. They are rarely found in gymnosperms (e.g. conifers) and cryptogams (e.g. ferns).

3.3 TERPENOIDS AND STEROIDS

Terpenoids are the largest group of secondary compounds (15 000–20 000 currently fully characterized). This group shows an incredible structural diversity of compounds the basic skeletons of which are all derived from mevalonic acid or a closely related precursor. This biosynthetic unity does not imply any functional unity nor any detailed unity in chemical properties. Most members of this group can be viewed as being built up of isoprene units (Fig. 3.2), linked together in various ways with different types of ring closure and varying in level of saturation and functional groups. Isoprene itself is emitted from many plants at high temperatures [172], but is otherwise rarely found in nature. Terpenoids can be classified according to the number of their constituent isoprene units (Table 3.2).

Most **monoterpenoids** are volatile compounds, mainly found as components of essential oils. They are responsible for the characteristic odours of many plant species. Monoterpenoids may be acyclic (with an open ring), e.g. geraniol (24), monocyclic, e.g. limonene (34), or bicyclic, e.g. pinene (46). In order to prevent autotoxicity monoterpenoids require specialized storage structures in the plant for sequestration.

The largest class of terpenoids comprises the **sesquiterpenoids**, also commonly found in essential oils. Well-known examples of the drimane-type aldehydes are polygodial (47) and warburganal (74), which act as feeding deterrents to a broad range of insect species

[197]. Sesquiterpene lactones possess a five-membered lactone ring, as exemplified by glaucolide-A (**26**). They occur frequently in the Compositae family, where they are localized in glandular hairs or in latex ducts. Gossypol (**28**) is a well-known phenolic sesquiterpene dimer found in cotton (*Gossypium* sp.) and related genera of the family Malvaceae. Monoterpene and sesquiterpene hydrocarbons are relatively weak odorants for humans, but often serve as important olfactory cues to insects.

Diterpenoids include resin acids in conifers (e.g. abietic acid (**1**)) and the clerodanes, such as clerodin (**12**) from the Indian bhat tree and ajugarin (**3**) from the leaves of *Ajuga remota*. The clerodanes are potent feeding deterrents to many insect species.

Triterpenoids are widespread and diverse, occurring in resins, cutins and corks. They include the limonoids (with azadirachtin (**6**) as one of the strongest insect feeding deterrents known [135]), the lantadenes and the cucurbitacins (e.g. cucurbitacin B (**14**)). The latter compounds, which taste intensely bitter to humans, deter feeding in many herbivorous insects. There are on the other hand also several insect species specialized on cucurbit plants, which use cucurbitacins as powerful host-recognition cues [131].

Saponins contain a polycyclic aglycone moiety of either triterpenoid (C_{30}) or steroid (C_{27}) structure attached to a sugar moiety. Aescin (**2**) and dioscin (**19**) are examples of the two types, occurring in horse chestnut (*Aesculus hippocastaneum*) and yams (*Dioscorea* spp.) respectively.

Insects are unable to synthesize the steroid nucleus in quantity and must obtain cholesterol or sitosterol (**62**) from their diet for the synthesis of steroid hormones such as the moulting hormone ecdyson (**21**). A number of plants produce ecdyson and closely resembling derivatives, which are called **phytoecdysteroids** (Fig. 3.2). In particular, some ferns and gymnosperms may contain concentrations up to five orders of magnitude above those occurring in insects. Rhizomes of the common fern (*Polypodium vulgare*), for instance, contain up to 1% β-ecdysone (i.e. the major insect ecdysteroid), and dry stems of *Diploclisia glaucescens* (Menispermaceae) have been reported to contain as much as 3.2% of this phytoecdysteroid. Because a true physiological role of ecdysteroids in plants is unknown, it seems attractive to postulate that they serve primarily as a defence mechanism against insect herbivores. The experimental evidence for this assumption is at the present time only meagre [115].

Some compounds that are of terpenoid origin but appear to have lost or gained carbon atoms include the gibberelins (**25**), which function as hormones in higher plants, the tocopherols (e.g. vitamin E (**73**)), which act as antioxidants in seed oils, and the active principles of marijuana, e.g. cannabidiol (**10**).

3.4 PHENOLICS

Phenolic compounds are ubiquitous in plants [83, 210]. Phenolics possess an aromatic ring with one or more hydroxyl groups, together with a number of other constituents. The name of this group derives from the simple aromatic parent substance phenol (**44**), but most contain more than one hydroxyl group (polyphenols). They are conveniently classified according to the number of carbon atoms in the basic skeleton (Table 3.3).

A group of relatively simple phenolics include the hydrobenzoic acids, e.g. vanillic acid (**72**), the hydroxycinnamic acids, e.g. caffeic acid (**8**), and the coumarins. Examples of the latter category are umbelliferone (**71**), widespread in the Umbelliferae, and scopoletin (**58**), commonly occurring in solanaceous plants but also present in other families.

By far the largest and most diverse group of plant phenolics are the **flavonoids**, which occur universally in higher plants. Usually, a plant contains several representatives of this group of compounds and almost every plant species possesses its own distinctive flavonoid profile. The flavonoids share a basic C_6–C_3–C_6

Table 3.3 The major classes of phenolics in plants. (Source: modified from Harborne, 1994)

Basic skeleton	Number of carbon atoms	Class	Examples
C_6	6	Simple phenols	Catechol, hydroquinone
		Benzoquinones	2,6-dimethoxybenzoquinone
C_6-C_1	7	Phenolic acids	p-hydroxybenzoic acid, salicylic acid
C_6-C_2	8	Acetophenones	3-Acetyl-6-methoxybenzaldehyde
		Phenylacetic acids	p-hydroxyphenylacetic acid
C_6-C_3	9	Hydroxycinnamic acids	Caffeic acid, ferulic acid
		Phenylpropenes	Myristicin, eugenol
		Coumarins	Umbelliferone, aesculetin
		Isocoumarins	Bergenin
		Chromones	Eugenin
C_6-C_4	10	Naphthoquinones	Juglone, plumbagin
$C_6-C_1-C_6$	13	Xanthones	Mangiferin
$C_6-C_2-C_6$	14	Stilbenes	Lunularic acid
		Anthraquinones	Emodin
$C_6-C_3-C_6$	15	Flavonoids	Quercetin, malvin
		Isoflavonoids	Genistein
$(C_6-C_3)_2$	18	Lignans	Podophyllotoxin
$(C_6-C_3-C_6)_2$	30	Biflavonoids	Amentoflavone
$(C_6-C_3)_n$	$9n$	Lignins	
$(C_6)_n$	$6n$	Catechol melanins	
$(C_6-C_3-C_6)_n$	$15n$	Flavolans (condensed tannins)	

structure (Fig. 3.2; e.g. kaempferol (**33**)). The flavonoid nucleus is normally linked to a sugar moiety to form a water-soluble glycoside. Most flavonoids are stored in the plant cell vacuoles. The flavonoids can be subdivided into flavones, e.g. luteolin (**36**), flavanones, e.g. naringenin (**40**), flavonols, e.g. kaempferol (**33**), anthocyanins and chalcones. Many flavones, flavanones and flavonols absorb light in the visible region and hence give flowers and other plant parts their bright yellow or cream colours. Many colourless representatives of these groups are often of considerable significance as feeding deterrents (e.g. catechin) or as insect toxicants (e.g. rotenone (**52**)). Rutin (**53**), the most widely distributed flavonol glycoside in plants, is a strong feeding deterrent to a number of polyphagous insects, such as *Schistocerca americana* [14], and phaseollin (**43**) is among the most potent feeding deterrents ever recorded. In tests with root-feeding larvae of the beetle *Costelytra zealandica* the FD_{50} value (the concentration at which feeding is reduced to 50% of the control) of this compound was as low as 0.03 ppm [117]. On the other hand several flavonoids have been found to be used by monophagous or oligophagous insect species to recognize their host plants and to stimulate feeding (Table 3.4).

Anthocyanins embrace most of the natural red and blue pigments in flowers, fruits and leaves. They are glycosides with glucose as the most common sugar moiety. Thus cyanin (**15**) is a glucose ester of cyanidin.

Tannins are polyphenolic compounds (molecular weight 500–20 000 daltons), that are found in all classes of vascular plants, often in high concentrations (Table 3.5).

They usually occur as soluble components in the sap of living cells. Tannins bind with their phenolic hydroxyl groups to almost all

Table 3.4 Flavonoids belonging to different classes that have been implicated as insect feeding stimulants (Source: modified from Harborne and Grayer, 1993)

Flavonoid class	Feeding stimulant flavonoid	Host plant/family	Insect/order	Reference
Flavonol O-glycosides	Isoquercitrin	Morus alba (Moraceae)	Bombyx mori (Lepidoptera)	79
	Kaempferol 3-O-xylosylgalactoside	Armoracia rusticana (Cruciferae)	Phyllotreta armoraciae (Coleoptera)	142
Flavone O-glycosides	6-Methoxyluteolin 7-O-rhamnoside	Alternanthera phylloxeroides	Agasicles sp. (Coleoptera)	223
Flavone C-glycosides	Schaftoside, neoschaftoside, carniloside, isoorientin 2'-glucoside, neocarniloside, isocoparin 2'-glucoside, and its 6'-p-coumaric and ferulic acid esters	Oryza sativa (Gramineae)	Nilaparvata lugens, Sogatella furcifera, Laodelphax striatellus (Homoptera)	15
Dihydroflavonols and flavonone	Taxifolin, dihydrokaempferol, pinocembrin	Prunus spp. (Rosaceae)	Scolytus mediterraneus (Coleoptera)	120
Dihydrochalcone O-glycoside	Phloridzin (45)	Malus spp. (Rosaceae)	Aphis pomi, Rhopalosiphum insertum (Homoptera)	112
Flavanol O-glycoside	Catechin 7-O-xyloside	Ulmus americanus (Ulmaceae)	Scolytus multistriatus (Coleoptera)	50

Table 3.5 Concentrations of some secondary compounds in plants

Compound	Class	Plant species	Concentration (% dry weight)	Reference
Vincristine	Alkaloid	Catharanthus roseus (leaf)	0.0002	225
Digitoxin	Cardenolide	Digitalis purpurea (leaf)	0.06	129
Bergaptan (a.o.)	Furanocoumarin	Pastinaca sativa (leaf)	0.1	221
Glucobrassicin	Glucosinolate	Brassica oleracea (leaf)	0.1	35
Hypericin	Quinone	Hypericum hirsutum (inflorescence)	0.3	155
Tomatine	Glycoalkaloid	Lycopersicum esculentum (leaf)	0.5–5.1	174
Amygdalin	Cyanogenic glycoalkaloid	Prunus amygdalus (seed)	3–5	67
Tannins	Polyphenol	Quercus robur (leaf)	0.6–6.0	59
		Englerina woodfordioides (leaf)	15	211
		Thea sinensis (leaf)	<30	206
L-dopa	Amino acid	Mucuna (seed)	5–10	9
Tremulacin (a.o.)	Phenolglycoside	Populus trichocarpa (leaf)	23	16, 192
Resin	Phenolic aglycones	Larrea cuneifolia (young leaves)	44	157

Table 3.6 Distribution of hydrolysable and condensed tannins in the plant kingdom (Source: reproduced from Swain, 1979, with permission)

Taxon	% of plant genera containing tannins	
	Hydrolysable	Condensed
Psilopsida	0	0
Lycopsida (club mosses)	0	0
Sphenopsida (horsetails)	0	28
Filicopsida (ferns)	0	92
Gymnosperms	0	74
Angiosperms	13	54
Monocotyledons	0	29
Dicotyledons	18	62

soluble proteins, producing insoluble copolymers. Enzymes complexed in this way show a marked reduction in activity. Also proteins bound to tannins cannot be degraded by enzymes in the digestive tract and tannins are therefore generally thought to decrease the nutritional value of plant tissues. Tannins may also cross-link with nucleic acids and polysaccharides, thereby impeding their physiological function. Tannins are commonly classified into two types: the hydrolysable and the non-hydrolysable or condensed tannins. Hydrolysable tannins are limited to angiosperms, but the condensed tannins occur widespread in the plant kingdom (Table 3.6).

The most common hydrolysable tannins are esters of gallic acid (**23**) and hexahydroxydiphenilic acid (**30**) with sugars. Condensed tannins are polymers of flavonoid units (Fig. 3.2), linked by carbon–carbon bonds that are not susceptible to hydrolysis. During plant tissue maturation, such as fruit ripening, tannins often polymerize further and become concomitantly less soluble. As a result their astringent taste disappears.

3.5 GLUCOSINOLATES

This is a small but well defined and basically coherent group of compounds. Their general formula is given in Figure 3.2. All glucosinolates, or mustard oil glucosides, contain sulphur as well as nitrogen atoms. They can be either acyclic, e.g. sinigrin (or allyl isothiocyanate (**61**)), or aromatic, e.g. sinalbin (**60**). Hydrolysis of glucosinolates is facilitated by the enzyme myrosinase, which leads to the formation of isothiocyanates (or mustard oils), nitriles and other compounds, depending on pH and other conditions. Hydrolysis occurs rapidly when plant tissue is ruptured, but probably also takes place, though at a much lower rate, during normal catabolism. Glucosinolates occur mainly, but not exclusively, in the Cruciferae plant family. Ever since Verschaffelt's [204] historical experiments with sinigrin and cabbage white butterflies this group of compounds has attracted much interest from students of insect–plant relationships [121]. Glucosinolates are unpalatable and toxic to many generalist feeders and several specialists living on non-cruciferous plants [33]. They are weakly deterrent to some crucifer-feeding insects, such as the cabbage looper, *Trichoplusia ni* [132]. Glucosinolates are strong feeding stimulants to many specialists on plants belonging to the Cruciferae family [34] (Fig. 3.4).

3.6 CYANOGENICS

Probably all plants have the ability to synthesize cyanogenic glycosides (general formula in Fig. 3.2), but in most species they are metabo-

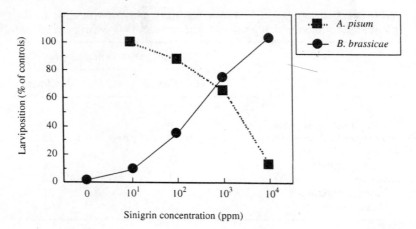

Figure 3.4 Production of larvae of pea aphids (*Acyrthosiphon pisum*) and cabbage aphids (*Brevicoryne brassicae*) on broad bean leaves systemically treated with sinigrin, a glucosinolate. Controls were untreated turnip leaves for *B. brassicae* and untreated broad bean for *A. pisum*. Sinigrin inhibits reproduction in the non-cruciferous species but stimulates the species specialized on crucifers, even when the compound occurs in non-host plants. (Source: redrawn from Nault and Styer, 1972.)

lized and not accumulated. Cyanogenic compounds are optically active because of the chirality of the hydroxylated C-atom. Thus prunasin (**48**), occurring in *Prunus* spp., is the stereoisomer of sambunigrin (**56**), typically found in *Sambucus* spp. The vacuoles of plant cells often serve as a storage place, e.g. in many Rosaceae. On damage of plant tissue the cyanogenic compounds are enzymatically hydrolysed and very toxic hydrogen cyanide (HCN) is formed [150]. When crushing the leaves of, for example, bird cherry (*Prunus padus*) its characteristic 'bitter almond smell' is easily discernible. In addition to a probable function as storage forms for reduced nitrogen, a protective role of cyanogenics against herbivores and pathogens is appealing [97]. The fact that HCN is a potent feeding deterrent to a diverse range of insects [218] supports its conjectural protective role.

3.7 LEAF SURFACE CHEMISTRY

The first physical contact between an insect and a plant occurs when the insect lands or otherwise touches the leaf surface. Chemical or structural characteristics of the plant surface may determine the insect's subsequent behaviour (Chapter 6). Therefore, the leaf surface, and especially its chemistry, merits special attention. The surface waxes or resins constitute the first line of resistance. The structural as well as chemical composition of the epicuticular wax layer differs among plant species (Fig. 3.5).

They show an extensive variation in their micromorphology [128], ranging from amorphous films to mixed arrays of wax tubes, rods and plates, and 14 major types have been distinguished [100]. The wax layer covering the cuticle consists of long-chain carbohydrates, alkylesters, primary alcohols and fatty acids [54]. Wax composition may vary considerably, even among congenic species, as was found to be the case in eight Papaveraceae species [102]. Frequently mixed with the wax are toxic constituents. When intact plants are briefly dipped into water or organic solvents [179] a wealth of compounds can be washed off that are present on the leaf exterior, mostly in small amounts. Such washings often contain not only sugars [46] and some amino acids [177]

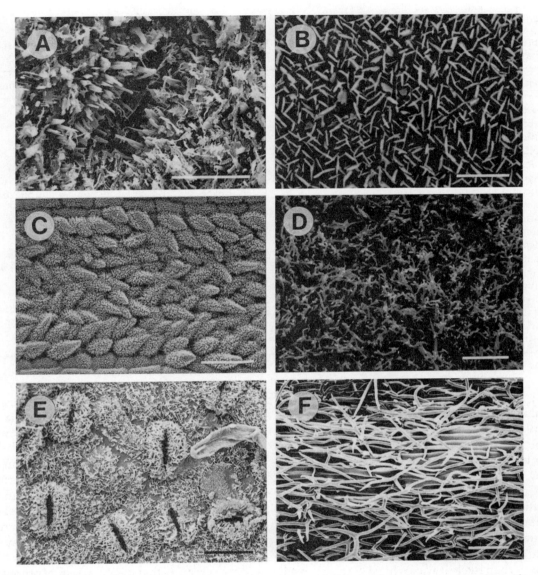

Figure 3.5 Scanning electron micrographs of plant surfaces. **(A)** *Brassica oleracea* var. capitata (Cruciferae). Abaxial leaf surface, showing a stomatal complex overarched by epicuticular wax crystals in the form of tapered, ridged tubes and dendritic plates. Scale = 10 μm. **(B)** *Festuca arundinacea* (Gramineae). Adaxial leaf surface, showing epicuticular wax on a cell on the top of an epidermal ridge. The wax crystals are in the form of plates, typical of the crystals formed by long-chain primary alcohols. The crystals stand on edge and are mutually aligned in three preferred orientations at 120°. Scale = 2 μm. **(C)** *Cyathodes colonsoi* (Epacridaceae). Abaxial leaf surface showing a band of short wax-crystal-encrusted epidermal trichomes overarching the stomatal complexes. Scale = 80 μm. **(D)** Sitka spruce, *Picea sitchensis* (Pinaceae). Adaxial leaf surface. Surface view of the epicuticular wax tubes, predominantly composed of *n*-nonacosan-10-ol, in the centre of the antitranspirant wax plug that fills the stomatal antechamber. Scale = 4 μm. **(E)** *Quercus pubescens* (Fagaceae). Stomatal complexes encrusted with primary-alcohol-rich wax crystals on the abaxial leaf surface. The upper (adaxial) surface lacks this thick epicuticular crust. Scale = 20 μm. **(F)** *Rosmarinus officinalis* (Labiatae). Abaxial leaf surface with dense indumentum, composed of many-branched trichomes. Scale = 100 μm. (Source: reproduced by courtesy of Dr C. E. Jeffree, University of Edinburgh.)

Table 3.7 Chemicals extracted and identified from leaf surfaces that have been found to affect insect behaviour

Chemical(s)	Plant species	Reference
Fructose, glucose, sucrose	Corn, sunflower	45
Amino acids	*Vicia faba*, *Beta vulgaris*	104
Amino acids	Corn, sunflower	45
Lipids	Cabbage and other species	54
Dulcitol (sugar alcohol)	*Euonymus europaea*	105
p-hydroxybenzaldehyde	Sorghum	220
Glucobrassicin (glucosinolate)	Cabbage	156, 203
Phloridzin (phenolic)	Apple	112
Anthraquinone (phenolic)	*Lolium perenne*	3
Luteolin, *trans*-chlorogenic acid (phenolics)	Carrot	60
Falcarindiol (polyacetylene)	Carrot	178
Sesquiterpenes	Wild tomato	107
Triterpeneol acetate	Sweet potato	144
Duvane diterpenes, α- and β-diols, saturated hydrocarbons	Tobacco	98
Tyramine (alkaloid), *trans*-chlorogenic acid	*Pastinaca sativa*	27
Naringin, hesperidin (flavanones), quinic acid	Citrus	96
Aristolochic acids	*Aristolochia* spp.	143

but also secondary plant substances such as phloridzin (**45**, a dihydrochalcon of apple leaves [112]), glucobrassicin (**27**, a glucosinolate from cabbage plants [203]), furanocoumarins [178] and alkaloids [102] (Table 3.7).

Contact with these surface chemicals often suffices to prevent insects from further investigation of the plant. Migratory locusts, for instance, may be inhibited from taking a test bite from an intact plant merely upon palpation. When the leaf waxes have been removed, however, the insects take one or a few test bites before deciding to reject these non-host plants (Fig. 3.6).

The other side of the coin is that leaf-surface chemicals may help some insects to recognize their specific host plants at an early stage. In several beetle species food intake is stimulated by the dominant wax components of their various host plants [1].

Plant leaves, stems and roots are often covered with hairs or trichomes, highly variable appendages of the epidermis, ranging from the simple unicellular type to arborized spiny structures (Fig. 3.7). They include glandular (or secretory; Fig. 3.8) and non-glandular hairs, scales and papillae.

They function among others as a structural or chemical resistance against small herbivores, including insects. Glandular trichomes are mainly found in the Labiatae, Solanaceae, Compositae and Geraniaceae. They contain highly specialized secretory cells, which synthesize and accumulate a large variety of terpene oils and other essential oils. More than 100 different mono-, sesqui-, and diterpenes have been identified in trichome secretions [161]. When ruptured, some types of trichome exude a sticky secretion in which small insects are trapped and killed [52, 164], whereas feeding by larger insects is hampered, thus limiting population development.

In view of its often decisive role in insect–plant relationships the plant surface, including its pubescent lining, has received relatively little attention. Plant breeders, however, have shown that foliar wax coatings,

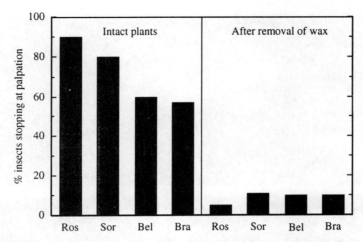

Figure 3.6 Rejection rates of four plant species at palpation by fifth instar nymphs of *Locusta migratoria*. Ros = *Rosa*; Sor = *Sorghum*; Bel = *Bellis*; Bra = *Brassica*. (Source: redrawn from Woodhead and Chapman, 1986.)

as well as various kinds of trichome, play an important role in the resistance of some crop cultivars to insect attack, strongly indicating the importance of surface characteristics in plant–insect relationships. The challenge is now to better understand the biological details of the boundary between an insect's mouthparts and its food, i.e. the plant interior. Our present knowledge is well covered by reviews by Juniper and Southwood [106], Chapman and Bernays [32] and Städler and Roessingh [179].

3.8 PLANT VOLATILES

Our knowledge of the chemistry of plant odours is still very incomplete because until recently it was difficult to collect sufficient amounts of volatiles from the atmosphere around intact plants for chemical identification. At the same time it is known that many secondary plant substances and several intermediates of primary metabolism have high enough vapour pressures to affect other organisms as a volatile. Many terpenoids, aromatic phenols, alcohols, aldehydes, etc. with molecular weights ranging from 100 to about 200 easily volatilize when exposed to the air and are indeed liberated when plant tissues are damaged. Intact plants also give off such volatile compounds, which permeate through open stomata, leaf cuticles and gland walls, but the release rate is much lower. In the past identification of plant volatiles began with extracts made of chopped or macerated plant material. During the last two decades 'headspace collection' methods have been developed to obtain the volatiles from the air around undamaged (or damaged) plants [62]. In combination with gas chromatography and mass spectrometry this technique obviously gives much more reliable information on the composition of naturally emitted blends of volatiles than the tissue extraction methods. For example, the headspace air of cotton plants contains 54 chemicals but only six of them also occur among the 58 compounds present in the essential oil of cotton buds [91].

Plant volatiles can be classified into general and specific volatiles. The commonly occurring 'green leaf volatiles' [207], which give damaged leaves a characteristic 'cut grass' smell, are six-carbon alcohols and aldehydes (Table 3.8). Some authors also include under

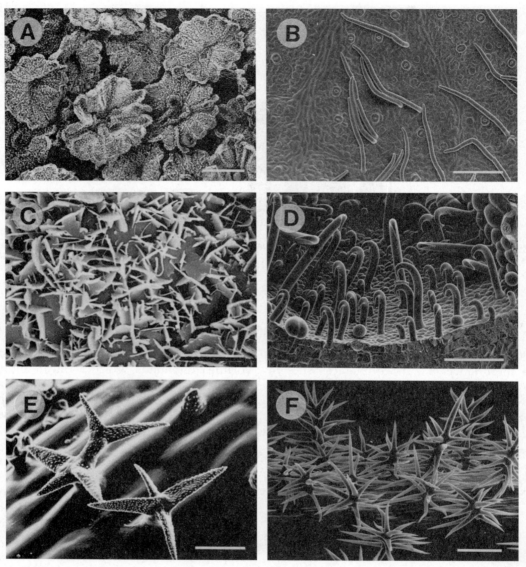

Figure 3.7 Scanning electron micrographs of trichomes. **(A)** *Rhododendron callostrotum* (Ericaceae). Waxy peltate trichomes on abaxial leaf surface. Scale = 100 μm. **(B)** Leaf of *Fagus sylvatica* (Fagaceae) that has just achieved full expansion in early summer, showing the deciduous clothing trichomes that confer a silky appearance on the expanding leaves. The epidermal cells are covered with a smooth wax film. Scale = 100 μm. **(C)** Abaxial leaf surface of *Quercus pubescens* (Fagaceae) showing detailed structure of the crystalline epicuticular wax plates. Scale = 4 μm. **(D)** Hooked trichomes on the abaxial surface of an expanding primary leaf of *Phaseolus vulgaris* (Leguminosae). The hooks catch in the tarsal joints of herbivorous arthropods, immobilizing them. Scale = 60 μm. **(E)** Branched trichomes of *Lavandula spicata* (Labiatae). The warty surface of the cells is produced by local enlargement of the cuticular layer of the cuticle. Scale = 30 μm. **(F)** Arboriform trichomes on a bud surface of kangaroo-paw, *Anigozanthus flavidus* (Amaryllidaceae). In the young buds the dense indumentum formed by these hairs may protect them from solar radiation. Scale = 200 μm. (Source: reproduced by courtesy of Dr C. E. Jeffree, University of Edinburgh.)

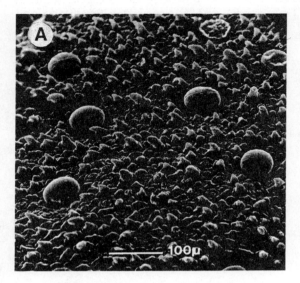
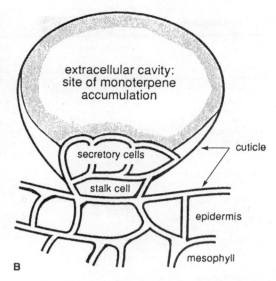

Figure 3.8 Glandular trichomes on leaf surfaces which accumulate monoterpenes. **(A)** Scanning electron micrograph of leaf surface of thyme (*Thymus vulgaris*). **(B)** Schematic cross-sectional view of a glandular trichome from the leaf surface of peppermint (*Mentha piperita*), showing secretory cells where monoterpenes are produced before being discharged in an extracellular cavity. (Source: reproduced from Gershenzon and Croteau, 1991, with permission).

Table 3.8 Green leaf volatiles synthesized from α-linolenic and linoleic acids (see Fig. 3.9)

Alcohols	Aldehydes
3-(*E*)-hexenol	3-(*E*)-hexenal
3-(*Z*)-hexenol	3-(*Z*)-hexenal
2-(*E*)-hexenol	2-(*E*)-hexenal
hexanol	*n*-hexanal

this heading some of their derivatives, e.g. acetates (Fig. 3.9).

They are generally produced, mostly in appreciable amounts, by oxidation of leaf lipids [88]. The precursor of the unsaturated aldehydes and alcohols, linolenic acid, often accounts for more than 1% of the dry weight of leaves. The relative amounts of the various green odour components emitted must be unique for a given plant species. Some insects can perceive these species specific variations and employ them to discriminate between host and non-host plants. Colorado potato beetles, for instance, respond positively to the mixture of green leaf volatiles produced by potato foliage, but when the natural combination is distorted by raising the concentration of one of its components, the response disappears [207].

Many or perhaps most plants also emanate taxon-characteristic volatiles, but so far they have been investigated with headspace techniques in only a limited number of cases (Table 3.9). The fact that many insects respond to a wide variety of plant-derived volatiles [130, 131] is an indication that they may use them as air-borne cues in finding or avoiding certain plants.

The number of volatile substances in the air around plants may run up to several hundreds, though often the blend is dominated by one or a few major compounds (Table 3.9). The air around corn leaves, for instance, contains at least 24 compounds, but the major fraction (75%) consists of only seven components [24].

Table 3.9 Headspace analysis of plants from five families (straw = strawberry; cab = cabbage); quantitative indications: tr = <0.1%; + = 0.1–1%; ++ = 1–5%; +++ = 5–20%; ++++ = >20% of total headspace volatile collection

Chemicals	Leguminosae		Rosaceae		Cruciferae		Solanaceae		Gramineae	
	Bean[1]	Clover[2]	Apple[3]	Straw[4]	Cab[5]	Mustard[6]	Tomato[7]	Potato[8]	Maize[9]	Oats[10]
Alcohols										
1-butanol	+		+							
1-penten-3-ol	+		+							
1-hexanol	++	+	tr	++						
(Z)-3-hexenol	++++	++++		++++	++	++			+++	+
(E)-2-hexenol		++								
2-ethyl-1-hexanol	++		+++		++					
1-octanol									++	
1-octen-3-ol	++	+		tr						
Others				3/+					1/tr	
Aldehydes										
hexanal	++		++		+				+++	
2-hexanal	++									
(E)-2-hexenal			++							
heptanal	+								+++	
others	3/+++		1/tr	1/tr	1/+					1/tr
Esters										
1-butyl acetate	+								++	
hexyl acetate	tr			tr	++					
(Z)-3-hexenyl acetate	+++	++++	+++	++++	+++	+++			++++	+
others	2/++								++++	++++
Terpenoids										
α-pinene					++		++		+++	
β-pinene					++		+		++	
α-thujene					++		++			
Sabinene					++					
Myrcene					++++					
β-phellandrene					++	+			++	tr
Limonene	++				+		++++			
1,8-cineole					+++		+++	+++	+++	tr
Linalool	++		+	+++	+++	+		++		
β-elemene					++				++	
(E)-β-ocimene	++++	++++	++		+		+			
(E)-β-farnesene			++			++				+
(E,E)-α-farnesene			++++			+++				
β-caryophyllene			tr	++						
α-terpinolene							+	++++		+
β-selinene							++++			
δ-cadinene								++++		
γ-cadinene								+++		
Others	2/++	2/++	2/++	2/++	1/+	2/+++	4/++	3/+		
Benzenoids										
Methyl salicylate	++	+	++	tr					++	tr
Benzyl alcohol				++						
2-phenylalcohol				++						
Benzaldehyde						++++				
Benzothiazole								++		tr
Others					2/+	1/++				
Nitrogen containing										
Benzonitrile								+++		
Phenylacetonitrile					+++					
Benzylisothiocyanate					++					
Indole					++					
Other compounds in other classes		1/+	2/tr		3/++	1/tr			3/++*	
Unidentified (%)	2.4		1.8							
Total number of identified compounds	23	10	20	16	21	14	12	11	16	10

[1] Excised leaves, greenhouse-grown (Ref. 189); [2] Excised leaves, field grown (Ref. 26); [3] Field grown (Ref. 188); [4] Excised and cut field-grown leaves from flowering plants (Ref. 80); [5] Intact potted plants, greenhouse-grown (Ref. 17); [6] Greenhouse-grown, whole shoots cut at ground level (Ref. 193); [7] Leaves (cut in strips) from freshly excised shoots grown in greenhouse (Ref. 5); [8] Intact potted plants, greenhouse-grown (Ref. 19); [9] Excised leaves, greenhouse-grown (Ref. 190); [10] Excised shoots, field-grown (Ref. 25)
* Three ketones were collected from maize, which together comprised 7.5% of the headspace collection

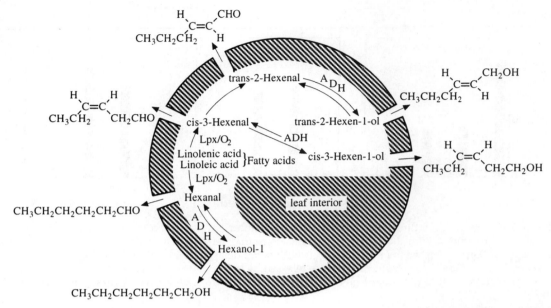

Figure 3.9 Green leaf volatiles are derived from fatty acids in the leaf interior and get into the surrounding air via the stomata. ADH = alcoholdehydrogenase; Lpx = lipoxygenase. (Source: redrawn from Visser and Avé, 1978.)

Regarding the headspace composition listed in Table 3.9, some trends can be noted. Cruciferous and solanaceous species produce very few or no alcohols and aldehydes in their headspace, while the legumes, the rosaceous species and maize emit them in large quantities. Cabbage and the two solanaceous plants release a variety of terpenoids in large amounts, while both the leguminous and rosaceous species, as well as oats, produce fewer compounds, which nevertheless can be major components. All plants figuring in this table, except the two solanaceous species, emit (Z)-3-hexenyl acetate as a major component of their headspace. Another interesting fact is that plant species belonging to the same family may show clear differences in their emitted volatiles. For example, while cabbage produces a greater variety of different terpenoids than mustard (11 *versus* six), the latter releases several isothiocyanates, which were not detected in the cabbage headspace. Sampling methodology may in part be responsible for such differences. Cabbage plants were sampled in intact, potted condition, whereas mustard plants were cut at the shoot base prior to sampling. In fact, most studies so far have employed excised plant tissues to obtain headspace collections. There is now convincing evidence that plants change their release profiles, often dramatically, upon damage [195] (Fig. 3.10), and that the bouquet emitted after mechanical damage differs from that induced by herbivore damage [194, 196].

The examples given show that large qualitative and quantitative differences often exist between the volatiles from different plant species. This may even be the case between the volatiles released by different cultivars. Thus, out of 43 compounds produced by three chrysanthemum cultivars only 14 were common to all three [182]. Although in this chapter the discussion is restricted to volatiles

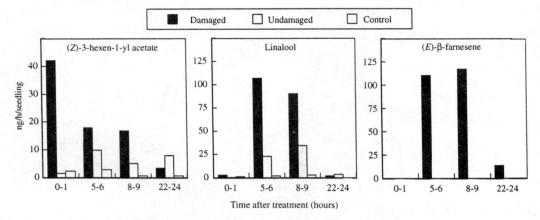

Figure 3.10 Amounts of three components of leaf volatiles emitted by corn seedlings after artificial damage and treatment with caterpillar regurgitant (to mimic herbivory) at various times after treatment ('damaged'). Some components are also, albeit with some delay, systemically released by 'undamaged' leaves of injured plants. As a 'control' volatiles released by unharmed plants were used. The composition of odour blends emitted upon damage changes with time. (Source: data from Turlings and Tumlinson, 1992.)

from vegetative plant parts only, it may be mentioned that flower fragrances may easily consist of 100 or more components [149].

Damage inflicted by herbivorous insects has been found to stimulate the emission of plant volatiles. It is noteworthy that the profile of the released *mélange* usually differs from that produced by the intact plant, with terpenoids often taking a prominent place. Two acyclic homoterpenes, E-4,8-dimethyl-1,3,7-nonatriene (**18**) and 4,8,12-trimethyl-1,3(E), 7(E),11-tridecatetraene (**69**) are of special interest, since they are often found in the headspaces of herbivore-infested plants. A curious detail is that the amounts of these compounds seem to vary with the herbivore species. Thus the headspace of apple leaves infested with the spider mite *Panonychus ulmi* contains 49% 4,8-dimethyl-1,3(E),7-nonatriene, while this is only 9% when another spider mite, *Tetranychus urticae*, infests the leaves. Interestingly, these differences suffice to attract different species of predatory mite [47]. Apparently these predators, like the Colorado potato beetles mentioned above, react to specific ratios of odour components.

In contrast to compounds that remain inside the plant and can be recycled when necessary, chemicals that are released into the air entail a permanent loss of energy. The energy the plant has to put into the production of compounds that are either purposely or unavoidably given off is correlated with the quantity as well as the types of compound produced. The limited data available suggest that in some cases the production costs are not negligible. For example, the production of volatile isothiocyanates by *Bretschneidera sinensis* during active growth may amount to 0.7% (expressed as a fraction of dry weight of total growth) per day [20]. At a different scale some figures are known of quantities of monoterpenes, which are emitted mainly by trees as isoprene and α-pinene. In the USA nationwide such emissions are estimated to amount to $3-6 \times 10^3$ kg carbon/km^2/year (anthropogenic hydrocarbon emissions are equivalent to 2×10^3 kg carbon/km^2/year) [31].

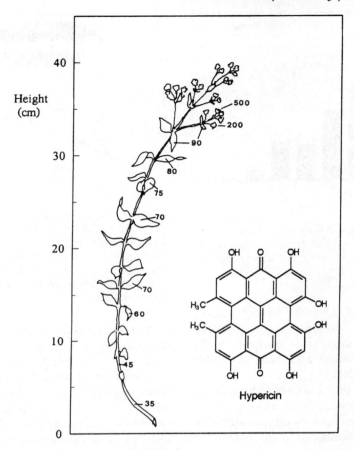

Figure 3.11 Hypericin concentration (μg/g fresh weight) of leaves and flowerheads of hairy St John's wort, *Hypericum hirsutum*. (Source: modified from Rees, 1969.)

3.9 CONCENTRATIONS OF SECONDARY PLANT SUBSTANCES

The relative amounts of secondary compounds found in plants not only vary spectacularly (Table 3.5) but are also not homogeneously distributed over the various plant parts. On the contrary. The latter point is very relevant to herbivorous insects, because they often feed on particular cells or certain tissues only (Chapter 2). From the plant's point of view it seems logical, when a protective function is attributed to secondary plant substances [65, 66], to allocate most of its defensive chemicals to those parts where insect damage would inflict the greatest losses in plant fitness [201]. Different plant parts then would store different levels of protectants, because damage to seeds, for instance, would have a greater impact on plant fitness than damage to old leaves. Fruits of wild parsnip (*Pastinaca sativa*), in accordance with this concept, harbour four times higher furanocoumarin concentrations than leaves and 800 times higher levels than roots [221]. Likewise, the flowers of *Hypericum hirsutum* contain five to ten times more hypericin than the leaves [155] (Fig. 3.11) and generally young leaves and other growing plant parts are better protected by secondary compounds than

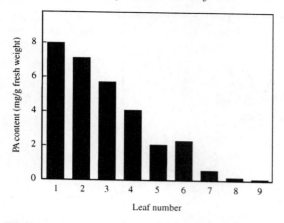

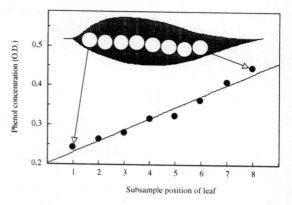

Figure 3.12 Pyrrolizidine alkaloids (PAs) concentration (mg/g fresh weight) per leaf of *Cynoglossum officinale* rosette plants. No. 1 is the youngest leaf. (Source: redrawn from van Dam *et al.*, 1994.)

Figure 3.13 Concentration of phenolics (O.D. = optical density) in eight subsamples of a single leaf blade of *Populus angustifolia*. Gall aphids (*Pemphigus betae*) prefer to form their galls at the base of the leaf, where the concentration of phenolics is lowest. (Source: redrawn from Whitham, 1983; data from Zucker, 1982.)

mature tissues [29]. Thus the youngest leaves of rosette plants of hound's tongue (*Cynoglossum officinale*) contain up to 53 times higher levels of pyrrolizidine alkaloids than older leaves [199] (Fig. 3.12).

Even the leaves of the same tree branch may differ in, for instance, polyphenol contents to such an extent that tree-dwelling caterpillars make extensive foraging trips within the canopy, feeding on some leaves only and avoiding others. Clearly many forest insects, like lepidopteran and sawfly larvae, make careful choices as they forage [69, 168].

Even within a single leaf local concentrations of protective chemicals may vary and in poplar leaves, for instance, gradually increase from the base to the leaf tip (Fig. 3.13). Colonizing gall aphids, therefore, do not settle randomly on a leaf, but nearly always attempt to form their galls at the base of that leaf, where the concentration of phenolics is lowest [224]. Another example in which different elements of a plant organ show strikingly different quantities of secondary compounds is provided by *Cola nitida*. The caffeine (9) content of its fruit is only a trace, that of the seed coat is 0.44% (dry weight) but that of the seed contents is 1.85% [146]. Not only different leaf parts, but also different leaf tissues often show considerable quantitative differences in resistance chemicals. Since neonate larvae, when starting feeding, often first encounter the contents of epidermal cells, it seems a good strategy for a plant to concentrate its chemical resistance in its epidermis. In Gramineae cyanogenic glucosides are concentrated in epidermal cells to such an extent that they represent 90% of the soluble carbohydrate content of the epidermal tissue [151]. The palisade layer in the mesophyll of holly leaves consists of 38% (dry weight) of saponins, whereas the remainder of the leaf contains on average only 1.3%. Despite these huge amounts of saponins in the palisade cells, which act as protease inhibitors in the guts of many herbivores, larvae of *Phytomyza ilicicola* tunnel exclusively in this particular tissue, which has a protein concentration of about ten times that of the remainder of the leaf tissues [111]. In contrast, some leaf-mining larvae of oak trees prefer to feed on the spongy mesophyll cells and rather avoid the palisade layer, which is in this case high in tannins [59].

In a number of cases secondary plant substances are produced in tissues other than where they accumulate. Some alkaloids, e.g. nicotine (41) in tobacco plants, are synthesized in the roots and transported *via* the xylem to the leaves. Alkaloids often occur at the highest concentrations in young growing plant tissues.

Of course the quantities of secondary compounds vary between individual plants. Often these differences are considerable, not only quantitatively but to some extent also qualitatively. There is a 20-fold interindividual variation in cyanogenic glycoside content in the foliage within a Costa Rican population of *Acacia farnesiana* occupying a few hectares, whereas the flavonoid content of these same leaves stays constant in kind and quantity [171]. Such variations are probably of great ecological significance, though it is only relatively recently that biologists realized that herbivores may behave quite differently on different plant individuals [108].

Given that a single plant can contain hundreds of constituents in its essential oils alone [78] the biosynthetic versatility of plants, which probably goes back to very ancient life forms [30], is difficult to encompass. Clearly the enormous diversity of secondary plant compounds is based upon the fact that plants generally contain mixtures of many compounds rather than just one or a few characteristic substances. Young tea shoots, for instance, contain more than 24 phenolic compounds [165] and the terpenoid mixtures in leaves of essential oil plants and in resin-producing trees are usually composed of 30–40 terpenoids of at least 1% concentration [118]. This biochemical richness emphatically indicates that plants, in addition to their form and age or season characteristics, possess an extra dimension: that of chemical composition.

3.10 PRODUCTION COSTS

Production of secondary plant substances requires matter and energy. Especially when plants contain appreciable quantities of secondary compounds, their synthesis and storage presumably exact a cost, but this has been hard to measure. The physiologist can express the costs of chemical resistance in biochemical terms, for instance the energy required to produce a certain quantity of secondary metabolites. Results for various groups of compounds, based on this method, are shown in Table 3.10.

Table 3.10 Costs of the formation of various primary and secondary plant compounds, expressed as grams of glucose per gram of compound. Average values are presented per group of compounds. For more detailed information, see Gershenzon, (1994b), on which this table is based

Primary compounds	
Carbohydrates	1.07
Organic acids	0.73
Lipids	3.10
Nucleotides	1.59
Amino acids	2.09
Secondary compounds	
Terpenoids	3.18
Phenolics	2.11
Alkaloids	3.24
Other nitrogenous secondary compounds	2.27

It shows that the production costs of secondary compounds are somewhat higher than for most primary metabolites. Terpenoids are especially expensive to produce because of their high level of chemical reduction and the often high number of enzymatic conversions involved in their formation. Several monoterpenes need nine steps, whereas the formation of the iridoid glycoside antirrhinoside, for instance, requires as many as 23 steps [71]. Of course the total costs of chemical resistance depend not only on costs of synthesis (which are relatively small), but also on the actual quantities of the chemicals present in the plant, their turnover rates and costs of transport and storage. Probably the costs of the 'handling' processes are appreciable. The figures presented in Table 3.10 compare remarkably well with the figures measured in

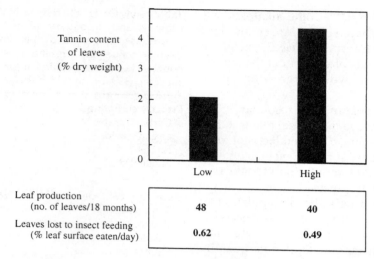

Figure 3.14 Comparison of leaf production and susceptibility to insect feeding on young *Cecropia peltata* trees with low and high tannin content. See text for further details. (Source: data from Coley, 1986.)

the plant *Diplacus aurantiacus* under natural conditions. A negative correlation was found between the amount of phenolic resins that coat the leaves of this species and growth rate. For each gram of resin produced there was a reduction in growth of 2.1 g dry weight shoot biomass [81]. Thus resin is more costly than growth itself.

Ecologists may compute the loss of fitness due to the commitment of resources to defensive chemicals and focus on the adaptive value of secondary compounds in terms of plant organ or tissue value, reduction in growth or seed production, apparency to herbivores, etc. This method was applied in a study on growth of young trees of *Cecropia peltata*, commonly occurring in the neotropics. An inverse relationship was found between growth rate and the concentration of secondary substances (tannins) present in the leaves (Fig. 3.14).

The antiherbivore effect of the tannins was demonstrated in the same study. When 18-month-old plants grown under standard conditions but varying in tannin content were placed in a large forest light gap and subjected for 10 days to naturally occurring herbivores, damage to low-tannin plants was significantly higher than damage to high-tannin plants. The investment in tannin production often appears to be quite substantial. In *Cecropia* the increase in tannin content from 1% to 6% translates into a more than 30% reduction in the rate of leaf production [37]. It should be remembered that secondary plant substances have seldom, or more probably never, one function. Reduced insect infestation, as measured in *Cecropia peltata* with increased tannin levels, is probably only one beneficial effect. It seems plausible that there is a balance between investment and profit, which may consist of several components.

Results from a number of studies indicate that secondary plant metabolites are costly to manufacture [72, 208], although some other studies present evidence that the costs of chemical resistance are small or even absent [200]. Such contradictory conclusions may be due to the use of different criteria to measure plant fitness. For instance, resistance in barley (*Hordeum vulgare*) to greenbugs appeared to be costly when plant biomass growth was used as a parameter, but when the number of leaf primordia differentiated on the apex, an indi-

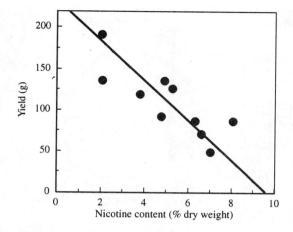

Figure 3.15 Relationship ($r = -0.87$) between the yield of foliage and nicotine content in eight native or primitive and two domesticated cultivars of *Nicotiana tabacum*. (Source: redrawn from Vandenberg and Matzinger, 1970.)

cator of future growth, was used, a reversed conclusion could be drawn [28]. Clearly more experimental data and theory development are needed to get a full grasp of the costs of resistance to herbivory [173].

It should be realized that plant growth is often limited by nitrogen availability rather than energy (glucose). In that case the production of N-containing compounds, such as alkaloids, comes at the expense of growth or reproduction (Fig. 3.15). This may explain why alkaloid levels are usually lower than those of, for instance, phenolics. The degree of 'protection' which a given chemical provides is another ecological factor which determines the amounts produced. In other words: when alkaloids are generally more toxic than phenolics one would expect lower alkaloid concentrations as compared to phenolics. Another point is that nitrogenous resistance compounds may simultaneously serve as nitrogen strategic reserves which, *via* metabolic degradation, can be reclaimed for growth or reproduction when environmental nitrogen supply falls short.

The topic of production costs is discussed in lucid reviews by Gershenzon [72] and Simms [173].

3.11 COMPARTMENTATION

One problem for a plant adopting a chemical resistance strategy against herbivores or pathogenic intruders is that any chemical toxic enough to be effective against a variety of organisms is likely to be self-toxic as well. This problem can be solved in two ways.

1. Instead of accumulating highly toxic compounds the plant stores less toxic precursors, which are transformed into toxins only when needed, for instance when damaged by herbivores.
2. Toxic chemicals are stored in cell compartments which are remote from metabolism, i.e. cell walls and vacuoles.

Both mechanisms do indeed commonly occur, often in combination with each other. Concentrations of toxic compounds in vacuoles are often extremely high. Berberine alkaloids (**7**), for example, occur at levels of more than 0.25 mol/l in cell vacuoles of the greater celandine (*Chelidonium majus*, Papaveraceae) [216]. Employing specialized storage sites requires a physiological machinery to transfer the compounds from their place of synthesis and the presence of specific membrane-carriers to accumulate them in the storage organs and to prevent them from 'leaking' away from these sites. Obviously these processes entail metabolic costs.

By binding to sugars the toxicity of many compounds is diminished and their solubility is increased, so that large amounts can be stored in the cell vacuoles. It is only upon leaf damage that such glycosides come together with specific degradation enzymes to produce the poison. Young shoots of sorghum (*Sorghum bicolor*) may contain up to 30% (dry weight) dhurrin (**16**), a cyanogenic glycoside. Most of it is stored, at still elevated concentrations, in the vacuoles of epidermal cells. The chloroplasts of the mesophyll cells contain the

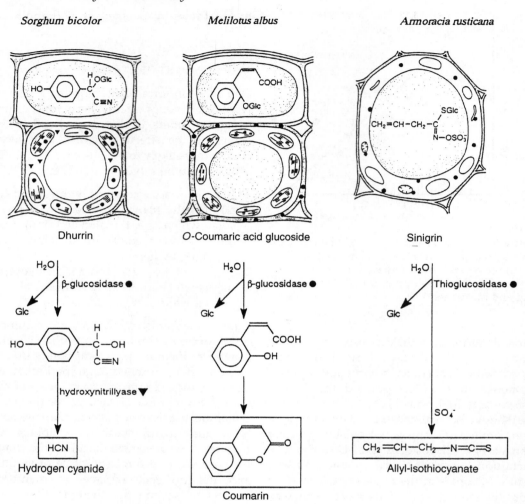

Figure 3.16 Compartmentation of precursors of toxic plant compounds and their degrading enzymes in three plant species: sorghum, white melilot and horseradish. The glucosidases (●) are present in all three plants, but localized in different cell compartments of mesophyll tissue. Another enzyme, hydroxynitrile lyase (▼), occurs in mesophyll cells of sorghum. (Source: redrawn from Matile, 1984.)

enzyme β-glucosidase and the cytosol of these cells holds hydroxynitrile lyase. When the leaves are damaged dhurrin is mixed with these two enzymes, resulting in immediate degradation and production of hydrogen cyanide [150] (Fig. 3.16).

This compound is well known for its general toxic effect on most living organisms. Likewise, coumarin is formed by hydrolysis when melilot leaves are injured (Fig. 3.16) and mechanical rupture of leaf tissues of crucifers causes enzymatic hydrolysis of glucosinolates. They are converted to mustard oils, which are deterrent and potentially toxic to many insect species (Fig. 3.16). In *Melilotus* self-toxicity is prevented by two membranes, which serve as barriers between substrate and enzyme: i.e. the tonoplast (vacuolar membrane) and the

Table 3.11 Half-lives of some secondary plant substances (Source: data from Adewusi, 1988 and Luckner, 1990)

Compound	Organism/plant part	Half-life
Ricinine (**51**)	*Ricinus communis*	4 hours
Quinolizidines	*Lupinus polyphyllus*	5 h
Morphine	*Papaver somniferum*	7.5 h
Nicotine (**41**)	*Nicotiana tabacum*	22 h
Marrubiin (**37**)	*Marrubium vulgare*	24 h
Mono- and diterpenes	*Pinus sylvestris* needles	46 h
	cortex	170 days
Gramine (**29**)	*Hordeum vulgare*	3.5 d
Tomatine (**66**)	*Lycopersicon* fruits	6 d
Dhurrin (**16**)	Sorghum seedlings	10 d

plasmalemma. In the case of the horseradish plant (Fig. 3.16) the distance between life and death is still shorter, i.e. only 7.5 nm, the thickness of the vacuolar membrane [123, 125]!

Another form of compartmentation is exemplified by the accumulation of low-molecular-weight terpenoids and other volatile oils in glandular hairs and idioblasts, resins in resin ducts and latex in cells called laticifers [42]. These specialized containment structures are wholly devoted to the storage and excretion of toxins.

Although many secondary plant substances are stored in vacuoles, they are not inert end products and disconnected from the plant's metabolic processes. Rather, the cell vacuole forms part of a dynamic environment from which metabolites can re-enter the cytoplasm, so this mechanism is not necessarily one of chemical disposal. It appears that many compounds are intimately involved with the plant's primary metabolic functions and may undergo rapid turnover, often having half-lives of less than one day (Table 3.11).

Large diurnal variations in the levels of secondary compounds support the view of a continuous metabolic involvement regulated by endogenic (e.g. developmental stage) and environmental factors (e.g. season, climate, amount of light). Recent evidence, however, indicates that turnover rates of alkaloids and terpenes may be much smaller than has been thought up to now [75, 87, 200].

Rapid metabolic turnover of secondary compounds may have a major impact on energy costs and nutrient investments. It therefore represents an important factor in theories on chemical plant resistance and plant evolution. Metabolic costs of production and storage of resistance substances have been calculated for nicotine synthesis. It was found that nicotine turnover in tobacco plants takes 17% of the daily photosynthesis yield, a startlingly high proportion of the net CO_2 fixation [160]. This high figure, however, has recently been disputed on methodological grounds [75].

3.12 TEMPORAL VARIABILITY

The chemical make-up of a plant is not a constant and fixed property but may show extensive temporal variability [113]. Mature plants differ in many respects from young plants and senescing plant tissues again show qualitative and quantitative changes as compared to full-grown life stages. In many parts of the world these changes parallel the seasons. This is quite obvious in the aging processes of annual species, but perennials that synchronize their reproductive periods and show morphological and physiological adaptations, such as leaf

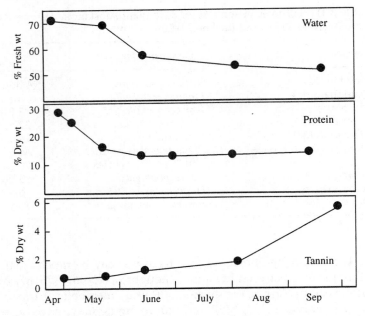

Figure 3.17 Seasonal variation of water, protein and tannin contents of sun leaves of oak, *Quercus robur*. (Source: redrawn from Feeny, 1970.)

abscission and nutrient storage, also continuously change their chemical profiles. Physical factors, such as leaf toughness [119], surface roughness [128] and water content [59], exhibit considerable changes with age but the changes in primary metabolites as well as secondary substances are often even more dramatic:

3.12.1 SEASONAL EFFECTS

Variations in leaf nutrients and allelochemics content with season are of paramount importance to insects feeding on them. Feeny, in a classical study [59], found that most insect species living on oak leaves concentrate their feeding in early spring because the nutritional value of leaves declines as they mature. Thus water and protein contents decrease and tannins accumulate during the summer (Fig. 3.17).

At the same time, increasing leaf toughness is an important factor, especially to smaller insects. The rapid increase of phenolics, reaching spectacularly high levels, in spring leaves of poplar (*Populus trichocarpa*; Fig. 3.18), undoubtedly has a physiological impact on its insect attackers.

Likewise, terpenoid levels in many plant species are much higher in young leaves than in mature, fully-expanded leaves. Similar patterns of change occur in other organs, such as stems and roots, with young organs often having terpenoid concentrations two to ten times as high as those of mature organs [73]. There is usually another surge in the production of secondary plant substances associated with the onset of flowering and seed production (Fig. 3.11). Many insects, however, grow better and attain higher fecundity levels when feeding on young leaves as compared to mature or senescent leaves of the same plant because of their higher nutritional value, although the opposite reactions can also be found [180]. Thus, larvae of *Pieris rapae* prefer to feed and grow better on young cabbage leaves than on mature leaves, while the

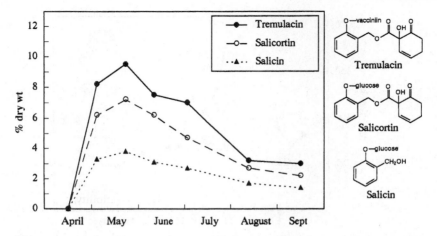

Figure 3.18 Seasonal variation in concentrations of three phenolics in the leaves of *Populus trichocarpa*. (Source: redrawn from Thieme and Benecke, 1971.)

cabbage looper (*Trichoplusia ni*) prefers mature leaves [21].

Sucking insects may encounter still larger variations in nutrient contents of their hosts than mandibulate species. A striking example is provided by willow trees. Their phloem sap contains 0.4% free amino acids during springtime, a value that decreases to about 0.05% during the summer.

Whereas many allelochemics classified as toxins or deterrents decline in concentration with leaf age, compounds designated as digestibility-reducing or quantitative resistance factors generally exhibit the opposite pattern. These substances, such as tannins and resins, increase in many instances with leaf age and may render leaves of all growth forms less suitable for herbivores. The resistance compounds of early-season foliage are often nitrogen-based substances such as alkaloids, cyanogenic compounds and non-protein amino acids. This may be related to the increased soil levels of nitrogen early in the season, and plants may use the nitrogenous compounds not only for protection but also to store nitrogen for later growth. At the same time, carbon is limiting in the young growing tissues but later in the year carbon supply can exceed the demand for growth, which permits the plant to produce carbon-based quantitative resistance compounds such as tannins. The 'resource availability hypothesis' [38] (see Chapter 10) seems an attractive explanation for many recorded seasonal changes in N-based and C-based resistance compounds.

3.12.2 DAY/NIGHT EFFECTS

On another time scale many secondary plant substances appear to fluctuate daily. Because of changes in photosynthetic and metabolic activity of their food plants, herbivorous insects are confronted with a diet during the night that differs markedly from that available during the daytime. The amounts of resistance compounds may fluctuate by as much as 35% during a day–night cycle, as has been reported for cyanogenic compounds in cassava [147]. Still larger diurnal variations may occur in certain plant reservoirs such as latex (Fig. 3.19).

Clearly the chemical composition of plants is not constant throughout the day, but varies markedly, not only in absolute amounts of particular secondary plant substances but also in the ratios of different compounds [159].

60 Plant chemistry: endless variety

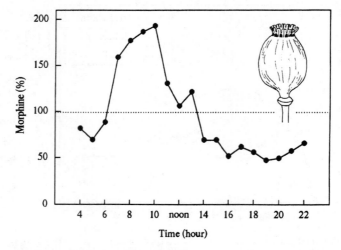

Figure 3.19 Diurnal changes in content of morphine in latex from developing capsules of *Papaver somniferum* as a percentage of daily average content. Dotted line = 100% level = daily average = 4.9% fresh weight. (Source: redrawn from Fairbairn and Wassel, 1964.)

Such diurnal fluctuations may provide a rationale for some of the ancient rules of drug plant harvesting. Theophrastos reports that the herb gatherers of his time (4th century BC) prescribed that 'some roots should be gathered at night, others by day and some before the sun strikes them' [159]. Looking at the graph for morphine (Fig. 3.19), it is obvious that the yield at 9 am could be four times the yield at 9 pm. Diurnal fluctuations of the kind documented above may also be one of the reasons why many insects are nocturnal feeders and other species choose to feed on different plant parts on different times of the day [110]. Larvae of the gypsy moth (*Lymantria dispar*), which normally forage at night, abandon their diurnal rhythm on nutritionally poor host plants and eat then intermittently throughout the day and night. When this insect is grown on an artificial diet feeding is also largely restricted to the night, except when the diet contains 2% tannin, in which case feeding also occurs during the daytime. The loss of the feeding rhythm is probably an adaptation to defoliation-induced changes in food quality such as would occur under population outbreak conditions [116].

3.12.3 INTERYEAR VARIATION

The role of external factors, such as climate and availability of nutrients, on chemical variation is manifested in interyear variations in the quality and quantity of allelochemics a perennial plant produces. This is exemplified by appreciable variations among years in the amounts of phenolics produced in three graminoid species (Fig. 3.20). In one grass species, *Andropogon scoparius*, the highest concentration observed was as much as 2.5 times that of the lowest [133].

3.13 EFFECTS OF LOCATION AND FERTILIZERS

Site factors that may greatly influence a plant's chemistry include exposure to direct sunlight and physical and chemical soil characteristics.

3.13.1 SUN AND SHADE

Light is a basic requirement of all green plants, which exist by virtue of their capacity to convert solar energy into organic matter. No wonder that light intensity generally affects

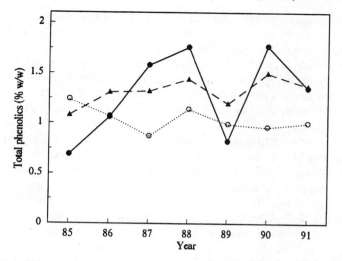

Figure 3.20 Average total phenolics content plotted by year for three grass species: *Andropogon scoparius* (●), *A. hallii* (▲), and *Carex heliophila* (○). (Source: redrawn from Mole and Joern, 1993.)

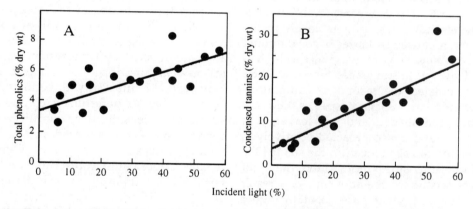

Figure 3.21 Foliar content of (**A**) total phenolics, and (**B**) condensed tannins against the percentage of available light that reached the leaves of *Acacia pennata* trees. Values along the y axes are not strictly comparable as in the phytochemical assays different standards were used. (Source: redrawn from Mole et al., 1988.)

plant primary and/or secondary metabolism. A reduction in light intensity will negatively affect photosynthesis with consequent decline in carbohydrate production. When mineral nutrient uptake is not affected, the result will be a net decrease in the C/N balance in shaded plants or plant parts. This will lead to lower levels of C-based metabolites, such as phenolics [51, 134] (Fig. 3.21).

Many instances are known of marked differences between plants growing in the open and in the shade. For example, bracken fern (*Pteridium aquilinum*) growing in shady areas may contain up to 50% more cyanogenic compounds than those in open sites, whereas for flavonoids and tannins higher concentrations are found in plants growing in sunny spots. These differences have noticeable effects on plant palatability to insect as well as mammalian herbivores [40]. Even different parts of the same plant sometimes show significant differences in their chemical composition due to differences in exposure to the sun [186]. As mentioned earlier (section 2.1), in a choice test larvae of the gypsy moth preferred alder leaves picked from the south side of a tree over those taken from the north side of the same tree (see Fig. 2.5). Likewise, eucalyptus trees are conspicuously more damaged on the sunny than on the shady side by the psyllid *Cardiaspina densitexta*, because the nymphs on the shaded side are exposed to a nutritionally suboptimal sap composition. This results in heavy mortality [213]. An example in which shading has markedly altered the primary metabolism of leaves is bittercress (*Cardamine cordifolia*). Analysis of leaf chemistry from plants growing in the sun and from individuals from the same clone that were shaded experimentally showed considerably higher sugar levels but lower protein levels in the sun. Damage caused by larvae of a leaf mining insect (*Scaptomyza nigrita*) was twice as high in shaded leaves as in sun leaves. Since the levels of glucosinolates, the secondary compounds characteristic of this plant species, did not differ between shaded and sun leaves, it is concluded that the insect reacted positively to the increased nitrogen levels in the shade [39]. High light intensities more often than not stimulate secondary metabolism. As a result whole plants, or plant parts, contain higher amounts of secondary metabolites in sunny areas than in the shade. A striking example is found in the concentration of diterpenoids, and also total diterpene resin content, in Scots pine (*Pinus sylvestris*), which may increase by 100% in insolated needles as compared to shaded foliage [77]. The existence of positive correlations between incident light and the production of secondary compounds has been confirmed by several other studies [209].

3.13.2 SOIL FACTORS

Other crucial environmental factors affecting plant growth are properties of the soil, including its mineral status. Numerous observations relate insect growth and abundance to the chemistry of the soil on which their host plants grow. In agriculture application of fertilizers is generally used to promote rapid, healthy plant growth and to increase yields. Fertilization primarily influences plant physiology but can also induce changes in plant morphology and phenology. Physiological responses are manifested by changes in nutrient composition, such as protein levels. Secondary metabolism is also affected, resulting in increased or decreased levels of secondary plant substances [70]. When the increase of secondary compounds is slower than the rate of increase of biomass, their concentrations decrease. In some cases the insect responds primarily to changes in the nutritive make-up of its host, whereas in other cases changes in allelochemics appear to dominate (Table 3.12).

Not only the nutritional status of a plant but also its leaf surface chemistry and its appearance are affected by fertilization. As a result insects searching for a host plant to oviposit may respond differently to fertilized plants as compared to unfertilized conspecifics. A spectacular sensitivity to fertilizer-induced changes has been observed in the cabbage white butterfly, *Pieris rapae*. Ovipositing females appear able to discriminate between fertilized and unfertilized host plants within 24 hours of fertilizer application [139]. This demonstrates not only that the plant may respond rapidly to the treatment but also that insects can perceive supposedly subtle differences between their host plants.

Many insects benefit from improved plant growth and other increased nutritional values. However, as so often in studies of insect–plant relationships, examples can also be cited in which plant fertilization negatively affected insect populations, e.g. by increased vigour or by shortening the stage of susceptibility to insect attack. The fact that generalizations in this area of research are often weakened by too many exceptions, seemingly negating the rule, is not evidence of poor science but is rather

Table 3.12 Changes in leaf chemistry of black mustard (*Brassica nigra*) grown under low sulphur (S) and low nitrogen (N) conditions, and weight of 7-day-old larvae of *Pieris rapae* and *Spodoptera eridania* grown on treated plants. *P. rapae* is a specialist on *Brassica* spp. and *S. eridania* is a generalist feeder that responds favourably to low glucosinolate levels. All values presented are expressed as a percentage of values for control plants and larvae. (Source: modified from Wolfson, 1982)

Treatment	Allylisothiocyanate	Protein	Weight of 7-day-old larvae	
			P. rapae	S. eridania
Low S	35	80	183	594
Low N	140	67	159	67

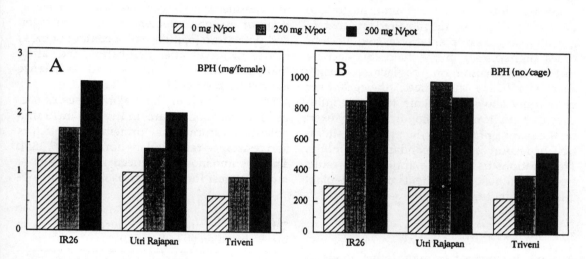

Figure 3.22 (**A**) Body weight of female brown planthoppers (*Nilaparvata lugens*) and (**B**) population development on three selected rice varieties grown in experimental cages under different nitrogen fertilization rates. IR26 is a susceptible cultivar, Utri Rajapan is tolerant and Triveni is moderately resistant. The brown planthopper (BPH) is one of the most devastating pest species on rice. (Source: redrawn from Heinrichs and Medrano, 1985.)

due to the incredible flexibility of insects, their enormous diversity in species and the great subtlety of the interactions between insects and plants.

Since nitrogen is the mineral that most often limits crop productivity, nitrogen fertilization of field crops often seems to stimulate insect populations as a result of increased consumption and higher utilization rates [138] (Fig. 3.22).

Scriber [170] has reported that in about 115 studies crop damage by pest insects increased with the nitrogen content of their host plants. Curiously, the opposite effects often occur after fertilizing forest trees. The reasons for this difference are uncertain. No generalizations can be made with respect to the effects of phosphorus, potassium or organic fertilizers on insect populations because they are variable. All that can be said is that in many cases fertilization practices evidently have profound effects on insect herbivores, although 'our present knowledge on the basic nature of soil minerals–plant–insect interactions is weak' [41].

Environmental effects on secondary metabolism show up in some dramatic differences that were observed when crop plants were grown simultaneously in the field and in greenhouses. The glucosinolate content of greenhouse-grown cabbage plants reached only 10% of the levels measured in plants grown under field conditions [35]. Opposite reactions, however, were found in tomato plants. In this case the alkaloid content of greenhouse-grown plants was two to four times higher than in the field [8]. Lacking further information on the factors causing such significant differences, the examples given only show once again the great influence that environmental conditions may exert on the plant's physiology, including its allocation of resources into secondary metabolites. This cautions us against extrapolating results from greenhouse experiments on insect–plant relations to natural situations.

3.14 INDUCED RESISTANCE

A plant, in contrast to many other foods, is alive all the time it is being eaten by insects and other small herbivores. Therefore it does make sense to step up its resistance measures upon wounding. Ever since Green and Ryan [76] showed that plants may undergo chemical changes following herbivory, thereby increasing their resistance to insect attack, this previously unrecognized protection system has captured the imagination of plant physiologists, entomologists and ecologists as well. As a result, a large body of literature grew up, which is summarized in a number of excellent reviews, including those by Tallamy and Raupp [191], Baldwin [6] and Dicke [47]. It is now widely recognized that the chemical composition of a plant is not only affected by abiotic factors but may to a considerable degree be influenced by previous mechanical damage or herbivory. Clearly, many if not all plants possess, in addition to their 'constitutive resistance system' (which is permanently present), an 'inducible resistance system'. Plant responses have been designated by different terms, which has led to some confusion in the literature. The use of an appropriate terminology (Table 3.13) may avoid misconceptions about mechanisms and functions of induced responses.

Silicon figures as a major mineral constituent of plants and it occurs in some species in quantities that exceed the tissue concentrations of essential elements such as nitrogen and potassium [55]. The silica content of grass blades increases after defoliation, indicating that silification is an induced resistance against herbivores [124].

Most if not all of the major classes of secondary metabolites are inducible, including volatile hydrocarbons, proteinase inhibitors and resistance-related plant hormones [6, 185]. The phenomenon of induced resistance is of much interest for two reasons. From a biolog-

Table 3.13 Terminology of induced responses in plants (based on [47] and [109])

1. As a result of stress or damage a plant may (or may not) show an **induced response**.
2. The induced response may (or may not) decrease herbivore preference or performance: **induced resistance**.
3. Reduced herbivore preference/performance may (or may not) increase plant fitness: **induced protection**.
4. **Constitutive resistance** factors are normally present in unstressed and undamaged plants, wheras **induced resistance** factors are absent in unstressed and undamaged plants.
5. By producing constitutive or induced deterrents, toxins and/or digestibility reducers plants possess a **direct protection** line against herbivores. Plants may also promote, either constitutively or only after induction, the effectiveness of carnivores that attack herbivores and thus show **indirect protection**.

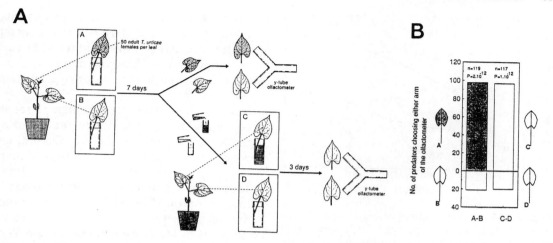

Figure 3.23 An experiment showing that a water-soluble elicitor mediates induction of systemic production of herbivore-induced signal factor (synomone). **(A)** One primary leaf of a lima bean plant is infested with spider mites (cage A) and the other serves as the control (cage B). After 7 days of incubation the leaves are tested in a Y-tube olfactometer for their attractiveness to predatory mites. The leaves that had been exposed to spider mites appear to be significantly more attractive to the predators (**B**, left column). The water from the vials in cage A is transferred to new vials, which are placed in cage C, and the water from the vials in cage B to new vials placed in cage D. Now uninfested primary lima bean leaves are incubated in the vials in cages C and D. After 3 days leaves from cages C and D are tested for their attractiveness to predators (**B**, right column). The (uninfested) leaves kept on water from infested leaves (cage C) appear to be more attractive to predatory mites than the control leaves from cage D kept on water from control cage B. (Source: reproduced from Dicke et al., 1993, with permission.)

ical point of view it is important because the plant can apparently mobilize some latent resistance capacity that, when unused, does not consume energy. Agronomists are interested because knowledge of induced resistance systems may be employed to mobilize resistance in crop plants in order to reduce their susceptibility to pest organisms or, by molecular technology, to introduce the basic machinery for resistance mechanisms from alien plant species.

Induced resistance may express itself on various time scales: a rapid (hours to days) and short-term (days to weeks) *de novo* synthesis of hitherto absent compounds or an elevation of the normal level of resistance compounds, and a long-term (years) response, such as that following defoliation in trees. The short-term reaction may be restricted to the damaged leaf or may spread systemically (i.e. internal transport *via* water phase) over the whole plant (Fig. 3.23). An example of a rapidly induced systemically transmitted change in secondary metabolism is the two-fold increase in nicotine in wild plants of *Nicotiana attenuata* after leaf damage (simulated herbivory; Fig. 3.24).

Induced resistance may involve a decrease both in insect feeding preference for and the nutritional value of induced foliage. It also can affect the oviposition responses of adult females [183].

The sensitivity of a plant to wounding is sometimes amazingly high. When larvae of *Epirrita autumnata*, a geometrid occurring on birch, were fed on undamaged leaves of branches from which a single leaf had been torn 2 days earlier, their developmental rate was significantly reduced [89]. This observation raises the intriguing question of whether

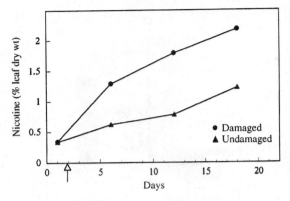

Figure 3.24 Nicotine content of undamaged leaves from *Nicotiana attenuata* plants, which were either artificially damaged by removing all but four to six basal leaves in one cutting (arrow indicates time of damage) or not (control). (Source: redrawn from Baldwin and Ohnmeiss, 1993.)

in nature a plant without some induced response must be considered to be an exception, because hardly any plant will escape some damage or other.

As mentioned before, a plant's constitutive chemical resistance system generally consists of a whole gamut of (predominantly related) chemicals. Induced chemical resistance usually differs qualitatively and/or quantitatively from the constitutive resistance chemicals. Thus in conifers the resin induced by bark beetle attack has a different monoterpenoid composition from the constitutive resin (Fig. 3.25).

Possibly the altered composition raises its effectiveness to deter or toxify the invading organisms. Furthermore, the volatiles emitted by spider-mite-infested leaves of several plant species appear to be quite different from those produced by undamaged or even artificially damaged leaves. Mite-infested plants produce some volatile terpenoids that do not normally occur in the headspace emanation of this plant [47]. The biological significance of these newly induced volatiles may be deduced from the observation that they are attractive to carnivorous mites that prey on spider mites. Populations of spider mites grow exponentially and rapidly overexploit their food plant unless sufficient predatory mites invade their populations in time. Thus the plant, at least under experimental conditions, is able to mobilize natural enemies of the pest organism by which it is victimized, and 'cries for help' [189]. In fact, the carnivores have evolved the ability to recognize the mite-infested plants by the spe-

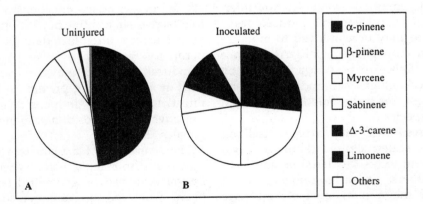

Figure 3.25 Comparison between constitutive monoterpenoid composition **(A)** and composition induced by inoculation with fungus vectored by bark beetles **(B)** in grand fir trees (*Abies grandis*). Values in % of total monoterpenoid content of phloem tissue. Wound reaction was determined 14 days after inoculation. (Source: data from Raffa and Berryman, 1982b.)

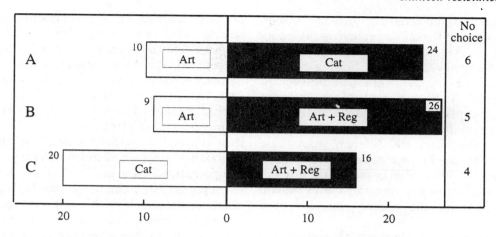

Preferences of *Cotesia marginiventris* females in a two-choice flight tunnel

Figure 3.26 Responses of 40 parasitic wasps (*Cotesia marginiventris*) when given choices between odours of corn seedlings that underwent various treatments one day before the tests. **(A)** Female wasps had the opportunity to choose between air streams from seedlings with artificial damage (Art) or caterpillar (*Spodoptera exigua*) feeding damage (Cat). **(B)** Artificial damage was compared with artificial damage treated with caterpillar regurgitant (to mimic feeding damage) (Art + Reg). Whereas the wasps prefer volatiles from caterpillar-damaged plants and regurgitant-treated damaged plants to artificially damaged plants, they do not discriminate caterpillar-damaged and regurgitant-treated damaged plants **(C)**. Total numbers of responding insects are given with each bar. Numbers at the right represent females that did not respond. (Source: modified from Turlings *et al.*, 1990.)

cific volatiles emanated. Parasitic wasps also appear to be able to smell the difference between plant volatiles emitted upon mechanical damage and damage due to feeding by caterpillars (Fig. 3.26). These natural enemies use the injury-induced chemicals to locate potential host insects [190, 196, 205].

An interesting side effect of plant volatiles as airborne signals is that they may induce resistance responses in conspecific plants. When a cotton seedling free of any pest organism is positioned downwind of a spider-mite-infested plant, it appears to be more attractive to predatory mites than mite-free plants grown upwind of the infested cotton plant. Concomitant with their increased attractiveness to predatory mites the downwind plants reduce in some way or other the oviposition rate of spider mites once they have settled on them. In this case of plant–plant interaction the nature of the volatiles emanating from the infested cotton seedling, which serve as a signal to conspecifics, remains to be elucidated [22].

Induced responses appear in many cases to be much greater and conclusive when caused by herbivore attack than after mechanical damage such as is caused by pruning or rubbing of leaves with carborundum powder. As a possible explanation it has been argued that artificial damage does not sufficiently mimic wounding as inflicted by mandibulate or piercing–sucking insects. Several researchers [126, 196], however, have shown that phytochemical induction is stimulated more intensely and more specifically by insect secretions, indicating that 'somehow trees can distinguish between sterile scissors and caterpillar mouthparts' [86]. Recently β-glucosidase, present in *Pieris brassicae* larval regurgitate and possibly in its saliva as well, has been shown to act as an elicitor of rapidly induced responses [127]. Moreover, there is evidence that plants can respond differentially to different herbivores and types of damage. The changes in two foliar proteinase inhibitors and three oxidative enzymes vary with the type of damage caused by chewing caterpillars, mining insects or tissue-fluid-sucking mites [184]. Another example of the dynamics in induced plant responses is the finding that volatiles produced upon damage inflicted by young caterpillars are different from those induced by late-instar conspecific caterpillars. Supposedly differences in the composition of the saliva between young and old larvae are responsible for the altered plant response [190].

All examples of induced responses discussed so far were elicited by injury inflicted by insect feeding or mechanical damage. Interestingly, a plant may also produce resistance chemicals without any wounding. Thus a cabbage plant on which a cabbage white butterfly (*P. brassicae*) has deposited one or a few egg masses produces some so far unidentified chemicals that act as an oviposition deterrent to conspecific butterflies [18]. Obviously, chemical communication between plants and insects *via* induced responses is a complex phenomenon the intricacies of which can as yet hardly be fathomed [63, 191].

The foregoing discussion should not lead to the wrong impression that all induced responses are synonymous with induced resistance. Several instances are known in which the attacking herbivore profits from the change in the plant's physiology. Out of 42 published studies on induced changes 27 have reported reduced fitness or feeding rates in herbivores. Increased fitness was observed in seven cases and no effect in eight cases [36]. Taking into account the fact that a number of no-effect cases probably remained unpublished, it is clear that the phenomenon of induced resistance, although widespread, is by no means universal. Moreover, since only few studies have revealed a protective function of induced compounds under field conditions [7], it seems premature to attribute a resistance role against herbivores as the primary function of this phenomenon. A tenable alternative view holds that the observed effects upon plant damage reflect generalized wound responses, which some herbivores coincidentally cannot tolerate [101, 198]. The observation that induced resistance in wild parsnip (*Pastinaca sativa*) is most pronounced in those plant parts (roots) that are least attacked by herbivores and that contain the smallest amounts of constitutive resistance compounds, whereas inducibility is lacking in the reproductive plant parts with high probability of attack and the highest levels of constitutive chemicals, is consistent with the idea that inducibility has a defence function. The tissues at risk are permanently protected; tissues that are seldom damaged have a dormant system [222].

The discussion so far has concerned rapid plant responses, which operate on relatively short time scales. Delayed induced responses have been found to occur in the next season's or later foliage of woody plants. One of the best documented examples concerns birch trees, which are abundant and widespread in the vast forests of Finland. E. Haukioja and his colleagues measured increased concentrations of phenolics in leaves when trees were exposed to caterpillar feeding. These changes negatively affect their nutritional value for folivores. In this case the chemical changes are manifested over two different time-scales: an increase in phenolic levels that builds up in hours to days and is short-lasting and a long-term response that may last for months to years [141]. When larvae of *Epirrita autumnata* are reared on the foliage of trees that were defoliated 2 or more years previously, growth and fecundity was significantly reduced (Fig. 3.27).

The fact that the nutritional quality of trees after complete defoliation may affect insect performance even after several years is of ecological relevance. It introduces a time lag into the negative feedbacks regulating the population dynamics of insect herbivores and may generate cyclic density fluctuations. Thus

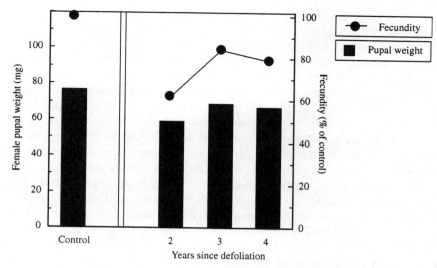

Figure 3.27 Effects of artificially defoliated mountain birch trees (*Betula pubescens*) on the growth (histograms = female pupal weight) and fecundity (dots) of the autumnal moth (*Epirrita autumnata*). Insects grown on foliage picked from trees that were defoliated 2–4 years earlier still show reduced growth and fecundity. (Source: data from Neuvonen and Haukioja, 1991.)

cyclic outbreaks of the larch budmoth (*Zeiraphera diniana*) may be partly explained by chemical and morphological changes in needles of its conifer host, *Larix decidua*, which persist for up to 4–5 years after defoliation [10].

Although many insect species appear to avoid feeding or ovipositing on damaged plants when they have access to undamaged conspecifics [90], it cannot be concluded that all plant responses to herbivory are necessarily detrimental to future folivores. Plant damage may alter not only secondary plant chemistry but also nutritional factors in remaining leaves or in reflushed foliage. Several instances are known in which conspecific or heterospecific insects grow better on damaged than on undamaged plants [56, 57, 145]. Most probably such interactions, although often subtle and indirect, have significant ecological implications.

Differences in resistance of plants are often correlated not necessarily with the amounts of constitutive resistance chemicals but with the capacity to respond with an induced chemical change. For example, lodgepole pines resistant to mountain pine beetles and their associated microorganisms appear to have a considerably stronger induced reaction than susceptible trees (Table 3.14) and induced responses can explain a major portion of the difference between resistant and susceptible individuals [152].

The phenomenon of induced resistance may have serious implications for the design of lab-

Table 3.14 Comparison of constitutive monoterpene content and induced monoterpene accumulation (mg/g dry weight phloem tissue) of lodgepole pines. Trees susceptible and resistant to mountain pine beetle were inoculated with fungi that are vectored by the beetle. (Source: reproduced from Raffa and Berryman, 1982a, with permission)

	Constitutive	Induced
Susceptible	1.1	13.4
Resistant	1.3	90.9

oratory experiments employing, for instance, excised leaves or leaf discs in bioassays, even if these experiments do not focus on induced changes [148]. Wounding effects may alter the taste of cut leaves to such an extent that insects show responses that may differ significantly from those under natural conditions, i.e. on intact plants [181]. One rather spectacular example may suffice as a warning. When examining the feeding preferences of two generalist leaf beetle species, *Diabrotica balteata* and *D. adelpha*, Risch [158] offered these beetles three acceptable plant species using three test methods; whole plants, whole detached leaves and leaf discs. The preference ranking for three plants, corn, bean and squash, appeared to be significantly affected by the method of testing. Whereas squash was preferred to bean and bean to corn when whole plants were used, the preference order turned out to be reversed in a leaf disc test. Now corn evoked the strongest feeding responses and squash the weakest! There were also some significant differences when responses to detached leaves were compared to whole plants [158]. Likewise, lettuce cultivars resistant to the lettuce aphid (*Nasonovia ribisnigri*) suddenly seem to be fully acceptable when leaf fragments are offered in a leaf disc test [49]. Again, the leaf discs cannot be considered to represent accurately an intact plant.

3.15 GENOTYPIC VARIATION

Quantitative and qualitative differences in secondary compounds among individuals in natural and domesticated plants are genetically controlled, although the environment exerts some modifying effects. The concentrations of most compounds are determined by genetic variation in excess of 50% and thus show substantial genetic influence over phenotypic variability [13, 113]. Not only are the levels of constitutive secondary metabolites under genetic control but obviously also those of the induced compounds synthesized upon plant damage [36], although few studies have quantified this trait. In the case of inducible pyrrolizidine alkaloids in hound's tongue (*Cynoglossum officinale*) the heritability accounted for 35% of the variation [198]. Quantities of secondary compounds occurring in plants exhibit continuous variation, and are usually polygenically controlled (i.e. in a manner involving several or many minor genes). Genotypic variation may be substantial in natural populations. Estimated salicin concentrations in leaves of willow clones, for example, ranged from 0.05% of dry weight to over 5%, a hundredfold range, whereas the standard deviation within clones varied less than twofold on average [175]. Since the concentrations of secondary plant substances are generally under tight genetic control, selection may easily modify the quantities produced. As a result striking differences are seen when the amounts of secondary chemicals in some man-made cultivars are compared to those found in wild relatives. Of course selection of low-allelochemics lines will change the plants' susceptibility to insect attack. Cotton (*Gossypium hirsutum*) varieties with reduced levels of gossypol, a phenolic sesquiterpene pigment (**28**), are a better food source to a number of insects than high-gossypol lines, since part of their natural resistance has been eliminated (Fig. 3.28).

The production of cucurbitacin (**14**), the triterpenoid that gives cucumbers their bitter taste, is controlled by a single gene. Breeding programmes have deliberately selected non-bitter varieties to suit human taste. Because this compound is a potent deterrent and also highly toxic to many herbivores, including man, low-cucurbitacin cultivars appear very susceptible to infestation by two-spotted spider mites (*Tetranychus urticae*) and several insect species.

Plant breeders employ the vast genetic variation in chemical and physical properties of natural plant species to develop cultivars with specific desirable traits. Different cultivars of almost any crop plant species appear to have different degrees of susceptibility to insect

Genotypic variation 71

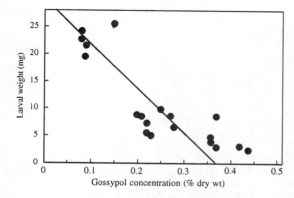

Figure 3.28 Growth of first instar tobacco budworm larvae over a 5-day period when kept on intact plants of 20 cotton cultivars, which vary in their gossypol concentrations. (Source: data from Hedin et al., 1983.)

attack, because of differences in their genetic make-up (Fig. 3.29).

Intense selection for fast plant growth and high reproductive output in the past generally resulted in a lowering of the plants' allocation to resistance. As a consequence, cultivated plants often became more vulnerable than their wild progenitors. Today plant breeders intend to select cultivars that still possess their natural chemical protection, except in the organs that are used as human food, e.g. fruits or seeds. This type of selection has been achieved with, for instance, potato tubers [103]. The topic of resistance breeding will be discussed in more detail in Chapter 12.

Although the concept of an individual does not cause confusion in most animals, including insects, it may be less clear in plants. Modular organisms make the definition of an individual more complex. Thus, a tree may be considered as a population of suborganismic units (modules), each of which develops and dies at its own time. A tree in this view can be described as a population of modules with associated stems and roots [212]. The concept of generation span then also becomes less clear. Some plant species have (in the usual sense) very long generation periods. Consequently, their rate of genetic recombination is very low compared to, for instance, insects. Broadleaf trees, which may reach an age of 200 years or more, may be considered as a monoculture in time, because of their long life span. Moreover, plant populations that essentially consist of clones are by nature very homogeneous. Some bracken fern clones (*Pteridium aquilinum*) are supposedly 1000 years or more

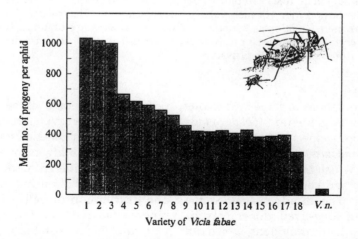

Figure 3.29 Mean number of progeny of a single aphid (*Aphis fabae*) when kept on 18 different cultivars of broad bean (*Vicia faba*). V.n. = *Vicia narbonensis*, a related species. (Source: data from Davidson, 1922.)

72 Plant chemistry: endless variety

Figure 3.30 Mosaic resistance in *Eucalyptus meliodora*. The Christmas beetle (*Anoplognatus montanus*) may, during an outbreak, defoliate trees completely. Some branches or sometimes whole trees are, however, immune because the compositions of their volatile oils are different. The resistant plant parts most probably developed from meristematic cells containing newly arisen somatic mutations. The resistant branches will produce seeds carrying the genes for resistance (Edwards *et al.*, 1990). (Source: drawing by P. Kostense after a photo kindly provided by P. B. Edwards.)

old. Similarly, populations of some herbaceous angiosperms such as goldenrod (*Solidago missouriensis*) or woody species such as aspen trees (*Populus tremuloides*) may essentially represent one clone, which may cover large areas (several hectares) and could date back to the Pleistocene [214]. The disadvantages of the genetic rigidity of long-lived clones may be compensated by somatic mutations, which can be inherited by naturally occurring mechanisms of sexual and asexual reproduction. The accumulation of somatic mutations may permit a plant or a clone to develop as a genetically diverse individual [53]. An example of such genetic diversity is provided by the susceptibility to infestation by a gall aphid, which differed markedly between different branches of an individual poplar tree. The distribution of galls, therefore, was not random but reflected the underlying mosaic pattern of host resistance (see Fig. 9.13). This high level of variation in susceptibility to gall aphids within an individual tree appeared to be of the same magnitude as the range of vari-

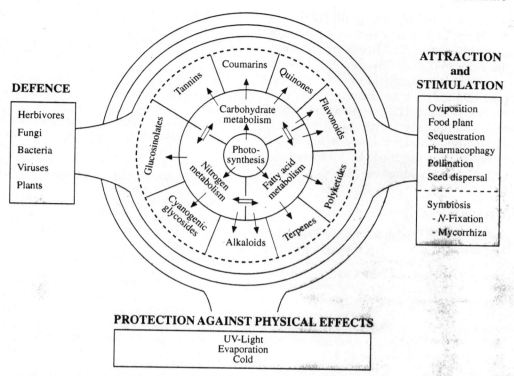

Figure 3.31 Secondary plant substances derive from primary metabolites. They show a multitude of functions and are involved in many biotic and abiotic environmental factors. (Source: redrawn from Hartmann, 1996.)

ation observed between extreme trees in the population [214]. Another example of 'mosaic resistance' is shown in Figure 3.30.

3.16 CONCLUSIONS

Superimposed on a relatively uniform primary metabolism plants produce a perplexingly wide spectrum of secondary compounds. The multifaceted roles of these chemicals are as yet poorly understood, but undoubtedly they provide protection against harsh environmental conditions, invading microorganisms and plant-eating animals, as summarized in Figure 3.31.

When searching for general principles in nature scientists often have to neglect slight variations that occur in measurements of all biological phenomena. This is done by averaging. In this chapter, however, emphasis has been laid on variations in the chemical composition of plants at the species level, within populations or within an individual plant, which may be modulated by insect attack, light conditions, nutrients in soil and atmosphere, etc. These spatial and temporal variations, caused by genotypic and environmental factors, cannot be considered as deviations from the 'normal' or standard but represent an essential feature of the strategy of plants to optimize their survival rates.

3.17 LITERATURE

There is an extensive literature on secondary plant compounds. A fine introduction is provided by Vickery and Vickery [206]. The chemotaxonomy of plants is covered by

Frohne and Jensen [67] and Smith [176]. Hegnauer [93] has produced a renowned standard work on the subject. The books by Rosenthal and Janzen [163] and Rosenthal and Berenbaum [162] contain much information on plant compounds relevant to insects. The well-known *Merck Index* [129] contains structural formulas and information on 10 000 chemicals, including plant substances and drugs. Other recent inventories are those by Harborne and Baxter [84] and Buckingham [23].

3.18 REFERENCES

1. Adati, T. and Matsuda, K. (1993) Feeding stimulants for various leaf beetles (Coleoptera, Chrysomelidae) in the leaf surface wax of their host plants. *Appl. Entomol. Zool.*, **28**, 319–324.
2. Adewusi, S. R. A. (1988) Turnover of dhurrin in green sorghum seedlings. *Plant Physiol.*, **94**, 1219–1224.
3. Allebone, J. E., Hamilton, R. J., Bryce, T. A. and Kelly, W. (1971) Anthraquinone in plant surface waxes. *Experientia*, **27**, 13–14.
4. Deleted in proof.
5. Andersson, B. Å., Holman, R. T., Lundgren, L. and Stenhagen, G. (1980) Capillary gas chromatograms of leaf volatiles. A possible aid to breeders for pest and disease resistance. *J. Agric. Food Chem.*, **28**, 985–989.
6. Baldwin, I. T. (1994) Chemical changes rapidly induced by folivory, in *Insect–Plant Interactions*, vol. 5, (ed. E. A. Bernays), CRC Press, Boca Raton, FL, pp. 1–23.
7. Baldwin, I. T. and Ohnmeiss, T. E. (1993) Alkaloidal responses to damage in *Nicotiana* native to North America. *J. Chem. Ecol.*, **19**, 1143–1153.
8. Barbour, J. D. and Kennedy, G. G. (1991) Role of steroidal glycoalkaloid α-tomatine in host-plant resistance of tomato to Colorado potato beetle. *J. Chem. Ecol.*, **17**, 989–1005.
9. Bell, E. A. and Janzen, D. H. (1971) Medical and ecological considerations on L-dopa and 5-HTP in seeds. *Nature*, **229**, 136–137.
10. Benz, G. (1974) Negative Rückkoppelung durch Raum- und Nahrungskonkurrenz sowie zyklische Veränderungen der Nahrungsgrundlage als Regelprinzip in der Populationsdynamik des Grauen Lärchenwicklers, *Zeiraphera diniana* (Guenee)(Lep., Tortricidae). *Z. Angew. Entomol.*, **76**, 196–228.
11. Berenbaum, M. R. (1995) Turnabout is fair play, secondary roles for primary compounds. *J. Chem. Ecol.*, **21**, 925–940.
12. Berenbaum, M. and Seigler, D. (1992) Biochemicals, engineering problems for natural selection, in *Insect Chemical Ecology*, (eds B. D. Roitberg and M. B. Isman), Chapman & Hall, New York, pp. 89–121.
13. Berenbaum, M. and Zangerl, A. R. (1992) Genetics of secondary metabolism and herbivore resistance in plants, in *Herbivores. Their Interactions with Secondary Plant Metabolites*, 2nd edn, vol. 2, (eds G. A. Rosenthal and M. R. Berenbaum), Academic Press, San Diego, CA, pp. 415–438.
14. Bernays, E. A., Howard, J. J., Champagne, D. and Estesen, B. J. (1991) Rutin, a phagostimulant for the polyphagous acridid *Schistocerca americana*. *Entomol. Exp. Appl.*, **60**, 19–28.
15. Besson, E., Dellamonica, G., Chopin, J., Markham, K. R., Kim, M., Koh, H. S. and Fukami, H. (1985) C-Glycosylflavones from *Oryza sativa*. *Phytochemistry*, **24**, 1061–1064.
16. Bingaman, B. R. and Hart, E. R. (1993) Clonal and leaf age variation in *Populus* phenolic glycosides, implications for host selection by *Chrysomela scripta* (Coleoptera, Chrysomelidae). *Environ. Entomol.*, **22**, 397–403.
17. Blaakmeer, A., Geervliet, J. B. F., van Loon, J. J. A., Posthumus, M. A., van Beek, T. A. and de Groot, Æ. (1994a) Comparative headspace analysis of cabbage plants damaged by two species of *Pieris* caterpillars, consequences for in-flight host location by *Cotesia* parasitoids. *Entomol. Exp. Appl.*, **73**, 175–182.
18. Blaakmeer, A., Hagenbeek, D., van Beek, T. A., de Groot, Æ., van Loon, J. J. A. and Schoonhoven, L. M. (1994b) Plant response to eggs vs. host marking pheromone as factors inhibiting oviposition by *Pieris brassicae*. *J. Chem. Ecol.*, **20**, 1657–1665.
19. Bolter, C. J., Dicke, M., van Loon, J. J. A., Visser, J. H. and Posthumus, M. A. (1997) Attraction of the Colorado potato beetle to herbivore-damaged plants during herbivory and after its termination. *J. Chem. Ecol.*, **23**, 1003–1023.
20. Boufford, D. E., Kjaer, A., Madsen, J. O. and Skrydstrup, T. (1989) Glucosinolates in Bretschneideraceae. *Biochem. Syst. Ecol.*, **17**, 375–379.
21. Broadway, R. M. and Colvin, A. A. (1992) Influence of cabbage proteinase inhibitors in situ on the growth of larval *Trichoplusia ni* and *Pieris rapae*. *J. Chem. Ecol.*, **18**, 1009–1024.
22. Bruin, J., Dicke, M. and Sabelis, M. W. (1992)

Plants are better protected against spider-mites after exposure to volatiles from infested conspecifics. *Experientia*, **48**, 525–529.
23. Buckingham, J. (1993) *Dictionary of Natural Products*, Chapman & Hall, London.
24. Buttery, R. G. and Ling, L. C. (1984) Corn leaf volatiles, identification using Tenax trapping for possible insect attractants. *J. Agric. Food Chem.*, **32**, 1104–1106.
25. Buttery, R. G., Ling, L. C. and Wellso, S. G. (1982) Oat leaf volatiles, possible insect attractants. *J. Agric. Food Chem.*, **30**, 791–792.
26. Buttery, R. G., Kamm, J. A. and Ling, L. C. (1984) Volatile components of red clover leaves, flowers, and seed pods, possible insect attractants. *J. Agric. Food Chem.*, **32**, 254–256.
27. Carter, M., Sachdev-Gupta, K. and Feeny, P. (1994) Tyramine, an oviposition stimulant for the black swallowtail butterfly from the leaves of wild parsnip, in *Abstracts of the 11th Annual Meeting of the International Society for Chemical Ecology, Syracuse, NY*, p. 55.
28. Castro, A. M., Rumi, C. P. and Arriaga, H. O. (1988) Influence of greenbug on root growth of resistant and susceptible barley genotypes. *Environ. Exp. Bot.*, **28**, 61–72.
29. Cates, R. G. (1980) Feeding patterns of monophagous, oligophagous, and polyphagous insect herbivores, the effect of resource abundance and plant chemistry. *Oecologia*, **46**, 22–31.
30. Cavalier-Smith, T. (1992) Origins of secondary metabolism, in *Secondary Metabolites, Their Function and Evolution*, (eds D. J. Chadwick and J. Whelan), John Wiley, Chichester, pp. 64–87.
31. Chameides, W. L., Lindsay, R. W., Richardson, J. and Kiang, C. S. (1988) The role of biogenic hydrocarbons in urban photochemical smog, Atlanta as a case study. *Science*, **241**, 1473–1475.
32. Chapman, R. F. and Bernays, E. A. (1989) Insect behavior at the leaf surface and learning as aspects of host plant selection. *Experientia*, **45**, 215–222.
33. Chew, F. S. (1988) Searching for defensive chemistry in the Cruciferae, or, do glucosinolates always control interactions of Cruciferae with their potential herbivores and symbionts? No!, in *Chemical Mediation of Coevolution*, (ed. K. S. Spencer), Academic Press, San Diego, CA, pp. 81–112.
34. Chew, F. S. and Renwick, J. A. A. (1995) Host-plant choice in *Pieris* butterflies, in *Chemical Ecology of Insects*, vol. 2, (eds R. T. Cardé and W. J. Bell), Chapman & Hall, New York, pp. 214–238.
35. Cole, R. A. (1994) Locating a resistance mechanism to the cabbage aphid in two wild Brassicas. *Entomol. Exp. Appl.*, **71**, 23–31.
36. Coleman, J. S. and Jones, C. G. (1991) A phytocentric perspective of phytochemical induction by herbivores, in *Phytochemical Induction by Herbivores*, (eds D. W. Tallamy and M. J. Raupp), John Wiley, New York, pp. 3–45.
37. Coley, P. D. (1986) Costs and benefits of defense by tannins in a neotropical tree. *Oecologia*, **70**, 238–241.
38. Coley, P. D., Bryant, J. P. and Chapin, T. (1985) Resource availability and plant antiherbivore defense. *Science*, **230**, 895–899.
39. Collinge, S. K. and Louda, S. M. (1988) Herbivory by leaf miners in response to experimental shading of a native crucifer. *Oecologia*, **75**, 559–566.
40. Cooper-Driver, G., Finch, S. and Swain, T. (1977) Seasonal variation in secondary plant compounds in relation to palatability of *Pteridium aquilinum*. *Biochem. Syst. Evol.*, **5**, 177–183.
41. Dale, D. (1988) Plant-mediated effects of soil mineral stresses on insects, in *Plant Stress–Insect Interactions*, (ed. E. A. Heinrichs), John Wiley, New York, pp. 35–110.
42. Data, E. S., Nottingham, S. F. and Kays, S. J. (1996) Effect of sweetpotato latex on sweetpotato weevil (Coleoptera, Curculionidae) feeding and oviposition. *J. Econ. Entomol.*, **89**, 544–549.
43. Davidson, J. (1922) Biological studies of *Aphis rumicis* Linn. Reproduction on varieties of *Vicia faba*. *Ann. Appl. Biol.*, **9**, 135–142.
44. Denno, R. F. and McClure, M. S. (1983) *Variable Plants and Herbivores in Natural and Managed Systems*, Academic Press, San Diego, CA.
45. Derridj, S., Fiala, V. and Boutin, J. P. (1991) Host plant oviposition preference of the European corn borer (*Ostrinia nubilalis* Hbn.). A biochemical explanation. *Symp. Biol. Hung.*, **39**, 455–456.
46. Derridj, S., Wu, B. R., Stammitti, L., Garrec, J. P. and Derrien, A. (1996) Chemicals on the leaf surface, information about the plant available to insects. *Entomol. Exp. Appl.*, **80**, 197–201.
47. Dicke, M. (1994) Local and systemic production of volatile herbivore-induced terpenoids, Their role in plant-carnivore mutualism. *J. Plant Physiol.*, **143**, 465–472.
48. Dicke, M., van Baarlen, P., Wessels, R. and Dijkman, H. (1993) Herbivory induces sys-

temic production of plant volatiles that attract predators of the herbivore, extraction of endogenous elicitor. *J. Chem. Ecol.*, **19**, 581–599.
49. Dieleman, F. L. (1991) Personal communication.
50. Doskotch, R. W., Mikhail, A. A. and Chatterjee, S. K. (1973) Structure of the water-soluble feeding stimulant for *Scolytus multistriatus*: a revision. *Phytochemistry*, **12**, 1153–1155.
51. Dudt, J. F. and Shure, D. J. (1994) The influence of light and nutrients on foliar phenolics and insect herbivory. *Ecology*, **75**, 86–98.
52. Dussourd, D. E. (1995) Entrapment of aphids and whiteflies in lettuce latex. *Ann. Entomol. Soc. Am.*, **88**, 163–172.
53. Edwards, P. B., Wanjura, W. J., Brown, W. V. and Dearn, J. M. (1990) Mosaic resistance in plants. *Nature*, **347**, 434.
54. Eigenbrode, S. D. and Espelie, K. E. (1995) Effects of plant epicuticular lipids on insect herbivores. *Annu. Rev. Entomol.*, **40**, 171–194.
55. Epstein, E. (1994) The anomaly of silicon in plant biology. *Proc. Natl Acad. Sci. USA*, **91**, 11–17.
56. Faeth, S. H. (1987) Community structure and folivorous insect outbreaks: the roles of vertical and horizontal interactions, in *Insect Outbreaks*, (eds P. Barbosa and J. C. Schultz), Academic Press, San Diego, CA, pp. 135–171.
57. Faeth, S. H. (1992) Do defoliation and subsequent phytochemical responses reduce future herbivory on oak trees? *J. Chem. Ecol.*, **18**, 915–925.
58. Fairbairn, J. B. and Wassel, G. (1964) The alkaloids of *Papaver somniferum* L. I. Evidence for a rapid turnover of the major alkaloids. *Phytochemistry*, **3**, 253–256.
59. Feeny, P. (1970) Seasonal changes in oak leaf tannins and nutrients as a cause of spring feeding by winter moth caterpillars. *Ecology*, **51**, 565–581.
60. Feeny, P., Sachdev, K., Rosenberry, L. and Carter, M. (1988) Luteolin 7-O-(6'-O-malonyl)-β-D-glucoside and *trans*-chlorogenic acid, oviposition stimulants for the black swallowtail butterfly. *Phytochemistry*, **27**, 3439–3448.
61. Fellows, L. E., Kite, G. C., Nash, R. J., Simmonds, M. S. J. and Schofield, A. M. (1989) Castanospermine, swainsonine and related polyhydroxy alkaloids, structure, distribution and biological activity. *Rec. Adv. Phytochem.*, **23**, 395–427.
62. Finch, S. (1986) Assessing host-plant finding by insects, in *Insect–Plant Interactions*, (eds J. R. Miller and T. A. Miller), Springer-Verlag, New York, pp. 23–63.
63. Firn, R. D. and Jones, C. G. (1995) Plants may talk, but can they hear? *Trends Ecol. Evol.*, **10**, 371.
64. Fitter, A. H. and Hay, R. K. M. (1987) *Environmental Physiology of Plants*, 2nd edn, Academic Press, London.
65. Fraenkel, G. S. (1959) The raison d'être of secondary plant substances. *Science*, **129**, 1466–1470.
66. Fraenkel, G. S. (1969) Evaluation of our thoughts on secondary plant substances. *Entomol. Exp. Appl.*, **12**, 473–486.
67. Frohne, D. and Jensen, U. (1973) *Systematik des Pflanzenreichs*, G. Fischer, Stuttgart.
68. Fung, S. Y. and Herrebout, W. M. (1988) Sorbitol and dulcitol in celastraceous and rosaceous plants, hosts of *Yponomeuta* spp. *Biochem. Syst. Ecol.*, **16**, 191–194.
69. Gall, L. F. (1987) Leaflet position influences caterpillar feeding and development. *Oikos*, **49**, 172–176.
70. Gershenzon, J. (1984) Changes in the levels of plant secondary metabolites and water and nutrient stress. *Rec. Adv. Phytochem.*, **18**, 273–320.
71. Gershenzon, J. (1994a) Metabolic costs of terpenoid accumulation in higher plants. *J. Chem. Ecol.*, **20**, 1281–1354.
72. Gershenzon, J. (1994b) The cost of plant chemical defense against herbivory: a biochemical perspective, in *Insect–Plant Interactions*, vol. 5, (ed. E. A. Bernays), CRC Press, Boca Raton, FL, pp. 105–173.
73. Gershenzon, J. and Croteau, R. (1991) Terpenoids, in *Herbivores. Their Interactions with Secondary Plant Metabolites*, 2nd edn, vol. 1, (eds G. A. Rosenthal and M. R. Berenbaum), Academic Press, San Diego, CA, pp. 165–219.
74. Gershenzon, J., Maffei, M. and Croteau, R. (1989) Biochemical and histochemical localization of monoterpene biosynthesis in the glandular trichomes of spearmint (*Mentha spicata*). *Plant Physiol.*, **89**, 1351–1357.
75. Gershenzon, J., Murtagh, G. J. and Croteau, R. (1993) Absence of rapid terpene turnover in several diverse species of terpene-accumulating plants. *Oecologia*, **96**, 583–592.
76. Green, T. R. and Ryan, C. A. (1972) Wound induced proteinase inhibitors in plant leaves. *Science*, **175**, 776–777.

77. Gref, R. and Tenow, O. (1987) Resin acid variation in sun and shade needles of Scots pine (*Pinus sylvestris*). *Can. J. For. Res.*, **17**, 346–349.
78. Guenther, E. (1948) *The Essential Oils*, D. van Nostrand & Co., New York.
79. Hamamura, Y., Hayashiya, K., Naito, K., Matsuura, K. and Nishida, J. (1962) Food selection by silkworm larvae. *Nature*, **194**, 754–755.
80. Hamilton-Kemp, T. R., Andersen, R. A., Rodriguez, J. G., Loughrin, J. H. and Patterson, C. G. (1988) Strawberry foliage headspace vapor components at periods of susceptibility and resistance to *Tetranychus urticae* Koch. *J. Chem. Ecol.*, **14**, 789–796.
81. Han, K. and Lincoln, D. E. (1994) The evolution of carbon allocation to plant secondary metabolites, a genetic analysis of cost in *Diplacus aurantiacus*. *Evolution*, **48**, 1550–1563.
82. Harborne, J. B. (1993) *Introduction to Ecological Biochemistry*, 4th edn, Academic Press, London.
83. Harborne, J. B. (1994) Phenolics, in *Natural Products. Their Chemistry and Biological Significance*, (eds J. Mann, R. S. Davidson, J. B. Hobbs et al.), Longman, Harlow, pp. 362–388.
84. Harborne, J. B. and Baxter, H. (1993) *Phytochemical Dictionary. A Handbook of Bioactive Compounds From Plants*, Taylor & Francis, London.
85. Harborne, J. B. and Grayer, R. J. (1993) Flavonoids and insects, in *The Flavonoids: Advances in Research Since 1986*, (ed. J. B. Harborne), Chapman & Hall, London, pp. 589–618.
86. Hartley, S. E. and Lawton, J. H. (1991) Biochemical aspects and significance of the rapidly induced accumulation of phenolics in birch foliage, in *Phytochemical Induction by Herbivores*, (eds D. W. Tallamy and M. J. Raupp), John Wiley, New York, pp. 105–132.
87. Hartmann, T. (1996) Diversity and variability of plant secondary metabolism, a mechanistic view. *Entomol. Exp. Appl.*, **80**, 177–188.
88. Hatanaka, A. (1993) The biogeneration of green odour by green leaves. *Phytochemistry*, **34**, 1201–1218.
89. Haukiója, E. and Niemelä, P. (1977) Retarded growth of a geometrid larva after mechanical damage to leaves of its host tree. *Ann. Zool. Fenn.*, **14**, 48–52.
90. Heard, T. (1995) Oviposition preferences and larval performance of a flower-feeding weevil, *Coelocephalapion aculeatum*, in relation to host development. *Entomol. Exp. Appl.*, **76**, 203–209.
91. Hedin, P. A., Thompson, A. C. and Gueldner, R. C. (1975) Survey of the air space volatiles of the cotton plant. *Phytochemistry*, **14**, 2088–2090.
92. Hedin, P. A., Jenkins, J. N., Collum, D. H., White, W. H. and Parrott, W. L. (1983) Multiple factors in cotton contributing to resistance to the tobacco budworm, *Heliothis virescens* F, in *Plant Resistance to Insects. American Chemical Society Symposium 208*, (ed. P. A. Hedin), American Chemical Society, Washington, DC, pp. 347–365.
93. Hegnauer, R. (1962–1994) *Chemotaxonomie der Pflanzen*, Birkhäuser, Basel.
94. Heidorn, T. and Joern, A. (1984) Differential herbivory on C_3 versus C_4 grasses by the grasshopper *Ageneotettix deorum* (Orthoptera, Acrididae). *Oecologia*, **65**, 19–25.
95. Heinrichs, E. A. and Medrano, F. G. (1985) Influence of N fertilizer on the population development of brown planthopper (BPH). *Int. Rice Res. Newsl.*, **10**(6), 20–21.
96. Honda, K. (1995) Chemical basis of differential oviposition by lepidopterous insects. *Arch. Insect Biochem. Physiol.*, **30**, 1–23.
97. Hruska, A. J. (1988) Cyanogenic glucosides as defense compounds: a review of the evidence. *J. Chem. Ecol.*, **14**, 2213–2217.
98. Jackson, D. M., Severson, R. F., Johnson, A. W., Chaplin, J. F. and Stephenson, M. G. (1984) Ovipositional response of tobacco budworm moths (Lepidoptera, Noctuidae) to cuticular chemical isolates from tobacco leaves. *Environ. Entomol.*, **13**, 1023–1030.
99. Janzen, D. H. (1979) New horizons in the biology of plant defenses, in *Herbivores. Their Interactions with Secondary Plant Metabolites*, (eds G. A. Rosenthal and D. H. Janzen), Academic Press, San Diego, CA, pp. 331–350.
100. Jeffree, C. E. (1986) The cuticle, epicuticular waxes and trichomes of plants, with reference to their structure, functions and evolution, in *Insects and the Plant Surface*, (eds B. Juniper and T. R. E. Southwood), Edward Arnold, London, pp. 23–64.
101. Jermy, T. (1984) Evolution of insect/host plant relationships. *Am. Nat.*, **124**, 609–630.
102. Jetter, R. and Riederer, M. (1996) Cuticular waxes from the leaves and fruit capsules of eight Papaveraceae species. *Can. J. Bot.*, **74**, 419–430.
103. Johns, T. and Alonso, J. G. (1990) Glycoalka-

loid change during domestication of the potato, *Solanum* section *Petota*. *Euphytica*, **50S**, 203–210.
104. Jördens-Röttger, D. (1979) Das Verhalten der schwarzen Bohnenblattlaus *Aphis fabae* Scop. gegenüber chemische Reizen von Pflanzenoberflächen. *Z. Angew. Entomol.*, **88**, 158–166.
105. Juniper, B. E. and Jeffree, C. E. (1983) *Plant Surfaces*, Edward Arnold, London.
106. Juniper, B. E. and Southwood, T. R. E. (1986) *Insects and the Plant Surface*, Edward Arnold, London.
107. Juvik, J. A., Babka, B. A. and Timmermann, E. A. (1988) Influence of trichome exudates from species of *Lycopersicon* on oviposition behavior of *Heliothis zea* (Boddie). *J. Chem. Ecol.*, **14**, 1261–1278.
108. Karban, R. (1992) Plant variation: its effect on populations of herbivorous insects, in *Plant Resistance to Herbivores and Pathogens*, (eds R. S. Fritz and E. L. Simms), University of Chicago Press, Chicago, IL, pp. 195–215.
109. Karban, R. and Myers, J. H. (1989) Induced plant responses to herbivory. *Annu. Rev. Ecol. Syst.*, **20**, 331–348.
110. Karowe, D. (1989) Facultative monophagy as a consequence of prior feeding experience, behavioural and physiological specialization in *Colias philodice* larvae. *Oecologia*, **78**, 106–111.
111. Kimmerer, T. W. and Potter, D. A. (1987) Nutritional quality of specific leaf tissues and selective feeding by a specialist leaf miner. *Oecologia*, **71**, 548–551.
112. Klingauf, F. (1971) Die Wirkung des Glucosids Phlorizin auf das Wirtswahlverhalten von *Rhopalosiphum insertum* (Walk.) und *Aphis pomi* de Geer (Homoptera, Aphididae). *Z. Angew. Entomol.*, **68**, 41–55.
113. Krischik, V. A. and Denno, R. F. (1983) Individual, populational, and geographic patterns in plant defense, in *Variable Plants and Herbivores in Natural and Managed Systems*, (eds R. F. Denno and M. S. McClure), Academic Press, San Diego, CA, pp. 463–512.
114. Kutchan, T. M. and Zenk, M. H. (1993) Enzymology and molecular biology of benzophenanthridine alkaloid biosynthesis. *J. Plant Res.*, **3**, 165–173.
115. Lafont, R., Bouthier, A. and Wilson, I. D. (1991) Phytoecdysteroids, structures, occurrence, biosynthesis and possible ecological significance, in *Proceedings of a Conference on Insect Chemistry and Ecology, Tábor 1990*, pp. 197–214.
116. Lance, D. R., Elkinton, J. S. and Schwalbe, C. P. (1986) Feeding rhythms of gypsy moth larvae; effect of food quality during outbreaks. *Ecology*, **67**, 1650–1654.
117. Lane, G. A., Biggs, D. R., Sutherland, O. W. R., Williams, E. M., Maindonald, J. M. and Donnell, D. J. (1985) Isoflavonoid feeding deterrents for *Costelytra zealandica*, Structure activity relationships. *J. Chem. Ecol.*, **11**, 1713–1735.
118. Langenheim, J. H. (1994) Higher plant terpenoids: a phytocentric overview of their ecological roles. *J. Chem. Ecol.*, **20**, 1223–1280.
119. Larsson, S. and Ohmart, C. P. (1988) Leaf age and larval performance of the leaf beetle, *Paropsis atomari*. *Ecol. Entomol.*, **13**, 19–24.
120. Levy, E. C., Ishaaya, I., Gurevitz, E., Cooper, R. and Lavie, D. (1974) Isolation and identification of host compounds eliciting attraction and bite stimuli in the fruit bark beetle, *Scolytus mediterraneus* (Col., Scolytidae). *J. Agric. Food Chem.*, **22**, 376–379.
121. Louda, S. and Mole, S. (1991) Glucosinolates, chemistry and ecology, in *Herbivores. Their Interactions with Secondary Plant Metabolites*, 2nd edn, vol. 1, (eds G. A. Rosenthal and M. R. Berenbaum), Academic Press, San Diego, CA, pp. 123–164.
122. Luckner, M. (1990) *Secondary Metabolism in Microorganisms, Plants, and Animals*, 3rd edn, Springer-Verlag, Berlin.
123. Luthy, B. and Matile, P. (1984) The mustard oil bomb, rectified analysis of the subcellular organization of the myrosinase system. *Biochem. Physiol. Pflanz.*, **179**, 5–12.
124. McNaughton, S. J. and Tarrants, J. L. (1983) Grass leaf silicification, natural selection for an inducible defense against herbivores. *Proc. Natl Acad. Sci. USA*, **80**, 790–791.
125. Matile, P. (1984) Das toxische Kompartiment der Pflanzenzelle. *Naturwissenschaften*, **71**, 18–24.
126. Mattiacci, L., Dicke, M. and Postumus, M. A. (1994) Induction of parasitoid attracting synomone in Brussels sprouts plants by feeding of *Pieris brassicae* larvae. Role of mechanical damage and herbivore elicitor. *J. Chem. Ecol.*, **20**, 2229–2247.
127. Mattiacci, L., Dicke, M. and Postumus, M. A.

(1995) β-Glucosidase: an elicitor of herbivore-inducible plant odors that attract host-searching parasitic wasps. *Proc. Natl Acad. Sci. USA*, **92**, 2036–2046.
128. Mechaber, W. L., Marshall, D. B., Mechaber, R. A., Jobe, R. T. and Chew, F. S. (1996) Mapping leaf surface landscapes. *Proc. Natl Acad. Sci. USA*, **93**, 4600–4603.
129. *Merck Index* (1996), 12th edn, Merck & Co., Rahway, NJ.
130. Metcalf, R. L. (1987) Plant volatiles as insect attractants. *CRC Crit. Rev. Plant Sci.*, **5**, 251–301.
131. Metcalf, R. L. and Metcalf, E. R. (1992) *Plant Kairomones in Insect Ecology and Control*, Chapman & Hall, New York.
132. Mitchell, B. K., Justus, K. A. and Asaoka, K. (1996) Deterrency and the variable caterpillar, *Trichoplusia ni* and sinigrin. *Entomol. Exp. Appl.*, **80**, 27–31.
133. Mole, S. and Joern, A. (1993) Foliar phenolics of Nebraska sandhills prairie graminoids, between-years, seasonal, and interspecific variation. *J. Chem. Ecol.*, **19**, 1861–1874.
134. Mole, S., Ross, J. A. M. and Waterman, P. G. (1988) Light-induced variation in phenolic levels of foliage of rain-forest plants. I. Chemical changes. *J. Chem. Ecol.*, **14**, 1–21.
135. Mordue (Luntz), A. J. and Blackwell, A. (1993) Azadirachtin: an update. *J. Insect Physiol.*, **39**, 903–924.
136. Morgan, M. E., Kingston, J. D. and Marino, B. D. (1994) Carbon isotopic evidence for the emergence of C_4 plants in the Neogene from Pakistan and Kenya. *Nature*, **367**, 162–165.
137. Mothes, K. (1955) Physiology of alkaloids. *Annu. Rev. Plant Physiol.*, **6**, 393–432.
138. Muthukrishnan, J. and Selvan, S. (1993) Fertilization affects leaf consumption and utilization by *Porthesia scintillans* Walker (Lepidoptera, Lymantridae). *Ann. Entomol. Soc. Am.*, **86**, 173–178.
139. Myers, J. H. (1985) Effect of physiological condition of the host plant on the ovipositional choice of the cabbage white butterfly *Pieris rapae. J. Anim. Ecol.*, **54**, 193–204.
140. Nault, L. R. and Styer, W. E. (1972) Effects of sinigrin on host selection by aphids. *Entomol. Exp. Appl.*, **15**, 423–437.
141. Neuvonen, S. and Haukioja, E. (1991) The effects of inducible resistance in host foliage on birch-feeding herbivores, in *Phytochemical Induction by Herbivores*, (eds D. W. Tallamy and M. J. Raupp), John Wiley, New York, pp. 277–291.
142. Nielsen, J. K., Larsen, L. M. and Sørensen, H. (1979) Host plant selection of the horseradish flea beetle *Phyllotreta armoraciae* (Coleoptera, Chrysomelidae), identification of two flavonol glycosides stimulating feeding in combination with glucosinolates. *Entomol. Exp. Appl.*, **26**, 40–48.
143. Nishida, R. and Fukami, H. (1989) Oviposition stimulants of an Aristolochiceae-feeding swallowtail butterfly, *Atrophaneura alcinous. J. Chem. Ecol.*, **15**, 2565–2575.
144. Nottingham, S. F., Son, K. C., Wilson, D. D., Severson, R. F. and Kays, S. J. (1989) Feeding and oviposition preferences of sweet potato weevils, *Cyclas formicarius elegantulus* (Summers), on storage roots of sweet potato cultivars with differing surface chemistries. *J. Chem. Ecol.*, **15**, 895–903.
145. Nuorteva, P. (1952) Die Nahrungspflanzenwahl der Insekten im Lichte von Untersuchungen an Zikaden. *Ann. Acad. Sci. Fenn. Ser. A, IV Biol.*, **19**, 1–90.
146. Ogutuga, D. B. A. (1975) *Ghana J. Agric. Sci.*, **8**, 121–125.
147. Okolie, P. N. and Obasi, B. N. (1993) Diurnal variation of cyanogenic glucosides, thiocyanate and rhodanese in cassava. *Phytochemistry*, **33**, 775–778.
148. Olckers, T. and Hulley, P. E. (1994) Host specificity tests on leaf-feeding insects, aberrations from the use of excised leaves. *Afr. Entomol.*, **2**, 68–70.
149. Pham-Delegue, M. H., Etievant, P., Guichard, E., Marilleau, R., Douault, P., Chauffaille, J. and Masson, C. (1990) Chemicals involved in honeybee–sunflower relationship. *J. Chem. Ecol.*, **16**, 3053–3065.
150. Poulton, J. E. (1990) Cyanogenesis in plants. *Plant Physiol.*, **94**, 401–405.
151. Pourmohseni, H., Ibenthal, W. D., Machinek, R., Remberg, G. and Vray, V. (1993) Cyanoglucosidases in the epidermis of *Hordeum vulgare. Phytochemistry*, **33**, 295–297.
152. Raffa, K. F. and Berryman, A. A. (1982a) Physiological differences between lodgepole pines resistant and susceptible to the mountain pine beetle and associated microorganisms. *Environ. Entomol.*, **11**, 486–492.
153. Raffa, K. F. and Berryman, A. A. (1982b) Accu-

mulation of monoterpenes and associated volatiles following fungal inoculation of grand fir with a fungus transmitted by the fir engraver *Scolytus ventralis* (Coleoptera, Scolytidae). *Can. Entomol.*, **114**, 797–810.
154. Rausher, M. D. (1992) Natural selection and the evolution of plant–insect interactions, in *Insect Chemical Ecology*, (eds B. D. Roitberg and M. B. Isman), Chapman & Hall, New York, pp. 20–88.
155. Rees, C. J. C. (1969) Chemoreceptor specificity associated with choice of feeding site by the beetle, *Chrysolina brunsvicensis* on its foodplant, *Hypericum hirsutum. Entomol. Exp. Appl.*, **12**, 565–583.
156. Renwick, J. A. A., Radke, C. D., Sachdev-Gupta, K. and Städler, E. (1992) Leaf surface chemicals stimulating oviposition by *Pieris rapae* on cabbage. *Chemoecology*, **3**, 33–38.
157. Rhoades, D. F. (1977) The antiherbivore chemistry of *Larrea*, in *Creosote Bush*, (eds T. J. Mabry, J. H. Hunziker and D. R. DiFeo), Dowden, Hutchinson & Ross, pp. 135–177.
158. Risch, S. J. (1985) Effects of induced chemical changes on interpretation of feeding preference tests. *Entomol. Exp. Appl.*, **39**, 81–84.
159. Robinson, T. (1974) Metabolism and function of alkaloids in plants. *Science*, **184**, 430–435.
160. Robinson, T. (1979) The evolutionary ecology of alkaloids, in *Herbivores. Their Interactions with Secondary Plant Metabolites*, (eds G. A. Rosenthal and D. H. Janzen), Academic Press, San Diego, CA, pp. 413–448.
161. Rodriguez, E., Healey, P. L. and Mehta, I. (eds) (1984) *Biology and Chemistry of Plant Trichomes*, Plenum Press, New York.
162. Rosenthal, G. A. and Berenbaum, M. R. (1991–1992) *Herbivores. Their Interactions with Secondary Plant Metabolites*, 2nd edn, Academic Press, San Diego, CA.
163. Rosenthal, G. A. and Janzen, D. H. (1979) *Herbivores. Their Interaction with Secondary Plant Metabolites*, Academic Press, San Diego, CA.
164. Ryan, J. D., Gregory, P. and Tingey, W. M. (1982) Phenolic oxidase activities in glandular trichomes of *Solanum berthaultii. Phytochemistry*, **21**, 1885–1887.
165. Sanderson, G. W. (1972) The chemistry of tea and tea manufacturing. *Rec. Adv. Phytochem.*, **5**, 247–316.
166. Schlee, D. (1992) *Ökologische Biochemie*, 2nd edn, G. Fischer Verlag, Jena.
167. Schoonhoven, L. M. (1972) Secondary plant substances and insects. *Rec. Adv. Phytochem.*, **5**, 197–224.
168. Schultz, J. C. (1983) Habitat selection and foraging tactics of caterpillars in heterogeneous trees, in *Variable Plants and Herbivores in Natural and Managed Systems*, (eds R. F. Denno and M. S. McClure), Academic Press, San Diego, CA, pp. 61–90.
169. Schultz, J. C. (1988) Many factors influence the evolution of herbivore diets, but plant chemistry is central. *Ecology*, **69**, 896–897.
170. Scriber, J. M. (1984) Nitrogen nutrition of plants and insect invasion, in *Nitrogen in Crop Production*, (ed. R. D. Hack), American Society of Agronomy, Madison WI, pp. 175–228.
171. Seigler, D. S. and Conn, E. E. (cited in ref. 99).
172. Sharkey, T. D. and Singsaas, E. L. (1995) Why plants emit isoprene. *Nature*, **374**, 769.
173. Simms, E. L. (1992) Costs of plant resistance to herbivory, in *Plant Resistance to Herbivores and Pathogens*, (eds R. S. Fritz and E. L. Simms), University of Chicago Press, Chicago, IL, pp. 392–425.
174. Sinden, S. L., Schalk, J. M. and Osman, S. F. (1978) Effects of daylength and maturity of tomato plants on tomatine content and resistance to the Colorado potato beetle. *J. Am. Soc. Hortic. Sci.*, **103**, 596–600.
175. Smiley, J. T., Horn, J. M. and Rank, N. E. (1985) Ecological effects of salicin at three trophic levels, new problems from old adaptations. *Science*, **229**, 649–651.
176. Smith, P. M. (1976) *The Chemotaxonomy of Plants*, Edward Arnold, London.
177. Soldaat, L. L., Boutin, J. P. and Derridj, S. (1996) Species-specific composition of free amino acids from the leaf surface of four *Senecio* species. *J. Chem. Ecol.*, **22**, 1–12.
178. Städler, E. and Buser, H. R. (1984) Defense chemicals in leaf surface wax synergistically stimulate oviposition by a phytophagous insect. *Experientia*, **40**, 1157–1159.
179. Städler, E. and Roessingh, P. (1991) Perception of surface chemicals by feeding and ovipositing insects. *Symp. Biol. Hung.*, **39**, 71–86.
180. Stamp, N. E. and Bowers, M. D. (1990) Phenology of nutritional differences between new and mature leaves and its effect on caterpillar growth. *Ecol. Entomol.*, **15**, 447–454.
181. Stamp, N. E. and Bowers, M. D. (1994) Effects of cages, plant age and mechanical clipping on plant chemistry. *Oecologia*, **99**, 66–71.
182. Storer, J. R., Elmore, J. S. and van Emden, H. F.

(1993) Airborne volatiles from the foliage of three cultivars of autumn flowering chrysanthemums. *Phytochemistry,* **34**, 1489–1492.
183. Stout, M. J. and Duffey S. S. (1996) Characterization of induced resistance in tomato plants. *Entomol. Exp. Appl.,* **79**, 273–283.
184. Stout, M. J., Workman, J. and Duffey, S. S. (1994) Differential induction of tomato foliar proteins by arthropod herbivores. *J. Chem. Ecol.,* **20**, 2575–2594.
185. Stout, M. J., Workman, K. V. and Duffey, S. S. (1996) Identity, spatial distribution, and variability of induced chemical responses in tomato plants. *Entomol. Exp. Appl.,* **79**, 255–271.
186. Suomela, J., Kaitaniemi, P. and Nilson, A. (1995) Systematic within-tree variation in mountain birch leaf quality for a geometrid, *Epirrita autumnata*. *Ecol. Entomol.,* **20**, 283–292.
187. Swain, T. (1979) Tannins and lignins, in *Herbivores. Their Interactions with Secondary Plant Metabolites*, (eds G. A. Rosenthal and D. H. Janzen), Academic Press, San Diego, CA, pp. 657–682.
188. Takabayashi, J., Dicke, M. and Posthumus, M. A. (1991) Variation in composition of predator attracting allelochemicals emitted by herbivore-infested plants: relative influence of plant and herbivore. *Chemoecology,* **2**, 1–6.
189. Takabayashi, J., Dicke, M. and Posthumus, M. A. (1994) Volatile herbivore-induced terpenoids in plant–mite interactions: variation caused by biotic and abiotic factors. *J. Chem. Ecol.,* **20**, 1329–1354.
190. Takabayashi, J., Takahashi, S., Dicke, M. and Posthumus, M. A. (1995) Developmental stage of the herbivore *Pseudaletia separata* affects the production of herbivore-induced synomone by corn plants. *J. Chem. Ecol.,* **21**, 273–287.
191. Tallamy, D. W. and Raupp, M. J. (1991) *Phytochemical Induction by Herbivores*, John Wiley, New York.
192. Thieme, H. and Benecke, R. (1971) Die Phenylglykoside der Salicaceen. Untersuchungen über die Glykosidakkumulation in einigen mitteleuropäischen *Populus*-Arten. *Pharmazie,* **26**, 227–231.
193. Tollsten, L. and Bergström, G. (1988) Headspace volatiles of whole plants and macerated plant parts of *Brassica* and *Sinapis*. *Phytochemistry,* **27**, 2072–2077.
194. Turlings, T. C. J. (1994) The active role of plants in the foraging successes of entomophagous insects. *Norw. J. Agric. Sci.,* **16**, 211–219.
195. Turlings, T. C. J. and Tumlinson, J. H. (1992) Systemic release of chemical signals by herbivore-injured corn. *Proc. Natl Acad. Sci. USA,* **89**, 8399–8402.
196. Turlings, T. C. J., Tumlinson, J. H. and Lewis, W. J. (1990) Exploitation of herbivore-induced plant odors by host-seeking parasitic wasps. *Science,* **250**, 1251–1253.
197. Van Beek, T. A. and de Groot, Æ (1986) Terpenoid antifeedants, part I. An overview of terpenoid antifeedants of natural origin. *Recl. Trav. Chim. Pays Bas,* **105**, 513–527.
198. Van Dam, N. M. and Vrieling, K. (1994) Genetic variation in constitutive and inducible pyrrolizidine alkaloid levels in *Cynoglossum officinale* L. *Oecologia,* **99**, 374–378.
199. Van Dam, N. M., Verpoorte, R. and van der Meijden, E. (1994) Extreme differences in pyrrolizidine alkaloid levels between leaves of *Cynoglossum officinale*. *Phytochemistry,* **27**, 1013–1016.
200. Van Dam, N. M., Witte, L., Theuring, C. and Hartmann, T. (1995) Distribution, biosynthesis, and turnover of pyrrolizidine alkaloids in *Cynoglossum officinale* L. *Phytochemistry,* **39**, 287–292.
201. Van Dam, N. M., de Jong, T. J., Iwasa, Y. and Kubo, T. (1996) Optimal distribution of defenses, are plants smart investors? *Funct. Ecol.,* **10**, 128–136.
202. Vandenberg, P. and Matzinger, D. F. (1970) Genetic diversity and heterosis in *Nicotiana*. III. Crosses among tobacco introductions and flue-cured varieties. *Crop Sci.,* **10**, 437–440.
203. Van Loon, J. J. A., Blaakmeer, A., Griepink, F. C., van Beek, T. A., Schoonhoven, L. M. and de Groot, Æ. (1992) Leaf surface compound from *Brassica oleracea* (Cruciferae) induces oviposition by *Pieris brassicae* (Lepidoptera, Pieridae). *Chemoecology,* **3**, 39–44.
204. Verschaffelt, E. (1910) The cause determining the selection of food in some herbivorous insects. *Proc. K. Ned. Akad. Wet.,* **13**, 536–542.
205. Vet, L. E. M. and Dicke, M. (1992) Ecology of infochemical use by natural enemies in a tritrophic context. *Annu. Rev. Entomol.,* **37**, 141–172.
206. Vickery, M. L. and Vickery, B. (1981) *Secondary Plant Metabolism*, Macmillan, London.
207. Visser, J. H. and Avé, D. A. (1978) General green leaf volatiles in the olfactory orientation of the Colorado potato beetle, *Leptinotarsa decemlineata*. *Entomol. Exp. Appl.,* **24**, 738–749.

208. Vrieling, K. and van Wijk, C. A. M. (1994) Cost assessment of the production of pyrrolizidine alkaloids in ragwort (*Senecio jacobaea* L.). *Oecologia*, **97**, 541–546.
209. Waterman, P. G. and Mole, S. (1989) Extrinsic factors influencing production of secondary metabolites in plants, in *Insect–Plant Interactions*, vol. 5, (ed. E. A. Bernays), CRC Press, Boca Raton, FL, 107–134.
210. Waterman, P. G. and Mole, S. (1994) *Analysis of Phenolic Plant Metabolites*, Blackwell, Oxford.
211. Waterman, P. G., Choo, G. M., Vedder, A. L. and Watts, D. (1983) Digestibility, digestion-inhibitors and nutrients of herbaceous foliage and green stems from an African montane flora and comparison with other tropical flora. *Oecologia*, **60**, 241–249.
212. Watson, M. A. (1986) Integrated physiological units in plants. *Trends Ecol. Evol.*, **1**, 119–123.
213. White, T. C. R. (1970) The nymphal stage of *Cardiaspina densitexta* (Homoptera, Psyllidae) on leaves of *Eucalyptus fasciculosa*. *Aust. J. Zool.*, **18**, 273–293.
214. Whitham, T. G. (1983) Host manipulation of parasites, within-plant variation as a defense against rapidly evolving pests, in *Variable Plants and Herbivores in Natural and Managed Systems*, (eds R. F. Denno and M. S. McClure), Academic Press, San Diego, CA, pp. 15–41.
215. Whittaker, R. H. (1970) The biochemical ecology of higher plants, in *Chemical Ecology*, (eds E. Sondheimer and J. B. Simeone), Academic Press, New York, pp. 43–70.
216. Wink, M. (1987) Physiology of accumulation of secondary metabolites with special reference to alkaloids, in *Cell Cultures and Somatic Cell Genetics of Plants*, vol. 4, (eds F. Constabel and I. K. Vasil), Academic Press, San Diego, CA, pp. 17–42.
217. Wolfson, J. L. (1982) Developmental responses of *Pieris rapae* and *Spodoptera eridania* to environmentally induced variation in *Brassica nigra*. *Envir. Entomol.*, **11**, 207–213.
218. Woodhead, S. and Bernays, E. A. (1977) Changes in release rates of cyanide in relation to palatability of *Sorghum* to insects. *Nature*, **270**, 235–236.
219. Woodhead, S. and Chapman, R. F. (1986) Insect behaviour and the chemistry of plant surface waxes, in *Insects and the Plant Surface*, (eds B. Juniper and T. R. E. Southwood), Edward Arnold, London, pp. 123–135.
220. Woodhead, S., Galeffi, C. and Marini Betollo, G. B. (1982) p-Hydroxybenzaldehyde is a major constituent of the epicuticular wax of seedling *Sorghum bicolor*. *Phytochemistry*, **21**, 455–456.
221. Zangerl, A. R. and Bazzaz, F. A. (1992) Theory and pattern in plant defense allocation, in *Plant Resistance to Herbivores and Pathogens*, (eds R. S. Fritz and E. L. Simms), University of Chicago Press, Chicago, IL, pp. 363–391.
222. Zangerl, A. R. and Rutledge, C. E. (1996) The probability of attack and patterns of constitutive and induced defense, a test of optimal defense theory. *Am. Nat.*, **147**, 599–608.
223. Zielske, A. F., Simons, J. N. and Silverstein, R. M. (1972) A flavone feeding stimulant in alligatorweed. *Phytochemistry*, **11**, 393–396.
224. Zucker, W. V. (1982) How aphids choose leaves, the role of phenolics in host selection by a galling aphid. *Ecology*, **63**, 972–981.
225. Svoboda, G. H. (1961) Alkaloids of *Vinca rosea* (*Catharanthus roseus*). IX. Extraction and characterization of leurosidine and leurocristine. *Lloydia*, **24**, 173–178.

PLANTS AS INSECT FOOD: NOT THE IDEAL

4

4.1	Insect feeding systems	83
4.2	Plants are suboptimal food	89
4.2.1	Nitrogen	90
4.2.2	Water	92
4.3	Artificial diets	93
4.4	Consumption and utilization	95
4.4.1	Food quantities eaten	95
4.4.2	Utilization	95
4.4.3	Suboptimal food and compensatory feeding behaviour	100
4.4.4	Allelochemicals and food utilization	102
4.4.5	Detoxification of plant allelochemicals	104
4.5	Symbionts	108
4.6	Host-plant effects on herbivore susceptibility to pathogens and insecticides	109
4.7	Plant-mediated effects of air pollution on insects	112
4.8	Conclusions	113
4.9	References	113

The subject of this chapter, plants as food for herbivorous insects, touches the heart of insect–plant biology. Its theme can be captured in two basic questions. First: what do plants offer to insects in the way of nutrition? And second: what do insects need for optimal growth and reproduction? Answering these two questions is seriously hindered by the fact that (1) the chemical composition of plants, as noted before, varies among species, as well as in space and time, and (2) the nutritional requirements of insects vary between species and with developmental stage and environmental conditions. Apart from these complications, clearly a major nutritional discrepancy exists between what plants provide and what insects require. Plants appear to supply food that is at best of marginal quality. What is the basis of this statement and how do we know?

Since insects, like all animals, need food as material for conversion into body substance and as a source of energy, it is appropriate to compare the chemical composition of insects with that of plants. Figure 4.1 shows the concentrations of some major elements in insects and plants.

None of the values found in insects, except that for magnesium, equals the average value measured in plants. Nitrogen deserves special attention because, relative to the other major components of living organisms, it is in short supply in a form available to insects. Whereas the nitrogen content of animals amounts to about 8–14% of their (dry) body weight, plants usually contain only 2–4% nitrogen (Fig. 4.2).

Likewise, the caloric value of animal tissue (in insects: 22.8 joules per milligram) exceeds that of plants (terrestrial plants on average: 18.9 J/mg). These ratios show that herbivorous insects must concentrate nitrogen when converting plant food into body tissue. Attaining the caloric value typical of insects requires less grading up. Therefore, the nutritional value of a plant for an insect (more so than for mammals, which grow much more slowly, but use more energy) is primarily determined by its nitrogen content, while its caloric value is of less importance (Table 4.1).

4.1 INSECT FEEDING SYSTEMS

The three salient features of feeding behaviour are food choice, method of feeding and quan-

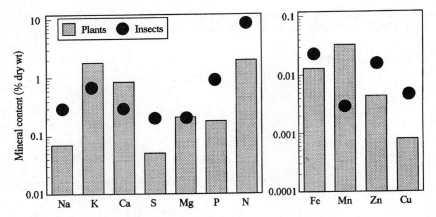

Figure 4.1 Average concentrations of elements in plant tissues (bars) as compared to those in insects (dots). It should be emphasized that the levels presented for plants in particular vary greatly between species. Environmental factors and plant (tissue) age cause further inter- and intraspecific variation. The vertical scale is logarithmic. (Source: data from Allen *et al.*, 1974.)

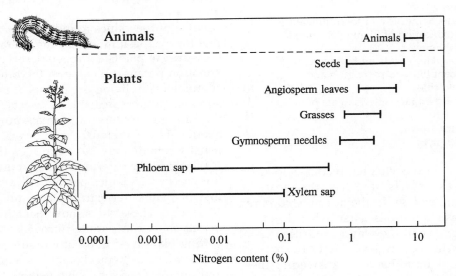

Figure 4.2 Variations in nitrogen concentration (% dry weight) of different plant parts as compared to animals. Xylem and phloem sap concentrations expressed as nitrogen weight/volume. (Source: redrawn from Mattson, 1980.)

titative food consumption. Insects use two feeding methods and either bite off and chew their food [46] or imbibe liquid nourishment [143]. Insects of the former type, known as **mandibulate**, possess the more general type of mouthparts. There are three pairs of appendages, which more or less oppose each other (Fig. 4.3).

The mandibles (or jaws) serve to cut and grind the food. They are equipped with tooth-like ridges to cut food and grinding surfaces to crush it. Below the mandibles are the max-

Table 4.1 Approximate optimal ratios of protein to available carbohydrate plus fat (expressed as grams of glucose) in the diets of some herbivorous insects as compared to some mammals. (Source: modified from Bernays, 1982)

Animal	Ratio of protein to glucose
Silkworm	1:3
Silkworm (artificial diet)	1:1.5
Locust (artificial diet)	1:1
Cabbage butterfly larva	1:1
Calf (very young)	1:4
Human child	1:6
Cow	1:7
Buffalo	1:10
Goat	1:15

Table 4.2 Relative leaf toughness or hardness in plants with different growth form, with leaves of herbaceous dicots standardized to 1. n = number of species tested. (Source: modified from Bernays, 1991)

Plant type	n	Relative toughness
Herbaceous dicots: all leaves	166	1.0
Woody plants: new leaves	25	1.7
C_3 grasses: all blades	42	3.1
C_4 grasses: all blades	34	6.2
Woody plants: fully expanded leaves	89	6.3
Palms: expanded fronds	8	9.8

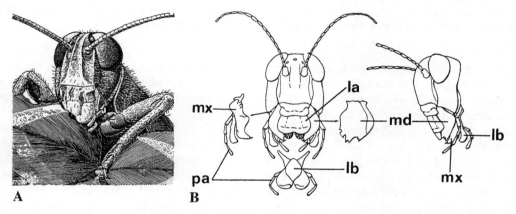

Figure 4.3 A grasshopper as an example of a mandibulate insect, feeding on clover (left) and frontal and lateral views of its mouthparts. mx = maxilla; md = mandible; pa = palps; la = labrum; lb = labium. (Source: **(A)** reproduced from Frost, 1959, with permission; **(B)** from *A Textbook of Entomology* by H. H. Ross. Copyright (©) 1948. Reprinted by permission of John Wiley & Sons, Inc.)

illae. Each maxilla bears a segmented appendage, the maxillary palp, which is equipped with chemosensory sensilla. The maxillae aid in manipulating the food and guiding it toward the mouth. The labrum, or upper lip, forms the roof of the preoral cavity and mouth. Its ventral surface, called the epipharynx, often contains taste sensilla. The labium, or lower lip, forms the floor of the preoral cavity. It has one pair of palps bearing mechanoreceptors.

Although the highly sclerotized mandibles of many insect species can be extremely hard, deriving extra hardness by the incorporation of zinc or manganese in the cuticle [60], there is often considerable wear caused by feeding on tough plant parts [46]. Plant leaves vary greatly in toughness and hardness, grasses, for instance, being three times tougher than an average herb [28] (Table 4.2).

Differences in leaf hardness probably affect insect feeding and growth more than is often

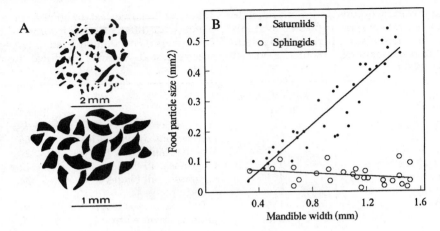

Figure 4.4 (A) Drawings of food particles from the midgut of (above) a sphingid (*Pachylia ficus*) and (below) a saturniid (*Rothschildia lebeau*) fifth instar caterpillar. **(B)** Relationship between mandible width and size of chewed food particles in the gut of 10 saturniid and 15 sphingid caterpillars (various instars). The saturniid species fed on tough and thick, mature leaves whereas the sphingids fed upon soft and flimsy leaves, both old and new. (Source: redrawn from Bernays and Janzen, 1988.)

thought. The polyphagous beet armyworm (*Spodoptera exigua*), for example, needs over three times longer to swallow food particles from celery (*Apium graveolens*) than from nettleleaf goose foot (*Chenopodium murale*), because the leaves of celery are 1.5 times as tough as those of goose foot [21]. The size of the leaf fragments swallowed by chewing insect species varies with size (instar) of the insect and hardness of the food. Thus saturniid caterpillars feeding on tough leaves have in their guts leaf particles that are relatively large and very regular in size, whereas sphingid larvae generally feeding on the soft leaves of herbaceous hosts bite off small leaf particles independent of caterpillar size (Fig. 4.4).

Since most insects digest cell walls only to a very limited degree [73], inefficient digestion would be expected for insects with the habit of taking only large bites. This is not the case. The frass of a lepidopteran (*Paratrytone melane*) contains leaf pieces with 76–86% uncrushed cells. Yet the approximate digestibility of soluble carbohydrates and protein averages 78 and 88% respectively. It is supposed that the nutrients are extracted from the uncrushed cells through plasmodesmata and cell wall pores after the cell membranes are digested [14].

Chewing off particles of tough plant tissues requires quite some energy and causes severe wear of mouthparts as compared to feeding on softer tissues. Cellulose, as an important component of cell walls, may thus act as a broad-spectrum resistance factor to insect herbivores and hence play a role in structuring terrestrial communities [1]. Wear can be especially excessive when feeding on plants with high silicon content (up to 10% dry weight), such as grasses and horsetails (Equisetales). Amorphous silica ($SiO_2 \cdot nH_2O$) particles deposited in cell walls and cell lumens [56] serve as a harsh abrasive that may cause complete loss of mandibular teeth during the feeding process, resulting in death by starvation. Increased silica content in rice and other gramineous crop plants contributes to resistance to several insect pest species [82, 115] (Fig. 4.5).

Haustellate mouthparts, which serve to pierce the plant tissues and suck liquid food,

Insect feeding systems   87

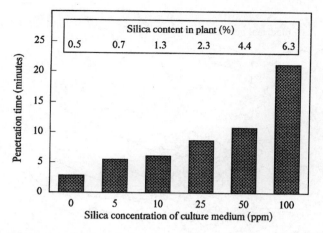

Figure 4.5 Penetration time required by newly hatched yellow stem-borer (*Scirpophaga incertulas*) larvae to enter stems of rice plants grown on nutrient solutions with different silica levels. (Source: redrawn from Khan and Ramachandran, 1989.)

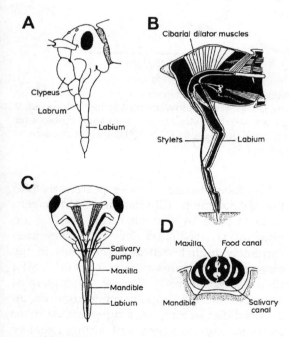

Figure 4.6 Haustellate mouthparts of Hemiptera. **(A)** Lateral and **(C)** frontal view of hemipteran head. **(B)** Longitudinal section of head and mouthparts of a pentatomid during feeding. **(D)** Schematic transverse section through stylet bundle. (Source: reproduced from Davies, 1988, after Weber, with permission.)

are polyphyletic in origin, having arisen independently in Hemiptera, Thysanoptera and adult Lepidoptera. The elongated suctorial mouthparts of butterflies and moths and some adult flies consist only of the maxillae, which fit together to form a proboscis or rostrum. In the Hemiptera the labium is developed into a pronounced structure. The mandibles and maxillae are styliform and the maxillary palps are atrophied. The labium is shaped into an anteriorly grooved sheath in which the two mandibular and the two maxillary stylets are enclosed. In most hemipteran plant feeders its distal end is equipped with taste sensilla [117] but in aphids they seem to be absent. The two maxillae are interlocked in such a way that a double-barrelled tube is formed. The upper channel in the stylet bundle serves to take up food and the lower one to deliver saliva (Fig. 4.6). The needle-like stylets can pierce the plant cuticle and cell walls and, once inside plant tissue, can be oriented into different directions in search of an acceptable feeding site (Fig. 4.7).

The degree of control exercised over the stylets allows movements towards a vascular bundle, possibly by following a chemical gradient. The stylets gradually penetrate in

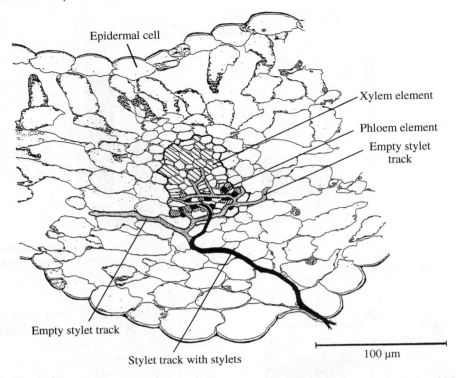

Figure 4.7 Stylet pathway of an aphid (*Aphis fabae*) feeding on a sieve element in the vein of a broad bean leaf. The stylet track shows many branches, representing earlier search movements during the process of phloem localization. The empty branches consist of salivary sheath material, which remains visible after the stylets are withdrawn. (Source: reproduced from Tjallingii and Hogen Esch, 1993, with permission.)

between mesophyll cells towards vascular elements. Cell punctures along the pathway allow the insect to imbibe protoplast or vacuole contents and chemosensory evaluation possibly takes place in the pharyngeal taste organ. Notwithstanding the occasional intracellular punctures, the stylet pathway from cuticle to phloem remains extracellular [143]. Sieve elements may be located within the vascular bundle on a hit-or-miss basis [68]. Locating a suitable feeding site within the plant is a tedious process. Aphids may need up to 4 hours to find and tap suitable phloem elements, a long period of time for a short-lived animal! Once they have hit an adequate sieve element they may tap it continuously for several days [145].

The food channel empties posteriorly into the cibarial cavity. Cibarial muscles may generate suction in the cibarium. Feeding on phloem is facilitated by the sometimes extremely high hydrostatic pressures in the sieve elements, ranging from about 0.2–1 MPa (2–10 atmospheres). Several hemipterans, however, are also capable of feeding on an artificial diet lacking plant turgor pressure, or in some cases on xylem with strong negative pressures. Apparently the cibarial pump then generates sufficient suction pressure to overcome the negative pressure [83].

Sucking insects utilize several kinds of plant fluids. Most aphids, mealybugs, leafhoppers and psyllids imbibe fluid from phloem cells, whereas many Heteroptera and some immature scale insects feed on the parenchyma. Thrips live on liquids extracted from individual epidermal or parenchymal cells.

Many plants support both mandibulate and haustellate insect species. For instance, 335 chewers and 88 sap-feeding insect species have been recorded as feeding upon two oak species [81]. The divergence in mouthpart structures allowing for the two feeding styles is an important prerequisite for food specialization and thereby for insect diversification. Obviously, both feeding methods have their advantages and disadvantages. The more delicate feeding strategy developed by sap-feeding insects places a restriction on larger size; sucking insects are generally smaller than chewers. They often inflict less mechanical injury to their host plants, thus exploiting their resources better than their chewing counterparts [102]. In other cases, however, sucking insects cause serious deformation and stunting of shoots, and pentatomid bugs may kill entire shoots (e.g. ears of grasses, wheat, etc.) by just a single feeding puncture (not to mention the transmission of viruses and mycoplasms by aphids and leafhoppers). The direct damage inflicted by aphids is often relatively small, but the impact of spittlebugs on their host can be more severe than that of leaf-eating species [99]. Mandibulate insects, on the other hand, cannot avoid ingesting, together with nutritive compounds, large amounts of the indigestible structural components of the plant, as well as toxic substances. Sap feeders can often avoid adulteration of their food with such compounds. Phloem fluid, for instance, has a lower ratio of allelochemicals to nutrients than most other plant tissues [109]. Furthermore, phloem feeders may derive additional protection by injecting salivary secretions into their food that detoxify some allelochemicals before ingestion [105]. Aphids may also escape changes in leaf palatability resulting from defoliation of their hosts, whereas chewing species are negatively affected [158].

Xylem forms a less suitable food source than phloem, since nitrogen concentrations of xylem sap (less than 0.01%) are typically ten times less than phloem concentrations and two orders of magnitude less than those of foliar tissues (Fig. 4.2); leafhoppers feeding on xylem must suck enormous amounts of sap to meet their nitrogen and carbohydrate demands. Feeding rates can be as high as 300–1000 times body weight per day [42]. Spittlebugs are therefore very selective in their host choice, depending on the way the plant transports fixed nitrogen in the xylem. They show a distinct preference for some subgroups of legumes but avoid others in order to secure their nitrogen requirements [142].

4.2 PLANTS ARE SUBOPTIMAL FOOD

The food of insect herbivores consists of dilute nutrients in a matrix of indigestible structural compounds, such as cellulose and lignin, and a variety of allelochemicals (which are often simply poisonous or deter feeding). To make things worse the quantitative ratios of nutrients present in the plant may differ greatly from those required by the insect. Qualitatively the nutritional requirements of insects are generally the same as for other animals, except that, unlike many other animals, insects lack the capacity to synthesize sterols. Therefore, they must extract sterols together with several other essential nutrients (amino acids, carbohydrates, lipids, fatty acids, vitamins, trace elements) from their food. The nutritional requirements of different insect species are often fairly specific and may allow for only small margins, qualitatively as well as quantitatively. Optimal growth, survival rates, growth and fecundity require certain protein:carbohydrate ratios, which may vary considerably among species. Polyphagous larvae of the corn earworm (*Helicoverpa zea*) grow best on an artificial diet with a protein:carbohydrate ratio of 79:21. Conversely, nymphs of the cockroach *Supella longipalpa* require a totally different ratio of 16:84. This striking contrast results from differences in organism characteristics [151]. Whereas corn earworms grow fast and therefore need protein-rich food, the cockroach is characterized by slow growth and high activity levels. The cockroach, as a consequence, needs a high intake of energy to supply its muscles with fuel.

When plants are truly suboptimal food due to inadequate nutrient ratios and the presence of allelochemics which need to be detoxified, it should be possible to develop artificial diets that support growth better than natural food plants. Young cutworm larvae (*Agrotis ipsilon*) raised on an artificial diet did indeed gain 12 times as much weight as those raised on tissues from 'susceptible' corn leaves, their natural food [110] (Fig. 4.8).

This observation and similar results from other insect species [13], which grew faster, attained higher pupal weights and showed better reproduction on artificial diets, prove that the most susceptible plants are in fact remarkably well defended against insect attack [110] and are poor food sources from a nutritional point of view. Given this conclusion Berenbaum [24] has presented an intriguing hypothesis based on the contention that nutritional inadequacy may be a major determinant of host-plant resistance. In that case the selective impact of herbivory may have been a driving factor in establishing a biosynthetic and structural diversity of primary metabolites that would render plants less suitable food sources for herbivores. Unfortunately, Berenbaum's hypothesis can not be proved at present. On the contrary, as will be discussed in section 10.4.2, there is growing evidence that the main characteristics of plant metabolism evolved before the appearance of terrestrial plants. Since insects probably had and have only limited impact on plant evolution, it seems unlikely that they have significantly affected the evolution of ancient plant traits such as the basic biochemical processes of metabolism.

4.2.1 NITROGEN

The importance of organic nitrogen for normal insect growth and reproductive success cannot be overemphasized. In spite of its general occurrence in the atmosphere, nitrogen is of all the elements essential to organic life on Earth the one that is least available in a usable form, i.e. combined with other chemicals. Proteins are the basic structural materials of insects, not only of soft tissues but also of the integument. Cuticular proteins usually make up more than 50% of cuticle by dry weight. In contrast the bulk of plant tissue consists of carbohydrates, since major components of cell walls include, in addition to lignin, cutin, silica and cell wall protein, cellulose and hemicellulose. Moreover, the balance of amino acids constituting plant proteins differs from the dietary requirements of insects (Fig. 4.9). Since large amounts of aromatic compounds serve to bind the cuticular proteins together, insects need considerably higher levels of aromatic amino acids, such as phenylalanine and tryptophan, than are present in plant proteins [26, 41].

The growth efficiency of a variety of insects is closely related with plant nitrogen content,

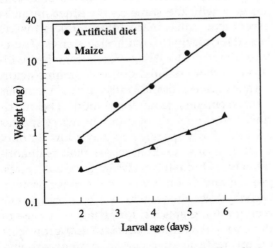

Figure 4.8 Larval growth of *Agrotis ipsilon* during six days after hatching on susceptible maize plants or on artificial diet. Note that the Y-axis (larval body weight) is a logarithmic scale. (Source: redrawn from Reese and Field, 1986.)

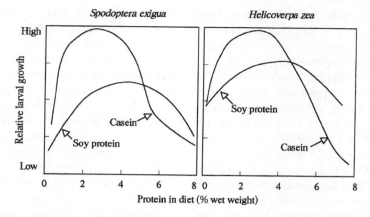

Figure 4.9 Ability of two dietary proteins to support larval growth of two noctuid species. Growth with the various protein regimens is relative to the growth of controls on a standard artificial diet containing 2.4% wet weight of casein. Soy protein is inferior to casein, an animal-derived protein. Higher soy protein levels are needed to obtain maximal growth than in the case of casein diets. (Source: redrawn from Duffey et al., 1986.)

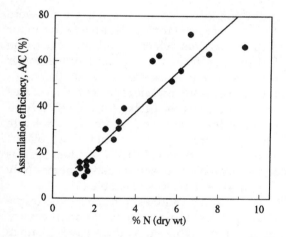

Figure 4.10 Assimilation efficiency as a function of total nitrogen levels in food plants of the plant bug *Leptoterna dolabrata*. (Source: redrawn from McNeill and Southwood, 1978.)

a correlate of protein content. As the nitrogen content of their food increases, insects become more efficient in converting plant material into body tissue (Fig. 4.10).

Thus insects on plants with 1% nitrogen require over three times more food than insects on plants containing 6% nitrogen [93]. Yet, despite its essential role in herbivore–plant relations, the total nitrogen content of a plant is frequently a poor index of its nutritional value. High nitrogen levels may occur together with metabolically useless nitrogen compounds, such as alkaloids, or with tannins, which in some cases reduce digestive efficiency (see section 4.4.4). In this respect phloem-feeding species are in a more privileged position than chewing insects, because nearly all nitrogen-containing compounds in phloem sap can be utilized.

Where nitrogen is truly an important indicator of food quality, or even the limiting factor [153] for optimal growth, the application of nitrogen fertilizer to plants can be expected to positively affect herbivore performance. Such effects do indeed often occur, but negative effects have also been reported. Scriber [123], on the basis of a literature survey, listed at least 115 studies in which insects grew better with increased plant nitrogen. On the other hand, at least 44 studies indicate a decrease in herbivore performance with high nitrogen concentration. These confused and often contradictory responses of insects to changes in nitrogen content have been explained by several mechanisms. Probably, insects are physiologically adapted to nitrogen levels that are normal (or slightly higher than normal) for their normal host(s). When the nitrogen content of fir trees was manipulated by growing seedlings on nutrient solutions that differed in nitrogen concentration, the nitrogen content of the needles ranged from 0.74–5.02% (dry weight). Western spruce budworms (*Choristoneura occidentalis*) performed best at levels around 2.5%, concentrations that are normally encountered in nature [40]. Similar results were obtained with two noctuid species on artificial diets with varying amounts of protein [55] (Fig. 4.9).

It has been conjectured that 'flush feeders', insect species that are adapted to high nitrogen levels in their food, would respond positively to an increased amount of nitrogen being transported to the growing tissues, while 'senescence feeders' would respond negatively to the decreased export of nitrogen from senescing tissues [153]. Moreover, nitrogen fertilization may cause many kinds of physiological and morphological alteration to plants and affect, among other things, secondary metabolism, resulting in increased production of defence substances. Susceptibility to plant pathogens and environmental factors such as microclimate and weed growth may also alter. Morphological changes may include an increase of leaf surface and leaf thickness, changes in length of internodes and toughness of veins, all of which could negatively affect herbivorous insects. Thus nitrogen-fertilization-induced changes naturally alter the value of the plant as a home for the herbivore and its natural enemies.

4.2.2 WATER

Water is the cradle of life. Insects, like other animals, need it and acquiring sufficient amounts of water is a major nutritional 'hurdle' for most herbivorous species [123, 139]. Even though water is according to traditional terminology not a nutrient and the water content of foliage varies from 45–95% of fresh weight, the amount of water in the food of many lepidopterous larvae provides a surprisingly useful index of its nutritional value and thus of growth performance. The importance of sufficient leaf water content for an insect has been shown by an experiment in which larval growth rates were determined on various legumes and alfalfa cultivars differing in water content. Significantly better growth occurred on plants with higher water content. The relevance of dietary water was confirmed in experiments with artificial diets varying in water content [123]. When the normal amount of water in foliage drops, its nutritional value decreases. When caterpillars belonging to 16 species were fed upon excised leaves without water supplementation *via* the petioles, their relative growth rates showed reductions of up to 40% even when the food did not show any indication of desiccation. Such effects were more pronounced for tree-leaf-feeders than for forb-leaf-feeders, which is probably due to the fact that the former group already has lower conversion efficiencies because of the naturally lower water and nitrogen contents of tree leaves compared to herbal foliage [122]. Acridids, on the other hand, more easily tolerate reduced water content in their host plants. Water stress even made 12 of 41 plant species more palatable to desert locusts, whereas only five of 41 became less palatable [31].

Water content and leaf nitrogen levels (especially protein and amino acids) often covary, both being higher in young leaves than in mature and senescing leaves. The nutritional value of herbs and grasses with water contents of 80–90% and nitrogen levels of 5–6% is higher than that of foliage from woody plants, which typically contains 60% water and 2% nitrogen [134] (see section 4.4.2). Superimposed on the differences between plant groups are seasonal changes in water (and nitrogen) content, with diminishing nutritional values as the summer progresses (Fig. 4.11).

Although herbivorous insects feed on material that contains a high proportion of water, many species can be seen to drink from dew

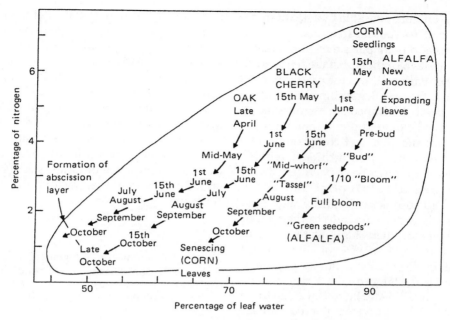

Figure 4.11 Nitrogen and water content of foliage of herbaceous plants and deciduous trees decreases as the season progresses and consequently the nutritional value of plants is reduced as compared to spring foliage. (Source: reproduced, with permission, from the *Annual Review of Entomology*, volume 26, © 1981, by Annual Reviews Inc.)

drops or other sources of free water. Caterpillars can perceive water from a distance of at least up to 2 cm and will readily drink when encountering water drops [51, 98]. In contrast to caterpillars, which easily evaporate water, grasshoppers normally need only an average water intake of about 60% of their food. The importance of water to this insect also is evident from the finding that their state of hydration influences food selection. Locusts after being fed for some time on drier food sources eat more than normally from plants with high water content to compensate for a water deficit [111]. Clearly, herbivores possess a behavioural regulation of water balance by selective feeding.

Drought stress in plants can be extremely detrimental to some herbivores. In other cases, however, effects are negligible or even beneficial for herbivore population increase. Periods of unusually warm, dry weather are often followed by outbreaks of insect pests in forests and rangeland. Their causes are not well understood. Probably drought stress affects the resistance mechanisms of plants negatively, while at the same time their nutritional value for insects increases. Drought, just as other kinds of stress, effects increased levels of soluble sugars and nitrogen in plant foliage, inner bark and sapwood. Foliar sugars in balsam fir, for instance, may increase 2.5-fold. Drought has also been found to disturb nitrogen metabolism in many woody as well as herbaceous plants, thereby influencing growth and fecundity of their insect fauna [94, 95]. Details of the causal relationships leading from water stress in plants to insect response are still hardly understood and await elucidation [74].

4.3 ARTIFICIAL DIETS

When studying behavioural responses to specific plant compounds or the nutritional role of certain plant components, artificial diets of known chemical composition have been shown to be an indispensable tool. Plant material is difficult to standardize, because individual plants and plant parts may vary greatly, with season, developmental stage, etc. Artificial diet formulation allows for precise control of nutritional factors. Beginning in the 1950s, diets have been developed for many species. This proved to be more difficult than might be expected once the chemical composition of a plant is known and the nutritional requirements of its herbivores have been listed. The difficulties arise from two plant traits which are difficult to copy in an artificial diet. First, plants provide a dry substrate to their herbivores in spite of their high water content, which may amount to 90% of total weight. Plant food is essentially a liquid packed in microcapsules (cells) giving it a dry outside. Second, the physical and chemical structure of the microcapsules prevents microorganisms invading the highly nutritive cell contents. To mimic the firm surface of plant parts artificial diets are given some rigidity by incorporating agar, cellulose or other nutritionally inert substances which add texture to the liquid food. It also provides roughage which aids the passage of the food material through the gut. To suppress bacterial decay the food has to be sterilized by heat and by adding antibiotics. These compounds, however, may also affect the feeder through their impact on gut microbes and detoxification enzymes or in other ways, and are often not tolerated. Therefore, finding an effective dose that is at the same time harmless to the insect consumer forms an important element in diet development. Because different species often differ slightly in their precise nutritional requirements even small changes in diet composition can have drastic effects on insect performance, including growth and reproduction. Moreover, origin and storage conditions of the ingredients and variations in diet preparation can seriously affect its quality, even when polyphagous and less finicky insects, for instance gypsy moth larvae (*Lymantria dispar*), are involved [80]. Currently several excellent books on artificial diets exist that have a 'cookbook' style, giving step-by-step rearing instructions for specific insect species [132]. Moreover, 'ready-to-eat' diets for several species are obtainable from commercial sources. Some artificial diets, lacking any host-plant specific chemical are suitable for a number of different insect species (Table 4.3). In the case of food specialists addition of host-plant material is part of many successful diets, due to the presence of either specific feeding stimuli or nutritional factors that are still unknown.

Table 4.3 Composition of a general purpose artificial diet. Wesson salts and Vanderzant vitamin mix are ready-made mixtures commercially available. (Source: modified from Singh, 1983)

Ingredients	Amount (g/100 g)
1. Protein	
Casein	3.5
Wheat germ (0.75 protein, 2.25 other)	3.0
2. Carbohydrates	
Sucrose	3.0
Glucose (+1.3 dextrose from vitamin mixture)	0.5
3. Lipids	
Linoleic acid	0.25
4. Sterols	
Cholesterol	0.05
5. Minerals	
Wesson salts	3.0
6. Vitamins	
Vanderzant vitamin mix (0.7 vitamin, 1.3 dextrose)	2.0
7. Gelling and bulking agents	
Agar	2.5
Cellulose	10.0
8. Microbial inhibitors	
Streptomycin	0.015
Benzylpenicillin sodium	0.015
Methyl-parahydroxy-benzoic acid	0.112
Sorbic acid	0.3
9. 4M potassium hydroxide	0.5
10. Water	72.17

Of course, artificial diets differ in many respects from natural food sources. The question arises whether insects reared on diets for many generations or even for only one generation exhibit changes in behaviour or physiology. When the European corn borer (*Ostrinia nubilalis*) was cultured continuously (i.e. for 46–108 generations) on an artificial diet, it would hardly feed anymore on normally very susceptible corn plants. However, when each time after seven generations on artificial diet one generation was raised on corn, the culture maintained its virulence for corn [64]. In another insect, however, vigour and food utilization on host plant remained unchanged after more than 275 generations on an artificial diet [48]. Short-term changes may also occur. Insects grown on a relatively soft artificial diet may, due to reduced physical exercise as compared to those feeding on tough plant leaves, develop differences in head musculature and dimensions. Such variations have been noted among caterpillars fed host plants differing in toughness [27]. Larvae of tobacco hornworms (*Manduca sexta*) raised on a diet are sluggish and less dextrous in holding themselves on vertical plant structures than conspecifics fed plant tissues. Therefore quality control is an essential part of all insect rearing procedures [8, 85].

4.4 CONSUMPTION AND UTILIZATION

4.4.1 FOOD QUANTITIES EATEN

Fast-growing insects consume large amounts of food. Their gut is shaped to process large food volumes and occupies in non-reproductive stages most of the body cavity. Food passage through the gut is fast and takes often only a few hours in leaf-feeding insects. Food transit time in aphids can be as short as 1 hour. Young caterpillars may consume plant tissues at a rate of up to six times their body weight per day, whereas adult locusts eat daily about their own weight of food. Sap feeding spittlebugs may even consume xylem sap in

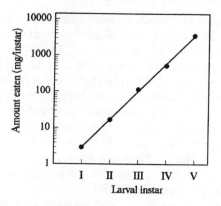

Figure 4.12 Food intake by silkworm (*Bombyx mori*) larvae over different instars. Note logarithmic scale of ordinate. (Source: data from Anantha Raman *et al.*, 1994.)

amounts ranging from 100–1000 times their body weight per day [137].

The stepwise growth of insects makes it convenient to determine food intake and digestive efficiency during larval growth [137]. Growth follows an exponential increase in weight, with often more than a doubling of weight occurring in each instar. The steep growth curve is paralleled by an exponential increase in food intake (Fig. 4.12).

As a result mature larvae often weigh several thousand times their weight at hatching. Full-grown silkworm (*Bombyx mori*) larvae, for instance, weigh 10 000 times the body weight of neonates. Lepidopterous larvae consume from 94–98% of all food during the penultimate and last stadia alone [7].

4.4.2 UTILIZATION

(a) Utilization as a factor driving host-plant use

A prime question in insect–plant studies is with what efficiency herbivores utilize their host plants nutritionally. To answer this, one needs to know which fraction of the total amount of food that is ingested is actually used for growth and ultimately reproduction.

A higher efficiency of food utilization can be seen as an indicator of a higher nutritional quality of a plant or plant part. In conjunction with sensory and behavioural factors determining host-plant specificity, differences in host-plant utilization efficiency may contribute to explain host-plant specificity. A considerable literature has accumulated in this field, often referred to as 'nutritional ecology' [136]. A landmark publication in the field of insect nutrition was the influential review article by Waldbauer [150], who summarized the earlier literature and proposed to rigorously standardize the quantitative methods and parameters employed in such approaches. The extensive literature which appeared since 1968 has been summarized in several reviews and books [134, 136, 137]. General conclusions from quantitative nutritional studies are that major herbivore guilds, such as tree and herb feeders, differ in their utilization efficiency (see below) and that, as stated earlier, water and nitrogen are primary determinants of nutritional quality irrespective of the group studied. These studies have also shown that, at the physiological level, for example when one is interested in assessing whether an insect species uses one plant species more efficiently than an alternative host plant, or whether a secondary plant substance affects utilization efficiency, accurate measurements are often more difficult to obtain than might be expected.

(b) Parameters of utilization and performance and their interrelationships

Waldbauer [150] defined three parameters of utilization, now commonly termed nutritional indices:

1. **approximate digestibility** (abbreviated as AD, also termed absorption efficiency);
2. **efficiency of conversion of ingested food** to body substance (ECI, also termed growth efficiency);
3. **efficiency of conversion of digested food** to body substance or utilization efficiency (ECD, also termed metabolic efficiency).

The prevalent method used to quantify food intake and utilization has been the **gravimetric method**, which involves weighing of food, body and faecal masses at the start and the end of the experimental period. It is based on the so-called **budget equation**, often given [106, 119] as: $C = G + R + FU$, where C is the amount of food consumed, G is insect biomass produced (i.e. somatic and reproductive growth and several secreted and excreted products that are not actually contributing to somatic growth *per se*, such as exuviae, silk and digestive enzymes), R is respiration (amount of carbon dioxide respired) and FU is faeces (urinary wastes and other metabolic waste products egested, together with undigested food as faecal constituents).

Budget items are routinely expressed in dry matter units, because water escapes from food, faeces and the insect body as vapour and the losses *via* these routes are very difficult to quantify. The amount of dry matter lost by respiration is in only few cases quantified directly [148]. As a consequence the accuracy of the budget can not be checked and this has been a matter of debate, especially in the wider field of ecological energetics [156]. An important source of error is an inaccurate determination of the dry matter content of the food, either plant food or artificial diet [11, 116, 147]. As plant tissues respire during the experiment and because plant species and tissues differ in their respiration rates, losses due to plant respiration should be taken into account for reliable measurements of nutritional indices. Errors become more serious when an excess of food is offered. For instance, when only half of the food presented to the feeding insect is consumed during the experiment and the percentage dry matter of e.g. leaf material is estimated only slightly incorrectly (e.g. at 14.5% while it actually is 14%), the value of

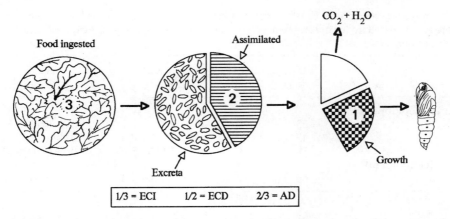

Figure 4.13 Partitioning of ingested food **(3)** between the fractions assimilated or absorbed **(2)** and excreta (faeces) and the subsequent partitioning between growth **(1)** and respiration (CO_2 and H_2O). Commonly used formulae to calculate utilization efficiencies (nutritional indices: ECI, ECD and AD) from the different fractions 1, 2 and 3 are given at the bottom.

ECD is calculated to be 40% when in reality it was 50%. Because of the spatial heterogeneity of dry matter content of leaves and other plant organs, these errors are of a random rather than a systematic nature and hence will hamper the experimenter from reliably measuring differences. When in addition leaf respiration rates are not appropriately accounted for, these errors may become even more serious and may lead to physiologically improbable degrees of variation in metabolic efficiency and consequently to erroneous conclusions [147]. Unfortunately, plant tissue respiration rates and proper controls have been taken into account in only a very few cases [148]. Details of methods and procedures of measurement are beyond the scope of this text and can be found elsewhere [85, 147]. The way in which AD, ECI and ECD values are calculated from the gravimetric measurement of C, G and FU is depicted in Figure 4.13. Some alternative methods to quantify food utilization exist, e.g. based on markers and elemental budgets, but these techniques are much less frequently used [85, 147].

Performance, i.e. the extent to which a herbivore is able to realize maximum growth and reproduction, is preferentially expressed as a rate parameter. The most commonly used parameter is the **relative growth rate** (RGR), expressed as amount of growth attained (mg dry matter) per unit of body weight (mg dry matter) per day. Relative growth rate is the product of **relative consumption rate** (RCR – mg/mg/d) and nutritional indices: RGR = RCR × AD × ECD = RCR × ECI.

This equation indicates that, on a certain food source, a higher RGR can be attained by either increasing food intake rate or increasing utilization. When consumption or growth rates are related to gravimetrically determined nutritional indices, inverse relationships are commonly found [123, 124, 137]. However, such relationships do not allow a distinction between cause and consequence: is growth rate reduced because of a lower metabolic efficiency or is metabolic efficiency reduced because of a lower growth rate? Growth rate of herbivorous insects is assumed to be nutrient-limited rather than limited by energy constraints [120, 135, 149, 155] (section 4.3.1). Thus, suboptimal availability of a limiting

Table 4.4 Average values (range in brackets) of performance and indices of nutritional utilization by mandibulate and haustellate herbivorous insects. n = no. of insect species investigated. (Source: data from Slansky and Scriber, 1985)

	AD %	ECD %	RCR mg/mg/d	RGR mg/mg/d	n
Mandibulates (Lepidoptera)					
Herbs	53	41	2.0 (0.27–6.0)	0.37 (0.03–1.5)	26
Grasses	43	45	2.0 (0.07–4.8)	0.29 (0.06–0.62)	6
Trees	39	37	1.5 (0.31–5.0)	0.17 (0.03–0.51)	82
Haustellates (Homoptera)					
Herbs	60	65	1.0 (0.90–1.6)	0.39 (0.11–0.67)	3

nutrient, often nitrogen or water, reduces growth rate, increases maintenance costs and causes a lower metabolic efficiency. Intake of the limiting nutrient can, however, be increased by compensatory feeding responses, that, as will be discussed below, by now are known to be well-developed in different herbivorous insects [129].

(c) Utilization of plant food by different feeding guilds

When looking at performance and utilization values large differences appear to exist between different feeding guilds, such as mandibulate feeders of herbs and forbs *versus* woody plants, or mandibulate *versus* haustellate (piercing–sucking) species (Table 4.4).

Tree-feeding species realize a much lower RGR, due to both a lower RCR and a lower AD. Haustellate species reach the highest RGR at a RCR that on average is twice as low as that of mandibulates, which can be ascribed to higher values of both AD and ECD. The differences in nutritional indices can largely be explained by the differences in nutritional quality of the respective tissues exploited by these guilds. Thus, piercing–sucking insects grow on average faster than leaf chewers. Obviously such differences are likely to have important ecological consequences (section 9.4.1).

(d) Changes in food utilization during development

During growth digestive performance values change. Values for approximate digestibility (AD, see section 4.4.2b) tend to decrease from early to late instars [7, 137]. This is probably related to increased feeding rate and increased gut size when larvae get older. Shorter retention times and larger food masses would make enzymatic degradation and nutrient absorption through the gut wall less efficient. Obviously this may have important consequences for an insect's ability to utilize a particular plant or plant part. It may explain why young insects are often more finicky eaters than older conspecifics. First instar larvae of *Helicoverpa virescens*, for instance, show a reduction in weight gain due to the presence of condensed tannins in their food at a concentration about ten times lower than that required to reduce growth of fifth instars [103].

The fact that food utilization efficiencies are typically higher for early-instar compared to late-instar larvae, together with the observation that levels of detoxifying enzymes are much lower in earlier than in later instars [3, 33], seems of crucial importance when investigating the suitability of a plant for a given insect species. Because variations in nutritive as well as secondary components of the food may have their greatest impact on early-instar

Table 4.5 Costs of growth for a holometabolous and a hemimetabolous insect species in their final larval stages for which continuous or repeated respirometric data as well as gravimetric growth data are available. Feeding took place on an optimal host plant. G = growth (mg dry matter); H = heat production, calculated from respirometric measurements. (Source: data from van Loon, 1991 and 1993)

Species	Diet	Duration (h)	G (mg)	RGR (mg/mg/day)	H (J)	H/G (J/mg)
Pieris brassicae	Cabbage leaves	90	88	0.640	1027	11.7
Locusta migratoria	Wheat	240	258	0.124	4536	17.6

larvae, studies on nutritive requirements and effects of allelochemicals should begin with early-instar larvae, in spite of the technical difficulties this may present [137].

(e) The cost of growth: factors determining metabolic efficiency

A relevant physiological question concerning host-plant utilization is how differences in utilization efficiency come about. The **metabolic load hypothesis** says that increased energetic processing costs (to be distinguished from maintenance costs) are a direct cause of lower growth rates, suggesting a trade-off between energy production for other purposes than anabolism and anabolism itself, resulting in growth. This idea has been put forward repeatedly, but experimental evidence is lacking [108, 118, 137]. Induction of the polysubstrate monooxygenases (PSMOs) enzyme system in response to allelochemicals (section 4.4.4) in the food has been one of the supposedly more important energy-requiring processes in the metabolic load hypothesis. However, an experimental test that was set up to quantify the cost of this induction gravimetrically did not yield proof for this, and the amount of enzymatic PSMO protein measured, although effective, was too small to expect any measurable cost [104].

Very few data are available on direct, longer-term measurements of metabolism (e.g. by respirometry), and it is premature to draw a reliable conclusion on the effect of dietary quality on processing costs. Due to the laboriousness of such measurements [146] few studies are available in which repeated or chronic respirometric measurements are combined with determinations of dry matter growth, allowing an actual check of the gravimetric budget [147]. Migratory locusts (*Locusta migratoria*), a hemimetabolous species, have distinctly lower growth rates than caterpillars, e.g. larvae of *Pieris brassicae* (Table 4.5), and clearly invest more energy per unit of dry matter growth.

The values of cost of growth for these two species, defined as the ratio between heat loss to the environment and growth (ratio H/G, Table 4.5), differ only by a factor of 1.5, in spite of the considerable differences in total amount of growth achieved, growth rate, body size and life style between the two. A similar comparison between two other species, a locust (*Melanoplus sanguinipes*) and a caterpillar (*Pseudaletia unipuncta*), when feeding on the same host plant (wheat) under identical conditions, reached the same conclusion: here also overall oxygen consumption required to double body mass was twice as high for the locust [27]. An important part of higher energy costs associated with the growth of the locusts results from extended duration of development *per se*, which is expected to result in a greater contribution of maintenance energy to total energy expenditure. An important fraction of the higher maintenance costs for the locusts is devoted to their cuticle, the mass of which is ten times greater than in caterpillars of similar size. From these physiological considerations it emerges that the large degree of variation in ECD reported for herbivorous insects based on gravimetric measurements was equivocal, especially in cases in which growth rates were hardly or not affected. Carnivores such as predators and parasitoids are nutritionally in a more comfortable position than herbivores, because the composition of their food closely fulfils their requirements for growth and development. As a result the approximate digestibility (AD) of animal tissue is higher than that of vegetable food. Whereas folivorous insects show AD values in the range of 40–50% (Table 4.4), carnivores generally reach values around 80% [137]. Nevertheless, vegetarians may grow faster since they have access to unlimited amounts of food. As a result their relative growth rate (RGR), although quite variable, is usually high, as exemplified for some groups in Table 4.4. The figures reached by herbivores, i.e. between 0.03 and 0.40, are in marked contrast to the values known for carnivorous species, which range between 0.01 and 0.03. Rapid growth combined with low energy expenditure on food acquisition mean that herbivorous insects pass twice as much of their assimilated food to the production of body tissues and eggs than predatory species (Fig. 4.14).

4.4.3 SUBOPTIMAL FOOD AND COMPENSATORY FEEDING BEHAVIOUR

(a) Extent and mechanisms of compensatory feeding

Herbivores use several mechanisms to maximize growth rates. First, they can leave a nutritionally poor plant that was selected previously to feed upon, and start searching for alternative food. This behaviour is essentially based on a nutritional feedback, whereby the insect resumes food-plant selection and may feed alternately from different host species [129]. This phenomenon, which can be studied in an experimental design employing a 'cafeteria' set-up, has been called 'dietary self-selection' [151] (section 7.6.2b). Secondly, they can increase food ingestion rate on the same plant, as mentioned above. This compensatory behaviour counteracts a reduced growth rate and concomitant higher

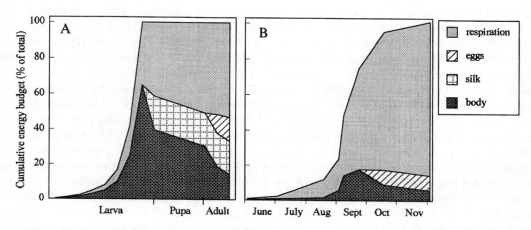

Figure 4.14 Cumulative energy budgets of **(A)** a herbivorous insect (silkworm, *Bombyx mori*) and **(B)** a predatory arthropod (a spider, *Oligolephus tridens*). Energy expenditure as a percentage of total assimilated energy is shown for the lifetime of a female. Note the differential allocation of energy that goes to cover maintenance metabolism compared to that stored in growth. The greater respiratory energy demands in the spider is related to hunting activity (Source: **(A)** data from Hiratsuka, 1920; **(B)** redrawn from Klekowski and Duncan, 1975, data supplied by J. Phillipson.)

maintenance costs, and may well be a functional response, assuming that the added costs of a higher feeding rate are smaller [91]. Unfortunately, very few data are available about the energetic demands of muscular activity involved in feeding, but those that are suggest that these costs are minor [4, 91]. Thirdly, the insect can, at least in theory, optimize utilization efficiencies, keeping consumption rate constant. Because of the probably spurious interactions between rates and efficiencies, however, very few convincing data are available for the latter option.

Compensatory feeding, adjusting feeding rate to approach or realize maximum growth rate has been found in a number of cases and is probably general among herbivorous insects [129]. When food quality is suboptimal and protein, for instance, is the limiting nutrient for growth, food consumption rate has been found to increase to 2.5–3 times higher levels [137, 138].

In experiments using artificial diets that allow levels of specific components to be diluted, increases in consumption rates by a factor of 7 have been noted (reviewed by Simpson and Simpson [129]). The physiological mechanisms allowing such a considerable span of variation in food consumption rates are complicated [129]. Recently, models have been formulated to integrate the various mechanosensory, peripheral and internal chemosensory feedbacks that operate in concert [128], but a discussion is beyond the scope of this text. The speed with which these feedbacks can operate is remarkable. One particularly well-studied case used *Locusta migratoria*. An injection into the haemocoel of a mixture of eight amino acids suffices to postpone the next meal significantly, suggesting that some as yet unknown internal chemosensors monitor haemolymph composition and provide a feedback response on feeding behaviour within minutes [2]. In this insect compensatory self-selection was also found to occur after only one nutritionally inadequate meal was taken [130].

(b) Constraints on compensatory feeding

Although several laboratory studies demonstrate that compensatory feeding can alleviate the effects of nutritional inadequacy of food sources, there are apparent constraints of both a physiological and an ecological nature. First, a trade-off probably exists between rate and efficiency. An increase in rate of consumption leads to a reduced retention time of food in the gut and this in turn will result in lower absorption efficiency [129]. Second, since, for example, protein and carbohydrate intake in the locust are regulated separately [127], an increased consumption rate to compensate for suboptimal availability of one nutrient may lead to an excess and thereby reduced utilization of the other, partly counteracting the effect of compensation. Third, increased consumption may cause intoxication because of the concomitant increased ingestion of allelochemicals: the detoxification system cannot keep up with the increased speed at which allelochemicals enter the body [138]. Herbivores may prevent intoxication, however, if they possess peripheral chemoreceptors that detect such allelochemicals (Chapter 6), or by avoiding such food sources through 'aversion learning' (Chapter 7).

Concerning ecological trade-offs it has been hypothesized that in herbivores short feeding periods will be selected for in order to expose them minimally to natural enemies [135, 141]. Indeed, dramatic differences in selection pressure on feeding *versus* resting insects have recently been demonstrated in a study on predation risks of caterpillars under field conditions. During feeding, the risk of being predated was 100 times higher than during non-feeding periods [29]. Indirect evidence comes from studies in which reduced nitrogen content of a crop increased predation rates of cabbage white caterpillars (*Pieris rapae*) [90]. This may be explained by assuming that compensatory feeding known to counteract suboptimal nitrogen ingestion rates in these caterpillars [135] led to increased exposure

4.4.4 ALLELOCHEMICALS AND FOOD UTILIZATION

Allelochemicals can negatively affect herbivores in three ways. They can reduce food intake by an inhibitory effect on feeding behaviour (see Chapter 6). Once ingested they can reduce the efficiency of food utilization or, third, they can poison the insect by interference with other physiological processes. Frequently, allelochemicals act through a combination of all three mechanisms. The various postingestive modes of action, which may operate in the gut or, after being absorbed, within other body parts, are often hard to separate [133], although a technique that permits a distinction between feeding inhibition and toxicity has been used with caterpillars [34]. Here we will discuss some effects on food utilization, because they figure as one of the main factors in plant resistance by reducing growth and development of the herbivore. Interference by allelochemicals with the intrinsic nutritional value a plant offers to herbivores is very probably not restricted to 'non-adapted' insects, which cannot grow on that plant species or cultivar, but also applies to insects that have adopted the plant as a normal host. This may explain why a number of insects fare better on artificial diets than on their natural host plants (Fig. 4.8). When food utilization indices on an artificial diet supplemented with allelochemicals are compared to control values negative (or positive) effects of the additive can be quantified. In such a study with a polyphagous insect five secondary plant compounds, when added to the diet, were all found to suppress growth, even when no reduction of food intake occurred. Different compounds had different effects on the various utilization parameters, which suggests that they interfere with different aspects of the digestion/absorption process [20]. These results, although indicative, need, in view of their important implications, to be repeated with modern rigid protocols [116].

Although utilization indices are rarely used to analyse physiological mechanisms involved in postingestive detrimental effects of allelochemicals (see section 4.4.2), the study by Beck and Reese [20] shows that they may provide indications on the type of processes affected. Other biochemical approaches can supply more detailed insight into the mode of action of particular secondary plant substances. Gossypol (28), a sesquiterpenoid typically occurring in cotton, inhibits feeding and growth in many insects. Larvae of *Spodoptera littoralis*, for instance, grow much faster on cotton leaves of low gossypol content than on high-gossypol strains. The proof that it is indeed gossypol that retards growth comes from an experiment with artificial diets. When larvae were fed gossypol-incorporated diets, protease and amylase activities in the gut decreased within one day (Table 4.6).

The affinity of gossypol to proteins in the gastrointestinal tract is well established. It may bind to the ingested dietary proteins, or to the digestive enzymes themselves. In both cases protein digestion will be hampered [97].

Interactions between secondary plant compounds and nutrients have been inferred from tests in which the nutritional content was varied, in combination with varying amounts

Table 4.6 Protease activity in 90–100-mg-weight larvae of *Spodoptera littoralis* fed for 2 days on an artificial diet containing various amounts of gossypol acetate. (Source: modified from Meisner et al., 1978. Reprinted by permission of the Entomological Society of America)

Gossypol acetate concentration (%)	Average larval weight after 2 days (mg)	Protease activity relative to control (%)
0 (control)	546	100
0.25	491	89
0.50	392	55

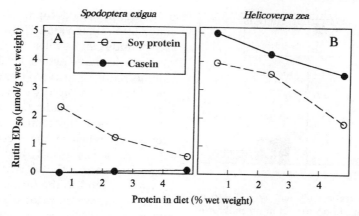

Figure 4.15 The effect of different regimens of dietary protein on the ability of rutin (**53**) to suppress larval growth in two noctuid species. Growth inhibition at various dietary regimens, as expressed by the dose of rutin required to reduce growth by 50% (= ED_{50}), is relative to the growth of control insects at 2.4% casein. Note that in *H. zea* growth on casein is suppressed at lower rutin concentrations than on soy protein, whereas in *S. exigua* the effect of protein type is reversed. (Source: redrawn from Duffey et al., 1986.)

of a particular allelochemical. Such an experiment (Fig. 4.15) showed that the detrimental effect of rutin (**53**), a widely distributed flavonoid, varies not only with the amount of protein in the food, but also with the kind of protein [55].

Feeny's classic paper [57] on differential growth of winter moth larvae on young and mature oak leaves (Fig. 9.3) initiated a lengthy debate on the role of tannins as digestibility reducers. Feeny [57] suggested that tannins complex either with leaf protein or with digestive enzymes in the gut, thereby reducing the efficiency of digestion and, as a consequence, retarding growth. Although the affinity of tannins for proteins is probably to some extent responsible for part of the detrimental effects, alternative mechanisms have come to light, such as inhibition of feeding, induction of midgut lesions, and pharmacological toxicity [15, 32, 58, 140]. The biochemical basis for the antinutritional effects of tannins seems complex and has not yet been fully elucidated.

Insects that are adapted to tannin-rich food are unaffected, and may even benefit from the presence of tannins in their food by stimulation of ingestion, among other factors [77] (Fig. 2.4). The tree locust *Anacridium melanorhodon* shows increased dry matter digestibility (AD) and growth efficiency (ECD) and a resultant 15% increase in growth rate when tannic acid is added to its diet [32]. Tannin-adapted insects possess several mechanisms to avoid their potentially harmful effects, including an alkaline gut pH [9], and absorption of tannins on to the peritrophic membrane [32, 57].

In Chapter 3 it was stated that plants never contain only one resistance compound, but rather have a variety of compounds. Plant chemicals may in various ways interact synergistically once they are inside the insect body [22]. As known from experience with synthetic insecticides, insects can relatively easily develop resistance to a specific group of chemicals. There is ample evidence that developing resistance to two or more groups with different modes of action is much more difficult. Consequently, adaptation to high levels of toxins in one host plant is often associated with a concomitant reduction in tolerance to compounds in other host plants [62]. In view of these insect response characteristics a plant probably cannot afford to produce only a single secondary chemical or even a single

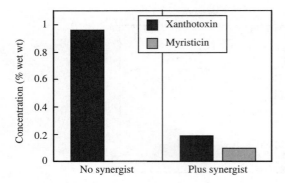

Figure 4.16 Concentration (% wet wt) of xanthotoxin (a furanocoumarin occurring in umbelliferous plants) which when added to artificial diets either in combination with myristicin (cooccurring in umbellifers and a synergist of organic insecticides) or not, cause 50% mortality in first instar larvae of *Helicoverpa zea*. (Source: data from Berenbaum and Neal, 1985.)

group of chemicals. In line with this reasoning wild parsnip produces secondary compounds from at least seven distinct biosynthetic pathways. The toxicity or deterrency of one group of compounds may be strongly affected by the presence or absence of other compounds [22]. Myristicin (**39**) is a lignan with a functional group characteristic of inhibitors of pivotal detoxification enzymes (cytochrome P_{450}). It commonly cooccurs with the phototoxic furanocoumarins in umbelliferous plants. Myristicin synergizes the toxicity of xanthocoumarin to the generalist caterpillar *Helicoverpa zea* almost fivefold. Thus, the production of 1 mg of myristicin can 'save' the plant producing 77 mg of furanocoumarin (Fig. 4.16).

This may represent a marked saving in production costs (energy) [25]. Undoubtedly, synergistic interactions between allelochemicals constitute an extremely important element in a plant's chemical protection.

4.4.5 DETOXIFICATION OF PLANT ALLELOCHEMICALS

Herbivores are confronted by relatively large amounts of noxious chemicals in their plant food and thus expose themselves to the hazard of being poisoned by every meal as aptly stated by Brattsten [36]. These potentially toxic compounds can be tolerated because herbivorous species have evolved various physiological mechanisms to avoid their harmful effects. They may either rapidly excrete the unwanted compounds or degrade them enzymatically, or otherwise neutralize such chemicals before they can reach pharmacologically active levels. As a last resort they have developed target-site insensitivity, that is failure of a toxicant to bind to the target because of alteration in the structure or accessibility of that target site [23]. Since insect herbivores consume huge amounts of food relative to their body weight (a caterpillar may eat five times its body weight of foliage per day) their detoxification system needs to be highly efficient. Insects indeed seem to exhibit a greater tolerance to, for instance, hydrogen cyanide (HCN) and alkaloids. HCN, toxic to all higher organisms, is released from plant cyanogens during digestion. Interestingly, polyphagous insects tend to be over 100 times less sensitive to HCN than mammals (Table 4.7).

Likewise, the toxicity of alkaloids to unspecialized insect species appears to be one or two orders of magnitude lower than for mammals (Fig. 4.17).

Specialized insects are often able to cope with still higher concentrations of those allelochemicals that typically occur in their food plants [26]. However, a comparison based on body weight is disputable. When toxicity is

Table 4.7 Oral toxicity of hydrogen cyanide in some polyphagous insect herbivores as compared to mammals. (Source: data modified from Bernays, 1982)

Animals	Oral LD_{50} (mg/kg)
Locusta migratoria (Orthoptera)	500
Zonocerus variegatus (Orthoptera)	1000
Spodoptera littoralis (Lepidoptera)	800
Spodoptera eridania (Lepidoptera)	1500
Mammals (general)	0.5–3.5

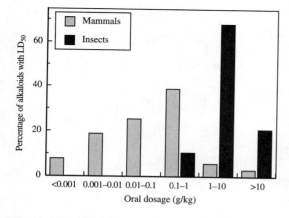

Figure 4.17 Alkaloid toxicity levels for unspecialized mammal herbivores and unspecialized insect herbivores, shown as the percentage of alkaloids with the LD_{50} in the different dosage ranges. (Source: redrawn from Bernays, 1982.)

compared on the basis of metabolic activity or body surface rather than of body weight, the differences between mammals and insects become less spectacular, or may even disappear.

(a) Physiological adaptations: rapid excretion

An effective way to prevent toxification consists of mechanisms that render target sites inaccessible. This can be accomplished by preventing potential toxins from passing the gut wall. Many secondary plant compounds are under physiological conditions either charged molecules, bulky or polar molecules, or hydrophilic compounds (such as glycosides), to which biomembranes are often almost impermeable. Such properties, in combination with the rapid intestinal passage normal for herbivorous insects, mean that many toxicants get little chance to enter the body cavity [157].

A striking example is found in the tobacco hornworm (*Manduca sexta*), which feeds on hosts containing nicotine, a traditional insecticide and deadly poison to other animals as well. Nicotine, in contrast to many other alkaloids, is lipophilic and therefore in most insects easily passes the gut epithelium. It derives its toxicity by a functional resemblance to acetylcholine, a pivotal neurotransmitter in the central nervous system of animals. By mimicking acetylcholine molecules nicotine disturbs delicate and basic functions of the central nervous system. Tobacco hornworm larvae have evolved a number of resistance mechanisms that protect them against this otherwise potent toxin. They rapidly excrete nicotine and other ingested alkaloids before a toxic dose can accumulate. In an experiment in which hornworms were fed food containing known quantities of nicotine, 93% of an ingested 0.5 mg dose was excreted in 2 hours, whereas in houseflies more than 90% of the administered dose remained in the insect's body for as long as 18 hours [125]. The small amounts of nicotine that do get into the haemolymph of tobacco hornworms cannot cross the ion-impermeable neural sheath and are eliminated *via* the Malpighian tubules. When, in spite of the physiological barriers which this species has developed, nicotine does reach the nerve cells in the central nervous system, these cells appear to tolerate this compound, demonstrating an example of target-site insensitivity.

This is a well-studied example of a multicomponent protection system in an adapted insect species against an allelochemical that is highly poisonous to all non-adapted animals. Not that rapid excretion is a physiological trait 'invented' only by specialists. The polyphagous larvae of the green hairstreak (*Callophrys rubi*), which feed on plants from ten different families, excrete all alkaloids from their host plant *Genista tinctoria* unchanged. They even do so when raised on a non-host, such as *Lupinus polyphyllus* [59].

(b) Enzymatic detoxification

Most herbivorous insects rely heavily on enzymatic degradation for neutralization of ingested plant allelochemicals. The most extensively studied enzymes that effectively metabolize a wide variety of toxicants (includ-

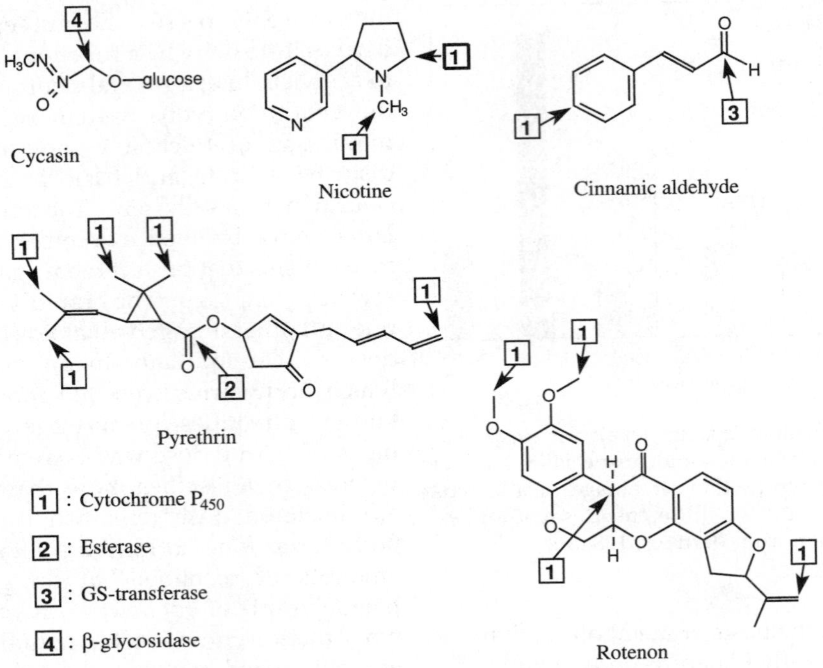

Figure 4.18 Sites of enzymatic attack on some plant allelochemicals. (Source: redrawn from Brattsten, 1988.)

1 : Cytochrome P_{450}
2 : Esterase
3 : GS-transferase
4 : β-glycosidase

ing man-made pesticides) are the polysubstrate monooxygenases (PSMOs; in the past often called mixed-function oxydases, MFOs). PSMOs constitute a three-tier system localized in the endoplasmic reticulum of cells in several tissues of all eukaryotic organisms. PSMOs convert allelochemicals into more polar, reactive compounds, which are further metabolized by secondary enzymes. One of its components is cytochrome P_{450}, a red-coloured enzyme that derives its name from its spectral absorption maximum around 450 nm. Activity levels of cytochrome P_{450} can differ greatly among herbivores. In a study of 58 caterpillar species the activity of this enzyme seemed to be related to the type of food plant. Species feeding on plants rich in monoterpenes, such as members of the Myrtaceae, Rutaceae or Solanaceae, tend to have considerably higher levels than those living on some other plant families, including Papilionoideae, Plantaginaceae and Poaceae [112].

Simultaneously with the PSMO system several other enzyme systems serve to detoxify allelochemicals. Toxicants can be metabolized not only by oxidations but also by hydrolytic cleavages and conjugations (Fig. 4.18).

The breakdown products of the toxic compound can be either recycled in the intermediary metabolic pathways or converted to products that are easily excreted. Although enzymatic degradation usually leads to nontoxic products, sometimes a breakdown product is more toxic than the parent molecule (Fig. 4.19). The primary (or phase 1) products are subsequently metabolized by other enzymes to harmless substances.

When insects are exposed to a novel toxin the levels of detoxifying enzymes, such as

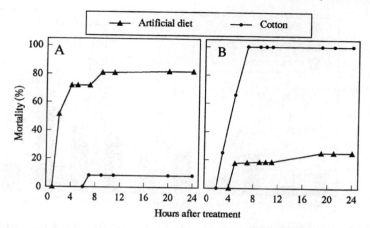

Figure 4.19 Effect of diet on the toxicity of two insecticides. Newly moulted sixth-instar larvae of *Spodoptera frugiperda* were fed artificial diet or cotton leaves for 2 days prior to insecticide treatment. **(A)** 2 μg diazinon per larva. **(B)** 15 μg isofenphos per larva. Rapid breakdown of diazinon in cotton-fed larvae to less toxic compounds reduces mortality. Rapid breakdown of isofenphos in cotton-fed larvae increases mortality, because the resulting metabolite is more toxic than the parent compound. (Source: redrawn from Yu, 1986.)

PSMOs, begin to increase within minutes. This phenomenon is termed 'induction' and depends on *de novo* synthesis of enzyme protein. Larvae of the noctuid *Peridroma saucia* show low PSMO activity when reared on an artificial diet. After being fed peppermint leaves their PSMO activity was up to 45 times higher. Enzyme induction was apparently due to high concentrations of monoterpenes in the peppermint leaves, because by feeding artificial diets with menthol or pinene the cytochrome P_{450} content of the midgut increased considerably [159]. When larvae of the tobacco cutworm (*Spodoptera litura*), a polyphagous species, were raised on different hosts belonging to 11 different plant families, their PSMO activity levels varied, depending on the kind of food experienced, within a 20-fold range [112]. Moreover, different plants affect different reactions controlled by the PSMO system differently, implicating the existence of isoenzymic forms of the cytochrome P_{450} (Fig. 4.20) [159].

This conclusion can explain why a particular insect species shows large differences in sensitivity to insecticide treatment depending on the crop plant species on which it occurs (Fig. 4.19). It also becomes clear why natural enemies of herbivores are commonly more susceptible to insecticide treatments than their hosts: they normally ingest little or no toxin at all with their meals. This applies to another time scale also, since during evolution parasitoids and predators have not been exposed to the plethora of secondary plant substances as intensively as plant-eating insects.

As soon as the inducing chemical is no longer present, enzyme activity begins to drop to preinduction levels. Because of this flexible induction mechanism herbivores can show highly variable activity levels depending on the food consumed and even on how long after a meal they were assayed. The phenomenon of induction suggests that there are costs in maintaining constantly high levels of detoxifying enzymes. However, there is no evidence that any significant energetic or nutritional costs are involved [133] (see section 4.4.2). Therefore the adaptive value of induction remains unclear.

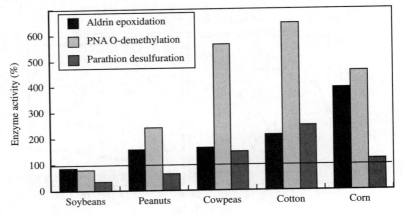

Figure 4.20 Effect of host plant on oxidase activities in the midgut of *Spodoptera frugiperda* larvae. Three different PSMO activities induced by five different food plants are compared to values measured in larvae fed on an artificial diet (=100%). These data on selective induction of various PSMO activities by different inducers strongly suggest a multiplicity of cytochrome P_{450}. (Source: redrawn from Yu, 1986.)

Only by evolving efficient detoxification mechanisms have insects been able to break the chemical protection line of plants, nature's most varied chemical repertoire. These mechanisms have been studied extensively in the recent past and are discussed in several authoritative reviews [38, 53, 88].

4.5 SYMBIONTS

The capacity to utilize plant food depends in a number of herbivores on the presence of symbiotic microbes. Evidence that microbes play any role in the digestion and nutrition of folivorous insect species is scarce [9], but certainly sap-feeding insects can hardly survive without them. Bacteria, yeasts and other unicellular fungi or protozoa aid in the degradation of plant food and the synthesis of nutritional requisites that plants do not provide at all or provide in insufficient quantities (sterols, some vitamins, ten of the 20 protein-amino acids). A third role assigned to symbionts is assistance in the detoxification of plant allelochemicals [19, 45, 52, 107].

Extracellular symbionts live in the alimentary tract, either free in the gut lumen or, more protected, in pockets (caeca) of the midgut or hindgut, as in a number of Coleoptera, Hemiptera and Orthoptera (Fig. 4.21).

Caeca are lacking in Lepidoptera. Approximately 10% of all insect species accommodate intracellular endosymbionts. They may occur in cells of the gut wall and be constantly set free into the lumen, as in the larvae of some cerambycids. Often, however, they are confined to specialized cells, **mycetocytes**, which are scattered singly throughout various tissues or are aggregated to form an organ-like structure called **mycetome**. Among herbivorous insects the most thoroughly studied groups with respect to endosymbionts are the Hemiptera and Coleoptera. Aphids, for instance, can exploit phloem sap in spite of its nutritional deficiencies because microbial 'brokers' help to overcome the nutritional hurdle [52]. Microorganisms in the gut of the green peach aphid (*Myzus persicae*) produce all the essential amino acids except four that are supplied by the host plant [100]. Several other studies on a variety of insect species have demonstrated that symbionts can supplement nutrients that the natural plant food does not provide at all, or provides at concentrations too low to support normal growth [45, 54]. In most of these studies the physiological role of

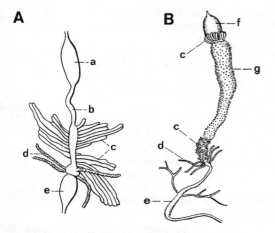

Figure 4.21 Alimentary canals of a sap feeding and a chewing insect species. A = gut of *Aphanus* sp. (Lygaeidae), feeding on seeds. B = gut of *Adoxus obscurus* (Chrysomelidae) feeding on *Epilobium angustifolium*. a, dilated part of midgut; b, narrow part of midgut; c, caeca with symbionts; d, Malpighian tubes; e, rectum; f, gizzard; g, midgut. (Source: **(A)** reproduced from Kuskop, 1923, with permission; **(B)** reproduced from Stammer, 1935, with permission.)

endosymbionts was inferred from reduced growth and poor survival of symbiont-free insects when fed, after treatment with antibiotics or heat exposure, on their normal host plants or on artificial diets.

Insects that tolerate host plants rich in toxic compounds must possess intrinsic mechanisms to prevent poisoning. In as yet a few cases detoxifying symbionts have been identified as actors in converting host-derived allelochemics to harmless compounds [54]. The cigarette beetle (*Lasioderma serricorne*), for example, houses a yeast that catabolizes a broad variety of xenobiotics, including many flavonoids and tannins. With its biochemical machinery this symbiont most probably contributes to the success of its polyphagous host in exploiting a diverse array of plant species [126].

An assessment of the significance of endosymbionts in the nutritional ecology of herbivorous insects is seriously hampered by the fact that symbionts cannot generally be cultured apart from their hosts, thereby limiting experimentation on their physiology. However, our increasing knowledge of the influence of symbionts on resource exploitation of plants by insects suggests that they frequently act as unseen but vital mediators in insect–plant interactions [19].

4.6 HOST-PLANT EFFECTS ON HERBIVORE SUSCEPTIBILITY TO PATHOGENS AND INSECTICIDES

A compelling body of evidence indicates that the food plant may influence an insect's susceptibility to entomopathogens such as bacteria [9, 103], viruses [121], fungi [65] and nematodes [16]. Effects ascribed to the plant can be either inhibition of the pathogen or potentiation of its toxicity and reproduction. Assuming a decisive role of allelochemicals in interactions between trophic levels most studies of this phenomenon have focused on the effects of plant compounds on pathogen efficacy. Such studies either related the concentration of certain allelochemicals in host plants with pathogenicity, or analysed the effects of pure compounds by adding them in conjunction with the pathogen to artificial diets. Rutin, for instance, was found by employing the latter method to markedly protect within a certain concentration range the larvae of *Trichoplusia ni* against the toxin produced by *Bacillus thuringiensis* [86]. Several classes of plant allelochemicals are now known to influence pathogens or the toxins they produce but other foliar factors, such as nutritional value, age and water content may be involved as well. Gypsy moth larvae show differences in susceptibility to a baculovirus depending on the kind of tree foliage they were offered before and during the test. After inoculation with a standard dose of the virus, mortality on foliage with low levels of hydrolysable tannins is higher than on high-tannin foliage [78] (Table 4.8). Because larval

Table 4.8 Host plant effects on mortality of gypsy moth (*Lymantria dispar*) larvae fed standard doses of a baculovirus. (Source: reproduced from Keating *et al.*, 1988, with permission)

Host plant	Mortality (%)	Hydrolysable tannins (% dry weight of leaf)
Black oak (*Quercus nigra*)	25	33.2
Red oak (*Q. rubra*)	47	36.6
Quaking aspen (*Populus tremuloides*)	79	1.4
Bigtooth aspen (*P. grandidentata*)	86	1.2

mortality is also correlated with differences in leaf tissue acidity (which is reflected in the pH of the insect's midgut) the interaction between host plant and pathogen susceptibility of the herbivore may be multifactorial [79] (see also section 9.4.3).

Likewise the susceptibility to insecticides varies in polyphagous insects with the plant species on which they happen to feed when treated. The migratory grasshopper *Melanoplus sanguinipes* is, when fed oats, killed by a dose of deltamethrin three times lower than when feeding on rye [71], and aphids (*Myzus persicae*) have even shown a 200-fold variation in insecticide susceptibility depending on host plants [6]. Differential insecticide susceptibility has also been linked to physiological variables occurring within one plant species. Green peach aphids (*Myzus persicae*) showed differences in insecticide susceptibility not only when reared on different Brussels sprout varieties but also on plants of the same cultivar that had been exposed to different nitrogen fertilization regimes. Thus host-plant condition can significantly affect the level of insecticide tolerance [101].

The physiological mechanisms responsible for changes in herbivore sensitivity to pathogens and insecticides remain largely unknown. However, most studies support the general hypothesis that the susceptibility to entomopathogens is inversely related to host-plant suitability. It seems likely that feeding upon a suboptimal host imposes a general stress on the herbivore that negatively influences its resistance to, for instance, microbial infections [96].

A plant's chemical composition may be altered significantly by the presence of mutualistic or plant pathogenic microorganisms. Obviously this will in turn affect herbivore performance and susceptibility to entomopathogens. Associations between endophytic fungi and various plants are classified as mutualistic, since these fungi have limited or no pathogenic effects, but may rather provide protection against herbivores (Fig. 4.22).

Their contribution to the greater vegetative vigour of their hosts consists mainly of the ability to produce alkaloids or other compounds that predispose their hosts against herbivory. Fungal endophytes occur in a wide range of grasses and are therefore of agricultural importance. This has stimulated a multitude of papers in recent years [39, 49]. As many grasses are relatively free of defensive chemicals the evolutionary *raison d'être* of endophytic fungi may lie in the mutualistic relationship they have established with their hosts. Fungal endophytes are not limited to grasses. There is increasing evidence that they are associated with many more angiosperms, including woody plants [114], than hitherto known. Hence our current thinking about insect–plant relationships needs to be expanded to the broader perspective of multitrophic interactions. As an interesting example the multitrophic system involving larvae of the Japanese beetle (*Popillia japonica*), feeding on the roots of fescue grasses, is given. When

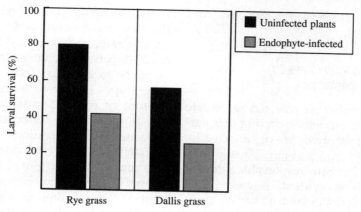

Figure 4.22 Survival of larvae to pupation of fall armyworms (*Spodoptera frugiperda*) on two grass species (*Lolium perenne* and *Paspalum dilatatum*), either or not infected with endophytic fungi. This noctuid species is a generalist herbivore on graminaceous plants. (Source: redrawn from Clay et al., 1986.)

host plants of this insect were infected by endophytic fungi, larval food intake was reduced because of the presence of feeding-deterrent alkaloids. As a result larval vigour was lowered, rendering them in turn more susceptible to entomopathogenic nematodes [63].

Plant diseases probably occur equally commonly in both natural and agricultural ecosystems [44]. Since diseased plants are, in addition to having changed physical characteristics, biochemically different from healthy plants, their nutritional suitability for herbivores will be changed. In diseased plants the concentrations and distribution of assimilates (sugars, amino acids, starch, etc.) and allelochemicals are often significantly modified [17]. Moreover, novel compounds, termed **phytoalexins**, are generated in response to invasion by microorganisms. During the last decade several hundreds of phytoalexins belonging to various chemical classes have been identified [67].

Insect herbivores are often negatively affected when their host plant is infested by some phytopathogen, although positive effects have been reported too [17, 18, 66]. Two examples may suffice. The chrysomelid beetle *Gastrophysa viridula*, a specialized feeder on dock (*Rumex* spp.), shows greater larval mortality, retarded development and reduced fecundity on plants infected by rust fungus as compared to healthy plants. Chemical analysis showed that infected leaves had lower nitrogen levels and higher oxalate concentrations than rust-free plants, which probably account for the deleterious effects. The interactions in this tripartite system, however, are bilateral since not only did the insects suffer from the presence of rusts but the beetles, by damaging the food plant, also elicited an induced resistance against the rust fungus. This plant response developed rapidly in the damaged leaf and was transferred, albeit to a limited degree, to undamaged plant parts [69].

Virus-infected plants have occasionally been found to be a better food source for insects than healthy plants. Aphids (*Aphis gossypii*) show increased population growth on squash plants (*Cucurbita pepo*) infected with zucchini yellow mosaic virus as compared to conspecifics on healthy plants. The change in population development was correlated with changes in the levels of specific amino acids from phloem exudate, although the concentrations of total amino acids did not show substantial differences between infected and healthy plants [35].

4.7 PLANT-MEDIATED EFFECTS OF AIR POLLUTION ON INSECTS

This relatively new field of research has to date focused primarily on demonstrating that particular insects either are or are not affected by pollutant-induced changes in their host plants, whereas the mechanisms responsible received less attention. It is evident, however, that structural characteristics such as surface morphology and toughness, as well as the levels of both primary metabolites and secondary compounds, can be affected by air pollutants, as clearly outlined in reviews by Hughes [75] and others [43, 70].

Air pollutants considered to be most important in terms of phytotoxicity include sulphur dioxide (SO_2), ozone, nitric oxide (NO) and nitrogen dioxide (NO_2). Evidence of a connection between air pollution and changes in insect attack on plants has been obtained by observational studies (outbreaks of forest insects in the vicinity of industry) and, more recently, by experimental studies. Many aphids and other sap-feeding insect species grow better on plants exposed to moderate concentrations of air pollutants, whereas among chewing insects some species are definitely favoured by air pollution and others show decreased population densities. Pollutants may affect herbivore populations by changes in host-plant quality or by affecting their natural enemies. There is a growing amount of evidence that the nutritional quality of plants can be altered significantly. In many instances the levels of free amino acids and reducing sugars are increased, while leaf protein content may either increase or decrease. These changes are reflected in changed nitrogen/carbohydrate ratios.

Contrary to what one would expect, the exposure of host plants to SO_2 often produces beneficial effects on herbivore performance [154]. Mexican bean beetles (*Epilachna varivestis*) prefer to feed on soybean foliage that has been exposed to SO_2, and show higher growth rates and increased fecundity on this food. A common change in plant leaves in response to oxidative pollutants, as well as other types of stress, is an increase in the amount of reduced glutathione. In soybean foliage glutathione concentration was found to change with fumigation in the same manner as insect growth. When non-fumigated foliage was enriched with glutathione by allowing excised leaves to imbibe a solution of this peptide through their petioles, insect growth was stimulated in the same way as by SO_2-treated plants, which suggests a pivotal role of this compound in pollutant-induced effects [75, 76] (Fig. 4.23). Although the evidence for air pollutants affecting plants and thereby insect herbivory is indisputable, our understanding of its consequences for population development or, at a larger scale, the functioning of ecosystems is still close to nil.

Acid precipitation is a phenomenon closely related to air pollution. It is caused primarily by oxides of sulphur and nitrogen. Acid rain probably does not affect plants directly. Most probably negative influences on plants are caused, especially in soils with poor buffering capacity, by indirect effects through alteration of soil properties and activity of soil microorganisms [75].

Although carbon dioxide (CO_2) is usually not considered an air pollutant, it is perhaps the most important atmospheric component changed worldwide by human activity. Global concentrations have increased by about 20% during the past 35 years, and a doubling of CO_2 levels is anticipated during the next 50–75 years. Most studies on the impact of elevated CO_2 levels have shown so far that in most plants, unaccountably, the nitrogen levels in their leaves are reduced [12]. To compensate for the lower nutritional quality of their food, several insects were observed to eat more on CO_2-treated plants than on control plants grown in ambient CO_2 levels [87]. In birch trees exposed to twice the normal CO_2 level,

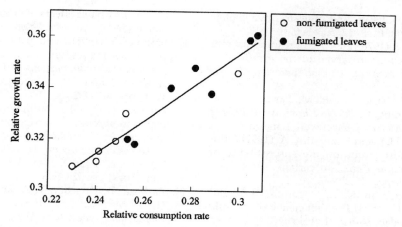

Figure 4.23 Food consumption and growth (expressed as relative consumption and relative growth rates) of Mexican bean beetle larvae fed on previously fumigated soybean leaves (0.3 ppm SO_2 for 24 h) or non-fumigated leaves enriched with glutathione. (Source: redrawn from Hughes and Voland, 1988.)

nitrogen concentration of their foliage was decreased by 23%, whereas concentrations were doubled for condensed tannins and tripled for starch. When the foliage was fed to three different lepidopterous insects the changes in chemical composition caused increased food consumption, unaltered or reduced growth, prolonged larval development and reduced food processing efficiency [89]. To what extent increased CO_2 levels acting directly or indirectly through insect herbivory contributes to changes in vegetation under natural conditions remains unknown [10]. It should also be realized that other factors that may be influenced by elevated CO_2 concentration, such as cloud cover, precipitation and temperature, could easily reverse the direct CO_2 effects on plant–herbivore interactions [12, 152].

4.8 CONCLUSIONS

Plant tissue is far from an optimal food for insects. It is low in nitrogen and high in allelochemicals. Naturally occurring large variations in plant chemical composition are augmented by environmental factors such as nutrient and water availability, plant pathogens, air pollution and other stress factors. Herbivores cope with nutritionally inadequate food by ingesting large amounts of it. Effective non-specific and inducible detoxification mechanisms neutralize (to what cost?) toxic food compounds and symbionts may assist in fulfilling nutritional requirements.

The balance between food quality offered by plants and food quality minimally needed by insects is a subtle one. It is, moreover, affected to a larger extent than hitherto supposed by other partners of an intricate network, such as pathogens and mutualistic organisms.

This field of research has produced a large body of literature. Much of the older literature is covered by thorough reviews by Slansky and Scriber [137] and Slansky and Rodriguez [136].

4.9 REFERENCES

1. Abe, T. and Higashi, M. (1991) Cellulose centered perspective on terrestrial community structure. *Oikos*, **60**, 127–133.
2. Abisgold, J. D. and Simpson, S. J. (1987) The physiology of compensation by locusts for changes in dietary protein. *J. Exp. Biol.*, **129**, 329–346.
3. Ahmad, S. (1986) Enzymatic adaptations of

herbivorous insects and mites to phytochemicals. *J. Chem. Ecol.*, **12**, 533–560.
4. Aidley, D. J. (1976) Increase in respiratory rate during feeding in larvae of the armyworm, *Spodoptera exempta*. *Physiol. Entomol.*, **1**, 73–75.
5. Allen, S. E., Grimshaw, H. M., Parkinson, J. A. and Quarmby, C. (1974) *Chemical Analysis of Ecological Materials*, Blackwell, London.
6. Ambrose, H. J. and Regupathy, A. (1992) Influence of host plants on the susceptibility of *Myzus persicae* (Sulz.) to certain insecticides. *Insect Sci. Appl.*, **13**, 79–86.
7. Anantha Raman, K. V., Magadum, S. B. and Datta, R. K. (1994) Feed efficiency of the silkworm *Bombyx mori* L. hybrid ($NB_4D_2 \times KA$). *Insect Sci. Appl.*, **15**, 111–116.
8. Anderson, T. E. and Leppla, N. C. (1992) *Advances in Insect Rearing Research and Pest Management*, Westview Press, Boulder, CO.
9. Appel, H. M. (1994) The chewing herbivore gut lumen, physicochemical conditions and their impact on plant nutrients, allelochemicals, and insect pathogens, in *Insect–Plant Interactions*, vol. 5, (ed. E. A. Bernays), CRC Press, Boca Raton, FL, pp. 209–223.
10. Arnone, J. A., Zaller, J. G., Ziegler, C., Zandt, H. and Körner, C. (1995) Leaf quality and insect herbivory in model tropical plant communities after long-term exposure to elevated atmospheric CO_2. *Oecologia*, **104**, 72–78.
11. Axelsson, B. and Ågren, G. I. (1979) A correction for food respiration in balancing energy budgets. *Entomol. Exp. Appl.*, **25**, 260–266.
12. Ayres, M. P. (1993) Plant defense, herbivory, and climate change, in *Biotic Interactions and Global Change*, (eds P. M. Kareiva, J. G. Kingsolver and R. B. Huey), Sinauer, Sunderland, MA, pp. 75–94.
13. Bailey, C. G. (1976) A quantitative study of consumption and utilization of diets in the bertha armyworm, *Mamestra configurata* (Lepidoptera, Noctuidae). *Can. Entomol.*, **108**, 1319–1326.
14. Barbehenn, R. V. (1992) Digestion of uncrushed leaf tissues by leaf-snipping larval Lepidoptera. *Oecologia*, **89**, 229–235.
15. Barbehenn, R. V. and Martin, M. M. (1994) Tannin sensitivity in larvae of *Malacosoma disstria* (Lepidoptera), roles of the peritrophic envelope and midgut oxidation. *J. Chem. Ecol.*, **20**, 1985–2001.
16. Barbercheck, M. E., Wang, J. and Hirsh, I. S. (1995) Host plant effects of entomopathogenic nematodes. *J. Invertebr. Pathol.*, **66**, 169–177.
17. Barbosa, P. (1991) Plant pathogens and non-vector herbivores, in *Microbial Mediation of Plant–Herbivore Interactions*, (eds P. Barbosa, V. A. Krischik and O. G. Jones), John Wiley, New York, pp. 341–382.
18. Barbosa, P. (1993) Lepidopteran foraging on plants in agrosystems, constraints and consequences, in *Caterpillars. Ecological and Evolutionary Constraints on Foraging*, (eds N. E. Stamp and T. M. Casey), Chapman & Hall, New York, pp. 523–566.
19. Barbosa, P., Krischik, V. A. and Jones, O. G. (1991) *Microbial Mediation of Plant–Herbivore Interactions*. Wiley, New York.
20. Beck, S. D. and Reese, J. C. (1976) Insect–Plant interactions, nutrition and metabolism. *Rec. Adv. Phytochem.*, **10**, 41–92.
21. Berdegue, M. and Trumble, J. T. (1996) Effects of plant chemical extracts and physical characteristics of *Apium graveolens* and *Chenopodium murale* on host choice by *Spodoptera exigua* larvae. *Entomol. Exp. Appl.*, **78**, 253–262.
22. Berenbaum, M. (1985) Brementown revisited, interactions among allelochemicals in plants. *Rec. Adv. Phytochem.*, **19**, 139–169.
23. Berenbaum, M. R. (1986) Target site insensitivity in insect–plant interactions, in *Molecular Aspects of Insect–Plant Associations*, (eds L. B. Brattsten and S. Ahmad), Plenum Press, New York, pp. 257–272.
24. Berenbaum, M. R. (1995) Turnabout is fair play, secondary roles for primary compounds. *J. Chem. Ecol.*, **21**, 925–940.
25. Berenbaum, M. and Neal, J. J. (1985) Synergism between myristicin and xanthotoxin, a naturally co-occurring plant toxicant. *J. Chem. Ecol.*, **11**, 1349–1358.
26. Bernays, E. A. (1982) The insect on the plant – a closer look, in *Proceedings of the 5th International Symposium on Insect–Plant Relationships, Wageningen, 1982*, (eds J. H. Visser and A. K. Minks), Pudoc, Wageningen, pp. 3–17.
27. Bernays, E. A. (1986) Evolutionary contrasts in insects, nutritional advantages of holometabolous development. *Physiol. Entomol.*, **11**, 377–382.
28. Bernays, E. A. (1991) Evolution of insect morphology in relation to plants. *Phil. Trans. Roy. Soc. Lond. B*, **333**, 257–264.
29. Bernays, E. A. (1996) Selective attention and

host-plant specialization. *Entomol. Exp. Appl.*, **80**, 125–131.
30. Bernays, E. A. and Janzen, D. H. (1988) Saturniid and sphingid caterpillars, two ways to eat leaves. *Ecology*, **69**, 1153–1160.
31. Bernays, E. A. and Lewis, A. C. (1986) The effect of wilting on palatability of plants to *Schistocerca gregaria*, the desert locust. *Oecologia*, **70**, 132–135.
32. Bernays, E. A., Cooper Driver, G. and Bilgener, M. (1989) Herbivores and plant tannins. *Adv. Ecol. Res.*, **19**, 263–302.
33. Berry, R. E., Yu, S. J. and Terriere, L. C. (1980) Influence of host plants on insecticide metabolism and management of variegated cutworm. *J. Econ. Entomol.*, **73**, 771–774.
34. Blau, P. A., Feeny, P. and Contardo, L. (1978) Allylglucosinolate and herbivorous caterpillars, a contrast in toxicity and tolerance. *Science*, **200**, 1296–1298.
35. Blua, M. J., Perring, T. M. and Madore, M. A. (1994) Plant virus-induced changes in aphid population development and temporal fluctuations in plant nutrients. *J. Chem. Ecol.*, **20**, 691–707.
36. Brattsten, L. B. (1979) Biochemical defense mechanisms in herbivores against plant allelochemicals, in *Herbivores. Their Interactions with Secondary Plant Metabolites*, (eds G. A. Rosenthal and D. H. Janzen), Academic Press, New York, pp. 200–270.
37. Brattsten, L. B. (1988) Enzymic adaptations in leaf-feeding insects to host-plant allelochemicals. *J. Chem. Ecol.*, **14**, 1919–1939.
38. Brattsten, L. B. (1992) Metabolic defenses against plant allelochemicals, in *Molecular Aspects of Insect–Plant Associations*, (eds L. B. Brattsten and S. Ahmad), Plenum Press, New York, pp. 176–242.
39. Breen, J. P. (1994) *Acremonium* endophyte interactions with enhanced plant resistance to insects. *Annu. Rev. Entomol.*, **39**, 401–423.
40. Brewer, J. W., O'Neill, K. M. and Deshon, R. E. (1987) Effects of artificially altered foliar nitrogen levels on development and survival of young instars of western spruce budworm, *Choristoneura occidentalis* Freeman. *J. Appl. Entomol.*, **104**, 121–130.
41. Brodbeck, B. and Strong, D. (1987) Amino acid nutrition of herbivorous insects and stress to host plants, in *Insect Outbreaks*, (eds P. Barbosa and J. C. Schultz), Academic Press, San Diego, CA, pp. 347–364.
42. Brodbeck, B. V., Mizell, R. F. and Andersen, P. C. (1993) Physiological and behavioral adaptations of three species of leafhoppers in response to the dilute nutrient content of xylem fluid. *J. Insect Physiol.*, **39**, 73–81.
43. Brown, V. C. (1995) Insect herbivores and gaseous air pollutants – current knowledge and predictions, in *Insects in a Changing Environment*, (eds R. Harrington and N. E. Stork), Academic Press, London, pp. 219–249.
44. Burdon, J. J. (1987) *Diseases and Plant Population Biology*, Cambridge University Press, Cambridge.
45. Campbell, B. C. (1989) On the role of microbial symbiotes in herbivorous insects, in *Insect–Plant Interactions*, vol. 5, (ed. E. A. Bernays), CRC Press, Boca Raton, FL, pp. 1–44.
46. Chapman, R. F. (1995) Mechanics of food handling by chewing insects, in *Regulatory Mechanisms in Insect Feeding*, (eds R. F. Chapman and G. de Boer), Chapman & Hall, New York, pp. 3–31.
47. Clay, F. K., Hardy, T. N. and Hammond, A. M. (1986) Fungal endophytes of grasses and their effect on an insect herbivore. *Oecologia*, **66**, 1–5.
48. Cohen, A. C. and Patana, R. (1984) Efficiency of food utilization by *Heliothis zea* (Lepidoptera, Noctuidae) fed artificial diets or green beans. *Can. Entomol.*, **116**, 139–146.
49. Dahlman, D. L., Eichenseer, H. and Siegel, M. R. (1991) Chemical perspectives of endophyte-grass interactions and their implications to insect herbivory, in *Microbial Mediation of Plant–Herbivore Interactions*, (eds P. Barbosa, V. A. Krischik and O. G. Jones), John Wiley, New York, pp. 227–252.
50. Davies, R. G. (1988) *Outlines of Entomology*, 7th edn, Chapman & Hall, London.
51. Dethier, V. G. and Schoonhoven, L. M. (1968) Evaluation of evaporation by cold and humidity receptors in caterpillars. *J. Insect Physiol.*, **14**, 1049–1054.
52. Douglas, A. E. (1992) Microbial brokers of insect–plant interactions, in *Proceedings of the 8th International Symposium on Insect–Plant Relationships*, (eds S. B. J. Menken, J. H. Visser and P. Harrewijn), Kluwer, Dordrecht, pp. 329–336.
53. Dowd, P. F. (1990) Detoxification of plant substances by insects, in *CRC Handbook of Natural Pesticides, vol. 4. Insect Attractants and Repellents*, (eds E. D. Morgan and N. B. Mandava), CRC Press, Boca Raton, FL, pp. 181–225.

54. Dowd, P. F. (1991) Symbiont-mediated detoxification in insect herbivores, in *Microbial Mediation of Plant–Herbivore Interactions*, (eds P. Barbosa, V. A. Krischik and O. G. Jones), John Wiley, New York, pp. 411–440.
55. Duffey, S. S., Bloem, K. A. and Campbell, B. C. (1986) Consequences of sequestration of plant natural products in plant–insect–parasitoid interactions, in *Interactions of Plant Resistance and Parasitoids and Predators of Insects*, (eds D. J. Boethel and R. D. Eikenbary), Ellis Horwood, Chichester, pp. 31–60.
56. Epstein, E. (1994) The anomaly of silicon in plant biology. *Proc. Natl. Acad. Sci. USA*, **91**, 11–17.
57. Feeny, P. (1970) Seasonal changes in oak leaf tannins and nutrients as a cause of spring feeding by winter moth caterpillars. *Ecology*, **51**, 565–581.
58. Feeny, P. (1992) The evolution of chemical ecology, contributions from the study of herbivorous insects, in *Herbivores. Their Interactions with Secondary Plant Metabolites*, 2nd edn, vol. 2, (eds G. A. Rosenthal and M. R. Berenbaum), Academic Press, San Diego, CA, pp. 1–44.
59. Fiedler, K., Krug, E. and Proksch, P. (1993) Complete elimination of hostplant quinolizidine alkaloids by larvae of a polyphagous lycaenid butterfly *Callophrys rube*. *Oecologia*, **94**, 441–445.
60. Fontaine, A. R., Olsen, N., Ring, R. A. and Singla, C. L. (1991) Cuticular metal hardening of mouthparts and claws of some forest insects of British Columbia. *J. Entomol. Soc. Br. Columbia*, **88**, 45–55.
61. Frost, S. W. (1959) *Insect Life and Insect Natural History*, 2nd edn, Dover Publications, New York.
62. Gould, F. (1991) Arthropod behavior and the efficacy of plant protectants. *Annu. Rev. Entomol.*, **36**, 305–330.
63. Grewal, S. K., Grewal, P. S. and Gaugler, R. (1995) Endophytes of fescue grasses enhance susceptibility of *Popillia japonica* larvae to an entomopathogenic nematode. *Entomol. Exp. Appl.*, **74**, 219–224.
64. Guthrie, W. D., Rathore, V. S., Cox, D. F. and Reed, G. L. (1974) European corn borer, virulence on corn plants of larvae reared for different generations on a meridic diet. *J. Econ. Entomol.*, **67**, 605–606.
65. Hajek, A. E. and St Leger, R. J. (1994) Interactions between fungal pathogens and insect hosts. *Annu. Rev. Entomol.*, **39**, 293–322.
66. Hammond, A. M. and Hardy, T. N. (1988) Quality of diseased plants as hosts for insects, in *Plant Stress–Insect Interactions*, (ed. E. A. Heinrichs), John Wiley, New York, pp. 381–432.
67. Harborne, J. B. (1993) *Introduction to Ecological Biochemistry*, 4th edn, Academic Press, London.
68. Harris, K. F. (1977) An ingestion–egestion hypothesis of noncirculative virus transmission, in *Aphids as Virus Vectors*, (eds K. F. Harris and K. Maramorosch), Academic Press, New York, pp. 165–219.
69. Hatcher, P. E. (1995) Three-way interactions between plant pathogenic fungi, herbivorous insects and their host plants. *Biol. Rev.*, **70**, 639–694.
70. Heliövaara, K. and Väisänen, R. (1993) *Insects and Pollution*, CRC Press, Boca Raton, FL.
71. Hinks, C. F. and Spurr, D. T. (1989) Effect of food plants on the susceptibility of the migratory grasshopper (Orthoptera, Acrididae) to deltamethrin and dimethoate. *J. Econ. Entomol.*, **82**, 721–726.
72. Hiratsuka, E. (1920) Researches on the nutrition of the silkworm. *Bull. Ser. Exp. Sta. Jpn.*, **1**, 257–315.
73. Hochuli, D. F. and Roberts, F. M. (1996) Approximate digestibility of fibre for a graminivorous caterpillar. *Entomol. Exp. Appl.*, **81**, 15–20.
74. Holtzer, T. O., Archer, T. L. and Norman, J. M. (1988) Host plant suitability in relation to water stress, in *Plant Stress–Insect Interactions*, (ed. E. A. Heinrichs), John Wiley, New York, pp. 111–137.
75. Hughes, P. R. (1988) Insect populations on host plants subjected to air pollution, in *Plant Stress–Insect Interactions*, (ed. E. A. Heinrichs), John Wiley, New York, pp. 249–319.
76. Hughes, P. R. and Voland, M. L. (1988) Increase in feeding stimulants as the primary mechanism by which SO_2 enhances performance of Mexican bean beetle on soybean. *Entomol. Exp. Appl.*, **48**, 257–262.
77. Karowe, D. N. (1989) Differential effect of tannic acid on two tree-feeding Lepidoptera, implications for theories of plant antiherbivore chemistry. *Oecologia*, **80**, 507–512.
78. Keating, S. T., Yendol, W. G. and Schultz, J. C. (1988) Relationship between susceptibility of

gypsy moth larvae (Lepidoptera; Lymantriidae) to a baculovirus and host plant foliage constituents. *Environ. Entomol.*, **17**, 952–958.
79. Keating, S. T., Schultz, J. C. and Yendol, W. G. (1990) The effect of diet on gypsy moth (*Lymantria dispar*) larval midgut pH, and its relationship with larval susceptibility to a baculovirus. *J. Invertebr. Pathol.*, **56**, 317–326.
80. Keena, M. A., Odell, T. M. and Tanner, J. A. (1995) Effects of diet ingredient source and preparation method on larval development of laboratory-reared gypsy moth (Lepidoptera, Lymantriidae). *Ann. Entomol. Soc. Am.*, **88**, 672–679.
81. Kennedy, C. E. J. and Southwood, T. R. E. (1984) The number of species of insects associated with British trees: a re-analysis. *J. Anim. Ecol.*, **53**, 455–478.
82. Khan, Z. R. and Ramachandran, R. (1989) Studies on the biology of the yellow stemborer, *Scirpophaga incertulas*. *ICIPE Annual Report*, 24–25.
83. Kingsolver, J. G. and Daniel, T. L. (1995) Mechanics of food handling by fluid-feeding insects, in *Regulatory Mechanisms in Insect Feeding*, (eds R. F. Chapman and G. de Boer), Chapman & Hall, New York, pp. 32–73.
84. Klekowski, R. Z. and Duncan, A. (1975) Physiological approach to ecological energetics, in *Methods for Ecological Bioenergetics*, (eds W. Grodzinski, R. Z. Klekowski and A. Duncan), Blackwell, Oxford, pp. 15–64.
85. Kogan, M. (1986) Bioassays for measuring quality of insect food, in *Insect–Plant Interactions*, (eds J. R. Miller and T. A. Miller), Springer-Verlag, New York, pp. 155–189.
86. Krischik, V. A. (1991) Specific or generalized plant defense: reciprocal interactions between herbivores and pathogens, in *Microbial Mediation of Plant–Herbivore Interactions*, (eds P. Barbosa, V. A. Krischik and O. G. Jones), John Wiley, New York, pp. 309–340.
87. Lincoln, D. E., Fajer, E. D. and Johnson, R. H. (1993) Plant–insect herbivore interactions in elevated CO_2 environments. *Trends Ecol. Evol.*, **8**, 64–68.
88. Lindroth, R. L. (1991) Differential toxicity of plant allelochemicals to insects: roles of enzymatic detoxification systems, in *Insect–Plant Interactions*, vol. 5, (ed. E. A. Bernays), CRC Press, Boca Raton, FL, pp. 1–33.
89. Lindroth, R. L., Arteel, G. E. and Kinney, K. K. (1995) Responses of three saturniid species to paper birch grown under enriched CO_2 atmospheres. *Funct. Ecol.*, **9**, 306–311.
90. Loader, C. and Damman, H. (1991) Nitrogen content of food plants and vulnerability of *Pieris rapae* to natural enemies. *Ecology*, **72**, 1586–1590.
91. McEvoy, P. B. (1984) Increase in respiratory rate during feeding in larvae of the cinnabar moth *Tyria jacobaeae*. *Physiol. Entomol.*, **9**, 191–195.
92. McNeill, S. and Southwood, T. R. E. (1978) The role of nitrogen in the development of insect plant relationships, in *Biochemical Aspects of Plant and Animal Coevolution*, (ed. J. B. Harborne), Academic Press, London, pp. 77–98.
93. Mattson, W. J. (1980) Herbivory in relation to plant nitrogen content. *Annu. Rev. Ecol. Syst.*, **11**, 119–161.
94. Mattson, W. J. and Haack, R. A. (1979a) The role of drought in outbreaks of plant-eating insects. *BioScience*, **37**, 110–118.
95. Mattson, W. J. and Haack, R. A. (1979b) The role of drought stress in provoking outbreaks of phytophagous insects, in *Insect Outbreaks*, (eds P. Barbosa and J. C. Schultz), Academic Press, San Diego, CA, pp. 365–407.
96. Meade, T. and Hare, J. D. (1994) Effects of genetic and environmental host plant variation on the susceptibility of two noctuids to *Bacillus thuringiensis*. *Entomol. Exp. Appl.*, **70**, 165–178.
97. Meisner, J., Ishaaya, I., Ascher, K. R. S. and Zur, M. (1978) Gossypol inhibits protease and amylase activity of *Spodoptera littoralis* Boisduval larvae. *Ann. Entomol. Soc. Am.*, **71**, 5–8.
98. Mellanby, K. and French, R. A. (1958) The importance of drinking water to larval insects. *Entomol. Exp. Appl.*, **1**, 116–124.
99. Meyer, G. A. (1993) A comparison of the impacts of leaf- and sap-feeding insects on growth and allocation of goldenrod. *Ecology*, **74**, 1101–1116.
100. Mittler, T. E. (1970) Uptake rates of plant sap and synthetic diet by the aphid *Myzus persicae*. *Ann. Entomol. Soc. Am.*, **63**, 1701–1705.
101. Mohamad, B. M. and van Emden, H. F. (1989) Host plant modification to insecticide susceptibility in *Myzus persicae* (Sulz.). *Insect Sci. Appl.*, **10**, 699–703.
102. Mullin, C. A. (1986) Adaptive divergence of chewing and sucking arthropods to plant allelochemicals, in *Molecular Aspects of Insect–Plant*

Associations, (eds L. B. Brattsten and S. Ahmad), Plenum Press, New York, pp. 175–209.
103. Navon, A., Hare, J. D. and Federici, B. A. (1993) Interactions among *Heliothis virescens* larvae, cotton condensed tannin and the CryIA(c) δ-endotoxin of *Bacillus thuringiensis*. *J. Chem. Ecol.*, **19**, 2485–2499.
104. Neal, J. J. (1987) Metabolic costs of mixed-function oxidase induction in *Heliothis zea*. *Entomol. Exp. Appl.*, **43**, 175–179.
105. Peng, Z. and Miles, P. W. (1988) Studies on the salivary physiology of plant bugs: function of the catechol oxidase of the rose aphid. *J. Insect Physiol.*, **11**, 1027–1033.
106. Petrusewicz, K. and MacFayden, A. (1970) *Productivity of Terrestrial Animals: Principles and Methods*, IBP Handbook 13, Blackwell, Oxford.
107. Phelan, P. L. and Stinner, B. R. (1992) Microbial mediation of plant–herbivore ecology, in *Herbivores. Their Interactions with Secondary Plant Metabolites*, 2nd edn, vol. 2, (eds G. A. Rosenthal and M. R. Berenbaum), Academic Press, San Diego, CA, pp. 279–315.
108. Rausher, M. D. (1982) Population differentiation in *Euphydras editha* butterflies, larval adaptation to different hosts. *Evolution*, **36**, 581–590.
109. Raven, J. A. (1983) Phytophages of xylem and phloem. *Adv. Ecol. Res.*, **13**, 135–234.
110. Reese, J. C. and Field, M. D. (1986) Defense against insect attack in susceptible plants: cutworm (Lepidoptera, Noctuidae) growth on corn seedlings and artificial diet. *Ann. Entomol. Soc. Am.*, **79**, 372–376.
111. Roessingh, P., Bernays, E. A. and Lewis, A. C. (1985) Physiological factors influencing preference for wet and dry food in *Schistocerca gregaria* nymphs. *Entomol. Exp. Appl.*, **37**, 89–94.
112. Rose, H. A. (1985) The relationship between feeding specialization and host plants to aldrin epoxidase activities of midgut homogenates in larval Lepidoptera. *Ecol. Entomol.*, **10**, 455–467.
113. Ross, H. H. (1948) *A Textbook of Entomology*, John Wiley, New York.
114. Saikkonen, N., Helander, M., Ranta, H. et al. (1996) Endophyte-mediated interactions between woody plants and insect herbivores? *Entomol. Exp. Appl.*, **80**, 269–271.
115. Salim, M. and Saxena, R. C. (1992) Iron, silica, and aluminum stresses and varietal resistance in rice: effects on whitebacked planthopper. *Crop Sci.*, **32**, 212–219.
116. Schmidt, D. J. and Reese, J. C. (1986) Sources of error in nutritional index studies of insects on artificial diet. *J. Insect Physiol.*, **32**, 193–198.
117. Schoonhoven, L. M. and Henstra, S. (1972) Morphology of some rostrum receptors in *Dysdercus* spp. *Neth. J. Zool.*, **22**, 343–346.
118. Schoonhoven, L. M. and Meerman, J. (1978) Metabolic costs of changes in diet and neutralization of allelochemics. *Entomol. Exp. Appl.*, **24**, 689–693.
119. Schroeder, L. A. (1981) Consumer growth efficiencies, their limits and relationships to ecological energetics. *J. Theoret. Biol.*, **93**, 805–828.
120. Schroeder, L. A. (1986) Protein limitation of a tree leaf feeding Lepidopteran. *Entomol. Exp. Appl.*, **41**, 115–120.
121. Schultz, J. C. and Keating, S. T. (1991) Host-plant-mediated interactions between the gypsy moth and a baculovirus, in *Microbial Mediation of Plant–Herbivore Interactions*, (eds P. Barbosa, V. A. Krischik and O. G. Jones), John Wiley, New York, pp. 489–506.
122. Scriber, J. M. (1979) Effects of leaf-water supplementation upon post-ingestive nutritional indices of forb-, shrub-, vine-, and tree-feeding Lepidoptera. *Entomol. Exp. Appl.*, **25**, 240–255.
123. Scriber, J. M. (1984) Host-plant suitability, in *Chemical Ecology of Insects*, (eds W. J. Bell and R. T. Cardé), Chapman & Hall, New York, pp. 159–202.
124. Scriber, J. M. and Slansky, F. (1981) The nutritional ecology of immature insects. *Annu. Rev. Entomol.*, **26**, 183–211.
125. Self, L. S., Guthrie, F. E. and Hodgson, E. (1964) Metabolism of nicotine by tobacco-feeding insects. *Nature*, **204**, 300–301.
126. Shen, S. K. and Dowd, P. F. (1991) Detoxification spectrum of the cigarette beetle symbiont *Symbiotaphrina kochii* in culture. *Entomol. Exp. Appl.*, **60**, 51–59.
127. Simpson, S. J. and Abisgold, J. D. (1985) Compensation by locusts for changes in dietary nutrients: behavioural mechanisms. *Physiol. Entomol.*, **10**, 443–452.
128. Simpson, S. J. and Raubenheimer, D. (1996) Feeding behaviour, sensory physiology and nutritional feedback: a unifying model. *Entomol. Exp. Appl.*, **80**, 55–64.
129. Simpson, S. J. and Simpson, C. L. (1990) The mechanisms of nutritional compensation by phytophagous insects, in *Insect–Plant Interactions*, vol. 2, (ed. E. A. Bernays), CRC Press, Boca Raton, FL, pp. 111–160.
130. Simpson, S. J., Simmonds, M. S. J. and Blaney,

W. M. (1988) A comparison of dietary selection behaviour in larval *Locusta migratoria* and *Spodoptera littoralis*. *Physiol. Entomol.*, **13**, 225–238.

131. Singh, P. (1983) A general purpose laboratory diet mixture for rearing insects. *Insect Sci. Appl.*, **4**, 357–362.
132. Singh, P. and Moore, R. F. (1985) *Handbook of Insect Rearing*, Elsevier, Amsterdam.
133. Slansky, F. (1992) Allelochemical-nutrient interactions in herbivore nutritional ecology, in *Herbivores. Their Interactions with Secondary Plant Metabolites*, 2nd edn, vol. 2, (eds G. A. Rosenthal and M. R. Berenbaum), Academic Press, San Diego, CA, pp. 135–174.
134. Slansky, F. (1993) Nutritional ecology: the fundamental quest for nutrients, in *Caterpillars. Ecological and Evolutionary Constraints on Foraging*, (eds N. E. Stamp and T. M. Casey), Chapman & Hall, New York, pp. 29–91.
135. Slansky, F. and Feeny, P. (1977) Stabilization of the rate of nitrogen accumulation by larvae of the cabbage butterfly on wild and cultivated food plants. *Ecol. Monogr.*, **47**, 209–228.
136. Slansky, F. and Rodriguez, J. G. (1987) *Nutritional Ecology of Insects, Mites, Spiders, and Related Invertebrates*. J. Wiley, New York.
137. Slansky, F. and Scriber, J. M. (1985) Food consumption and utilization, in *Comprehensive Insect Physiology, Biochemistry and Pharmacology*, vol. 4, (eds G. A. Kerkut and L. I. Gilbert), Pergamon Press, New York, pp. 87–163.
138. Slansky, F. and Wheeler, G. S. (1992) Caterpillars' compensatory feeding response to diluted nutrients leads to toxic allelochemical dose. *Entomol. Exp. Appl.*, **65**, 171–186.
139. Southwood, T. R. E. (1972) The insect/plant relationship – an evolutionary perspective. *Symp. Roy. Entomol. Soc. Lond.*, **6**, 3–30.
140. Steinly, B. A. and Berenbaum, M. (1985) Histopathological effects of tannins on the midgut epithelium of *Papilio polyxenes* and *Papilio glaucus*. *Entomol. Exp. Appl.*, **39**, 3–9.
141. Tabashnik, B. E. and Slansky, F. (1987) Nutritional ecology of forb foliage-chewing insects, in *Nutritional Ecology of Insects, Mites, Spiders and Related Invertebrates*, (eds F. Slansky and J. G. Rodriguez), John Wiley, New York, pp. 71–103.
142. Thompson, V. (1994) Spittlebug indicators of nitrogen-fixing plants. *Ecol. Entomol.*, **19**, 391–398.
143. Tjallingii, W. F. (1995) Regulation of phloem sap feeding by aphids, in *Regulatory Mechanisms in Insect Feeding*, (eds R. F. Chapman and G. de Boer), Chapman & Hall, New York, pp. 190–209.
144. Tjallingii, W. F. and Hogen Esch, T. (1993) Fine structure of aphid stylet routes in plant tissues in correlation with EPG signals. *Physiol. Entomol.*, **18**, 317–328.
145. Tjallingii, W. F. and Mayoral, A. (1992) Criteria for host-acceptance by aphids, in *Proceedings of the 8th International Symposium on Insect–Plant Relationships*, (eds S. B. J. Menken, J. H. Visser and P. Harrewijn), Kluwer, Dordrecht, pp. 280–282.
146. Van Loon, J. J. A. (1988) A flow-through respirometer for leaf chewing insects. *Entomol. Exp. Appl.*, **49**, 265–276.
147. Van Loon, J. J. A. (1991) Measuring food utilization in plant-feeding insects – toward a metabolic and dynamic approach, in *Insect–Plant Interactions*, vol. 3, (ed. E. A. Bernays), CRC Press, Boca Raton, FL, pp. 79–124.
148. Van Loon, J. J. A. (1993) Gravimetric vs. respirometric determination of metabolic efficiency in caterpillars of *Pieris brassicae*. *Entomol. Exp. Appl.*, **67**, 135–142.
149. Van't Hof, H. M. and Martin, M. M. (1989) The effect of diet water content on energy expenditure by third-instar *Manduca sexta* larvae (Lepidoptera, Sphingidae). *J. Insect Physiol.*, **35**, 433–436.
150. Waldbauer, G. P. (1968) The consumption and utilization of food by insects. *Adv. Insect Physiol.*, **5**, 229–288.
151. Waldbauer, G. P. and Friedman, S. (1988) Dietary self-selection by insects, in *Endocrinological Frontiers in Physiological Insect Ecology*, (eds F. Sehnal, A. Zabza and D. L. Denlinger), Wroclaw Technical University Press, Wroclaw, pp. 403–422.
152. Watt, A. D., Whittaker, J. B., Docherty, M., Brooks, G., Lindsay, E. and Salt, D. T. (1995) The impact of elevated atmospheric CO_2 on insect herbivores, in *Insects in a Changing Environment*, (eds R. Harrington and N. E. Stork), Academic Press, London, pp. 197–217.
153. White, T. C. R. (1993) *The Inadequate Environment. Nitrogen and the Abundance of Animals*, Springer-Verlag, Berlin.
154. Whittaker, J. B. and Warrington, S. (1995) Effects of atmospheric pollutants on interactions between insects and their food plants, in *Pests, Pathogens and Plant Communities*, (eds J. J. Burdon and R. S. Leather), Blackwell, Oxford, pp. 97–110.

155. Wiegert, R. G. and Petersen, C. E. (1983) Energy transfer in insects. *Annu. Rev. Entomol.*, **28**, 455–486.
156. Wightman, J. A. (1981) Why insect energy budgets do not balance. *Oecologia*, **50**, 166–169.
157. Wink, M. and Schneider, D. (1990) Fate of plant-derived secondary metabolites in three moth species (*Syntomis mogadorensis*, *Syntomeida epilais*, and *Creatonotus transiens*). *J. Comp. Physiol.*, **B160**, 389–400.
158. Wratten, S. D., Edwards, P. J. and Dunn, I. (1984) Wound-induced changes in palatability of *Betula pubescens* and *B. pendula*. *Oecologia*, **61**, 372–375.
159. Yu, S. J. (1986) Consequences of induction of foreign compound-metabolizing enzymes in insects, in *Molecular Aspects of Insect–Plant Associations*, (eds L. B. Brattsten and S. Ahmad), Plenum Press, New York, pp. 153–174.

HOST-PLANT SELECTION: HOW TO FIND A HOST PLANT   5

5.1	Terminology	121
5.2	Host-plant selection: a catenary process	122
5.3	Searching mechanisms	126
5.4	Orientation to host plants	129
5.4.1	Optical *versus* chemical cues	129
5.4.2	Visual responses to host-plant characteristics	131
5.4.3	Olfactory responses to host plants	136
5.4.4	Flying moths and walking beetles: two cases of olfactory orientation	136
5.5	Chemosensory basis of host-plant odour detection	138
5.5.1	Morphology of olfactory sensilla	138
5.5.2	Sensitivity of olfactory sensilla	142
5.5.3	Specificity and olfactory coding	143
5.6	Host-plant searching in nature	146
5.7	Conclusions	148
5.8	References	148

One of the most notable features of herbivorous insects that has emerged from the previous chapters is that most species are very selective feeders and meticulously choose the plants on which they deposit their eggs. Recent research on several species has shown that they select not only certain plant species but also specific plant organs. For several herbivorous species it has been shown that they may even prefer particular individuals within a host-plant population. At the outset of this chapter on selection behaviour it is important to note that the host-plant range of a certain insect species does not necessarily include all plant species that appear under laboratory testing conditions behaviourally acceptable or nutritionally adequate; under natural circumstances it is often more restricted. Also, host selection behaviour may change with the developmental phase of the insect and different life stages often differ in their host-plant preference or their ability to use a plant species as a host. Despite the fact that neonate insect larvae have a small body size and consequently possess limited energy reserves, they are capable of leaving the plant on which they hatched if they appear to judge it unsuitable.

There are a number of situations that make it necessary for a herbivorous insect to search for a host plant. For instance, eclosion of adults from pupae that overwintered in the soil may occur far from potential food or oviposition plants if these are annuals. The arrival in a novel habitat after migration or dispersal and local exhaustion of food plants are other examples of such circumstances. Food specialization requires the ability to find and recognize host plants, which in natural habitats often grow in mixed and complex vegetations.

5.1 TERMINOLOGY

It is useful first to carefully define terms that are generally used to describe or categorize host-plant selection behaviour.

Searching. Whenever an insect is remote from a potential foodplant, it needs to search for and find that plant. To locate a host plant, it needs to move towards it and contact it, or at least to arrive and stay in the proximity of it in order to further examine its characteristics. The observation that the insect contacts the plant, however, gives no information on the mechanism used in establishing this contact. The term 'searching' means 'to look carefully in a place in an effort to find something'. 'Finding' (sometimes unfortunately used as a synonym [68]) may rather be the end result of searching. As searching has a connotation of directionality, it is important to note that the movement pattern of an insect may vary from random, resulting in contact by chance, to oriented and strongly directed movements (see below).

Selection. In the strict sense of the word, 'to select' means to choose from among alternatives. In order to do this, it is necessary that differential sensory perception of alternative food plants occurs. Selection thus implies a weighing of alternatives. From a methodological point of view it is difficult to prove that comparison of alternatives is being made during selection behaviour, especially if contacts with potential hosts occur sequentially. Sequential contacting occurs more frequently than simultaneous contacting and this implies that a short-term memory must be invoked to enable comparisons over time. In cases in which alternatives have been assessed before final acceptance occurs, either at a distance by approaching and turning away again or by actual contact-testing, the term 'selection behaviour' is appropriate.

Acceptance. Acceptance of a plant is said to occur when either sustained feeding or oviposition occurs. 'Acceptance' is a term devoid of the assumptions implied by the term 'selection'. For example, when a beetle is released in the middle of a monoculture of beans and is observed to initiate sustained feeding after climbing a bean plant, it cannot be concluded that the beetle selected the bean plant as a host plant, since no alternatives were available. It can only be said that the bean plant is accepted by the insect.

Preference. When in dual or multiple choice assays an insect consistently feeds or oviposits more often on one of the alternative plants, it is said to 'prefer' that plant over the others.

This may also be observed under field conditions when the degree of feeding or oviposition on a certain plant species is higher than would be predicted from its relative abundance. Clearly, preference is a relative concept and applicable only to the set of plant species or genotypes that were actually available to the insect.

Recognition. This term is often used in connection with acceptance. It means 'to know again' and implicitly refers to a neural process. It implies that there is an internal standard or 'image' of the plant(s) sought for. This image is present in one or another form in the central nervous system (CNS) of the insect. The profile of incoming sensory information on plant cues is compared to this stored image and when it matches sufficiently the plant is recognized as a host. The putative image is genetically fixed, but can be modified by experience to quite some extent (see Chapter 7).

From the above it appears that the term 'acceptance' is the most straightforward one, while searching, selection, preference and recognition implicitly refer to complex behavioural processes, the neural mechanisms of which are being elucidated (see below) but as yet are only partly understood.

It is also important at this point to relate the behavioural terms defined above to the classification of semiochemicals ('chemicals that mediate interactions between organisms') that may affect host-plant selection behaviour. For this purpose we adopt the terminology proposed by Dethier et al. [32], which is summarized in Table 5.1. These terms will be used in this and the following chapters.

5.2 HOST-PLANT SELECTION: A CATENARY PROCESS

Insects are often said to show 'programmed behaviour' and stereotyped, predictable sequences of behavioural acts, so-called **reaction chains** [6]. This means that more or less

Table 5.1 Chemical designations in terms of insect responses [32]

Attractant	A chemical that causes insects to make oriented movements towards its source
Repellent	A chemical that causes insects to make oriented movements away from its source
Arrestant	A chemical that may slow the linear progression of an insect by reducing actual speed of locomotion or by increasing turning rate
Feeding or ovipositional stimulant	A chemical that elicits feeding or oviposition in insects ('feeding stimulant' is synonymous with 'phagostimulant')
Deterrent	A chemical that inhibits feeding or oviposition when present in a place where insects would, in its absence, feed or oviposit

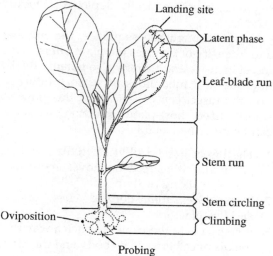

Figure 5.1 Complex behaviour patterns involve a sequence of stimulation and response steps, as exemplified by oviposition behaviour in the cabbage root fly (*Delia radicum*). An airborne gravid female fly may land in response to yellow-green wavelengths (500–600 nm), as reflected by green foliage. During the 'latent phase' she walks along the leaf, pausing now and then to groom or to make short flights. During the next phase, the 'leaf-blade run', she walks continuously, often along the leaf edge and frequently changing direction. With taste hairs on her tarsi she assesses the suitability of the plant. If she contacts the appropriate chemical stimuli she moves on to a stem or a midrib of a leaf, which is quickly followed ('stem run'). At the stem base she moves around it sideways ('stem circling'), keeping her head downwards. During the 'climbing phase' she walks around close to the cabbage stem and occasionally climbs up the stem a few centimetres. She then starts 'probing' the soil with her ovipositor, probably testing soil particle size and water content. When again the adequate stimuli are perceived, she finally lays her eggs in the soil close to the stem. (Source: redrawn from Zohren, 1968.)

distinct behavioural elements follow each other in a fixed order. The insect shows appropriate reactions to a succession of stimuli (Fig. 5.1).

When the outcome of a sensory evaluation is rejection of a particular plant or plant part as food or oviposition site, the herbivore 'jumps back' to one of the earlier steps in the reaction sequence. Modification of selection behaviour as a result of previous experience (Chapter 7) leads to faster decision making or to changes in preference, but the sequence remains the same. As we will see from the examples presented below, such sequences of behavioural phases and of elements within each phase can be quite long and elaborate.

In the process of host-plant selection two main consecutive phases may be distinguished, delimited by the intermittent decision to stay in contact with the plant: (1) searching and (2) contact-testing. The first phase may end with the event of finding, the second phase ends by acceptance or rejection. Acceptance is a crucial behavioural decision as it results in ingestion of plant material or deposition of eggs, with possible negative consequences for fitness. A host-plant selection sequence is schematically depicted in Figure 5.2(A).

Going through the sequence, the number and intensity of the cues the plant offers to the insect increase and thereby also potentially the intensity and modalities of sensory information the insect can collect about the plant. A standardized host-plant selection sequence can be described as follows.

1. The insect has no physical contact with a plant and either rests or moves about randomly, walking or flying.
2. It perceives plant-derived cues, optical and/or olfactory.
3. It responds to these cues in such a way that the distance between its body and the plant decreases.
4. The plant is found, i.e. it is contacted by either touching or climbing it or landing on it.
5. The plant surface is examined by contact-testing (e.g. palpation of leaf surface).
6. The plant may be damaged and the content of tissues released by nibbling or test-biting (in the case of biting–chewing species), probing (piercing–sucking species) or puncturing with the ovipositor.
7. The plant is accepted (as evidenced by one or more eggs being laid or continued feeding) or is rejected, resulting in the insect's departure.

During each of these steps the insect may decide to turn away from the plant before contacting it or to leave it after contact. When it arrives in a patch of potential host plants, it may exhibit repetition of the same sequence with respect to different plant individuals of the same or other species. In the end it may return to and select the plant that was examined first but was left after that initial contact.

In this and the next chapters, host-plant selection behaviour will be discussed using this sequential framework. The focus will be on the different plant cues affecting selection behaviour and the sensory apparatus *via*

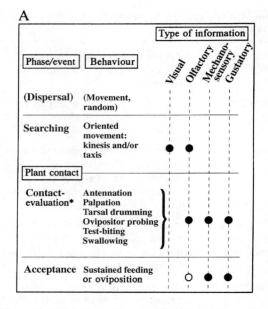

Figure 5.2 (A) Generalized sequence of host-plant selection behaviour of herbivorous insects. Left column: behavioural phase or event. Middle column: common behavioural elements occurring within a behavioural phase. Right column: main plant-derived stimuli affecting the behaviour. ● = well documented plant cues for several species; ○ = suggested or probable; * = examples of behavioural elements displayed by many species; not all elements occur in a particular species and not necessarily in this sequence. Between brackets at the top, dispersal is indicated as a preceding behavioural phase with its behavioural elements (which do not belong to the host selection sequence). **(B)–(E)** Host selection behaviour sequences of representatives of the four major herbivorous orders, following the scheme of **(A)**, with specific elements and terms. **(B)** Alate aphids (*Myzus, Aphis* spp.). **(C)** Adult bark beetles (*Dendroctonus, Ips* spp.). # = progressive colonization by gallery elongation occurs when repellents or deterrents are absent. **(D)** Adult herbivorous flies (*Delia, Rhagoletis* spp.); for optomotor anemotaxis, visual cues are ground pattern movements, mechanosensory cues are air streams, both not plant-derived. **(E)** Adult nocturnal moths (*Heliothis* spp., *Manduca sexta*); optomotor anemotaxis: as **(D)**. (Source: compiled from various sources.)

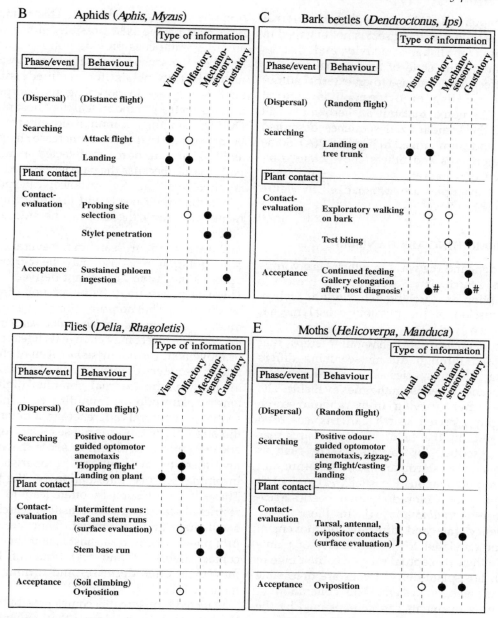

Figure 5.2 (*Continued*)

which these are perceived and affect selection behaviour. The crucial decision to accept or to reject a plant is based not only on sensory information of plant cues but also on the insect's physiological status (satiety, sexual maturity, egg maturation, etc. [8]). The integration of these two variables, together with information on previous experiences stored in the insect's memory, occurs in the central nervous system (CNS) [30]. For the purpose of this chapter we will assume that the internal status is such that the insect is not engaged in migration or dispersal activity and that its motivation (the state of an internal variable that affects the response to an external stimulus) for feeding or oviposition is high. It should be noted that not all herbivores will follow the standardized sequence described above and summarized in Figure 5.2(A). Some take short cuts and others show more complicated or elaborate sequences. Some well-studied examples are schematized in Figure 5.2(B)–(E).

5.3 SEARCHING MECHANISMS

To understand the ways in which herbivorous insects search, it is necessary to present a description of searching behaviour as well as a discussion of the possible causal mechanisms involved.

The sequence of behavioural steps that is passed through during searching differs among insect species and developmental phases and depends on the cues available. The whole range, varying from random search to highly directed search patterns, has been observed. In the field, random search has been described for various insects, such as polyphagous caterpillars [31], immature and mature polyphagous locusts [2, 72] and adult oligophagous Colorado potato beetles (*Leptinotarsa decemlineata*) [54]. In these cases, frequency, rate and direction of movement appear unrelated to the suitability of plants within their perceptual range, i.e. the range in which host-plant-derived cues are detectable by the sensory system. The generation of random movements can be explained by the functioning of so-called 'central motor programmes' located in the CNS. When an insect becomes motivated to search for food, e.g. because blood trehalose levels fall below a certain level (an internal state parameter), these programmes are activated and as a result the insect may start a random walk. Only internally stored (e.g. in memory) and proprioceptive information is used. This searching type may be the best possible, either when environmental cues provide no directionality or when the sensory capacity of the insect is insufficient to obtain the required stimuli. Such is the case in polyphagous caterpillars moving on the ground in search of host plants. During searching, scanning movements may be performed that serve to increase the probability that a resource is detected along the path, mainly because the path is widened. This is often seen in, for instance, caterpillars, which raise their heads and first thoracic segments and sway these from one side to the other.

During random searching, several types of orientation response may be performed upon stimulation by plant-derived cues. These responses may be either non-directed or directed. The non-directional changes in random movement are classified as **kineses** [56, 97]. The insect may change its linear speed of movement (orthokinesis) or it may change rate or frequency of turning (klinokinesis). The intensity of the external stimulus (light intensity, plant odours, humidity, etc.) and the spatial or temporal differences in it determine the strength of these responses. One (unilateral) receptor is sufficient to sense the stimulus intensity by temporal comparisons of incoming sensory information by the CNS. These kinetic responses often lead to area-restricted search, i.e. an intensified search in a small area, and arrestment. They are most prominent close to a host plant or upon contact (Figs 5.3 and 5.4). Rate of linear movement often decreases and turning rates increase [71].

Directed movement becomes possible when the host plant emits signals that, either alone or in combination with a second cue, allow directionality to be perceived by the sensory system of the searching insect. Movements in this case are directed by sensory information on external cues but may still be under influence of central motor programmes (see below). When a distinct directionality towards

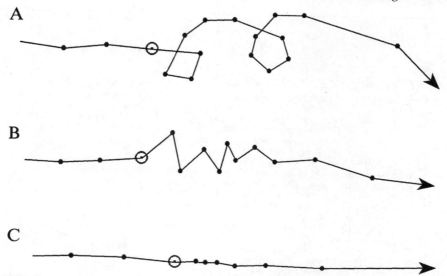

Figure 5.3 Searching patterns used where resources are aggregated. In these cases it may be advantageous for an insect to search an area more thoroughly once it has already encountered a host plant. This strategy increases its chance of finding another host plant. Mechanisms used for restricting the area of search include: **(A)** periodic increases in turning tendency, generating looping or circling; **(B)** alternation in turning direction, generating zigzags; **(C)** adjustments in length of moves between stops. Dots indicate landings, circled dots represent landings on host plants followed by egg laying. (Source: redrawn from Bell, 1991.)

the food plant results from the analysis of movement patterns, such oriented movements relative to an external source of stimulation are termed **taxes**, and may be toward the source (positive) or away from the source (negative). Orientation to visual or chemical cues or to their combination is common to many insects. Over short distances in relatively undisturbed, still air, insects may respond to plant odour gradients by positive chemotaxis. This may be achieved either by temporal comparisons of information coming from the olfactory receptors (klinotaxis) or by comparing sensory input coming simultaneously from a bilateral pair of (olfactory) receptors and trying to obtain equal stimulation of both sides (tropotaxis; symmetrical orientation). A third type of orientation is menotaxis, i.e. maintenance of a constant angle with stimulus direction by preserving a non-symmetrical distribution of sensory stimulation.

Two special cases of menotaxis, i.e. **anemotaxis** and **photomenotaxis**, need special attention because they have been found to operate in herbivorous insects. Anemotaxis and photomenotaxis mean oriented movement by maintaining a set angle to the prevailing wind direction or light direction respectively. Wind or light direction, perceived as air flow by mechanoreceptors or photon flow by photoreceptors, may be sampled successively at the left and right sides of the body by serial counterturning movements. Wind direction is detected mechanically by walking insects but mainly visually in the case of flying insects. Anemotactic behaviour, influenced by plant odours, is seen in a number of herbivorous insects under laboratory conditions. In contrast to what might be expected, odorous cues do not exhibit a gradient, required for chemotaxis, at distances greater than a few centimetres (section 5.4.4). Photomenotaxis, or light

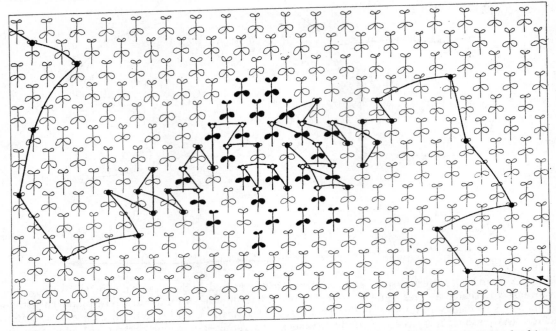

Figure 5.4 Schematized search behaviour in egg-laying females of *Cidaria albulata*, a specialist herbivore on *Rhinanthus* spp. The moths fly shorter distances between alightings and show more turning flight near a host-plant stand, thereby increasing the chance of alighting on a host plant. Turning of flight path and alighting (at least the latter) are stimulated by host-plant odour. Total number of plants = 252; no. of *Rhinanthus* plants = 25 (10%). Total no. of alightings = 45; number of alightings on *Rhinanthus* = 15 (33%). (Source: reproduced from Douwes, 1968, with permission.)

compass orientation, is a main mechanism for insects walking on the ground [97]. While it is difficult to demonstrate anemotaxis in the field, because of lack of control over wind direction, and the ubiquitous occurrence of air turbulence, which prevents a consistent directionality and is prominent especially in the boundary layer over the soil surface, the use of photomenotaxis can be investigated relatively simply. One method is Santschi's 'mirror test' [97] and a second method to demonstrate photomenotaxis is the 'turntable test' [54].

Although the descriptions of movement types and the way in which plant-derived cues may be used are useful to demonstrate the existence of different searching strategies, the number of documented cases for which the orientation mechanism has been fully analysed is small, especially under field conditions. It has repeatedly been suggested that combinations of mechanisms, rather than a single one, operate under natural circumstances. Alternative or additional classifications of searching patterns can be found in the literature [10, 122, 124]. Models of searching behaviour indicate that, contrary to what one might expect, random walking can be a very effective search strategy and that the rate of random movement is an important factor in determining the success of non-random search [71]. Directed orientation is often viewed as adaptive, as it improves the efficiency of search, i.e. it produces a higher success ratio per unit of time and energy invested in searching behaviour.

5.4 ORIENTATION TO HOST PLANTS

5.4.1 OPTICAL *VERSUS* CHEMICAL CUES

Two important types of stimulus that could be used as directionality cues by herbivorous insects are optical and odorous characteristics of plants. The relative importance of the two varies between species, as becomes particularly noticeable when diurnal and nocturnal species are compared. The two types of stimulus are often used in an integrated way (section 5.6).

The nature of optical and chemical plant-derived cues differs in some important aspects. The unit of light energy, the photon, moves self-propelled at the speed of light. The spectral reflectance pattern of a plant is not substantially altered by air movements and is relatively constant at varying distances from the plant. In contrast, volatile compounds emanating from plants move slowly. In still air they move by diffusion and in all dimensions, but in moving air their concentration in space is highly variable (see below). Odour concentrations rise sharply when the plant is approached. Absolutely still air and complete absence of turbulence are very rare, if not completely lacking, under natural circumstances and wind speeds are mostly greater than the linear speed of diffusion of organic molecules. In moving air, the normal situation, volatiles are carried away from the source with the prevailing direction of air flow.

In the literature mention is often made of an odour-filled space which, based on Sutton's model of diffusion, has a semi-ellipsoidal shape in moving air. More recently, however, through the use of ion-detectors with a short response time, it has become clear that the odour occurs in a stochastic fashion as pockets or filaments of molecules in a meandering plume (Fig. 5.5).

Outside the plume boundary, which can be visualized by the use of smoke, no odour pockets occur. When moving upwind, the insect may contact spatially separated pockets of odour molecules at concentrations only slightly lower than those found close to the plant. Most information on the spatial distribution of odorous molecules comes from studies on distribution patterns of sex pheromones, which are released from the insect body, virtually a point source. Chapman [23] has stressed the fact that point sources produce odour plumes different from those emanating from big plants or plant patches; clearly, the form of the food source may shape the plume.

In summary, when considering abiotic factors, optical plant characteristics are relatively constant with respect to their distribution and largely independent of temperature and wind speed, but of course they depend on light intensity. Odours emanating from plants have a spatially highly variable distribution and concentration, which depends on wind speed, on temperature and to some extent also on light intensity. Moreover, the quality and quantity of emitted plant volatiles may vary with the plant's physiological state and wounding effects [13, 103]. Apart from these abiotic factors, the main issues to be considered regarding the relative usability of optical and odorous cues are their specificity and their 'active space', 'effective zone' or 'effective attraction radius' [17].

Quite often it has been assumed implicitly that optical cues cannot be used to recognize host plants, for the reason that 'all plants are green' (i.e. the dominant reflectance–transmission hue is 500–580 nm). In apparent contrast, several plant species have been found to emit volatile chemicals or chemical blends that appear to be taxon-specific, either qualitatively (unique compounds) or quantitatively (characteristic ratios) [122]. This has probably led to the greater attention paid in the literature to odours as guiding factors in host-plant searching, especially in the case of specialized herbivores. In contrast to the low variability of spectral composition of light reflected by foliage, however, intensity of reflected light may differ more pronouncedly

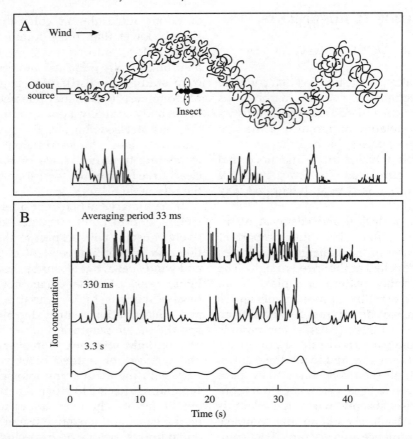

Figure 5.5 (A) Schematic drawing of an undulating and meandering odour plume and an odour signal encountered over time when an insect moves upwind in a straight line to a small odour source. **(B)** Signal amplitudes generated by a stationary ion probe located in an odour plume when different averaging periods are used. Varying amounts of odour, caused by irregular bursts of odour production at the source, pass the odour receiver. Upon increasing the averaging period, the signal amplitude differences decrease, which will lead to a decreased resolution of concentration differences by olfactory receptors. However, even at an average time of 3.3 s the signal is still intermittent and the major bursts of the original can clearly be distinguished. (Source: redrawn from Murlis, 1986.)

between species, because of the presence of wax crystals or trichomes on the leaf surface, or because of biotic (age, nutrient status) and abiotic (density, incident light intensity, background) factors.

The maximum distance over which plant cues can guide an insect to its host plant is another important factor, related to the concept of **active space**. Active space is defined as the space within which the intensity of a stimulus or cue is above the threshold for a behavioural response. In the absence of visual cues, behavioural responses to plant odours have been demonstrated at distances of 5–30 m for several oligophagous species, with a maximum of 100 m reported for the onion fly (*Delia antiqua*; Table 5.2).

The fact that some insects can be lured to scented traps suggests that volatile plant compounds may under field conditions attract herbivorous insects, sometimes over large distances. Tephritid fruit flies and diabroticite

Table 5.2 Distances over which odorous or optical plant cues have been shown to elicit positive taxis-type responses from herbivorous insect species

Insect species	Distance (m)	Reference
Odorous cues		
Leptinotarsa decemlineata	0.6	54
	6	33
Ceutorhynchus assimilis	20	38
Delia radicum	24	41
Dendroctonus spp.	30	127
Pegomya betae	50	2
Delia antiqua	100	55
Optical cues		
Delia brassicae	2	87
Empoasca devastans	3.6	95
Leptinotarsa decemlineata	8	115
Rhagoletis pomonella	10	4

rootworm beetles can be attracted in large numbers to traps baited with specific blossom aroma components. This applies also to some polyphagous species, such as corn earworms [49] and Japanese beetles. The latter may be attracted in open areas to such traps from a distance of up to 400 m. In these cases, traps appear to be an effective and sensitive tool for monitoring insect densities [66].

The significance of figures on linear distances and conclusions about active spaces under natural conditions heavily depend on both the biomass and complexity of the vegetation, factors that have not been varied extensively in field studies on insect host-plant searching. The integrity (unmixed character) of the stimulus produced by an individual host plant or a patch of host plants in a mixed plant stand is thought to be preserved over relatively short distances only [105], although in some instances odours may remain attractive despite mixing with other plant volatiles. Thus gravid beet flies (*Pegomya betae*) are attracted by the odour of young beet leaves over distances of up to 50 m, even if these odours have passed non-host plants [92]. Optical contrasts in a mixed plant stand may be perceived over distances of a few metres, especially in flying insects. At present, few firm data exist on the size of active spaces based on either optical or odorous signals, and the conclusion that the active space of odorous signals is greater than that of optical cues [12, 87] seems premature. Indeed, under field conditions they always occur together and it will be shown below (section 5.6) that insects use combinations of signals, which may enable them to overcome the disadvantage inherent in relying solely on either one.

5.4.2 VISUAL RESPONSES TO HOST-PLANT CHARACTERISTICS

Three optical characteristics of plants may influence host selection behaviour: spectral quality, dimensions (size) and pattern (shape) [87]. The spectral sensitivity of insect compound eyes ranges from 350–650 nm (near-ultraviolet to red) and thus includes shorter wavelengths than that of the human eye (Fig. 5.6).

The ommatidium, the basic photoreceptor and image-formation unit of the insect compound eye, is of a fixed-focus type. This results in maximum acuity at very close range while at greater distances perception of shape is poor. For a more detailed discussion of the characteristics of photoreceptors and the sophisticated visual system of insects, the reader is referred to other texts [12, 29]. Although the size of plants or plant parts and their shapes show considerable variation between and within plant species, this variation presumably aids plant selection only at close distances.

To illustrate visual discrimination some examples of insect responses to optical host-plant cues, such as shape and colour, will be presented.

(a) Lepidoptera

The responsiveness of day-foraging butterflies to colours has been relatively well studied. When artificial green paper leaves are offered

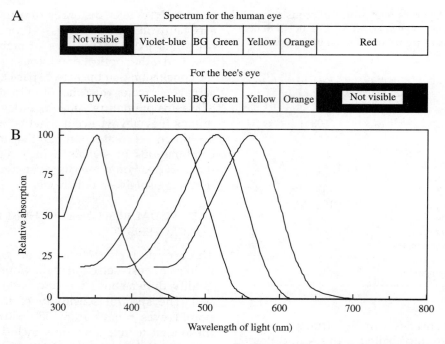

Figure 5.6 (**A**) Comparison of the wavelength spectra (nm) perceived by humans and honeybees. (**B**) Spectral sensitivity curves of a tetrachromatic insect eye (*Spodoptera* sp.). The absorption of each pigment is expressed as a percentage of the maximum for that pigment. (Source: redrawn from Langer *et al.*, 1979.)

to oligophagous cabbage white butterflies (*Pieris brassicae* and *P. rapae*), naive individuals show landing responses, albeit at much lower frequencies than with cabbage leaves. Immediately upon alighting on the substrate they start to 'drum' it for a few seconds, even though specific host-plant chemicals are absent. For *P. brassicae*, true colour vision and wavelength specific behaviour have been found (Fig. 5.7), and *P. rapae* clearly showed landing preferences for differently coloured artificial substrates.

In both *P. brassicae* and *P. rapae* associative learning (see Chapter 7) in response to different shades of green has been demonstrated [112, 113, 119]. These butterflies switch their colour preference for landing responses from green leaf colour to flower colours (yellow, blue and violet), depending on their motivation for oviposition or nectar feeding re-

spectively. In the papilionid butterfly *Battus philenor*, discrimination of leaf shape has been demonstrated and this butterfly uses leaf shape as an associatively learned signal for preferential landing on host plants [80].

Despite the fact that single rhabdome stemmata of caterpillars are very simple organs compared to the compound eye of the adult butterfly, caterpillars are able to discriminate object sizes and colours, enabling them to orient towards plant silhouettes after dropping on the ground [65, 90, 94].

(b) **Diptera**

In the case of herbivorous flies among the families Tephritidae (fruit flies) and Anthomyiidae (root maggots), the use of visual cues has been amply demonstrated [86]. For a flying *Rhagoletis pomonella* female in search of oviposition

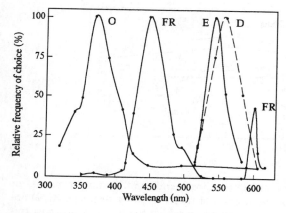
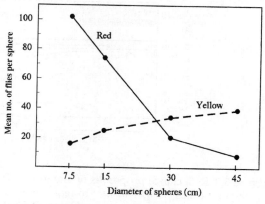

Figure 5.7 Relative effectiveness of different wavelengths (abscissa = nm) in eliciting behavioural responses from *Pieris brassicae* butterflies. The ordinate gives the relative frequency of choice (%), normalized to the maximally visited wavelength (= 100%). A so-called open space reaction (O, lack of response to plants and tendency to increase flight altitude) is induced by wavelengths in the ultraviolet; a feeding reaction (FR, extension of the proboscis) is maximally induced by blue and to a lesser extent by yellow, egg-laying (E) and drumming (D) by slightly different wavelengths in the green part of the spectrum. (Source: redrawn from Scherer and Kolb, 1987.)

Figure 5.8 Visual responses of apple maggot flies (*Rhagoletis pomonella*) to red and yellow odourless sticky spheres of increasing diameter under orchard conditions. Visual preferences were measured on the basis of catches of flies on the spheres. A red sphere of 7.5 cm diameter matches the size and colour of a ripe apple. The higher number of flies caught on larger yellow spheres is interpreted as a response to a supernormal substitute stimulus for the green of leaves, on which the flies search for aphid honeydew as a source of energy. (Source: redrawn from Prokopy, 1968.)

sites, i.e. apple fruits, the sequence of visually oriented behaviour can be described as a series of consecutive steps. At a distance of about 10 m, a single tree is perceived as a silhouette contrasting against the background. Perception of colour is unlikely at this stage, especially when the insect is facing direct sunlight, as is the perception of details of shape, because of its limited visual acuity. When the fly is at a distance of a few metres or less from the plant and finds itself either in front, under or above the tree crown, spectral quality and intensity of the reflected light are the main cues evoking alightment on e.g. foliage, fruits or trunk. At still closer range (1 m or less), as a third step, detailed discrimination on the basis of size or shape becomes possible (Fig. 5.8).

In the cabbage root fly, *Delia radicum*, visually-based landing responses occur when the flies are offered artificial leaves that have been painted with colours mimicking host-plant leaf reflectance profiles (Fig. 5.9).

In contrast to *Pieris* butterflies, no landing preference was displayed when spectrally matched artificial leaves of three different host plants were offered simultaneously with the real leaves. The flies shifted their preferences with plant age. The overriding preference for radish in the mature plant stage was much less pronounced in the young plant stage and this correlated with smaller differences in reflectance properties between the three host plants. After alightment, during the next phases of host selection, leaf shape does not seem to influence oviposition, but artificial leaves possessing a stem are clearly preferred over those lacking one (Fig. 5.10).

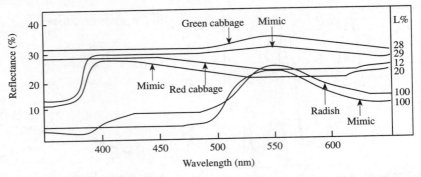

Figure 5.9 Reflectance properties of leaves of different cruciferous host plants and landing responses of cabbage root flies (*Delia radicum*) to real leaves or artificial mimics of these. Landing responses (L%) are expressed as the percentage of landings relative to radish, the plant on which the flies landed most frequently in a multiple choice test. Alternative host plants were green cabbage and red cabbage. In direct comparisons of real leaves and their mimics, flies landed with equal frequencies on both. Vertical axis = % reflectance of incident light. (Source: redrawn from Prokopy et al., 1983a.)

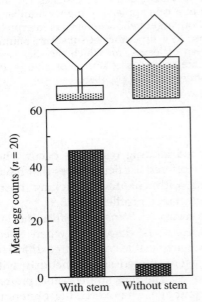

Figure 5.10 Effect of the presence of a stem as a morphological feature of artificial leaves on oviposition preference of cabbage root flies (*Delia radicum*). Artificial leaves (13×13 cm pieces of green paper dipped in paraffin and sprayed with a surface extract of cabbage leaves) of each type were offered together in the same test arena. The type with a stem is significantly preferred for oviposition at its base. (Source: redrawn from Roessingh and Städler, 1990.)

When the flies were allowed to choose between different sizes of artificial leaf, the one that was four times as big was also landed on four times as often and received 2.5 times as many eggs [89, 91]. Clearly plant colour, shape and size play important roles in the host selection behaviour of these herbivorous flies, which belong to the best-studied species in this respect.

(c) Homoptera

The attraction to the colour of foliage has been studied extensively in aphids and whiteflies [25, 58, 70]. These small insects can generate only small motoric forces and at wind speeds exceeding 1 m/s, they are unable to maintain their airspeed against the wind direction. This has led to the opinion that these insects move only passively. Alate (i.e. the winged morph) aphids can still exert control over their transport, however, by active taking-off (spreading their wings) and alighting (by folding the wings). The main factor that elicits an alighting response is the perception of plant colours. Thus *Brevicoryne brassicae* and *Myzus persicae* alight in the field preferentially on leaves

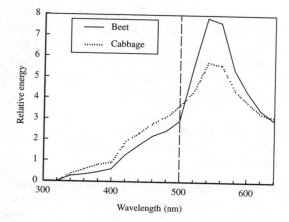

Figure 5.11 Relative energy curves of light reflected from the upper surfaces of mature leaves of sugar beet and cabbage in July under direct sunlight. The ratio between the areas under the curve to the right and left of the dividing line at 500 nm (the 'long/short ratio') is 3.2 for beet and 2.1 for cabbage. (Source: redrawn from Kennedy et al., 1961.)

reflecting a greater proportion of long-wave energy, with little or no regard for the taxonomic status of the plants. Since sugar beet leaves have a higher 'long/short-wave ratio' than cabbage leaves (Fig. 5.11), more cabbage aphids alight on sugar beet leaves than on cabbage, although the former is not one of its hosts. 'Long/short-reflectance ratios' change with leaf age and water status. The colour attraction of these 'yellow-sensitive' aphid species serves to bias their landings towards plants of the appropriate physiological type rather than to recognize their host-plant species [58].

Likewise, *Aphis fabae*, which alights three times as frequently on *Beta* plants as on reed (*Phragmites communis*), has a preference for saturated yellow, which more closely resembles the reflectance profile of *Beta* leaves (Fig. 5.12).

The mealy plum aphid *Hyalopterus pruni* displays so-called host alternation (section 7.4.1) between its summer host *Phragmites* and its winter host *Prunus* spp. Alates, which

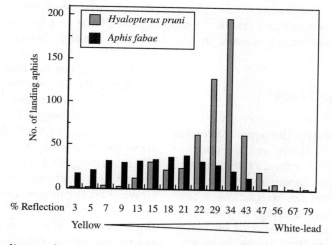

Figure 5.12 Visual landing preferences of two aphid species (*Hyalopterus pruni*, the mealy plum aphid, and *Aphis fabae*, the black bean aphid). Preference was measured as the number of alate aphids that landed on each of 16 plates, which together constituted a graded series from yellow to white colours with increasing reflection in the short wavelength band (decreasing saturation from left to right). (Source: redrawn from Moericke, 1969.)

search for *Phragmites* in the spring, alight twice as often on reed plants than on adjacent non-host beet (*Beta vulgaris*) plants [70]. Discrimination between these two plant species is done in this case on the basis of a lower degree of saturation of the yellow reflectance of the *Phragmites* blades as compared to that of *Beta* leaves. Thus visually-based response to colours and reflectance intensity is species-specific. Whiteflies avoid settling in the presence of short-wavelength illumination (400 nm), but will alight on green light (550 nm) [25].

Not only homopterans but also insects belonging to other taxonomic groups use differences in reflectance intensity between plant species, or between leaves or organs within a plant as a visual selection criterion for more nutritious tissues. These are often younger tissues, which display a relatively strong reflection in the yellow region. In fact, most diurnal insects are attracted to yellow. In many cases yellow surfaces act as a 'supernormal' stimulus, because they emit peak energy in the same bandwidth as foliage, but at greater intensity.

Although there is a large body of information on the mechanisms of insect photoreception, our knowledge on the visual performance of herbivorous species in the field is meagre.

5.4.3 OLFACTORY RESPONSES TO HOST PLANTS

When attempting to test the separate role of visual stimuli, test insects are exposed to objects with controlled optical characteristics, which are odourless. Conversely, to assess the effect of odours alone on orientation to host plants, the visual surroundings in which the odour tests are carried out should be homogeneous. For experiments in the laboratory several setups have been developed that allow quantitative studies of orientation responses to odours [40] (see Appendix C). As discussed above (section 5.4.1), control over an odorous stimulus in terms of concentration and distribution is usually less exact than is often assumed. We will discuss two examples of orientation to odours as demonstrated under laboratory circumstances, one of a flying insect and one of a walking insect.

5.4.4 FLYING MOTHS AND WALKING BEETLES: TWO CASES OF OLFACTORY ORIENTATION

When a flying female tobacco hornworm moth (*Manduca sexta*) is searching for a host plant, she displays positive anemotaxis, i.e. she flies upwind using the prevailing direction of air flow as a cue. Mechanoreceptors located on her antenna and serving as anemoreceptors provide this directional information (either by klino- or tropotaxis; section 5.3). Her flight path can be described as a regular zigzag (a series of counterturns) of limited amplitude.

How does the odour emitted by the tobacco plant come into play? Firstly, the host-plant odour may have acted as an activator (arousing agent) for flight to occur, by inducing the moth to take off from a resting or walking condition. Once in flight, she may pick up an odour plume emanating from one or a group of host plants and her subsequent flight path is then mainly determined by trying to prevent loss of the odour plume. When over a certain minimum time interval olfactory receptor cells do not detect odour, a so-called 'casting' response ensues. The moth reduces speed and increases the amplitude of the counterturns, thereby flying more across wind and regressing in a downwind direction. When during casting odour molecules are picked up again by the olfactory sensilla, upwind zigzagging is resumed. This sequence of behavioural acts may be reiterated until final approach of the host plant. Closer to the odour source the intervals between counterturns decrease. This host-searching mechanism is designated as odour-conditioned (or odour-modulated) positive anemotaxis.

The female's host-plant searching behaviour

is in fact very similar to odour-modulated upwind flight of male moths in search of a female [7]. In the latter case the odorous signal is a sex pheromone emitted by the female. A present view of the mechanisms steering this behaviour maintains that the serial counterturning is controlled by a motor programme in the central nervous system that is set in motion by olfactory activity, but afterwards is continued automatically (self-steered) [129]. The switch from zigzagging to casting, however, is controlled by olfactory information: absence of activity changes in the odour receptors over a certain minimum time span causes casting behaviour. Upwind progress is made possible by optomotor feedback, that is the flow of visual images of the surroundings, mainly the ground, controls the motor response *via* a feedback loop.

The female is able to maintain the parameters of its flight path (ground speed, track angle) and counterturning frequency close to some apparently preferred values over a range of wind speeds. Odour-conditioned anemotactic flight enables directed flight to an odour source and is basically different from the relatively straightforward chemotactic orientation to odour gradients. It probably has evolved because, as we have seen, such gradients do not exist over any distance in the field. It may be noted that in contrast to the vast literature on behavioural responses to sex pheromones in male Lepidoptera, information is still scanty for members of this and other insect orders on orientation mechanisms to plant odours in the field [129].

One of the best studied cases of the ability of a walking insect to orient to host-plant odours is the Colorado potato beetle, *Leptinotarsa decemlineata* [122]. This specialist on solanaceous plants has a strong preference for the cultivated potato, *Solanum tuberosum*, on which it is one of the most devastating insect pests. During the first 7 days of adult life the beetles need to feed in order to fully develop their flight muscles and, as a consequence, host-plant location is done by walking. To quantify its walking behaviour a 'locomotion compensator' in combination with a wind tunnel has been used. This instrument allows detailed and automated recording of walking tracks without the insect contacting any obstacles (Appendix C).

When clean air is blown over a hungry beetle, it shows a menotactic response to the wind (anemotaxis), maintaining a relatively constant angle to the wind direction (Fig. 5.13). The walking track shows circling by making turns of 360°. When the airstream carries the odour of intact potato plants, the straightness of the path increases dramatically. Now circling is absent, average walking speed is increased and the beetles spend more time walking upwind. This response can be classified as positive (i.e. upwind) odour-conditioned anemotaxis. When the odour of non-hosts, for instance cabbage plants, is offered, the track parameters are similar to those recorded for clean air. When the odour of potato plants is combined with that of cabbage plants, the orientation response to potato is neutralized and the walking tracks of the beetles cannot be distinguished from those performed in clean air (Fig. 5.13).

Somewhat unexpectedly, similar effects were found when the odour of another solanaceous plant, wild tomato (*Lycopersicon hirsutum* f. *glabratum*) was offered. This plant is an unsuitable food for the beetle (Fig. 5.14). Despite the taxonomic relatedness of tomato to potato and therefore a presumed similarity in their odours, mixtures of their volatiles were not attractive to the beetles. The phenomenon that the presence of tomato odour prevents the beetles from orienting to their host plants has been termed 'odour-masking' [106]. It has been suggested that this phenomenon plays a role in the reduced population levels of herbivorous insects in mixed cropping systems (Chapter 12).

Positive odour-conditioned optomotor anemotaxis and olfactory-induced visual orientation are presently considered as the main mechanisms used during host-plant searching

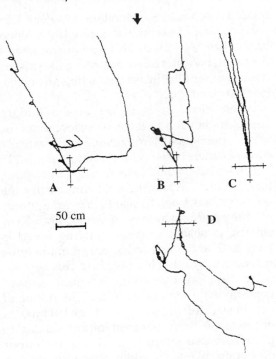

Figure 5.13 Walking tracks of an individual female Colorado potato beetle (*Leptinotarsa decemlineata*) during four consecutive periods **(A)–(D)** of 10 min. The stimulus situations were: **(A)** clean air stream; **(B)** air stream carrying the odour of potted cabbage (*Brassica oleracea*) plants; **(C)** air stream carrying the odour of potato (*Solanum tuberosum*) plants (the favourite host plant of the beetle); **(D)** air stream carrying a mixture of odours emanating from cabbage and potato. The arrow indicates the direction of the air stream. The plotter reset the position of the beetle to the origin (centre of cross) after a certain maximum distance had been travelled. Total distance travelled and track straightness are for **(C)** significantly higher than for the other three situations, which do not differ from each other. (Source: reproduced from Thiéry and Visser, 1986, with permission.)

in herbivorous insects, both in specialized and polyphagous species [100, 122]. In addition, there is evidence that chemotaxis occurs within ranges of a few centimetres from the host plant, as has been demonstrated for several caterpillars and various root feeding insects [59, 76]. Table 5.3 presents a summary of data on behavioural responses to plant odours in adults of herbivorous species belonging to four major orders. In each order food specialists have been found to respond to identified odours specific to their host plant.

5.5 CHEMOSENSORY BASIS OF HOST-PLANT ODOUR DETECTION

Insects heavily rely upon chemoreception when searching for food, oviposition sites and mating partners, as well as for social communication. In this context it is often stated that 'insects live in a chemical world'. Chemoreception refers to the classical senses of smell (olfaction, organs for detecting volatile chemical stimuli) and taste (gustation, or 'contact chemoreception' for the detection of dissolved or solid chemicals; Chapter 6). The distinction between the two is not absolute as insect taste sensilla have occasionally been found to respond also to odours [101].

5.5.1 MORPHOLOGY OF OLFACTORY SENSILLA

Olfactory chemoreceptor cells are associated with so-called **sensilla** (singular: sensillum), organs consisting of neurons, accessory cells and a cuticular structure (Fig. 5.15).

The cell bodies (perikarya) of the neurons are located at the base of the sensillum. Typically there are two to three neurons per sensillum, but examples of more (up to 30) neurons innervating one sensillum have also been reported [75, 130]. The dendrites are usually located in specialized cuticular structures, which are classified on the basis of external form. They include hairlike varieties (sensilla trichodea), pegs and cones (sensilla basiconica), pegs or cones sunk in shallow depressions (sensilla coeloconica) and pore plate organs (sensilla placodea). Chemosensory neurons are mostly bipolar and their axons run to the CNS *via* peripheral nerves without intermittent synapses. A filament-like

Chemosensory basis of host-plant odour detection 139

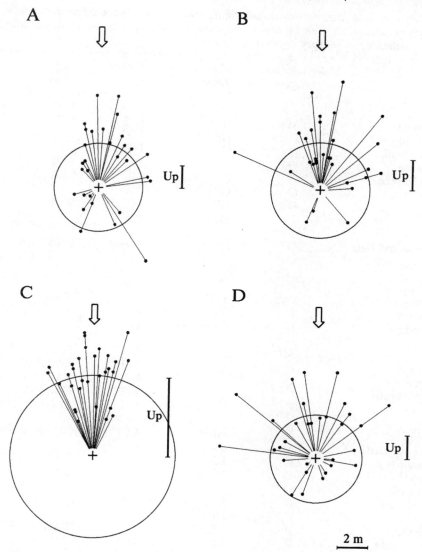

Figure 5.14 Distribution of displacement vectors of 31 female Colorado potato beetles after 10 min of walking upon exposure to four successive stimulus situations: **(A)** wind control (clean air); **(B)** odour released by tomato plants (*Lycopersicon hirsutum*); **(C)** potato (*Solanum tuberosum*) odour **(D)** odour blend of tomato and potato. The mean vector length in each situation is shown as the radius of the circle, Up = the upwind component of the mean vector; arrows indicate wind direction. Displacement vectors are a way to summarize the walking tracks shown in Figure 5.13. Vector length, straightness and upwind length are significantly higher in **(C)** than in the other three situations. (Source: redrawn from Thiéry and Visser, 1987.)

Table 5.3 Selected cases of adult herbivores belonging to four major insect orders that display behavioural responses to plant odours; the insect's host-plant specificity, the type of odour source, the test environment and the availability of sensory data are indicated

Order species	Specialization category	Odour source	Type of test environment	Sensory data	Reference
Hemiptera					
Phorodon humuli	M	G	L/F	SCR	19
Cryptomyzus korschelti	O	HP	L		126
Cavariella aegopodii	O	G	F(T)		24
Lipaphis erysimi	O	S	L	SCR	78
Brevicoryne brassicae	O	S	L(F−)	SCR	78, 82
Rhopalosiphum padi	O	G	L(F−)		81
Aphis fabae	P	HP	L(F−)	SCR	78, 57
Aphis gossypii	P	HP	F		84
Coleoptera					
Leptinotarsa decemlineata	O	HP/G	L(F−)	EAG/SCR	63, 120, 105
Anthonomus grandis	O	G	L/F	EAG/SCR	34, 35, 36
Ips typographus	O	G	L/F	SCR	69, 110
Phyllotreta spp.	O	S	L/F		83
Ceutorhynchus assimilis	O	S/HP	L/F	SCR	14, 38
Popillia japonica	P	G	F		1
Listroderes obliquus	P	G/S*	L		64
Diptera					
Psila rosae	M	S	L/F	EAG	44, 45, 77
Delia antiqua	O	S	L/F	EAG/SCR	43, 51, 55
Delia radicum	O	S	L/F	EAG	43, 48, 77
Rhagoletis pomonella	O	G	L/F	EAG	39, 42
Dacus dorsalis	P	G	L/F	EAG	67
Lepidoptera					
Heliothis subflexa	M	HPE	L		109
Acrolepiopsis assectella	M	S	L		104
Plutella xylostella	O	HPE	L	EAG	79
Manduca sexta	O	G/HPE	L	EAG	107
Papilio polyxenes	O	G	L	EAG	9
Heliothis virescens	P	HPE	L		52, 53, 108
Trichoplusia ni	P	HP	L		61
Ostrinia nubilalis	P	HP/G	L	EAG	20, 114
Spodoptera littoralis	P	HP	L		93

M = monophagous; O = oligophagous; P = polyphagous; HP = intact (host) plants; HPE = host-plant extract; G = generally occurring green leaf volatiles; S = volatile(s) specific to the host plant taxon; L = behavioural test in the laboratory, in an olfactometer or a wind tunnel; F = field test, either trap catches (F(T)) or direct observations; (F−) = behavioural responses to the odour source attractive under laboratory conditions could not be demonstrated under field conditions; EAG data on sensory perception of volatiles from the odour source available on the electroantennogram (EAG); SCR data on sensory perception of volatiles from the odour source available at single-cell level
*The specific volatiles were isothiocyanates, which are characteristic for Cruciferae, one of the preferred host-plant families

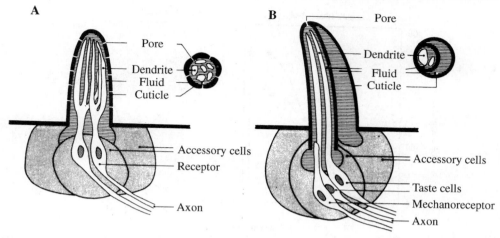

Figure 5.15 Schematic drawing of a longitudinal and a transverse section of **(A)** an insect olfactory hair and **(B)** an insect taste hair. The olfactory hair is innervated by two bipolar neurons; the taste hair is innervated by two chemoreceptors and one mechanoreceptor. (Source: reproduced by courtesy of Dr F. W. Maes, Groningen State University, The Netherlands.)

extension of the neuron that protrudes into the sensillum cavity, the dendrite, is specialized to respond to the chemical stimulus with a graded potential called the receptor potential. When this potential reaches a value above a certain threshold, it gives rise to a train of action potentials.

There are some important structural differences between olfactory and gustatory sensilla. Olfactory sensilla are multiporous, the entire sensillum wall or plate is perforated by up to thousands of minute pores (diameter about 10–50 nm) and dendrites are often branched. In contrast, gustatory sensilla are uniporous, the pore (diameter 200–400 nm) mostly being located at the very tip of a peg-, hair-, or papilla-like sensillum (Fig. 5.15). In both cases the dendritic tips are close to the pores, but are protected from desiccation by receptor lymph, which is secreted into the sensillum lumen by the tormogen and trichogen cells at the sensillum base. Olfactory sensilla are predominantly present on antennae but may also occur on palpi and ovipositor. The number of olfactory sensilla and the olfactory receptor cells associated with them is quite variable between species. Larvae of holometabolous insects have only small numbers of olfactory cells (for instance, less than ten neurons for beetle larvae and about 100 for caterpillars). For female adults this number amounts to up to a few hundreds in Hemiptera, while for Lepidoptera it varies between 6500 and 177 000 (in female *Manduca sexta*) per antenna [22]. The sensilla that house olfactory receptor cells may be multimodal, i.e. they may also contain thermo-, hygro- and mechanoreceptors.

The transduction process, i.e. the process by which quality and quantity of the chemical stimulus is converted into a receptor potential and eventually into action potentials, is thought to be similar for gustatory and olfactory receptor cells. It involves a sequence of steps, starting with the diffusion of stimulus molecules into the sensillum lumen *via* the pore(s) in the sensillum wall and, when the magnitude of the receptor potential exceeds a threshold, resulting in the generation of action potentials. When receptor potentials occur simultaneously in many olfactory cells, a summated receptor potential, the so-called elec-

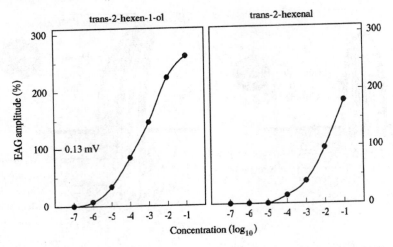

Figure 5.16 Relationship between concentration of two green leaf volatiles and EAG response intensity evoked in antennae of female Colorado potato beetles. Concentration is expressed as the dilution (v/v) in paraffin oil. EAG response is expressed relative to the response to a standard dose (10^{-3} or $1\,\mu l/ml$) of another green leaf volatile, cis-3-hexen-1-ol. (Source: redrawn from Visser, 1979.)

troantennogram (EAG; Appendix C) may be recorded. Action of stimulus molecules is most probably terminated by deactivation molecules present in the sensillar lymph [102].

5.5.2 SENSITIVITY OF OLFACTORY SENSILLA

Like most sensory cells, chemoreceptors are especially responsive to changes in stimulus intensity, i.e. changes in the concentrations of chemicals. Olfactory cells have been shown to handle up to ten odour pulses per second [74], which means that they can resolve the temporal pattern of odour bursts in a plume quite well (Fig. 5.5). Responses to constant stimuli generally show sigmoid concentration–response relationships at the level of EAGs as well as single cell recordings (Figs 5.16 and 5.17).

Upon increasing the odour concentration by one order of magnitude, EAG amplitude and frequency of action potentials typically become 1.5–3 times higher until saturating concentrations are reached, above which no further increase occurs. The discrimination of concentration differences is optimal in the range between threshold and saturating concentrations, i.e. the rising phase of the dose–response curves (Fig. 5.16 and 5.17). This in principle enables the insect to sense odour gradients, on the basis of which it may perform tropotactic behaviour.

Sensitivity of detection is enhanced enormously by the neural phenomenon of **convergence**. The axons running from olfactory receptor cells make synaptic contacts with a limited number of first-order interneurons in the antennal lobe of the brain, which means that they converge [50]. This leads to an amplification of the signal (an interneuron receives inputs from many receptor cells simultaneously and its threshold may therefore be reached at a lower concentration than that necessary to depolarize a receptor cell) and improved signal–noise ratio by separating background activity from the presence of a volatile signal. Thus, a 100–1000-fold amplification of the signal can be measured in the deuterocerebral interneurons responding to green leaf volatiles in the Colorado potato beetle, as compared to the sensitivity of its antennal receptors [26].

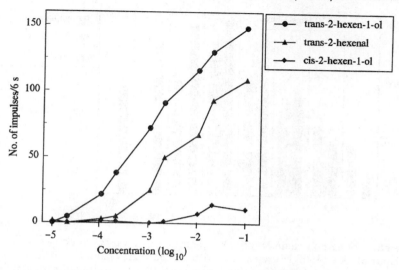

Figure 5.17 Relationship between concentration of three green leaf volatiles and the response of single olfactory neurones innervating sensilla basiconica on the antenna of female Colorado potato beetles. Concentration is expressed as the dilution (v/v) in paraffin oil. (Source: redrawn from Ma and Visser, 1978.)

The interneurons are organized in so-called glomeruli. When the number of output neurons that send their axons from out of the glomeruli to other brain centres are compared to the number of olfactory cells, a convergence ratio can be calculated, higher values of which are predicted to be associated with higher sensitivity. For *Locusta migratoria* this ratio is 60, for the omnivorous cockroach *Periplaneta americana* 750 and for *Manduca sexta* 3000 (see Visser [122] for review).

5.5.3 SPECIFICITY AND OLFACTORY CODING

EAG recordings (for techniques see Appendix C) from several species of herbivorous insect have shown that generally occurring plant volatiles, such as green leaf alcohols, aldehydes, acetates and terpenoids evoke different responses [122]. The responses increase in amplitude with increasing carbon-chain lengths of the alcohols and aldehydes until an optimum is reached at or near C_6-compounds, which at the same time are the most abundant chemicals in leaf headspaces (Fig. 5.18).

Studies in which EAG response profiles to a number of generally occurring plant volatiles have been correlated with the degree of host-plant specificity for several related or unrelated insect species have failed to demonstrate such a correlation [121]. Thus, no relationships between host preferences and general plant volatiles were found in five unrelated lepidopterans [111], four unrelated locust species [128] or nine related *Yponomeuta* spp. [116]. Conversely, three oligophagous dipterans (*Delia antiqua*, *Delia brassicae* and *Psila rosae*) that use host-plant-specific volatiles to locate their hosts did show EAG responses to the volatiles characteristic of each of their respective host plants that correlate well with their behavioural specificity (Fig. 5.19).

The EAG technique can also be used to demonstrate that previous exposure to odours has left a (quantitative) effect on olfactory sensitivity [118]. Furthermore, EAGs can be employed to analyse whether or not closely related species choose different host plants because of different messages or because of differences in interpretation of the same

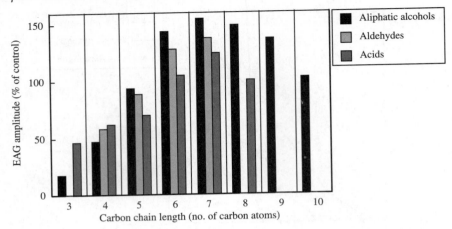

Figure 5.18 The effect of carbon chain length on strength of mean EAG responses to saturated aliphatic alcohols (black bars), aldehydes (light grey bars) and acids (dark grey bars) in females of seven *Yponomeuta* species and *Adoxophyes orana*. Missing bars = not measured. (Source: redrawn from van der Pers, 1981.)

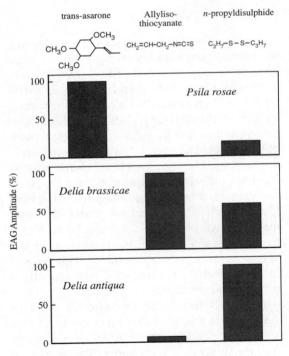

Figure 5.19 EAG responses of three herbivorous Diptera to plant volatiles expressed as a percentage of the response to the respective host-plant attractant. (Source: redrawn from Guerin and Städler, 1982.)

sensory information. For instance, an EAG study of two sibling species, *Rhagoletis pomonella* and *R. mendax*, and two *R. pomonella* host-plant races has revealed differences in olfactory sensitivity to their respective host fruit extracts and identified volatiles (Fig. 5.20).

How do olfactory receptors encode the multitude of volatile chemical stimuli present in the outside world into a message that will increase the chance of finding a host plant? Rather than total antennal responses (as in EAGs) electrophysiological studies based on recordings of the activity of individual olfactory neurons, so-called 'single-cell-recording' (Appendix C), offer the possibility to determine which individual plant chemicals evoke changes in chemosensory activity (of either the excitation or inhibition type) and how mixtures of chemicals are coded. The olfactory system functions as a filter because olfactory receptors are sensitive to only a limited array of stimuli. For both olfactory and gustatory receptors two main categories are recognized: 'specialist' and 'generalist' chemoreceptors. A specialist cell responds to only a small number of structurally related compounds. Among

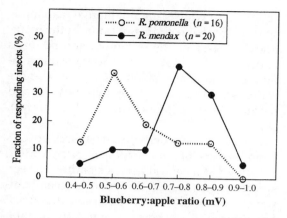

Figure 5.20 Frequency distribution of electroantennogram responses of two sibling fruit fly species to pentane extracts of apples and blueberries, respective host plants of *Rhagoletis pomonella* and *R. mendax*. Overlap of response patterns is caused by some individuals showing responses that are typical for their sibling species. (Source: redrawn from Frey and Bush, 1990.)

olfactory receptors, sex pheromone receptors are the classical example of specialist receptors. They have been studied more extensively than host-plant odour receptors or, *de facto*, any olfactory receptor in the animal kingdom. Recordings of neural responses evoked in single olfactory cells, obtained from some oligophagous species, have revealed the presence of specialist-type olfactory neurons that are sensitive to certain host-plant specific volatiles only. Such specialist receptors have been found, among others, in coleopterans [14], lepidopterous larvae [123], and aphids [78], but not in all species studied.

Volatiles emanating from plants in most cases excite generalist-type odour receptors. The response spectra of generalist receptors are by definition fairly broad and vary from cell to cell, often with overlapping patterns (Fig. 5.21).

The underlying molecular mechanism is unknown, but it has been suggested that such cells possess in their receptor membranes variable numbers of several types of receptor molecule with limited specificity [27, 102]. The

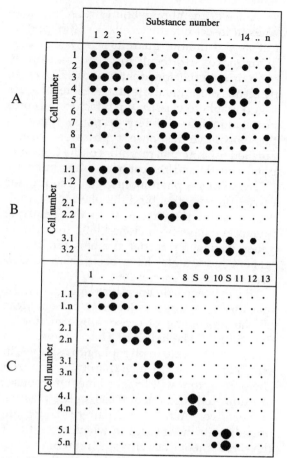

Figure 5.21 Schematic representation of the different sensitivity spectra of olfactory cells of insects. Each line represents the series of quantitative responses exhibited by the same cell to a number of substances ($1 \ldots n$). Size of dots indicates reaction strength. **(A)** Each cell has its own individual sensitivity profile, but overlaps with other cells occur (e.g. bark beetles). **(B)** Three cell types can be recognized; many cells of each type may be present. Overlaps do not occur, but within each type there is variation (e.g. honeybee). **(C)** Three overlapping cell types are discerned which, however, display little variation within one type. Cell types 4 and 5 show relatively narrow spectra and are used in labelled-line coding and may include highly sensitive and specific cells (S; e.g. Colorado potato beetle). (Source: redrawn from Boeckh and Ernst, 1983.)

existence of membrane-bound receptor molecules in insect olfactory cells is still hypothetical as the minute quantities in which they occur makes it difficult to isolate and purify enough material for identification. However, by means of multivariate statistical techniques it is possible to classify the olfactory receptor population into different response types. In this way, five response types were distinguished for antennal receptors of the Colorado potato beetle and the cabbage white butterfly [27, 63]. For sex pheromones, genes coding for carrier proteins that bind the pheromone molecules and transport them to the dendritic membrane have been cloned [60]. In addition genes coding for carrier proteins binding to 'general odourants' have been characterized in several moth species [60].

The existence of specialist and generalist olfactory receptors is reflected in present ideas on the processing of sensory information in the CNS. 'Labelled-line' codes have been inferred to operate in oligophagous species, in which the activity of such (highly) specialized chemoreceptors will trigger kinetic responses or odour-induced anemotaxis, either positive or negative. 'Across-fibre pattern' codes are operating in those cases in which the ratio of the simultaneous activity of a number of cells, be it specialists, generalists or both, is the crucial characteristic of the code. The across-fibre pattern can be visualized as a key that has to fit a hypothetical 'lock' (the image), which is localized in the brain [98]. In the relatively few instances studied in detail, across-fibre patterning seems to operate in oligophagous and polyphagous species, while a combination of labelled-line and across-fibre patterning seems to be used in the few oligophagous species for which the presence of specialist odour receptors has been demonstrated. Although both coding modes are sometimes described as mutually exclusive, they more probably represent extremes of a continuum [99]. As many plant species release a complex blend of generally occurring green leaf volatiles (section 3.8) and terpenoids into the atmosphere which lack qualitative taxonomic specificity, neural coding of ratios of the quantities released of these compounds becomes critical, as these ratios may contain information on the plant taxon [122]. Across-fibre codes are better suited for this purpose and require fewer receptors to accomplish this task [15].

In only a few herbivorous insect species have aspects of olfactory specificity, coding principles and central nervous pathways been studied in any detail. Progress has been hampered by a lack of knowledge on the actual headspace volatile composition, both qualitatively and quantitatively, which is essential for offering relevant odours at natural concentrations. In fact, the minimal blend of identified volatiles causing attraction has been successfully formulated for relatively few species [66].

The situation is further complicated by the often non-additive nature of responses to mixtures [126]. At a behavioural level, generally occurring green leaf volatiles may synergize with each other and also with taxonomically specific volatiles or with pheromones [18, 39]. Likewise, at the olfactory receptor level, interactions have been shown to occur between host-plant odour components as well as between host-plant odours and pheromones [15, 34, 35, 46, 117, 126].

5.6 HOST-PLANT SEARCHING IN NATURE

When a herbivorous insect is searching in the field, it meets a multitude of stimuli, which are distributed heterogeneously. Inherent to the field situation is a lack of control over abiotic parameters and the stimulus situation. It is therefore difficult to assess the relative importance of the two main stimulus modalities, optical and odorous plant cues. For several insect species it has been shown that significant stimulus interactions occur. During searching for food or oviposition sites, the importance of different types of stimuli may change with distance to the plant. Stimulus interactions may be one of the causes of the discrepancies indicated in Table 5.3, in which behavioural responses to odours observed in the laboratory could not be confirmed in the field.

The Colorado potato beetle, for instance, is well able to perform directed orientation in response to odours alone and uses odour-conditioned positive anemotaxis, as has been convincingly demonstrated in laboratory studies [105, 120]. Behavioural observations on host-plant searching in the field, however, have given variable results with respect to the role of odours in host-plant location. De Wilde [33] found upwind menotactic responses in the field at distances less than 6 m from a plot of potato plants (of unstated size). Jermy and coworkers [54], however, found only a low proportion of beetles moving upwind in the field and even in these cases their walking tracks did not reveal directed movement towards potato plants. The beetles showed photomenotaxis and a high directionality of movement based on light-compass orientation rather than on odour-induced anemotaxis. In the vicinity of potato plants interruptions of straight paths occurred, accompanied by an increased rate of turning. Jermy and coworkers estimated that the maximum distance at which a walking beetle could detect a single potato plant was about 60 cm, based on either olfactory cues, visual cues or a combination. However, only one of every two beetles that came within this radius of detection was attracted to the plant. Odour masking is likely to be one of the causes of the small radius of detection in a complex natural vegetation. It was concluded [54] that, under natural conditions, where individual potato plants may be scattered between non-hosts, host-plant finding is a chance event when the beetle starts at a distance of more than 60 cm from a potato plant. These findings fit well into the model of 'alternating random and non-random (kinetic arrestment-type) search strategies' formulated by Morris and Kareiva [71].

Of all herbivorous insects in which host searching behaviour is studied, the apple maggot fly (*Rhagoletis pomonella*) is probably the one analysed in most detail [3, 4, 5, 86]. Its visually-guided host searching behaviour has been described above (5.4.2). These flies are highly responsive to particular visual stimuli, but only after they have been 'activated' by apple odour. They show preferences for either yellow or red, depending on the size of the object and their motivational state (section 5.2). Spherical red objects of a limited diameter are preferred when the fly is searching for oviposition sites. In order to acquire carbohydrates, the flies feed on aphid honeydew, which is present on apple leaves. Larger yellow spheres are preferred over red ones when the motivation for carbohydrate ingestion is high. Yellow serves as a supernormal substitute stimulus for the green hue of apple leaves. Apple odour elicits upwind flight and odour-induced anemotaxis allows the flies to locate an apple-bearing tree within a patch of trees devoid of apples by a series of tree-to-tree displacements. In the same way they can find a synthetic odour source outside an odourless patch. Once at a tree bearing apples, selection of individual fruits by size or colour is done mainly visually. However, when there are few fruits or when they are green instead of red and therefore lack contrast with the leaves, odorous cues are used to aid the selection process (Fig. 5.22).

Field studies to date on host-plant searching behaviour have logically been focused on larger species, which, by virtue of their visual conspicuousness, can be directly observed and followed for some time while moving from plant to plant. As a result the oviposition behaviour of butterflies has been studied in quite some detail [12]. The picture that emerges from these studies is a predominant role of vision and associative learning involving optical and contact-chemosensory cues, promoting time- and energy-optimization of host selection behaviour.

In contrast to the situation mentioned above for fruit flies, host selection behaviour of bark beetles in forest ecosystems is largely

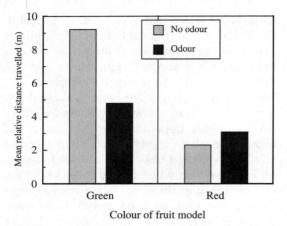

Figure 5.22 Demonstration of the interaction between olfactory and visual information in host selection behaviour of the apple maggot fly, *Rhagoletis pomonella*; the effect of host fruit odour (a synthetic blend of six esters), released at 500 µg/h, on finding green or red fruit models in a tree carrying 16 models. The response to colour of a fruit model (green or red) was significantly affected by the presence of odour. (Source: redrawn from Aluja and Prokopy, 1993.)

governed by chemical cues. Highly intricate chemical communication systems are operating based on complicated interactions between host-tree odours, aggregation pheromones produced by the beetles or associated microorganisms, and interspecific inhibitory semiochemicals [16].

5.7 CONCLUSIONS

Although our knowledge of the plant characteristics influencing host-plant searching and the ways in which insect herbivores detect and use them to their own advantage is increasing, the general picture is built upon information from a small number of relatively well-studied species. Clearly, plant factors that affect insects over some distance are difficult to manipulate experimentally in the field. In many specialized herbivores no evident orientation can be demonstrated when they are at some distance from their host plant and it appears that in order to find a suitable plant they must literally bump into it. Searching then is essentially a random process, in which the chance of an encounter is largely determined by spatial factors [21, 28, 54]. On the other hand, several specialized insect species have been observed under natural conditions to be perfectly able to integrate information from different cues, and have shown that the outcome of this integration, manifested as searching behaviour, is more complex than expected from a mere summation of responses across sensory modalities [11, 23, 47].

5.8 REFERENCES

1. Ahmad, S. (1982) Host location by the Japanese beetle: evidence for a key role for olfaction in a highly polyphagous insect. *J. Exp. Zool.*, **220**, 117–120.
2. Aikman, D. and Hewitt, G. (1972) An experimental investigation of the rate and form of dispersal in grasshoppers. *J. Appl. Ecol.*, **9**, 807–817.
3. Aluja, M. and Prokopy, R. (1992) Host search behaviour by *Rhagoletis pomonella* flies: inter-tree movement patterns in response to wind-borne fruit volatiles under field conditions. *Physiol. Entomol.*, **17**, 1–8.
4. Aluja, M. and Prokopy, R. (1993) Host odor and visual stimulus interactions during intratree host finding behavior of *Rhagoletis pomonella* flies. *J. Chem. Ecol.*, **19**, 2671–2696.
5. Aluja, M., Prokopy, R. J., Buonaccorsi, J. P. and Cardé, R. T. (1993) Wind tunnel assays of olfactory responses of female *Rhagoletis pomonella* flies to apple volatiles: effect of wind speed and odour release rate. *Entomol. Exp. Appl.*, **68**, 99–108.
6. Atkins, M. D. (1980) *Introduction to Insect Behavior*, Macmillan, New York.
7. Baker, T. C. (1988) Pheromones and flight behavior, in *Insect Flight*, (eds G. Goldsworthy and C. Wheeler), CRC Press, Boca Raton, FL, pp. 231–255.
8. Barton Browne, L. (1993) Physiologically induced changes in resource oriented behavior. *Annu. Rev. Entomol.*, **38**, 1–25.
9. Baur, R., Feeny, P. and Städler, E. (1993) Oviposition stimulants for the black swallowtail but-

terfly: identification of electrophysiologically active compounds in carrot volatiles. *J. Chem. Ecol.*, **19**, 919–937.

10. Bell, W. J. (1984) Chemo-orientation in walking insects, in *Chemical Ecology of Insects*, (eds W. J. Bell and R. T. Cardé), Chapman & Hall, New York, pp. 93–109.
11. Bell, W. J. (1991) *Searching Behaviour. The Behavioural Ecology of Finding Resources*, Chapman & Hall, London.
12. Bernays, E. A. and Chapman, R. F. (1994) *Host-Plant Selection by Phytophagous Insects*, Chapman & Hall, New York.
13. Blaakmeer, A., Geervliet, J. B. F., van Loon, J. J. A., Posthumus, M. A., van Beek, T. A. and de Groot, Æ. (1994) Comparative headspace analysis of cabbage plants damaged by two species of *Pieris* caterpillars: consequences for inflight host location by *Cotesia* parasitoids. *Entomol. Exp. Appl.*, **73**, 175–182.
14. Blight, M. M., Pickett, J. A., Wadhams, L. J. and Woodcock, C. M. (1989) Antennal responses of *Ceutorhynchus assimilis* and *Psylliodes chrysocephala* to volatiles from oilseed rape. *Aspects Appl. Biol.*, **23**, 329–334.
15. Boeckh, J. and Ernst, K. D. (1983) Olfactory food and mate recognition, in *Neuroethology and Behavioral Physiology*, (eds F. Huber and H. Markl), Springer-Verlag, Berlin, pp. 78–94.
16. Byers, J. A. (1995) Host-tree chemistry affecting colonization of bark beetles, in *Chemical Ecology of Insects*, vol. 2, (eds R. T. Cardé and W. J. Bell), Chapman & Hall, New York, pp. 154–213.
17. Byers, J. A., Anderbrant, O. and Löfqvist, J. (1989) Effective attraction radius: a method for comparing species attractants and determining densities of flying insects. *J. Chem. Ecol.*, **15**, 749–765.
18. Byers, J. A., Birgersson, G., Löfqvist, J., Appelgren, M. and Bergström, G. (1990) Isolation of pheromone synergists of bark beetle, *Pityogenes chalcographus*, from complex insect–plant odors by fractionation and subtractive-combination bioassay. *J. Chem. Ecol.*, **16**, 861–876.
19. Campbell, C. A. M., Pettersson, J., Pickett, J. A., Wadhams, L. J. and Woodcock, C. M. (1993) Spring migration of damsonhop aphid *Phorodon humuli* (Homoptera, Aphididae), and summer host plant-derived semiochemicals released on feeding. *J. Chem. Ecol.*, **19**, 1569–1576.
20. Cantelo, W. W. and Jacobson, M. (1979) Corn silk volatiles attract many pest species of moths. *J. Environ. Sci. Health*, **A14**, 695–707.
21. Cappuccino, N. and Kareiva, P. M. (1985) Coping with capricious environment: a population study of a rare pierid butterfly. *Ecology*, **66**, 152–161.
22. Chapman, R. F. (1982) Chemoreception: the significance of sensillum numbers. *Adv. Insect Physiol.*, **16**, 247–356.
23. Chapman, R. F. (1988) Odors and the feeding behavior of insects, in *ISI Atlas of Science, Animal and Plant Sciences*, pp. 208–212.
24. Chapman, R. F., Bernays, E. A. and Simpson, S. J. (1981) Attraction and repulsion of the aphid, *Cavariella aegopodii*, by plant odors. *J. Chem. Ecol.*, **7**, 881–888.
25. Coombe, P. E. (1982) Visual behaviour of the greenhouse whitefly *Trialeurodes vaporariorum*. *Physiol. Entomol.*, **7**, 243–251.
26. De Jong, R. and Visser, J. H. (1988) Integration of olfactory information in the Colorado potato beetle brain. *Brain Res.*, **447**, 10–17.
27. Den Otter, C. J., Behan, M. and Maes, F. W. (1980) Single cell responses in female *Pieris brassicae* (Lep., Pieridae) to plant volatiles and conspecific egg odours. *J. Insect Physiol.*, **26**, 465–472.
28. Dethier, V. G. (1959) Foodplant distribution and density and larval dispersal as factors affecting insect populations. *Can. Entomol.*, **91**, 581–596.
29. Dethier, V. G. (1963) *The Physiology of Insect Senses*, Methuen, London.
30. Dethier, V. G. (1982) Mechanisms of host plant recognition. *Entomol. Exp. Appl.*, **31**, 49–56.
31. Dethier, V. G. (1989) Patterns of locomotion of polyphagous arctiid caterpillars in relation to foraging. *Ecol. Entomol.*, **14**, 375–386.
32. Dethier, V. G., Barton Browne, L. and Smith, C. N. (1960) The designation of chemicals in terms of the responses they elicit from insects. *J. Econ. Entomol.*, **53**, 134–136.
33. De Wilde, J. (1976) The olfactory component in host-plant selection in the adult Colorado beetle (*Leptinotarsa decemlineata* Say). *Symp. Biol. Hung.*, **16**, 291–300.
34. Dickens, J. C. (1984) Olfaction in the boll weevil, *Anthonomus grandis* Boh. (Coleoptera, Curculionidae), electroantennogram studies. *J. Chem. Ecol.*, **10**, 1759–1785.
35. Dickens, J. C. (1986) Orientation of boll weevil *Anthonomus grandis* Boh. (Coleoptera, Curculionidae), to pheromone and volatile host

compound in the laboratory. *J. Chem. Ecol.*, **12**, 91–98.

36. Dickens, J. C. (1990) Specialized receptor neurons for pheromones and host plant odors in the boll weevil, *Anthonomus grandis* Boh. (Coleoptera, Curculionidae). *Chem. Senses*, **15**, 311–331.
37. Douwes, P. (1968) Host selection and host finding in the egg-laying female *Cidaria albulata* L. (Lep. Geometridae) *Opusc. Entomol.*, **33**, 233–279.
38. Evans, K. A. and Allen-Williams, L. J. (1993) Distant olfactory response of the cabbage seed weevil, *Ceutorhynchus assimilis*, to oilseed rape odour in the field. *Physiol. Entomol.*, **18**, 251–256.
39. Fein, B. L., Reissig, W. H. and Roelofs, W. L. (1982) Identification of apple volatiles attractive to the apple maggot, *Rhagoletis pomonella*. *J. Chem. Ecol.*, **8**, 1473–1487.
40. Finch, S. (1986) Assessing hostplant finding by insects, in *Insect–Plant Interactions*, (eds J. R. Miller and T. A. Miller), Springer-Verlag, New York, pp. 23–63.
41. Finch, S. and Skinner, G. (1982) Upwind flight by the cabbage root fly, *Delia radicum*. *Physiol. Entomol.*, **7**, 387–399.
42. Frey, J. E. and Bush, G. L. (1990) *Rhagoletis* sibling species and host races differ in host odor recognition. *Entomol. Exp. Appl.*, **57**, 123–131.
43. Guerin, P. M. and Städler, E. (1982) Host odour perception in three phytophagous Diptera: a comparative study, in *Proceedings of the 5th International Symposium on Insect–Plant Relationships, Wageningen, 1982*, (eds J. H. Visser and A. K. Minks), Pudoc, Wageningen, pp. 95–105.
44. Guerin, P. M. and Städler, E. (1984) Carrot fly cultivar preferences: some influencing factors. *Ecol. Entomol.*, **9**, 413–420.
45. Guerin, P. M. and Visser, J. H. (1980) Electroantennogram responses of the carrot fly *Psila rosae*, to volatile plant components. *Physiol. Entomol.*, **5**, 111–119.
46. Hansson, B. S., van der Pers, J. N. C. and Löfqvist, J. (1989) Comparison of male and female olfactory cell response to pheromone compounds and plant volatiles in the turnip moth, *Agrotis segetum*. *Physiol. Entomol.*, **14**, 147–155.
47. Harris, M. O. and Foster, S. P. (1995) Behavior and Integration, in *Chemical Ecology of Insects*, vol. 2, (eds R. T. Cardé and W. J. Bell), Chapman & Hall, New York, pp. 3–46.
48. Hawkes, C. and Coaker, T. H. (1976) Behavioural responses to host plant odours in adult cabbage root fly (*Erioischia brassicae* (Bouché)). *Symp. Biol. Hung.*, **16**, 85–89.
49. Hesler, L. S., Lance, D. R. and Sutter, G. R. (1994) Attractancy of volatile non-pheromonal semiochemicals to northern corn rootworm beetles (Coleoptera, Chrysomelidae) in eastern South Dakota. *J. Kansas Entomol. Soc.*, **67**, 186–192.
50. Hildebrand, J. G. and Montague, R. A. (1986) Functional organization of olfactory pathways in the central nervous system of *Manduca sexta*, in *Mechanisms in Insect Olfaction*, (eds T. L. Payne, M. C. Birch and C. E. J. Kennedy), Clarendon Press, Oxford, pp. 279–285.
51. Honda, I., Ishikawa, Y. and Matsumoto, Y. (1987) Electrophysiological studies on the antennal olfactory cells of the onion fly, *Hylemya antiqua* Meigen (Diptera, Anthomyiidae). *Appl. Entomol. Zool.*, **22**, 417–423.
52. Jackson, D. M., Severson, R. F., Johnson, A. W., Chaplin, J. F. and Stephenson, M. G. (1984) Ovipositional response of tobacco budworm moths (Lepidoptera, Noctuidae) to cuticular chemical isolates from green tobacco leaves. *Environ. Entomol.*, **13**, 1023–1030.
53. Jackson, D. M., Severson, R. F., Johnson, A. W. and Herzog, G. A. (1986) Effects of cuticular duvane diterpenes from green tobacco leaves on tobacco budworm (Lepidoptera, Noctuidae). *J. Chem. Ecol.*, **12**, 1349–1359.
54. Jermy, T., Szentesi, Á. and Horváth, J. (1988) Host plant finding in phytophagous insects: the case of the Colorado potato beetle. *Entomol. Exp. Appl.*, **49**, 83–98.
55. Judd, J. G. R. and Borden, J. (1989) Distant olfactory response of the onion fly, *Delia antiqua*, to host-plant odour in the field. *Physiol. Entomol.*, **14**, 429–441.
56. Kennedy, J. S. (1986) Some current issues in orientation to odour sources, in *Mechanisms in Insect Olfaction*, (eds T. L. Payne, M. C. Birch and C. E. J. Kennedy), Clarendon Press, Oxford, pp. 11–25.
57. Kennedy, J. S., Booth, C. O. and Kershaw, W. J. S. (1959) Host finding by aphids in the field. II. *Aphis fabae* Scop. (gynoparae) and *Brevicoryne brassicae* L., with a reappraisal of the role of host finding behaviour in virus spread. *Ann. Appl. Biol.*, **47**, 424–444.

58. Kennedy, J. S., Booth, C. O. and Kershaw, W. J. S. (1961) Host finding by aphids in the field. III. Visual attraction. *Ann. Appl. Biol.*, **49**, 1–21.
59. Klingler, J. (1958) Die Bedeutung der Kohlendioxyd Ausscheidung der Wurzeln für die Orientierung der Larven von *Otiorhynchus sulcatus* F. und anderer bodenbewohnender phytophager Insektenarten. *Mitt. Schweiz. Entomol. Ges.*, **31**, 205–269.
60. Krieger, J., Ganzle, H., Raming, K. and Breer, H. (1993) Odorant binding proteins of *Heliothis virescens*. *Insect Biochem. Mol. Biol.*, **23**, 449–456.
61. Landolt, P. J. (1989) Attraction of the cabbage looper to host plants and host plant odor in the laboratory. *Entomol. Exp. Appl.*, **53**, 117–124.
62. Langer, H., Hamann, B. and Meinecke, C. C. (1979) Tetrachromatic visual system in the moth *Spodoptera exempta* (Insecta, Noctuidae). *J. Comp. Physiol.*, **129**, 235–239.
63. Ma, W. C. and Visser, J. H. (1978) Single unit analysis of odour quality coding by the olfactory antennal receptor system of the Colorado beetle. *Entomol. Exp. Appl.*, **24**, 520–533.
64. Matsumoto, Y. (1970) Volatile organic sulfur compounds as insect attractants with special reference to host selection, in *Control of Insect Behavior by Natural Products*, (eds D. L. Wood, R. M. Silverstein and M. Nakajima), Academic Press, New York, pp. 133–160.
65. Meisner, J. and Ascher, K. R. S. (1973) Attraction of *Spodoptera littoralis* larvae to colours. *Nature*, **242**, 332–334.
66. Metcalf, R. L. and Metcalf, E. R. (1992) *Plant Kairomones in Insect Ecology and Control*, Chapman & Hall, New York.
67. Metcalf, R. L., Ferguson, J. E., Lampman, R. and Andersen, J. F. (1987) Dry cucurbitacin-containing baits for controlling diabroticite beetles (Coleoptera, Chrysomelidae). *J. Econ. Entomol.*, **80**, 870–875.
68. Miller, J. R. and Strickler, K. L. (1984) Finding and accepting host plants, in *Chemical Ecology of Insects*, (eds W. J. Bell and R. T. Cardé), Chapman & Hall, New York, pp. 127–157.
69. Moeckh, H. A. (1981) Host selection behavior of bark beetles (Col, Scolytidae) attacking *Pinus ponderosa*, with special emphasis on the western pine beetle, *Dendroctonus brevicomis*. *J. Chem. Ecol.*, **7**, 49–83.
70. Moericke, V. (1969). Hostplant specific colour behaviour by *Hyalopterus pruni* (Aphidiidae). *Entomol. Exp. Appl.*, **12**, 524–534.
71. Morris, W. F. and Kareiva, P. M. (1991) How insect herbivores find suitable host plants: the interplay between random and nonrandom movement, in *Insect–Plant Interactions*, vol. 3, (ed. E. A. Bernays), CRC Press, Boca Raton, FL, pp. 175–208.
72. Mulkern, G. B. (1969) Behavioral influences on food selection in grasshoppers (Orthoptera, Acrididae). *Entomol. Exp. Appl.*, **12**, 509–523.
73. Murlis, J. (1986) The structure of odour plumes, in *Mechanisms in Insect Olfaction*, (eds T. L. Payne, M. C. Birch and C. E. J. Kennedy), Clarendon Press, Oxford, pp. 27–38.
74. Murlis, J., Elkinton, J. S. and Cardé, R. T. (1992) Odor plumes and how insects use them. *Annu. Rev. Entomol.*, **37**, 505–532.
75. Mustaparta, H. (1984) Olfaction, in *Chemical Ecology of Insects*, (eds W. J. Bell and R. T. Cardé), Chapman & Hall, New York, pp. 37–70.
76. Nordenhem, H. and Nordlander, G. (1994) Olfactory oriented migration through soil by root-living *Hylobius abietis* (L.) larvae (Col., Curculionidae). *J. Appl. Entomol.*, **117**, 457–462.
77. Nottingham, S. F. (1987) Effects of nonhost-plant odors on anemotactic response to host-plant odor in female cabbage root fly, *Delia radicum*, and carrot rust fly, *Psila rosae*. *J. Chem. Ecol.*, **13**, 1313–1318.
78. Nottingham, S. F., Hardie, J., Dawson, G. W., Hick, A. J., Pickett, J. A., Wadhams, L. J. and Woodcock, C. M. (1991) Behavioral and electrophysiological responses of aphids to host and nonhost plant volatiles. *J. Chem. Ecol.*, **17**, 1231–1242.
79. Paliniswamy, P. and Gillot, C. (1986) Attraction of diamondback moths, *Plutella xylostella* (L.) (Lepidoptera, Plutellidae), by volatile compounds of canola, white mustard, and faba bean. *Can. Entomol.*, **118**, 1279–1285.
80. Papaj, D. R. and Prokopy, R. J. (1989) Ecological and evolutionary aspects of learning in phytophagous insects. *Annu. Rev. Entomol.*, **34**, 315–350.
81. Pettersson, J. (1970) Studies on *Rhopalosiphum padi* (L.). I. Laboratory studies on olfactometric responses to the winter host *Prunus padus* L. *Lantbrukshoegsk. Ann.*, **36**, 381–399.
82. Pettersson, J. (1973) Olfactory reactions of *Brevicoryne brassicae* (L.) (Hom., Aph.). *Swed. J. Agricult. Res.*, **3**, 95–103.
83. Pivnick, K. A., Lamb, R. J. and Reed, D. (1992) Responses of flea beetles, *Phyllotreta* spp., to

mustard oils and nitriles in field trapping experiments. *J. Chem. Ecol.*, **18**, 863–873.

84. Pospisil, J. (1972) Olfactory orientation of certain phytophagous insects in Cuba. *Acta Entomol. Bohemoslov.*, **69**, 7–17.

85. Prokopy, R. J. (1968) Visual responses of apple maggot flies, *Rhagoletis pomonella* (Diptera, Tephritidae), orchard studies. *Entomol. Exp. Appl.*, **11**, 403–422.

86. Prokopy, R. J. (1986) Visual and olfactory stimulus interaction in resource finding by insects, in *Mechanisms in Insect Olfaction*, (eds T. L. Payne, M. C. Birch and C. E. J. Kennedy), Clarendon Press, Oxford, pp. 81–89.

87. Prokopy, R. J. and Owens, E. D. (1983) Visual detection of plants by herbivorous insects. *Annu. Rev. Entomol.*, **28**, 337–364.

88. Prokopy, R. J., Collier, R. and Finch, S. (1983a) Leaf color used by cabbage root flies to distinguish among host plants. *Science*, **221**, 190–192.

89. Prokopy, R. J., Collier, R. and Finch, S. (1983b) Visual detection of host plants by cabbage root flies. *Entomol. Exp. Appl.*, **34**, 85–89.

90. Roden, D. B., Miller, J. R. and Simmons, G. A. (1992) Visual stimuli influencing orientation by larval gypsy moth *Lymantria dispar* (L.). *Can. Entomol.*, **124**, 287–304.

91. Roessingh, P. and Städler, E. (1990) Foliar form, colour and surface characteristics influence oviposition behaviour in the cabbage root fly *Delia radicum*. *Entomol. Exp. Appl.*, **57**, 93–100.

92. Röttger, U. (1979) Untersuchungen zur Wirtswahl der Rübenfliege *Pegomya betae* Curt. (Diptera, Anthomyidae). I. Olfaktorische Orientierung zur Wirtspflanze. *Z. Angew. Entomol.*, **87**, 337–348.

93. Salama, H. S., Rizk, A. F. and Sharaby, A. (1984) Chemical stimuli in flowers and leaves of cotton that affect behaviour in the cotton moth, *Spodoptera littoralis* (Lepidoptera, Noctuidae). *Entomol. Gen.*, **10**, 27–34.

94. Saxena, K. N. and Khattar, P. (1977) Orientation of *Papilio demoleus* larvae in relation to size, distance, and combination pattern of visual stimuli. *J. Insect Physiol.*, **23**, 1421–1428.

95. Saxena, K. N. and Saxena, R. C. (1975) Patterns of relationships between certain leafhoppers and plants. Part III. Range and interaction of sensory stimuli. *Entomol. Exp. Appl.*, **18**, 194–206.

96. Scherer, C. and Kolb, G. (1987) Behavioral experiments on the visual processing of color stimuli in *Pieris brassicae* L. (Lepidoptera). *J. Comp. Physiol. A.*, **160**, 645–656.

97. Schöne, H. (1984) *Spatial Orientation: The Spatial Control of Behavior in Animals and Man*, Princeton University Press, Princeton, NJ.

98. Schoonhoven, L. M. (1987) What makes a caterpillar eat? The sensory code underlying feeding behavior, in *Perspectives in Chemoreception and Behavior*, (eds R. F. Chapman, E. A. Bernays and J. G. Stoffolano), Springer-Verlag, New York, pp. 69–97.

99. Smith, B. H. and Getz, W. M. (1994) Nonpheromonal olfactory processing in insects. *Annu. Rev. Entomol.*, **39**, 351–375.

100. Städler, E. (1992) Behavioral responses of insects to plant secondary compounds, in *Herbivores. Their Interactions with Secondary Plant Metabolites*, 2nd edn, vol. 2, (eds G. A. Rosenthal and M. R. Berenbaum), Academic Press, New York, pp. 45–87.

101. Städler, E. and Hanson, F. E. (1975) Olfactory capabilities of the gustatory chemoreceptors of the tobacco hornworm larvae. *J. Comp. Physiol.*, **104**, 97–102.

102. Stengl, M., Hatt, H. and Breer, H. (1992) Peripheral processes in insect olfaction. *Annu. Rev. Physiol.*, **54**, 665–681.

103. Takabayashi, J., Dicke, M. and Posthumus, M. A. (1994) Volatile herbivore-induced terpenoids in plant–mite interactions: variation caused by biotic and abiotic factors. *J. Chem. Ecol.*, **20**, 1329–1354.

104. Thibout, E., Auger, J. and Lecomte, C. (1982) Host plant chemicals responsible for attraction and oviposition in *Acrolepiopsis assectella*, in *Proceedings of the 5th International Symposium on Insect–Plant Relationships, Wageningen, 1982*, (eds J. H. Visser and A. K. Minks), Pudoc, Wageningen, pp. 107–115.

105. Thiéry, D. and Visser, J. H. (1986) Masking of host plant odour in the olfactory orientation of the Colorado potato beetle. *Entomol. Exp. Appl.*, **41**, 165–172.

106. Thiéry, D. and Visser, J. H. (1987) Misleading the Colorado potato beetle with an odor blend. *J. Chem. Ecol.*, **13**, 1139–1146.

107. Tichenor, L. H. and Seigler, D. S. (1980) Electroantennogram and oviposition responses of *Manduca sexta* to volatile components of tobacco and tomato. *J. Insect Physiol.*, **26**, 309–314.

108. Tingle, F. C. and Mitchell, E. R. (1992) Attraction of *Heliothis virescens* (F.) (Lepidoptera,

Noctuidae) to volatiles from extracts of cotton flowers. *J. Chem. Ecol.*, **18**, 907–914.
109. Tingle, F. C., Heath, R. R. and Mitchell, E. R. (1989) Flight response of *Heliothis subflexa* (Gn.) females (Lepidoptera, Noctuidae) to an attractant from groundcherry, *Physalis angulata* L. *J. Chem. Ecol.*, **15**, 221–231.
110. Tommerås, B. A. and Mustaparta, H. (1987) Chemoreception of host volatiles in the bark beetle *Ips typographus*. *J. Comp. Physiol. A*, **161**, 705–710.
111. Topazzini, A., Mazza, M. and Pelosi, P. (1990) Electroantennogram responses of five Lepidoptera species to 26 general odorants. *J. Insect Physiol.*, **36**, 619–624.
112. Traynier, R. M. M. (1984) Associative learning in the ovipositional behaviour of the cabbage butterfly, *Pieris rapae*. *Physiol. Entomol.*, **9**, 465–472.
113. Traynier, R. M. M. (1986) Visual learning in assays of sinigrin solution as an oviposition releaser for the cabbage butterfly, *Pieris rapae*. *Entomol. Exp. Appl.*, **40**, 25–33.
114. Valterova, I., Bolgar, T. S., Kalinova, B., Kovalev, B. G. and Vrkoc, J. (1990) Host plant components from maize tassel and electroantennogramme responses of *Ostrinia nubilalis* to the identified compounds and their analogues. *Acta Entomol. Bohemoslov.*, **87**, 435–444.
115. Van der Ent, L. J. and Visser, J. H. (1991) The visual world of the Colorado potato beetle. *Proc. Exp. Appl. Entomol., N. E. V. Amsterdam*, **2**, 80–85.
116. Van der Pers, J. N. C. (1981) Comparison of electroantennogram response spectra to plant volatiles in seven species of *Yponomeuta* and in the tortricid *Adoxophyes orana*. *Entomol. Exp. Appl.*, **30**, 181–192.
117. Van der Pers, J. N. C., Thomas, G. and den Otter, C. J. (1980) Interactions between plant odors and pheromone reception in small ermine moths (Lepidoptera, Yponomeutidae). *Chem. Senses*, **5**, 367–371.
118. Van Loon, J. J. A. and Frentz, W. H. (1991) Electroantennogram responses to plant volatiles in two species of *Pieris* butterflies. *Entomol. Exp. Appl.*, **62**, 253–260.
119. Van Loon, J. J. A., Everaarts, T. C. and Smallegange, R. C. (1992) Associative learning in host-finding by female *Pieris brassicae* butterflies: relearning preferences, in *Proceedings of the 8th International Symposium on Insect–Plant Relationships*, (eds S. B. J. Menken, J. H. Visser and P. Harrewijn), Kluwer, Dordrecht, pp. 162–164.
120. Visser, J. H. (1979) Electroantennogram responses of the Colorado beetle, *Leptinotarsa decemlineata*, to plant volatiles. *Entomol. Exp. Appl.*, **25**, 86–97.
121. Visser, J. H. (1983) Differential sensory perceptions of plant compounds by insects, in *Plant Resistance to Insects. American Chemical Society Symposium 208*, (ed. P. A. Hedin), American Chemical Society, Washington, DC, pp. 215–230.
122. Visser, J. H. (1986) Host odor perception in phytophagous insects. *Annu. Rev. Entomol.*, **31**, 121–144.
123. Visser, J. H. (1987), cited in reference 98.
124. Visser, J. H. (1988) Host-plant finding by insects: orientation, sensory input and search patterns. *J. Insect Physiol.*, **34**, 259–268.
125. Visser, J. H. and Taanman, J. W. (1987) Odour-conditioned anemotaxis of apterous aphids (*Cryptomyzus korschelti*) in response to host plants. *Physiol. Entomol.*, **12**, 473–479.
126. Visser, J. H. and de Jong, R. (1988) Olfactory coding in the perception of semiochemicals. *J. Chem. Ecol.*, **14**, 2005–2018.
127. Vité, J. P. and Gara, R. I. (1962) Volatile attractants from ponderosa pine attacked by bark beetles. (Coleoptera, Scolytidae). *Contrib. Boyce Thompson Inst.*, **21**, 251–273.
128. White, P. R. and Chapman, R. F. (1990) Olfactory sensitivity of gomphocerine grasshoppers to the odours of host and nonhost plants. *Entomol. Exp. Appl.*, **55**, 205–212.
129. Willis, M. A. and Arbas, E. A. (1991) Odor-modulated upwind flight of the sphinx moth, *Manduca sexta* L. *J. Comp. Physiol. A*, **169**, 427–440.
130. Zacharuk, R. Y. and Shields, V. D. (1991) Sensilla of immature insects. *Annu. Rev. Entomol.*, **36**, 331–354.
131. Zohren, E. (1968) Laboruntersuchungen zu Massenanzucht, Lebensweise, Eiablage und Eiablageverhalten der Kohlfliege, *Chortophila brassicae* (Bouché) (Diptera, Anthomyiidae) *Z. Angew. Entomol.*, **62**, 139–188.

HOST-PLANT SELECTION: WHEN TO ACCEPT A PLANT

6.1	Contact phase: plant evaluation	155
6.2	Physical plant features acting during contact	156
6.2.1	Surface morphology	157
6.2.2	Surface texture	157
6.2.3	Surface wax structure	159
6.3	Plant chemistry: contact-chemosensory evaluation	159
6.4	The importance of plant chemistry for host-plant selection: a historical intermezzo	160
6.5	Stimulation of feeding and oviposition	162
6.5.1	Primary plant metabolites	162
6.5.2	Plant secondary metabolites promoting acceptance: sign stimuli	165
6.5.3	Generally occurring secondary plant metabolites acting as stimulants	169
6.6	Inhibition of feeding and oviposition	169
6.6.1	Deterrence as a general principle in host-range determination	170
6.6.2	Deterrence as a mechanism to avoid herbivore competition	170
6.7	Plant acceptability: a balance between stimulation and deterrence	172
6.8	Contact-chemosensory basis of host-plant selection behaviour	172
6.8.1	Contact chemoreceptors	172
6.8.2	Gustatory coding	174
6.8.3	Caterpillars as models for coding principles	174
6.8.4	Sign stimulus receptors: unsurpassed specialists	177
6.8.5	Sugar and amino acid receptors: detectors of nutrients	178
6.8.6	Deterrent receptors: generalist taste cells	180
6.8.7	Peripheral interactions	181
6.8.8	Oviposition preference	184
6.8.9	Host-plant selection: a three-tier system	185
6.9	Conclusions	186
6.10	References	187

6.1 CONTACT PHASE: PLANT EVALUATION

When engaged in host-plant finding, a herbivorous insect that touches a plant may enter what we will call the 'contact phase' of host-plant selection (section 5.3). This phase consists of a series of behavioural elements that serve to evaluate physical and chemical plant traits that could not be perceived from a distance. After initial plant contact, locomotion is often halted rather suddenly. This phenomenon has been called **arrestment**; the insect tends to restrict its movements to a small area. For example, after a first brief landing a flying insect may take off and immediately thereafter alight again on the same or a neighbouring leaf. A walking insect may start climbing along the plant stem and start moving in small circles over the plant surface. Caterpillars often sway their heads, probably facilitating orientation to odours. Plant structures like leaf edges, veins or stems seem to guide walking movements in this phase. During movement intermittent evaluation is performed, which shows itself as repetitive contacting of legs, antennae, mouthparts or ovipositor with the plant surface: scratching and drumming with tarsi, antennating, palpating and ovipositor-dragging are commonly observed types of behaviour. These movements are a direct response to physical and chemical contact cues offered by the plant. At the same time, volatile plant compounds that occur at relatively high concentrations in the leaf boundary layer can affect behaviour as well [1, 3, 133]. It is important to note that many species base their initial behavioural decision either to proceed with evaluation or to reject the plant individual just contacted on the physical and/or chemical surface characteristics [6, 38, 134].

As a next step in the evaluation sequence, the insect may damage the plant and thereby release chemicals from the plant interior, comprising a complex mixture of primary and secondary metabolites. Injury is often inflicted by the insect's mouthparts and is designated **test biting**, or **probing** in the case of piercing–sucking insects. A test bite is often smaller than a regular bite and the plant material may be kept longer in the preoral cavity than during regular food intake. When the sensory information gathered during contact evaluation is judged positively by the central nervous system, **acceptance**, the final decision taken in the host-plant selection process, is confirmed and food intake or oviposition is started. The amount of sensory information gathered during the entire sequence has reached its maximum. Acceptance of food is normally expressed as a certain minimal bout of food intake. Acceptance of an oviposition substrate is evident from the deposition of one or more eggs. It should be noted that the actual amount of food intake or the number of eggs laid is highly variable and depends not only on the outcome of the sensory evaluation, but also on the physiological status of the individual (such as deprivation, egg load, age) and experience (Chapter 7). From an evolutionary perspective, acceptance can be considered as the crucial decision taken during host-plant selection, as it has direct consequences for the acquisition of nutrients and energy or, in the case of oviposition, for the survival of progeny.

6.2 PHYSICAL PLANT FEATURES ACTING DURING CONTACT

Upon contact with the plant the insect obtains additional information on plant quality that was not yet accessible during previous phases of host selection: tactile (mechanosensory) and contact chemosensory (taste or gustatory) stimuli. Physical features of plant organs or tissues can drastically influence host-plant selection behaviour. The presence of trichomes and wax crystal structures on the plant surface, leaf thickness and toughness, sclerotization and high silica content may cause avoidance behaviour and such plant traits are often assumed to fulfil a defensive function. Many examples of this morphologically based type of resistance can be found in the plant resistance literature for a wide range of insects belonging to different groups (Table 6.1).

Insects are equipped with numerous

Table 6.1 Selected examples of physical plant characteristics that affect host-plant selection by members of three insect orders: Lepidoptera, Hemiptera and Coleoptera

	Plant species	Insect affected	Larva/Adult	Ref.
Trichomes (non-glandular)	Soybean	Cabbage looper (*Trichoplusia ni*) (Lep.)	L	62
	Cotton	Western lygus bug (*Lygus hesperis*) (Het.)	L + A	4
	Strawberry	Black vine weevil (*Otiorhynchus sulcatus*) (Col.)	A	35
Tissue thickness				
Pod	Soybean	Pod borer (*Grapholita glycinivorella*) (Lep.)	L	87
Stems	Tomato	Potato aphid (*Macrosipum euphorbiae*) (Hom.)	A	93
Leaf	Mustard	Mustard beetle (*Phaedon cochleariae*) (Col.)	L	140
Wax microstructure	Cabbage	Small white (*Pieris rapae*) (Lep.)	L	137
	Raspberry	Raspberry aphid (*Amphorophora rubi*) (Hom.)	A	66
	Mustard	Mustard beetle (*Phaedon cochleariae*) (Col.)	A	138

mechanosensory sensilla on all parts of their body [69], and these probably perceive the relevant information on plant surface structure and texture. Taking plant features as a starting point, a few examples are presented in more detail to illustrate to what extent physical features of plants can affect host-plant selection. The primary interface in plant–insect contact is the plant surface: a plant does not suffer damage until the surface is penetrated [19], and we will examine its features first.

6.2.1 SURFACE MORPHOLOGY

(a) Trichomes

Plant surfaces are often covered with various types of hair or trichome, which may be either glandular or non-glandular (section 3.7). These structures may hinder normal behaviour in insects, especially smaller ones. Intraspecific variation in trichome type or density has been successfully exploited in resistance breeding against some pest insects. In several cases the extent of pubescence is determined by one or two genes, which makes selection relatively easy [89, 128].

The effects of non-glandular trichomes on host-plant selection are varied. Attachment and movement can be negatively influenced by long trichomes, as is the case in pubescent soybeans and the Mexican bean beetle, *Epilachna varivestis* [143]. Pubescent bean plants also impair oviposition of an agromyzid fly [24]. Impaired insect movement and attachment are determined by length, density and alignment of the trichomes. Heavy pubescence can prevent small piercing–sucking species from reaching the epidermis with their mouthparts [126, 129]. *Helicoverpa zea* female moths, on the other hand, can get a better grip on the upper pubescent surface than on the glabrous underside of corn leaves and consequently lay more eggs on the upper side. These females display similar within-plant preference behaviour for the most pubescent leaves of millet and cotton [98].

In glandular trichomes ('sticky hairs') we find a sophisticated combination of morphological and chemical plant resistance against insect colonization. The contents of glands associated with trichomes are liberated by mechanical damage caused by the moving insect, or are continuously produced. Gland secretions may be repellent, deterrent and/or toxic or may effectively glue smaller species to the surface, after which they will succumb to starvation [45] (Fig. 6.1).

In larger species, active avoidance of plant species or cultivars carrying glandular trichomes on the basis of the allelochemicals they release has been demonstrated. A particularly well studied case is that of the Colorado potato beetle, which avoids the wild potato *Solanum berthaultii*. Adult beetles prefer to feed on the cultivated potato *Solanum tuberosum* in a choice situation, with *S. berthaultii* as the alternative. When *S. berthaultii* leaflets are appressed to *S. tuberosum* leaflets, these are avoided, indicating that deterrent chemicals are exuded from the trichomes of *S. berthaultii*. Removal of trichomes [45] rendered *S. berthaultii* leaf material just as acceptable as *S. tuberosum* (Fig. 6.2) [151]. When acetone leaf rinses of *S. berthaultii* were applied to *S. tuberosum* leaf discs, the non-volatile fraction was highly deterrent. Several different active compounds are involved, but their exact nature is as yet unknown.

6.2.2 SURFACE TEXTURE

Surface morphology may be quite important to female insects searching for an acceptable oviposition site. The diamondback moth, *Plutella maculipennis*, prefers rough over smooth artificial surfaces (Fig. 6.3) and females deposit eggs mainly along leaf veins and cavities of leaves or stems.

The cabbage root fly, *Delia radicum*, lays 2.5 times more eggs at the basis of artificial leaves with vertical folds compared to leaf models with horizontal folds. Moreover, the transition from leaf-blade exploration (Fig. 5.1) to stem

158   Host-plant selection: when to accept a plant

Figure 6.1 Trichomes on the leaves of a wild potato species, *Solanum berthaultii*, release when ruptured a clear liquid exudate, which upon exposure to air turns into a viscous sticky substance that acts as a natural glue trap for small insects. (Source: reproduced from Gregory *et al.*, 1986, with permission.)

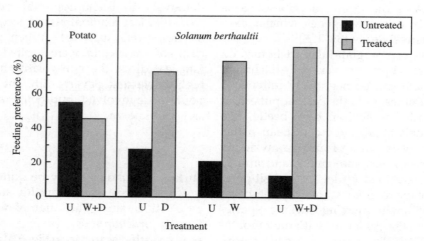

Figure 6.2 Effect of trichome removal of susceptible potato and resistant *Solanum berthaultii* by dipping (D = 95% ethanol dip), wiping (W = soft bristle-brush wipe) or combined dipping/wiping (W+D). Preference for treated versus untreated (U) leaves in adult Colorado potato beetles was determined in paired-choice experiments. It is seen that the combined wipe/dip treatment has no effect on potato while all three treatments to remove trichome-produced substances from *S. berthaultii* result into a preference for the treated leaflets. (Source: redrawn from Yencho and Tingey, 1994.)

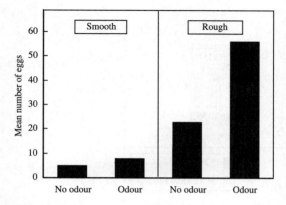

Figure 6.3 Effects of combinations of mechanosensory and olfactory cues on oviposition by the diamondback moth, *Plutella xylostella*. Smooth or rough plastic caps were offered as an oviposition substrate, with or without 10 ppm allylisothiocyanate as the odour (this compound is a major volatile released by host plants of this crucifer specialist). A rough surface baited with odour is by far the most stimulatory substrate; a rough substrate stimulates oviposition more strongly than a smooth substrate baited with odour. (Source: data from Gupta and Thorsteinson, 1960.)

run is more likely to occur on leaves with vertical folds [99] (Fig. 6.4).

The related anthomyid fly *Delia antiqua*, oligophagous on *Allium* spp., has been shown to take into account size, shape and orientation of artificial plants. Integration of mainly mechanosensory information on these physical plant features enables the fly to select substrates that closely resemble its natural host plant. Numbers of eggs deposited at the basis of plant models are synergistically enhanced when a volatile characteristic of its host plants (dipropyldisulphide) is present [48].

6.2.3 SURFACE WAX STRUCTURE

A large variation in micromorphology of plant epicuticular waxes (section 3.7) occurs both among and within plant species. Micromorphology is correlated with chemical composition. When wax crystals are of a type that is easily abraded by insect tarsal setae, this can cause problems for the insect in adhering to the plant surface and it may abandon further examination. On normal glaucous (wax-rich) cabbage, mustard beetles (*Phaedon cochleariae*) have problems adhering, while on a so-called glossy (i.e. wax-poor) mutant, which differs in its degree of branching of wax crystals, the beetles adhere very well and can still walk hanging upside down [138] (Fig. 6.5). Often, but not always, glossy crop varieties suffer less insect damage than genotypes with normal wax [38] (Table 6.2).

Of course variations in mechanical features of plant surfaces are important not only to herbivorous insects but also to their arthropod natural enemies, which search the plant surface and are likewise affected [118].

6.3 PLANT CHEMISTRY: CONTACT-CHEMOSENSORY EVALUATION

The previous sections clearly demonstrate that physical plant traits can affect host selection behaviour to an important extent. When we turn back to the high degree of host-plant specialization observed in herbivorous insects (Chapter 2), it is evident, however, that the behavioural responses to physical plant features do not offer a satisfactory explanation for this taxonomic specialization. The main reason is that taxonomic patterns in physical and morphological features are absent [54], which is in marked contrast with the taxonomic patterns observed in plant chemistry. Indeed, many plant families are characterized by secondary metabolites that do not occur in other families (Chapter 3). Genera within plant families have also been found to contain either qualitatively specific or quantitatively dominant compounds that belong to the characteristic secondary chemistry of the family. Such chemotaxonomic patterns in the plant kingdom potentially provide a basis for host-plant specificity of herbivorous insects and it is now firmly established that this potential has indeed been utilized to an impressive

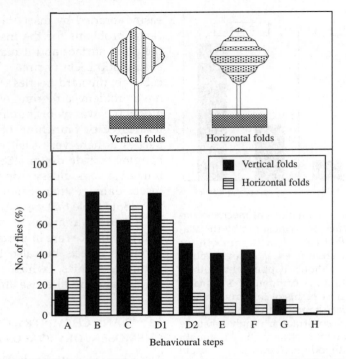

Figure 6.4 Influence of mechanosensory quality (horizontal *versus* vertical folds) of paper model leaves on oviposition behaviour of the cabbage root fly, *Delia radicum*. For each behavioural element (A–H) the number of flies performing this step is displayed. A = short visit, no exploration of leaf; B = rest, grooming; C = leaf run with exploration of surface; D1 = straight run on leaf borders or veins; D2 = straight geotactic run on stem; E = horizontal circular run around stem, heading towards ground; F = walk from stem to ground, probing sand surface; G = oviposition attempts; H = climbing back on stem, dragging ovipositor. Fewer flies complete the behavioural sequence on horizontal than on vertical folded surrogate leaves. The difference is associated with the transition from leaf exploration to stem run (D1 to D2) and significantly fewer females proceed to stem run and oviposition. (Source: redrawn from Roessingh and Städler, 1990.)

degree of refinement [6, 108, 132]. We will expound on this paradigm in what follows.

6.4 THE IMPORTANCE OF PLANT CHEMISTRY FOR HOST-PLANT SELECTION: A HISTORICAL INTERMEZZO

Mechanism and function of the botanical specificity shown by most herbivorous insects has historically been a challenging phenomenon to biologists. It was almost 200 years ago that the Swiss botanist A. P. de Candolle [26] implied that plant chemistry was the decisive factor in host-plant selection. J. H. Fabre [39] used the term 'botanical sense', referring to a sensory basis for behavioural specialization [112]. A tip of the veil over selection mechanisms was lifted by the Dutch botanist E. Verschaffelt [149], who demonstrated that mustard oil glucosides (glucosinolates), which are taxonomically characteristic for cruciferous plants, are decisive factors for plant acceptance by caterpillars of the cabbage white butterflies *Pieris brassicae* and *P. rapae*. The chemosensory basis of this behaviour was revealed only much later by the discovery of taste cells on the maxilla of the caterpillars that are specifically sensitive to these glucosides

Table 6.2 Susceptibility of crops with glossy phenotypes to insect attack; for references see Eigenbrode and Espelie, 1995, on which the table is based (Source: reproduced, with permission, from the *Annual Review of Ecology and Systematics*, volume 40, © 1995, by Annual Reviews Inc.)

Crop host	Pest insect	Effects of glossy phenotype	Beneficial trait
Allium cepa	*Thrips tabaci*	Lower infestation	Yes
Brassica napus	*Lipaphis erysimi*	Resistance	Yes
B. campestris	*L. erysimi*	Lower populations	Yes
B. oleracea	*L. erysimi*	Susceptibility	No
	Phyllotreta albionica	More susceptible	No
	P. cruciferae	Greater feeding damage	No
	Brevicoryne brassicae	Lower populations	Yes
	Erioischia brassicae	Fewer eggs	Yes
	Aleyrodes brassicae	Lower infestations	Yes
	Bemisia tabaci	Lower populations	Yes
	Thrips tabaci	Less damage	Yes
	Myzus persicae	Sometimes higher populations	No
	Plutella xylostella	Less damage, fewer eggs, lower larval survival	Yes
	Pieris rapae	Less damage and lower populations	Yes
Brassica spp.	*Phyllotreta nemorum*	Reduced leaf mining	Yes
Glycine max	*Epilachna varivestis*	Resistance	Yes
Hordeum vulgare	Four aphid species	Higher combined populations	No
Sorghum bicolor	*Schizaphis graminum*	Less preferred	Yes
	Spodoptera frugiperda	Less damaged	Yes
	Atherigona soccata	Resistance	Yes
	Chilo partellus	Resistance	Yes
Triticum aestivum	*Sitobion avenae*	Lower populations	Yes

[106]. Dethier [29] demonstrated the role of terpenoids contained in essential oils of Apiaceae in host-plant acceptance of black swallowtail (*Papilio polyxenes*) caterpillars, specialized feeders on this plant family. Fraenkel [41], in a seminal article entitled 'The *raison d'être* of secondary plant substances' brought together evidence that the food specificity of insects is based solely on the presence or absence of secondary metabolites and that several oligophagous species exploit taxon-specific secondary plant metabolites (posing effective defensive barriers against non-adapted species) as 'sign stimuli' that serve to identify their host plant. Jermy has drawn attention to the role of deterrents, secondary plant substances inhibiting feeding [55, 56] or oviposition [58], and advocated the view that host-plant selection is mainly based on avoidance of deterrents present in non-hosts. To counterbalance all the attention paid to secondary plant compounds, Kennedy and Booth [60] pointed to the combined importance of both secondary and primary plant metabolites in their 'dual discrimination' concept of host-plant selection. These concepts have all contributed tremendously to our current understanding of host-plant selection behaviour. They encompass the involvement of both primary and secondary compounds and also their stimulatory and inhibitory effects on herbivore behaviour.

In the following we will deal with the proximate mechanisms employed by plant feeding insects in selecting plants primarily on the basis of their chemistry. In this chapter we will focus on non-volatile (sapid) compounds that are perceived by gustatory receptors. A pos-

162   Host-plant selection: when to accept a plant

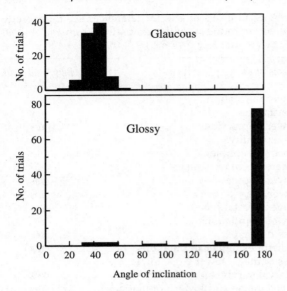

Figure 6.5 Effect of wax crystal microstructure of cabbage on attachment of the mustard beetle, *Phaedon cochleariae*. A single beetle was placed on a horizontal leaf. Once the insect started walking the angle of inclination was slowly increased from 0° to 180°. If the beetle fell off, the angle of inclination was recorded. Frequency distributions are shown, indicating at what angle with the horizontal beetles fell off. While on glaucous (normal) cabbage leaves, the majority of beetles fell off when the angle increased over 50°, on the glossy leaves (mutant wax crystal type) beetles could adhere while hanging upside down. (Source: redrawn from Stork, 1980.)

sible role of odours present at or near a feeding site has been much less studied, but there are indications that, during the contact phase also, volatiles play a role. Many plant chemicals are often confined to intra- or extracellular compartments (section 3.11). An extracellular 'compartment' that is particularly relevant for each discussion of host-plant selection mechanisms is the plant cuticle. As mentioned before, chemicals present at the plant's surface may affect selection behaviour prior to any injury that would release cell contents, either as an innate response or as a result of experience [21, 63]. Several groups of cuticular non-polar compounds, such as longer chain alkanes and esters, probably occur only on the surface [38, 54]. However, it is now known that sugars, amino acids and secondary metabolites, polar or non-polar, taxon-specific or generally occurring, also occur on plant surfaces (Table 3.7). We will indicate in the following discussion when behavioural responses have been found to surface-borne compounds.

6.5   STIMULATION OF FEEDING AND OVIPOSITION

6.5.1   PRIMARY PLANT METABOLITES

All plants contain carbohydrates and amino acids as primary metabolites resulting from their photosynthetic activity. There is ample evidence that most if not all herbivorous insects use carbohydrates especially as feeding stimulants (Table 6.3).

In most species studied, the disaccharide sucrose and its constituent monosaccharides fructose and glucose are the most powerful stimulants. These sugars are present at quite high concentrations (i.e. 2–10% dry weight, which roughly corresponds to 10–50 mmol/l) in green leaves, and even higher in fruits. They generally stimulate feeding in a dose-dependent way (Fig. 6.6).

Naturally they are also important nutrients needed to synthesize body tissue and to serve as energy sources (Chapter 4).

Although the protein content of plants is generally a limiting factor for the optimal growth of animals, protein molecules have never been found to stimulate feeding, but it must be noted that few explicit attempts have been made to demonstrate this. The protein building blocks, amino acids, however, do act as feeding stimulants (Table 6.4).

However, the stimulatory action of the 20 naturally occurring amino acids may at the sensory level vary significantly between even closely related species [108, 147]. Generally, ten amino acids are nutritionally essential for insects, but these are not necessarily stronger stimulants than non-essential amino acids, nor stimulatory to more species. Taste receptor

Table 6.3 Comparative stimulatory effectiveness of various sugars for some herbivorous insects; for references see Bernays and Simpson, 1982, on which the table is based (+++++ = highly stimulating; + = weakly stimulating; − = no effect; • = not tested) (Source: reprinted from *Advances of Insect Physiology* 16, E. A. Bernays and S. J. Simpson, Control of food intake, 59–118, 1982, by permission of the publisher Academic Press Limited London.)

	Locusts		Beetles		Caterpillars	
	Locusta migratoria	Schistocerca gregaria	Hypera postica	Leptinotarsa decemlineata	Pieris brassicae	Spodoptera spp.
Pentoses						
L-arabinose	+	•	•	−	−	−
L-rhamnose	−	−	•	−	−	−
D-ribose	−	−	•	−	−	−
D-xylose	−	−	•	−	−	•
Hexoses						
D-fructose	+++++	+++++	++++	+	−	+++++
D-galactose	++	+	•	+	−	++
D-glucose	+++	++++	+	+	++	++
D-mannose	−	+	++	−	++	++
L-sorbose	+	+	•	−	−	−
Disaccharides						
D-cellobiose	−	+	•	−	−	−
D-lactose	+	+	•	−	−	+
D-maltose	+++++	++++	++	−	−	+++
D-melibiose	+++	+++	•	−	−	++
D-sucrose	+++++	+++++	+++++	+++++	+++++	+++++
D-trehalose	+	++	+	+	−	−
Trisaccharides						
D-melizitose	++++	+	+++	++	−	++
D-raffinose	++++	+++	•	−	−	+++
Alcohols						
Inositol	+	•	•	−	−	−
Sorbitol	+	+	•	−	−	−
Mannitol	+	+	•	−	•	−

cells for sugars and amino acids have been found in many species and the ranking of chemosensory response intensities evoked by sugars or amino acids corresponds well with their behavioural effectiveness [67, 75] (section 6.8.5). Although less well studied, other substances taking part in plant primary metabolism, such as phospholipids and nucleotides and also minerals and vitamins (both nutritionally essential), are known to affect food acceptance in several species [7, 51].

Sugar and amino acid concentrations in different plant parts are spatially and temporally quite variable, variations that may be used as important cues for an insect when selecting a feeding site (Chapter 3). The significance of sugars and amino acids as feeding stimulants can only be quantified satisfactorily by incorporation into a neutral substrate (such as an agar-based artificial substrate or filter paper), which in itself elicits little or no feeding and is devoid of deterrents. In this way their relative stimulatory effectiveness can be assessed. Such an approach has been systematically

Host-plant selection: when to accept a plant

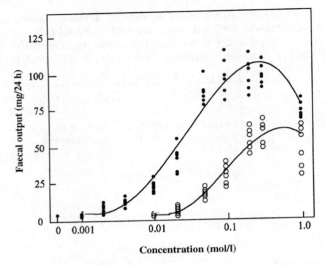

Figure 6.6 Behavioural response of *Pieris brassicae* larvae to two sugars, sucrose (●) and glucose (○) incorporated in an agar-based gel medium (a mixture of agar, water and cellulose). The parameter on the ordinate is dry weight of faecal output produced by six larvae over 24 hours, a fair indicator of the amount of food intake. At lower sugar levels, sucrose is a considerably stronger feeding stimulant than glucose. (Source: redrawn from Ma, 1972.)

Table 6.4 Comparative stimulatory effectiveness of various amino acids for some herbivorous insects; for references see Bernays and Simpson, 1982, on which the table is based (++ = stimulating; − = no effect; • = not tested; (+) = stimulating only in the presence of sucrose). (Source: reprinted from *Advances of Insect Physiology* 16, E. A. Bernays and S. J. Simpson, Control of food intake, 59–118, by permission of the publisher Academic Press Limited London.)

	Locusta migratoria	Leptinotarsa decemlineata	Hypera postica	Pieris brassicae	Aphis fabae	Acyrthosiphon pisum	Oncopeltus fasciatus
Glycine	−	+	•	−	•	•	++
D-alanine	−	•	•	•	•	•	++
L-serine	+	++	−	−	++	•	+
GABA	−	++	−	−	•	•	+
L-valine	+	+	−	•	•	++	•
L-threonine	+	−	•	−	•	++	+
L-leucine	−	+	•	•	•	•	•
L-isoleucine	−	−	•	•	•	−	•
L-cysteine	•	+	•	•	•	•	•
L-cystine	−	−	•	•	−	•	•
L-methionine	−	−	•	(+)	++	++	++
L-aspartic acid	−	−	•	•	•	•	•
L-asparagine	−	+	•	•	•	−	+
L-glutamic acid	−	−	•	•	−	•	•
L-glutamine	−	−	•	•	•	•	•
L-lysine	−	−	•	•	•	−	•
L-arginine	•	−	•	•	++	−	+
L-histidine	•	−	•	−	−	++	+
L-phenylalanine	−	−	•	•	•	•	•
L-tyrosine	•	•	•	•	−	++	•
L-tryptophan	−	−	•	•	•	•	•
L-proline	+	++	−	+	(++)	•	•
L-hydroxyproline	•	+	•	•	•	•	•

carried out for only few species [51]. In a no-choice situation, sucrose at concentration levels as occur in plants may induce on its own a maximum feeding rate on artificial substrates without any further compounds added. However, how these rates relate to those achieved on plant tissues has not been directly compared and they are therefore not directly indicative of the role of sugars in host selection behaviour. For example, larvae of several species, even after being raised for four instars on an artificial medium, still preferred plant tissue when this was offered together with the diet in a dual choice situation (van Loon, unpublished observations).

Several problems arise when attempting to compare feeding stimulation by an intact plant with that offered by plant chemical constituents presented in an artificial diet. Firstly, it is technically not possible to rule out differences in preference due to the obvious mechanosensory differences between the two. Secondly, in such studies artificial substrates generally contain a sugar and only one or two additional compounds and are therefore nutritionally deficient. When feeding rate is measured indirectly by weight of faecal pellets or substrate consumed over several hours, each comparison with feeding rates on plant tissues is questionable, because feeding rate on a deficient diet may also be affected by physiological feedback resulting from decreased nutrient levels in the haemolymph.

Sugars have also been shown to promote oviposition in, for instance, the polyphagous European corn borer (*Ostrinia nubilalis*) [27, 28]. Like most other ovipositing insects, the female moths do not seem to injure tissues and their oviposition response must be based on their perception of sugars present on the leaf surface. The dominant lipophilic constituents of leaf surfaces (alkanes, esters, fatty acids), to be considered as primary metabolites, are known to promote test biting or probing and subsequent feeding and oviposition in many insects, ranging from aphids to locusts (reviewed by Städler [131], Bernays and Chapman [6], and Eigenbrode and Espelie [38]).

Although primary plant substances, notably sugars and amino acids, do affect host-plant acceptance, the fact that they occur in all plants and that their concentrations vary greatly with plant developmental stage, age, physiological condition and environmental factors makes it unlikely that host-plant specificity can be explained by selection solely based on these categories of substances; in fact, no example is known. This notion leads us to consider the role of sapid plant secondary chemicals.

6.5.2 PLANT SECONDARY METABOLITES PROMOTING ACCEPTANCE: SIGN STIMULI

As noted in Chapter 3, plants offer a huge diversity of secondary metabolites to herbivores. In this diversity taxonomic patterns are discernible: a chemically distinct group of substances often occurs in one or a few related plant families only. Some other categories of secondary metabolites, however, have a wide distribution among unrelated plant families, notably many phenolics and flavonoids.

The number of instances in which particular taxon-specific secondary metabolites act as feeding or oviposition stimulants to monophagous or oligophagous species has grown considerably since Verschaffelt's [149] days. Table 6.5 lists examples of feeding or oviposition activity governed by secondary plant substances in a number of food specialists belonging to different orders.

In some cases the active compounds were found along an analogy approach (they had been found to be active for other insects feeding on the same plants), in other cases bioassay guided fractionation (Appendix C) led to their identification. Especially for oviposition, the degree of stimulation by one or a few identified compounds was similar or nearly so to the response to total extracts of the host plant, or even to the intact host plant itself. These substances are good examples of

Table 6.5 Monophagous and oligophagous herbivorous insects of different orders that use taxon-specific chemicals as token stimuli for host-plant acceptance, their host plant, the sign stimulus and the chemical class to which it belongs; all cases where sign-stimulus receptors have been identified are indicated

Insect species	Host plant	Sign stimulus	Chemical class	Ref.	Receptor identified	Ref.
Lepidoptera – feeding						
Pieris spp.	Brassica spp.	Sinigrin	Glucosinolate	149	Yes	106
Bombyx mori	Morus spp.	Morin	Flavonoid	107		
Euphydryas chalcedona	Plantago	Catalpol	Sesquiterpene	18		
Lepidoptera – oviposition						
Pieris spp.	Brassica spp.	Glucobrassicin	Glucosinolate	97, 148	Yes	37, 136
Papilio polyxenes	Daucus carota	Luteolin-glycoside	Flavonoid	40	Yes	100
Battus philenor	Aristolochia	Aristolochic acid	Iridoid glycoside	105		
Junonia coenia	Plantago	Aucubin + catalpol	Iridoid glycoside Sesquiterpene	90		
Coleoptera – feeding						
Phyllotreta armoraciae	Brassica spp.	Sinigrin + flavonoid glycos.	Glucosinolate Flavonoid	85		
Plagioderma versicolora	Salix spp.	Salicin	Phenolic	71		
Chrysolina brunsvicensis	Hypericum	Hypericin	Quinone	95	Yes	95
Diabrotica spp.	Cucurbita spp.	Cucurbitacins	Steroids (saponins)	92	Yes	83
Diptera – oviposition						
Delia radicum	Brassica spp.	Glucobrassicin + 'CIF'	Glucosinolate Indole derivative	84, 102	Yes Yes	130, 102
Psila rosae	Daucus spp.	Falcarindiol + bergapten, etc.	Polyacetylene Furanocoumarins	131	Yes	134
Delia antiqua	Allium spp.	n-propyl disulphide	Disulphide	70	Yes	134
Homoptera – feeding						
Brevicoryne brassicae	Brassica spp.	Sinigrin	Glucosinolate	150		
Aphis pomi	Malus	Phloridzin	Chalcone	81		
Acyrthosiphon spartii	Cytisus	Sparteine	Alkaloid	127		

what Fraenkel [41] called 'sign stimuli' (or 'token stimuli'): their occurrence is restricted to certain plant taxa and chemoreception of such compounds allows unambiguous recognition of the species' host plant. The best studied insect–plant interactions conforming to this principle are those between lepidopteran, dipteran and coleopteran herbivores of Cruciferae, Apiaceae (Umbelliferae) and Alliaceae [132].

The complexity of the stimulatory chemical signal comprising secondary metabolites may differ considerably. In two species of cabbage white butterfly (Pieris spp.), a single glucosinolate isolated from the surface of cabbage leaves elicits a strong oviposition response when sprayed on artificial leaves or some non-host plants, such as Phaseolus lunatus [97, 148]. Some other glucosinolates clearly differ in their stimulatory effect (Fig. 6.7). A much more complex situation has been revealed in swallowtail butterflies (Papilio spp.), where mixtures of compounds, only some specific to the host-plant taxon, were required to elicit a full behavioural response [40, 50, 86] (Fig. 6.8).

Table 6.5 also demonstrates that for different oligophagous species sharing the same host plants, the sign stimuli may be qualitatively different. Examples of this are the carrot root fly (Psila rosae) and the black swallowtail

Stimulation of feeding and oviposition 167

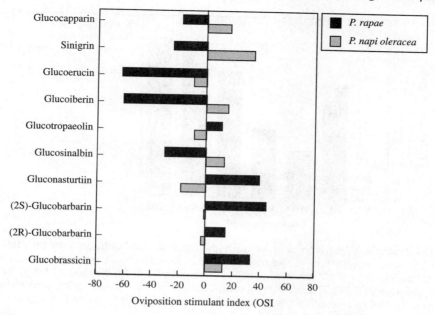

Figure 6.7 Stimulation of oviposition in *Pieris rapae* and *P. napi oleracea* by pure glucosinolates when sprayed on the non-host Lima bean (2 ml of a 0.1 mmol/l solution in water). The oviposition stimulant index (OSI) signifies the degree of preference in a dual choice situation relative to a Lima bean plant sprayed with 2 ml of a 0.1 g leaf equivalent/ml cabbage extract, in which the major glucosinolate is glucobrassicin. A negative OSI means that the females preferred the cabbage-extract-treated bean plant. Glucosinolates differ in their effectiveness to stimulate oviposition within each species and both species differ in their preference hierarchy. (Source: redrawn from Huang and Renwick, 1994.)

(*Papilio polyxenes*), both living on carrot, the flea beetle *Phyllotreta armoraciae* and caterpillars of *Plutella* and *Pieris*, living on cabbage, and the leek moth (*Acrolepiopsis assectella*) and the onion fly (*Delia antiqua*), specialists of Alliaceae (reviewed by Städler [132]). When specific compounds have been shown to exert an appreciable stimulatory activity, as is the case for the examples cited above, often no further attempts have been made to identify additional compounds, despite the fact that the full behavioural response as occurs to intact plants was not obtained. An intriguing example is the cabbage root fly, *Delia radicum*, for which glucosinolates act as taxon-specific oviposition stimulants [101, 130]; these were assumed to be the prime phytochemicals on which host-plant specificity in this species was based.

When later a classical bioassay-guided isolation procedure was carried out on leaf surface extracts, a non-glucosinolate compound was quite unexpectedly found to be a much more powerful stimulant, evoking equal stimulation at 100 times lower concentrations than the most stimulatory glucosinolate [102] (Fig. 6.9).

In the case of surface-borne compounds especially, the concentration actually available to the gustatory sensilla when these contact an intact plant surface remains an important unknown. Concentration values based on phytochemical extraction (assumed to be exhaustive) and quantities of surface-borne compounds can be expressed as micromoles per unit of surface area, but it is unclear which fraction of this quantity enters the taste sensilla and, consequently, what concentration is

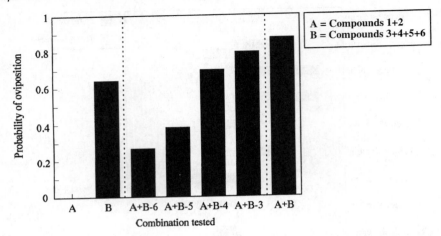

Figure 6.8 The probability of oviposition by individual females of *Papilio protenor* on filter-paper discs treated with different combinations of compounds isolated from the host plant *Citrus unshiu*. Compounds tested were (1) naringin 0.1%, (2) hesperidin 0.05%, (3) proline 0.2%, (4) synephrine 0.1%, (5) stachydrine 0.2% and (6) quinic acid 0.2%. The mixture of compounds 1 and 2 is inactive, the combination of A + B acts synergistically. Deletion of either compound 4 (i.e. A + B – 4), 5 or 6 results in a significant reduction of stimulatory activity. (Source: data from Honda, 1990.)

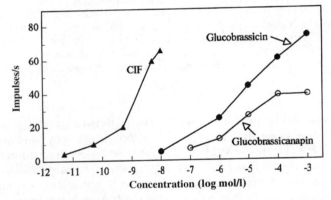

Figure 6.9 Dose – response curves of neural activity (number of action potentials in the first second after contact with the sensillar tip) in taste hairs on the fifth tarsomere of cabbage root flies (*Delia radicum*) for the glucosinolates glucobrassicin and glucobrassicanapin, the strongest glucosinolate oviposition stimulants for this species, and for 'CIF'. The latter chemical, occurring in surface extracts of cabbage leaves, is a much stronger oviposition stimulant than the two glucosinolates, but it does not belong to this chemical class. (Source: redrawn from Roessingh et al., 1992b.)

perceived. It is also remarkable that several insect species can be stimulated by polar chemicals present in the plant's epicuticle [38]. In view of the methods necessary to extract them from the surface these compounds seem to be anchored quite firmly in the epicuticular wax layer [134]. Possibly, taste sensilla possess as yet unknown mechanisms to release them from this substrate [148] and it would be interesting to investigate this aspect in more detail.

6.5.3 GENERALLY OCCURRING SECONDARY PLANT METABOLITES ACTING AS STIMULANTS

There are a growing number of insect species for which secondary plant metabolites found in unrelated plant families act as feeding stimulants. This is particularly true for some phenolic acids and flavonoids (see Table 3.4). For example, both caffeic acid (8) and its quinic acid ester chlorogenic acid (11) stimulate feeding in the silkworm, *Bombyx mori*, oligophagous on Moraceae, while the latter compound also stimulates feeding in the Colorado potato beetle [51], specialized on some solanaceous plants. Both the silkworm and the cotton boll weevil (*Anthonomus grandis*) are stimulated by the flavone-glycoside isoquercitrin (quercetin-3-glucoside) [107]. Polyphagous species also may be stimulated by the presence of flavonoids in their food. The ubiquitous quercetin glycoside rutin (53) has been documented as a feeding stimulant for both the desert locust (*Schistocerca americana*) [8] and *Helicoverpa virescens* caterpillars [107]. In view of the general occurrence of these secondary metabolites, the same reasoning applies as to nutrient chemicals, i.e. it would be difficult to conceive how, for specialized species, these compounds could constitute an unambiguous signal for acceptance.

For several well-studied oligophagous species, in particular for the Colorado potato beetle and the tobacco horn moth (*Manduca sexta*), both Solanaceae feeders, repeated attempts to identify token stimuli have been unsuccessful [57] (but these studies were made at a time when HPLC and HPLC-MS were not yet available). Therefore, an alternative mechanism of host recognition in these species was proposed: that host plants are acceptable because they lack compounds that inhibit feeding (at least in any appreciable amount), whereas non-host plants are rejected because of the presence of deterrents [55, 56, 57]. The problem here, of course, is that it is possible that host-specific sign stimuli do exist but so far they have defied chemical isolation. It should be realized that overall only few insect–plant combinations have been scrutinized in depth for the involvement of token stimuli and that isolation and identification of such compounds is a time-consuming process.

6.6 INHIBITION OF FEEDING AND OVIPOSITION

Fraenkel [41] pointed out that secondary plant substances are defensive substances that inhibit food intake in the majority of plant-feeding insects, except for some specialized species, which may exploit these chemicals with a limited taxonomic occurrence as sign stimuli enhancing acceptance. Relatively few studies have addressed rejection as a mechanism of host-plant specificity in a systematic way. Jermy [55, 56] clearly demonstrated that rejection of non-hosts by various insects is due to the presence of feeding inhibitors (feeding deterrents; Fig. 6.10).

A 'sandwich' test was used in which a disc of the test plant species was offered between two discs of a host plant. This rarely used method allows exclusion of the absence of feeding stimulants as a reason for rejection or low preference of a non-host plant. Another detailed study was performed on two locust species, *Locusta migratoria*, a Poaceae specialist, and the polyphagous *Schistocerca gregaria*, which led to similar conclusions. Acceptance is one criterion for identifying host plants and non-hosts. Meal size is another, and this makes it possible to discern more grades of difference in the acceptability of plants. When meal size on a stimulatory artificial wheat flour substrate was used as a measure for acceptance, *Locusta* was seen to take full meals on (and thus to fully accept) Poaceae, but to take only small meals on non-hosts. All the non-hosts contained deterrents, as also did several less acceptable species of Poaceae. *Schistocerca*, on the other hand, showed much more variability in meal size. All plant species on which

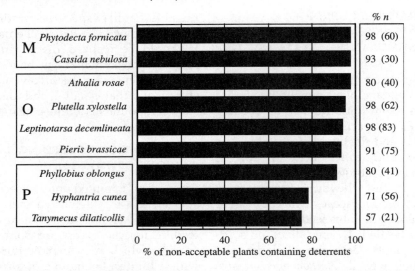

Figure 6.10 The percentage of rejected plants containing deterrents to monophagous (M), oligophagous (O) and polyphagous (P) species from different orders, from the evidence of the sandwich leaf disc test. Right: percentages of the total number of plant species tested that were only slightly nibbled or not consumed at all and the total number of plant species tested. (Source: data from Jermy, 1966.)

small meals were taken contained deterrents [5].

6.6.1 DETERRENCE AS A GENERAL PRINCIPLE IN HOST-RANGE DETERMINATION

Comparative research on many herbivorous insects has uncovered several general principles guiding their responses to feeding deterrents. First, non-hosts commonly contain deterrents. Second, monophagous and oligophagous species are generally more sensitive to deterrents from non-hosts than polyphagous species (Table 6.6). This has been documented for locusts [5, 6] and several caterpillar species [15]. Third, deterrents have not only been found in non-hosts, but in several instances also in acceptable plants, where their effect is apparently neutralized by the simultaneous presence of stimulants [23, 52, 56]. For several monophagous and oligophagous species for which sign stimuli have been identified in their host plants, lack of stimulation together with possible deterrence offers an explanation for rejection of non-hosts, as infusion or coating with sign stimuli renders some non-hosts acceptable and apparently overrides putative (weak) deterrents [68, 91, 149].

By now a vast literature is available on the effects of many hundreds of secondary metabolites inhibiting insect feeding [82]. The accumulation of these data has been promoted by an interest in identifying plant derived compounds with the prospect of their potential use in crop protection against insects [43] (section 12.4). Much less work has been done on oviposition deterrents [96], but the information available suggests that, as in food-plant recognition, deterrence is in many insects an important mechanism in host-plant selection.

6.6.2 DETERRENCE AS A MECHANISM TO AVOID HERBIVORE COMPETITION

Gravid females in pursuit of an acceptable oviposition site are, after landing, not only influenced by the chemical make-up of the

Table 6.6 Deterrent effects of compounds belonging to the major chemical classes of secondary plant substances to an oligophagous (O) lepidopteran and a polyphagous (P) lepidopteran or homopteran species (thr = threshold concentration)

Compound	Chemical class	Insect species	Host-plant specificity	Effective concentration (ppm)	Inhibition (%)	Ref.
Sinigrin	Glucosinolate	Papilio polyxenes	O	900	66	16
		Mamestra configurata	P	3100	50	119
Chlorogenic acid	Phenolic acid	Pieris brassicae	O	570	40	59
		Myzus persicae	P	3500	50	36
Phloridzin	Flavonoid	Schizaphis graminum	O	200	50	36
		Myzus persicae	P	4360	100	114
Strychnine	Alkaloid	Pieris brassicae	O	30	100	67
		Mamestra brassicae	P	3900	75	17
Tomatine	Steroid alkaloid	Pieris brassicae	O	40	100	67
		Myzus persicae	P	1034	(0)	114
Ajugarin	Diterpenoid	Spodoptera exempta	O	100	thr	64
		Spodoptera littoralis	P	300	thr	64
Azadirachtin	Triterpenoid	Pieris brassicae	O	7	50	65
		Spodoptera frugiperda	P	315	50	94

plant exterior but also by insect-produced compounds left by earlier visitors. Females of several butterfly, beetle and fly species secrete, concomitantly with egg deposition, substances that inhibit oviposition by conspecific females and oviposition behaviour of females arriving later [103, 111]. These substances have been termed 'host marking pheromones' or 'epideictic pheromones'. From the few cases in which the chemical structure of such signal compounds has been elucidated it appears that their chemical structures vary greatly. Host-marking is a well-known phenomenon in, for instance, many fruit flies. Female cherry fruit flies, *Rhagoletis cerasi*, drag their ovipositor over the fruit surface after an egg has been inserted under the skin of a cherry. During this dragging behaviour marking substances are deposited on the fruit surface. Other females landing on an 'occupied' fruit perceive these compounds with tarsal chemoreceptors. Investigations with synthetic analogues of the natural compound have shown that at the sensory level distinct structure–activity relationships exist [135], suggesting that the marking pheromone stimulates a specialized receptor.

In the case of two cabbage white butterflies (*Pieris brassicae* and *P. rapae*), egg washes were found to strongly deter oviposition, both intra- and interspecifically. This indicated the involvement of a chemical marker substance that causes avoidance [104]. Some avenanthramide alkaloids could be isolated from the egg washes that produced potent effects and were responsible for the activity of the crude egg wash. These compounds were found only in eggs of the genus *Pieris*, not in those from two other Pieridae nor in eggs from five non-pierid lepidopterans [9], a specificity reminiscent of sex pheromones. *Pieris* butterflies do not exhibit dragging behaviour after egg deposition on the underside of a leaf. Leaves that carry egg batches are avoided for oviposition after landing on the upperside and translocation of the identified putative marking substances was therefore investigated. Further studies could not demonstrate a translocation of the active principles of egg washes. Interestingly, however, fractions from surface extracts of leaves that had carried eggs were obtained that deterred oviposition but did not contain the egg alkaloids [10]. In contrast to the cherry fruit fly, where the marking substance is solely produced by the insect, in the case of *Pieris* butterflies there is a role for the plant. Apparently, contact with *Pieris* eggs induces a change in the plant's surface chemistry and as yet unknown substances are produced that act as strong deterrents to ovipositing females. Although the sources of the deterrent compounds are different, in both the cases cited they have the same effect on the insect, i.e. to reduce competition for food between offspring.

6.7 PLANT ACCEPTABILITY: A BALANCE BETWEEN STIMULATION AND DETERRENCE

The stimulatory and inhibitory effects that plant chemicals, either primary or secondary, exert on the host-plant selection behaviour of herbivorous insects counteract each other and their balance determines the possible outcomes of the decision-making process: rejection or variable degrees of acceptance, manifested as preference in choice situations [6, 32, 74]. When looking at the different categories of host-plant specialization this 'balance model' is a useful concept in understanding selection behaviour. In polyphagous species, several ubiquitous primary metabolites suffice to stimulate feeding on many plant species and rejection occurs only of those plants that produce deterrents of such a quality or in such a quantity that feeding stimulation is negated. A similar principle seems to govern the host-plant range of those oligophagous species that do not seem to use taxon-specific sign stimuli for host-plant recognition. A third category includes oligophagous and monophagous species that do require sign stimuli (Table 6.5) for acceptance. For this category, the stimulatory signal is a taxon-specific secondary metabolite with a definite chemical identity, often perceived by specialized taste receptors (section 6.8.4).

The view emerges that the mechanisms of host-plant selection employed in the different specialization categories are largely a matter of gradation rather than clearly definable and different modalities. In the third group, the association with a particular plant taxon has apparently given rise to a sensory specialization in the insect, constituting an overriding and unambiguous signal for recognition. It should be noted, however, that the balance between inhibitory and stimulatory chemicals is clearly asymmetrical. In other words, the effect of feeding inhibitors can be counterbalanced by feeding stimulants only to some degree. Above a certain level of inhibition no stimulants can evoke feeding. This is convincingly shown by sandwich tests, where the host-plant leaf discs do not neutralize the antifeeding effect of many or even most non-host-plant leaf discs.

In the following section we will discuss how insects use their contact-chemosensory apparatus to determine the balance of stimulatory and inhibitory plant chemicals.

6.8 CONTACT-CHEMOSENSORY BASIS OF HOST-PLANT SELECTION BEHAVIOUR

6.8.1 CONTACT CHEMORECEPTORS

The behavioural responses to plant substances described above are based on the perception of these substances by gustatory neurons. Like olfactory cells, taste cells have their cell bodies located just below the cuticle and send a dendrite into a hair-, cone- or papilla-like sensillum that has one terminal pore at its tip (see Fig. 5.15). Gustatory sensilla are predominantly located in the preoral cavity (e.g. the epipharyngeal sensilla) and on mouthparts, tarsi, ovipositor and antennae (Fig. 6.11).

Extremities equipped with sensilla can often be seen to move in such a way that the sensilla make brief intermittent contacts with the plant surface or interior during contact evaluation behaviour. The numbers of contact chemoreceptive sensilla differ markedly between species and between developmental stages within a species; in holometabolous insects especially, larvae have fewer than adults [20]. In grasshoppers, a trend is seen towards

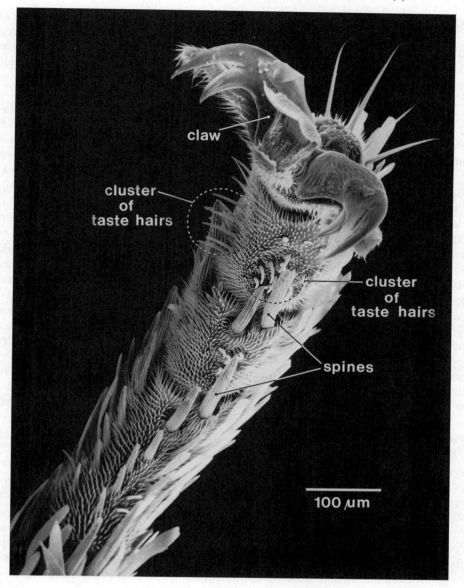

Figure 6.11 Scanning electron micrograph of the ventral side of the two distal tarsomeres of the prothoracic leg of a female *Pieris rapae* butterfly. Clusters of chemosensory hairs occur close to larger, non-innervated spines. (Source: reproduced by courtesy of Dr E. Städler, Wädenswil, Switzerland.)

decreasing numbers of taste sensilla in more specialized feeders [22]. Monophagous acridids that feed on plants with high deterrent properties have the fewest sensilla [11]. In all cases, three to five taste neurons are typically associated with a taste sensillum, whereas most sensilla contain in addition a mechanoreceptive neuron (see Fig. 5.15).

6.8.2 GUSTATORY CODING

Insect gustatory receptors are, like olfactory receptors, said to 'code' the complex chemistry of a plant by transducing the quality and quantity of the mixture of plant compounds into trains of action potentials (or 'spikes'), the electrical carriers of information. The number of action potentials per unit of time and temporal details of spike trains (such as distribution of intervals between them) contain information in an encoded form, which travels without intermittent synapses to the first relay station, located in the suboesophageal ganglion of the central nervous system [61, 79]. It is this ganglion that houses the motor neurons of the mandibular muscles and that therefore ultimately governs feeding activity [12]. Complex stimuli such as plant saps often evoke such trains in several cells innervating either the same sensillum or different sensilla simultaneously, and their axons converge in the suboesophageal ganglion. Here integration occurs by merging with other incoming information from either peripheral or internal receptors and with input from other parts of the brain. After integration has taken place (a process that may take only a fraction of a second), feeding may or may not occur. A complicating factor is that the sensory message conveyed to the brain is by no means constant but varies with age, time of day, physiological state and other biotic and abiotic parameters [14].

The common way to unravel the sensory code is by analysing so-called 'input–output' relationships: the input (trains of action potentials) is quantified electrophysiologically by stimulating identified gustatory sensilla and behaviour (the output) is quantified on the basis of either absolute amounts of food consumed or degree of preference for different feeding (or oviposition) substrates. On the basis of correlations between input and output, coding principles are inferred. In such studies, the sensillum rather than identified cells is often taken as the neurophysiological unit of response. This has a methodological rationale: in the extracellular recordings obtained by the standard tip-recording method a separation of the extracellularly recorded spike trains arising from several taste neurons is technically difficult, despite recently developed methods of computer-assisted spike-train analysis (Appendix C). A second reason is that in only few cases is the specificity of neurons innervating a sensillum so well known that the designation of a cell as 'sugar-best', 'salt-best' or 'water' cell is justified [31]. Indeed, the study of the chemical band-width or tuning of cells is an enterprise in itself and has been carried out in relatively few cases for the eight-cell caterpillar taste system located in the maxillary taste hairs [110], and to a limited extent for tarsal sensilla of *Pieris* butterflies [37, 136] and *Delia* flies [101, 124] in adult herbivorous insects. Theoretically, there is no need to know these specificities in any detail in order to derive gustatory codes. This notion sets the stage for the different starting points of the two most frequently discussed concepts of coding: labelled-line and across-fibre patterning.

6.8.3 CATERPILLARS AS MODELS FOR CODING PRINCIPLES

Caterpillars, many species of which are very specialized feeders, have been favourite models for both sensory coding and behavioural studies. This is because several species were found in ablation studies to require only two maxillary hairs, each with four taste cells, for complete host-plant discrimination behaviour (Fig. 6.12).

The eight taste cells represent about 10% of the total chemosensory complement (reviewed by Schoonhoven [110]). One of the prime questions about chemosensory coding has been whether or not obvious differences exist between codes for the extreme decisions taken during selection behaviour: acceptance and rejection. Dethier's study [30] on seven specialized caterpillar species (including both

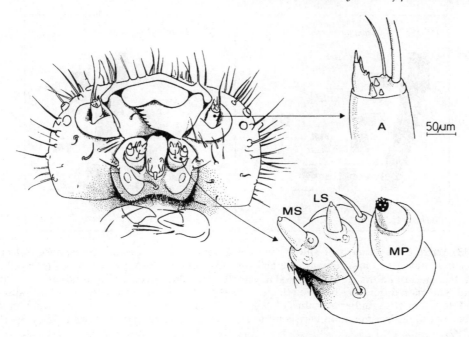

Figure 6.12 Diagram of the head of a caterpillar seen from below with enlargements of an antenna (A) and a maxilla. MP = maxillary palp; LS and MS = lateral and medial sensilla styloconica.

congeneric and unrelated species) led him to conclude that 'there is no universal difference between sensory patterns for acceptance and those for rejection'. This suggests that the nervous system bases its decisions for behavioural output on the combined input from several taste cells by reading synchronously across all afferent axons (fibres). This idea was formalized in the 'across-fibre' patterning concept of gustatory coding put forward in the vertebrate literature [32]. In an earlier study, the sensitivity spectra of the maxillary taste cells of the seven species had been characterized to some extent and little evidence for specialized taste neurons had been found [34]. In both the oligophagous species *Manduca sexta* and polyphagous *Spodoptera* and *Helicoverpa* caterpillars, the ratio of firing between lateral and medial maxillary sensilla styloconica correlated with acceptability [115, 122]. In *Manduca sexta*, across-fibre patterning has been proposed to function as the most proba-

ble mechanism of coding [33, 115], without detailed knowledge of gustatory cell specificities (see above). Evidently, it is the combined input from the two maxillary styloconic sensilla (and thus the across-fibre pattern generated by them) that determines the considerable subtlety in host-plant preference behaviour of these caterpillars [110, 122]. A detailed study of coding of preference behaviour in *Manduca sexta* in response to three solanaceous plants pointed to the role of temporal patterning as another coding principle, which is superimposed on the across-fibre patterning. As a result of different adaptation rates of gustatory cells, the ratios of firing across different cells changes with time [109] and therefore it is important to relate behavioural responses to the relevant time domain of the sensory response.

Most investigations on chemosensory physiology and discrimination behaviour of caterpillars made in concert have focused on

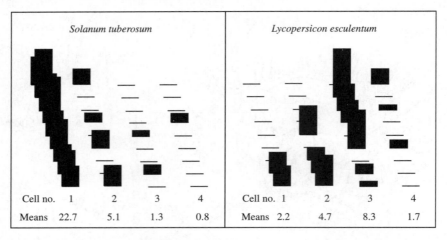

Figure 6.13 Across-fibre patterns of nine individual Colorado potato beetles in response to leaf saps of potato (*Solanum tuberosum*) and tomato (*Lycopersicon esculentum*). The activity rates of four cells in taste sensilla on the galea of adults are represented as bars (mean values over nine individuals are indicated at the bottom). The main differences between the responses to potato and tomato are the little or absent activity of cell 1, together with higher activities of cells 2 and 3 in response to tomato sap, which provide the basis for behavioural discrimination between the two plants. (Source: redrawn from Haley Sperling and Mitchell, 1991.)

the eight taste neurons located on the maxillary galea. Recent studies demonstrated, as might be expected, that input from antennae and maxillary palps also contributes to food-plant discrimination [25]. Clearly, these organs merit more attention than they have received so far.

Additional taste organs are located in the preoral cavity. Caterpillars commonly have two placoid sensilla on the epipharyngeal surface of the labrum, though none are present in *Mamestra brassicae*. These sensilla have three chemosensitive neurons each. Information from these sensilla may be involved in swallowing responses [110]. Colorado potato beetles also possess epipharyngeal sensilla [76], whereas acridids have whole groups on the epipharyngeal face of the labrum and on the hypopharynx [20].

Adult insects have considerably more sensilla and taste neurons at their disposal than larvae [20]. This is especially true in the Lepidoptera and Coleoptera, in which the difference is at least tenfold. Most probably these increased receptor numbers relate to the more complex behavioural tasks of adults. Whereas larvae are predominantly interested in eating the right food, adult insects represent the dispersal phase and must find, besides food, mating partners and, when female, oviposition sites. Despite the technical drawback of dealing with large receptor numbers, successful attempts have been made to analyse the coding of food preference in adult beetles [47, 80] and moths [13]. By recording responses from a limited sample of the galeal sensilla of Colorado potato beetles (*Leptinotarsa decemlineata*) it appeared that saps from three host-plant species elicited a much more consistent response in the taste neurons than those from non-hosts. Preference among different solanaceous host plants is most probably based on neural messages coded in across-fibre patterns, but there are also indications for the use of labelled-line coding (Fig. 6.13).

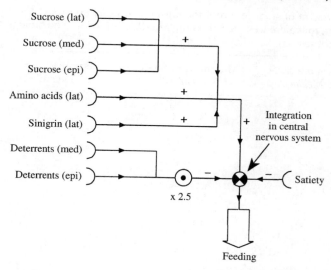

Figure 6.14 Schematic representation of how the inputs from different mouthpart chemoreceptors might be integrated within the central nervous system to regulate feeding in the caterpillar of *Pieris brassicae*. Impulses from the sucrose, amino acid and glucosinolate cells in the lateral (lat) and medial (med) sensilla styloconica on the galea and those from the epipharynx (epi) would have positive effects (+) tending to stimulate feeding, whereas inputs from the deterrent cells would have negative effects (−) tending to inhibit feeding. Satiety, representing a physiological parameter, would inhibit feeding when the gut is full. 'Feeding' or 'not-feeding' depends on the arimethric ratio between positive and negative inputs, i.e. nerve impulse frequencies. (Source: redrawn from Schoonhoven, 1987.)

6.8.4 SIGN STIMULUS RECEPTORS: UNSURPASSED SPECIALISTS

An important event in the study of the chemosensory basis of host-plant specialization was the discovery of taste cells highly sensitive to secondary plant substances in caterpillars of the large white butterfly (*Pieris brassicae*), a crucifer specialist [106]. These cells are located in both sensilla styloconica on the galea of each maxilla and respond to a number of glucosinolates, which are characteristic of Cruciferae. The two cells have overlapping but not identical sensitivity spectra. A certain minimal level of activity in these cells is required to signal acceptability of plant material. Such a chemoreceptor cell can be designated as a 'labelled line', i.e. a line (axon) along which information is transferred to the brain that correlates quantitatively with the strength of the behavioural response. The influence of these labelled-line-type receptors for sign stimuli can be neutralized, however, by deterrents like alkaloids or phenolic acids, which are perceived by so-called deterrent receptors [113, 144] (Fig. 6.14).

Since then more examples have been found of taste neurons that are specifically sensitive to a group of secondary plant metabolites (Table 6.5). Such chemosensory cells seem to be quite special for specialized herbivorous insects as they have not been documented for other animal groups, such as vertebrates, the taste system of which has been studied most extensively. This parallels the notion that the degree of host-plant specialization met in herbivorous insects is not equalled in other groups of herbivores, including vertebrates. In several monophagous or oligophagous species for which a sign stimulus was identified through combined phytochemical and behavioural investigations, electro-

Table 6.7 Sensitivity spectra of sucrose receptors of four caterpillar species in the medial or lateral galeal sensillum styloconicum; for references see Schoonhoven, 1987, on which the table is based (+ = sensitive; − = insensitive; • = not tested)

	Pieris brassicae Medial	Mamestra brassicae Lateral	Bombyx mori Lateral	Dendrolimus pini Medial
Pentoses				
D-arabinose	−	+	+	+
L-arabinose	−	+	+	+
L-rhamnose	•	−	+	•
D-xylose	−	•	+	+
Hexoses				
D-fructose	+	+	+	+
D-galactose	−	+	+	+
D-glucose	+	+	+	+
D-mannose	−	−	+	+
L-sorbose	+	•	+	+
Disaccharides				
Lactose	−	−	+	−
Maltose	−	+	+	−
Sucrose	+	+	+	+
Trehalose	−	−	+	•
Cellobiose	•	+	+	•
Alcohols				
Inositol	−	•	−	+
Sorbitol	−	•	+	•

physiological analyses revealed the presence of a corresponding sign-stimulus receptor. Stimulation of these cells is a signal to the brain: accept this food or oviposition site. For all cases documented so far such specialist cells detect stimulatory chemicals.

It should be noted that the across-fibre patterns and labelled-line concepts are not mutually exclusive. The two concepts can be amalgamated into one model in which across-fibre patterning (i.e. many cells, each with different but overlapping sensitivity spectrum) participate in coding complex stimuli (such as plant saps). However, some cells with a narrow and well circumscribed sensitivity spectrum (labelled-line cells) may have a more pronounced or dominant influence and even play a decisive role in behavioural decisions. Likewise, deterrent cells may take up a domi-

nant or indeed veto position in the decision process. The presence of one or more dominant information channels does not rule out the function of the other taste cells. The latter contribute to the decision process with more subtle details from the sensory evaluation of a plant's chemistry.

6.8.5 SUGAR AND AMINO ACID RECEPTORS: DETECTORS OF NUTRIENTS

In section 6.5.1 we discussed the general importance of primary metabolites as feeding stimulants. In caterpillars, some taste cells sensitive to primary plant metabolites (e.g. sugars) that stimulate feeding are also specialized: they can be excited only by a narrow range of sugars, but not by e.g. amino acids, or secondary plant metabolites (Table 6.7).

Table 6.8 Sensitivity spectra of amino acid receptors of lepidopterous larvae and coleopterous larvae or adults in the galeal sensilla styloconicum; for references see Schoonhoven, 1981, and Van Loon and Van Eeuwijk, 1989, on which the table is based (P.b. = Pieris brassicae; A.o. = Adoxophyes orana; H.z. = Heliothis zea; E.a. = Estigmene acrea; M.a. = Malacosoma americana; D.p. = Danaus plexippus; M.b. = Mamestra brassicae; P.p. = Papilio polyxenes; L.d. = Lymantria dispar; C.e. = Calpodes ethlius; P.r. = Pieris rapae; L.d. = Leptinotarsa decemlineata; E.a. = Entomoscelis americana; ++++ = strong response; + = weak response; − = no response; • = not tested; inh = inhibition as compared to control)

	Caterpillars											Beetles	
	P.b.	A.o.	H.z.	E.a.	M.a.	D.p.	M.b.	P.p.	L.d.	C.e.	P.r.	L.d.	E.a.
GABA	++	•	•	•	•	•	•	•	•	•	++	+++	+
Proline	+++	•	+	+++	+++	++++	•	+	+	+++	+++	++	++
Alanine	++	•	+	++++	++++	+++	•	+	inh	•	+++	+++	+++
Serine	++	•	+	++++	+	++++	•	+++	inh	•	+++	++	++
Glycine	−	•	+	+++	−	+	•	inh	+	•	−	+	++
Cysteine	++	•	•	+++	+	•	•	−	+	+++	−	•	•
Cystine	•	•	+++	+	++	++	•	−	+	++	•	•	•
Tryptophan	+++	++	+	++	+	+	•	++	++	•	+	•	+
Tyrosine	−	•	++	++	+	+	•	+	−	•	−	•	•
Arginine	−	•	+	+++	++	−	•	−	−	+++	−	•	++
Methionine	+++	++	+++	++	+	+++	++	−	−	+++	++++	•	++
Leucine	+++	++++	+++	+++	+	+	•	−	−	•	++++	+	+
Isoleucine	+++	++++	+	+	+	+	•	−	++	•	+++	+	+
Aspartic acid	−	•	+	−	++	+++	•	−	+	+++	−	•	++
Glutamic acid	−	•	+	+++	+++	−	•	−	+	+++	−	•	++
Histidine	++++	++	+	+	+	•	•	++	+	•	++	•	+
Valine	+++	+++	−	++	+++	+++	•	+++	+	•	+++	+	•
Phenylalanine	++++	+	++	+++	+	+	++	+	+	•	++	+	+
Threonine	++	+	+	++	+	+++	•	+++	++	•	•	+	+

In *Pieris* caterpillars, out of the eight taste cells present in the maxillary styloconic sensilla two are 'sugar-best' cells with overlapping but different sensitivity spectra [67]. Stimulation of these cells is essential to induce adequate feeding rates. Amino-acid-sensitive taste cells have been found in various insect species (Table 6.8).

Sometimes perception of sugars and amino acids occurs *via* the same cell. In the adult Colorado potato beetle a maxillary taste neuron sensitive to sugars also responds to two amino acids, GABA and alanine, which are known to stimulate feeding [78]. Moreover, in larvae of the red turnip beetle (*Entomoscelis americana*) the sucrose-best cell responds to some sugars, e.g. sucrose and maltose, as well as to some amino acids [77], whereas curiously in the adult insect this cell appears to be unresponsive to amino acids [139]. Clearly, the sensitivity spectra of taste cells differ among species and may even vary between developmental stages of the same species. The most thoroughly investigated insect 'sugar-best' cells are those on the proboscis of several adult Diptera that are saprophagous. These cells generally combine sensitivity to sugars and amino acids, although separate receptor sites have been postulated [83]. In contrast, many Lepidoptera use separate cells to mediate information on the presence of sugars and amino acids [147].

Another category of cell responding to generally occurring compounds is the 'inositol cell'. Several caterpillar species possess specialized receptor cells for sugar alcohols that stimulate feeding, like inositol (**32**) [108, 120]. In *Yponomeuta* species different taste cells have been found for the two stereo-isomeric sugar alcohols dulcitol (**20**) and sorbitol (**64**), which constitute strong feeding stimulants to the caterpillars: a rosaceous non-host can be rendered acceptable to the celastraceous specialist *Yponomeuta cagnagellus* by impregnating *Prunus* foliage with dulcitol, the sugar alcohol that typically occurs at high concentrations in Celastraceae [91].

6.8.6 DETERRENT RECEPTORS: GENERALIST TASTE CELLS

In many caterpillar species one or more taste cells have been identified that respond to a range of secondary plant substances occurring in non-hosts. Stimulation of these cells caused rejection of otherwise perfectly acceptable host plants after they had been treated with such compounds (reviewed by Schoonhoven *et al.* [117]). These cells are designated 'deterrent receptors'. They can be considered to be generalist taste cells in view of their sensitivity to a wide range of chemically unrelated classes of secondary plant compounds. The term 'generalist' does not mean, of course, that they respond to everything, e.g. sugars, or to all secondary plant compounds. For this cell type also, different caterpillar species display different sensitivity profiles [109]. How deterrent cells are able to express this broad sensitivity is poorly understood but, on the basis of electrophysiological and genetic findings, there is evidence that different receptor sites tuned to, for instance, phenolic or alkaloid compounds, are involved [44]. *Pieris* caterpillars have both a generalist as well as a more specialized deterrent cell in their maxillary taste hairs [144]. The specialist cell in the lateral sensillum (Fig. 6.12) is a 'cardenolide-best' receptor by virtue of its extreme sensitivity to cardenolides (threshold about 10^{-8} mol/l). These compounds act as powerful steroidal deterrents and their presence in certain members of the insect's host-plant family Cruciferae make these confamilial plant species unacceptable. The same cell also responds to phenolic acids and flavonoids, but only at a concentration more than 1000 times higher. The generalist deterrent neuron in the other hair, the medial sensillum, is also stimulated by cardenolides but at a concentration more than 10 times higher [145]. At present the cardenolide-sensitive cell is the only known example of a specialized deterrent cell. It can be envisaged that it has evolved from a generalist deterrent cell by loss of receptor sites for other classes of deterrent such as alkaloids.

If we could demonstrate the existence of receptor sites, which we currently think are proteinaceous membrane-bound macromolecules probably encoded by a single gene, it would significantly advance our understanding of deterrence and thereby the basis of specialized feeding habits [146] (Chapter 10). Although few data on the genetic basis of chemoreceptor sensitivity in herbivorous insects are available, a study on interspecific hybrids of two *Yponomeuta* species provided evidence that sensitivity to a feeding deterrent, a chalcone glycoside, is inherited *via* a single dominant gene [142]. Host-plant shifts in some individuals may be based on reduced sensitivity to deterrents present in a particular non-host species. This has possibly been an important factor in the evolution of, for instance, *Yponomeuta*. Phylogenetic reconstruction of this genus suggests that Celastraceae comprise the ancestral host-plant family and that a shift occurred to Rosaceae (Chapter 10). One species, *Yponomeuta malinellus*, feeding on the rosaceous genus *Malus*, and a second species, *Y. rorellus*, that has made a shift to yet another plant family, the Salicaceae, both lack sensitivity at the chemoreceptor level to compounds specifically occurring in *Malus* and *Salix* respectively, whereas these substances act as deterrents to the other species studied (Table 6.9).

The converse process may also occur, leading to a narrowing of host range. A better characterization of number and specificity of receptor sites is needed to support such scenarios. If a gene coding for a deterrent receptor molecule were to be expressed in a taste neuron sensitive to stimulants such as sugars, it would explain how sign-stimulus receptors originated. Indeed, that this can occur has

Contact-chemosensory basis of plant selection

Table 6.9 Chemosensory sensitivities in galeal styloconic taste receptors in four *Yponomeuta* species (Yponomeutidae), specialized feeders associated with host plants that are chemotaxonomically unrelated (Source: data from van Drongelen, 1979) + = receptor sensitive; − = receptor insensitive; n.t. = not tested

		Taste receptor specificities in lateral/medial sensilla styloconica			
Species	Host plant (family)	Dulcitol	Sorbitol	Phloridzin	Salicin
Yponomeuta cagnagellus	*Euonymus europaeus* (Celastraceae)	+/+*	−/−	−/+	n.t./+
Yponomeuta padellus	*Prunus/Crataegus* spp. (Rosaceae)	±/−†	+/−*	−/+	+/+
Yponomeuta malinellus	*Malus* spp. (Rosaceae)	−/−	+/−	−/−*	+/+
Yponomeuta rorellus	*Salix* spp. (Salicaceae)	−/−	−/−	−/+	−/+*

*compound present in hostplant mentioned
†dulcitol is present in some rosaceous host plants in low concentrations (about 10% of the levels found in Celastraceae)

been found in a taste mutant of *Drosophila melanogaster* [2]. The putative taste receptor proteins have not yet been isolated from insects, although recent reports suggest that insect sugar receptor molecules may become known soon [88].

Several recent studies have shown that so-called deterrent neurons in caterpillars act as 'labelled lines': the degree to which certain deterrent compounds coated on acceptable food causes rejection compared to untreated controls correlates with the firing rates of deterrent receptors in several caterpillar species [72, 92, 125] (Fig. 6.15).

6.8.7 PERIPHERAL INTERACTIONS

From the foregoing separate discussions of stimulant and deterrent receptors, a model emerges in which information on feeding stimulants and feeding deterrents is detected by independent chemoreceptor neurons and is transmitted separately to the brain; the subsequent weighing of inputs at the central level can conceivably occur according to arithmetical rules. Although relatively simple arithmetical rules could be derived for *Pieris* and *Mamestra* caterpillars feeding on artificial diets [113], from electrophysiological studies on other caterpillars, beetles and grasshoppers a growing number of examples have come to

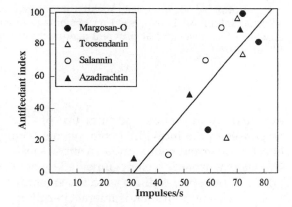

Figure 6.15 Relationship between antifeedant index (as determined by dual choice tests) and spike frequencies of a deterrent receptor cell in the medial sensillum styloconicum of *Pieris brassicae* larvae. Impulse frequencies in response to three different concentrations of azadirachtin, salannin, toosendanin and Margosan-O® have been plotted against antifeedant indices, at equimolar concentrations of the same compounds. A significant correlation is found between the intensity of the deterrent cell response and the antifeedant index. (Source: redrawn from Luo *et al.*, 1995.)

light in which interactions occur at the chemosensory periphery that do not conform to linear arithmetic: the presentation of mixtures to a sensillum produces responses from one or several taste cells that would not be

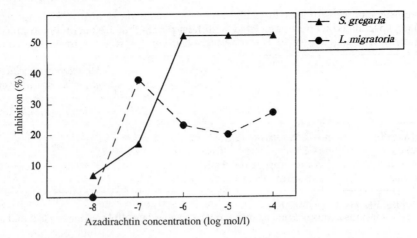

Figure 6.16 Inhibition of chemosensory responses to azadirachtin caused by the presence of sucrose (at a concentration of 50 mmol/l) in maxillary palps of two locust species, *Locusta migratoria* and *Schistocerca gregaria*. Inhibition is expressed as percentage of normal neural responses to azadirachtin, i.e. in the absence of sucrose, and at different azadirachtin concentrations. (Source: redrawn from Simmonds and Blaney, 1996.)

expected from simple adding up of the responses to the individual components. The effect of deterrent compounds on sugar-sensitive taste neurons is well documented by now [42, 117], but species differ as to the extent in which the same compounds interact peripherally [123] (Fig. 6.16).

An example is the effect of an anthocyanin on the sugar-best cell in *Pieris* caterpillars. This flavonoid compound not only excites both the lateral and medial deterrent cell in galeal taste hairs but also inhibits the sucrose sensitive cell present in both sensilla (Fig. 6.17). The reverse effect also occurs when stimulants suppress the response of deterrent receptors [121].

Interactions at the sensory level are not necessarily inhibitory as in the examples discussed so far. They may also be of the synergistic type. For example, the sinigrin-sensitive cell in the polyphagous larva of *Isia isabella* is synergized by sucrose, which when applied singly stimulates only the sugar cell [34] (Fig. 6.18).

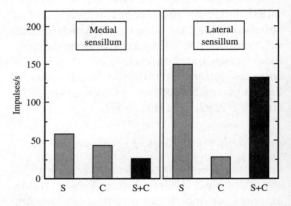

Figure 6.17 Inhibitory effects of the anthocyanin cyanin chloride on sugar responses in the two maxillary sensilla styloconica of *Pieris brassicae* larvae. Responses are presented as total impulse frequencies when stimulated with 15 mmol/l sucrose (S), 2.5 mmol/l cyanin chloride (C) and a mixture of these two stimuli (S+C). Neural activity in response to the mixtures is significantly lower than would be expected from adding up the values for single compounds in both sensilla. (Source: redrawn from van Loon, 1990.)

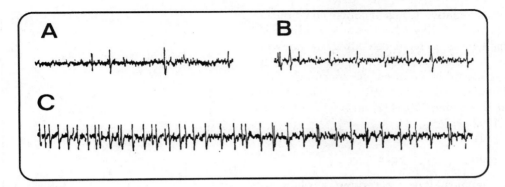

Figure 6.18 Synergistic receptor responses in the medial sensillum styloconicum on the maxilla of *Isia isabella* larvae. **(A)** Response to 0.001 mol/l sinigrin. **(B)** Response to 0.1 mol/l sucrose. **(C)** Response to a mixture of sinigrin and sucrose. The cell that preferentially responds to sinigrin alone shows a greatly increased response to the mixture. (Source: reproduced from Dethier and Kuch, 1971, with permission.)

This differs from the case in which two compounds both stimulate the same cell but evoke in combination an increased reaction in comparison to the response to either alone. An example of the latter is known from the maxillary taste cells of *Dendrolimus pini* caterpillars, which are responsive to a number of carbohydrates. Mixtures of glucose and inositol elicit much stronger reactions than either compound alone [110].

Peripheral interactions have been revealed in a growing number of cases since the attention has shifted from studying the stimulatory effects of pure compounds to the responses to binary mixtures of chemicals and to plant saps that represent natural but chemically undefined complex stimuli. Clearly, knowledge of responses to plant saps is very important to the understanding of the chemosensory basis of selection among different host plants. Studying interactions in responses to binary mixtures may lead to results that are not representative of the complex stimulus situation of a leaf sap. The triterpenoid toosendanin is a powerful deterrent to *Pieris brassicae* larvae.

It excites the medial deterrent neuron and inhibits sucrose and glucosinolate neurons, both of which mediate feeding stimulation [116]. The triterpenoid azadirachtin also excites the medial deterrent cell, but to a weaker extent and it does not affect the responses of the stimulant receptor cells [65]. When the deterrent effects of toosendanin and azadirachtin are compared in a bioassay employing host-plant leaf discs, the response of the deterrent cell alone correlates well with the level of deterrence and the putative contribution of the suppression of stimulant receptors by toosendanin seems to be minor if any. The occurrence and importance of peripheral interactions should therefore be studied by approaching the stimulus situation encountered during feeding or oviposition as closely as possible [146].

It is unknown how peripheral interactions of different kinds arise. Very probably competitive or allosteric interactions occur at receptor sites in the membrane [42, 83], but as yet no direct proof for this is available.

When deterrent compounds affect stimulant

receptors negatively, this of course contributes to the neural coding of deterrence. Additional mechanisms of deterrence coding are known, e.g. deterrents causing irregular firing in sucrose-sensitive neurons. A systematic discussion of the various coding principles can be found in Frazier, 1992 [42] and Schoonhoven et al., 1992 [117].

6.8.8 OVIPOSITION PREFERENCE

Adult females, when accepting a plant to oviposit on, make a choice that is of crucial importance to the survival chances of their offspring, as the mobility and energy reserves of many first instar larvae are so limited that their opportunities of finding a suitable host on their own are minimal. In two species of Delia flies (Diptera: Anthomyiidae), oligophagous on Cruciferae, egg-laying is induced when the female contacts glucosinolates. Females show a distinct order of preference for different glucosinolates. The neural responses of specific chemoreceptors located in sensory hairs on the tarsi, elicited by glucosinolates, correlate well with behavioural responses to these compounds (Fig. 6.19). It is concluded that tarsal sensilla play an important role in host-plant recognition.

The two butterflies Pieris rapae and P. napi oleracea each display their own preference hierarchy for different glucosinolates (Fig. 6.7). Electrophysiological studies on tarsal taste sensilla showed that here also the behaviourally most preferred compounds elicit the highest activity in glucosinolate-sensitive receptor cells [136]. Actually, it is surprising that such input–output relationships can be found, as the sample of sensory input quantified (the number of cells from which recordings were made relative to the total number of taste neurons present) comprises only 1–2% of the receptors available to the female. These findings, like those described above for caterpillars, indicate that the sensory characteristics vary among congenic butterflies. Presumably the sensory system of each species is adapted

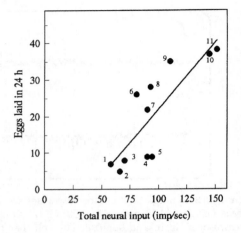

Figure 6.19 Relationship between summed neural input (impulses in the first second of stimulation) from two different receptor types on the legs and from labellar sensilla in the turnip root fly (Delia floralis) (X axis) and oviposition behaviour (number of eggs laid over a 24 h period in a no-choice situation) for 11 different glucosinolates sprayed at 10^{-2} mol/l on an artificial leaf (Y axis). A significant correlation is found between neural input and behavioural output. 1 = glucoerucin; 2 = glucoiberin; 3 = progoitrin; 4 = sinalbin; 5 = neoglucobrassicin; 6 = sinigrin; 7 = gluconapin; 8 = glucotropaeolin; 9 = gluconasturtiin; 10 = glucobrassicanapin; 11 = glucobrassicin. (Source: redrawn from Simmonds et al., 1994.)

to the host-plant selection typical of that particular species. Even within a species, i.e. between subspecies, differences in sensory responses have been observed, indicating an evolutionary flexibility of the system. This is exemplified by two subspecies of Pieris napi that show consistent differences in responses to glucosinolates [37] (Fig. 6.20).

Cardenolides, deterrents to larvae, have also proved to be powerful oviposition deterrents to both Pieris species (reviewed by Chew and Renwick [23]). They stimulate one cell, but do not affect the 'glucosinolate-best' cell. The preference hierarchy for glucosinolates is determined by the ensemble firing of the 'glucosinolate-best' neuron (positively correlated with higher preference) and the 'cardenolide-

Contact-chemosensory basis of plant selection

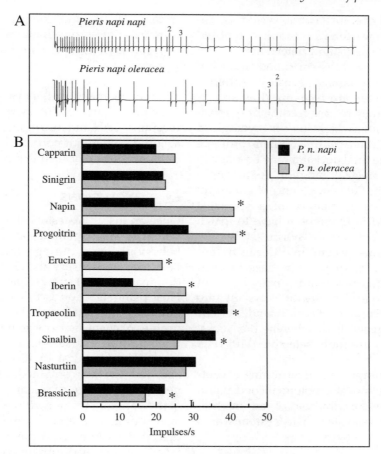

Figure 6.20 (A) Recordings of electrophysiological activity from taste hairs on tarsi of female *Pieris napi oleracea* and *P. napi napi* in response to the glucosinolate gluconapin at 10 μg/ml. In *P. napi napi* a second cell (designated '2') fires much more frequently than in *P. napi oleracea*. **(B)** Response profiles to ten different glucosinolates (the response strength is expressed as the number of one spike type (indicated by '3' in **(A)**) in the first second of stimulation); significant differences were found between both subspecies for seven compounds (indicated by *). (Source: redrawn from Du *et al.*, 1995.)

best' cell (negatively correlated with preference): the code is made up of a balance of two labelled lines, which is the most elementary across-fibre pattern. This example clearly shows the continuum that exists between the labelled-line and across-fibre pattern concepts. When a female alights upon a crucifer that carries a mixture of glucosinolates and cardenolides on its surface, both neurons will be excited and the balance of activity between the two will determine acceptability or rejection.

6.8.9 HOST-PLANT SELECTION: A THREE-TIER SYSTEM

Host-plant selection involves three major elements:

1. a peripheral chemoreceptive system, sensitive to multiple chemical stimuli, composed of phagostimulants and deterrents;
2. a central nervous system tuned in such a way as to recognize sensory patterns. Certain patterns are recognized as acceptable, that is they release feeding or oviposition behaviour (which may be synergized by a 'motivation centre'), others promote rejection. The final decision may be taken in the suboesophageal ganglion. As a simplified model the 'key–lock' concept is a useful one. The sensory pattern of a specialist feeder would in this model have to match more closely a certain norm set by the central nervous system, in order to trigger feeding activity, than is the case in food generalists. In other words, many different receptor activity profiles or 'keys' fit into the CNS template ('lock') and release feeding in generalists, whereas the 'locks' of specialists are more selective [110] (Fig. 6.21);
3. as a third component determining acceptance or rejection of a potential food plant, there is the contribution of an internal chemosensitive system. This system warns the CNS if food composition differs too much from physiological requirements, resulting in a change of food selection (see section 4.4.3).

Of course the three-tier system of host-plant selection, with its interacting elements of receptors, CNS and nutritional feedback, is not a closed system but perpetually interacts with numerous ecological constraints [112].

6.9 CONCLUSIONS

Once an insect has established contact with a potential host plant, elaborate evaluation behaviour ensues during which the insect uses both mechanosensory and chemosensory (predominantly taste) stimuli offered by the plant. Host-plant selection is to a large extent governed by a central neural evaluation of the profiles of chemosensory activity generated by the multitude of taste stimuli presented by the plant. The chemical quality of the plant is encoded in the combined activity of taste neurons that have different degrees of specificity, ranging from highly specialized (e.g. sign-stimulus receptors) to generalists (e.g. deterrent receptors). At the behavioural level it has been amply documented that acceptance is determined by the balance between stimulatory and inhibitory compounds. Only recently it has been demonstrated that this balance can be traced, partially at least, to activity at the chemosensory level as the ratio of identifiable stimulatory and inhibitory inputs. This ratio often seems to determine preference hierarchies in a straightforward way. In other cases, however, the codes have not been cracked and it is clear that uncovering the physiological basis of the often intricate discriminatory ability of plant-feeding insects is a continuing challenge. Because more and more peripheral interactions are being found in response to binary mixtures, the study of chemosensory activity profiles in response to plant saps, the natural stimuli, is implicated as the best way to account for the

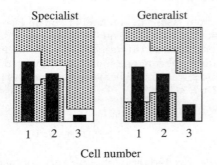

Figure 6.21 Model of CNS processing of sensory input in a food specialist and a generalist. The black bars represent action potential frequencies in three chemoreceptors (1–3) when stimulated by an acceptable food plant. The white space of the 'lock' reflects the variation allowed to the sensory input while still being interpreted as acceptable. Cell 3 is a deterrent receptor. (Source: redrawn from Schoonhoven, 1987.)

possibly large numbers of interactions occurring under field conditions.

Clearly our still limited knowledge of insect taste receptors permits the conclusion that herbivorous insects possess a highly sensitive system that allows them to detect subtle chemical differences between plants and between plant parts. Another important conclusion is that each species, perhaps even each biotype, is equipped with a species-specific sensory system that is optimally equipped to discriminate between host plants and non-hosts, as well as among different hosts.

The existence of highly specialized taste receptors in several specialized feeders together with evidence for the existence of several receptor sites with monogenic inheritance on generalist deterrent neurons are observations relevant to understanding the evolution of specialization and the probability of host shifts. As the activity of such receptors is the basis of acceptance or rejection decisions, mutational changes at the receptor level will affect behaviour. When, for instance, the sensitivity to a (class of) deterrents is lost by a mutation in the respective receptor site, a host shift may occur. Previously unacceptable plants containing such deterrents may then become acceptable and the host range is expanded when the deterrents involved are not lethally toxic (and many of them are not). Support for this scenario comes from the lepidopterous genus *Yponomeuta*.

The evolution of feeding specialization so characteristic for herbivorous insects may thus be determined to a considerable degree by neural constraints, either at the sensory or at the central level [146] (see also Chapter 10).

6.10 REFERENCES

1. Ahmad, S. (1982) Host location by the Japanese beetle: Evidence for a key role for olfaction in a highly polyphagous insect. *J. Exp. Zool.*, **220**, 117–120.
2. Arora, K., Rodrigues, V., Joshi, S., Shanbhag, S. and Siddiqi, O. (1987) A gene affecting the specificity of the chemosensory neurons of *Drosophila*. *Nature*, **330**, 62–63.
3. Baur, R., Feeny, P. and Städler, E. (1993) Oviposition stimulants for the black swallowtail butterfly: Identification of electrophysiologically active compounds in carrot volatiles. *J. Chem. Ecol.*, **19**, 919–937.
4. Benedict, J. H., Leigh, T. F. and Hyer, A. H. (1983) *Lygus hesperus* (Heteroptera, Miridae) oviposition behavior, growth and survival in relation to cotton trichome density. *Environ. Entomol.*, **12**, 331–335.
5. Bernays, E. A. and Chapman, R. F. (1978) Plant chemistry and acridoid feeding behaviour, in *Biochemical Aspects of Plant and Animal Coevolution*, (ed. J. B. Harborne), Academic Press, London, pp. 91–141.
6. Bernays E. A. and Chapman, R. F. (1994) *Host-Plant Selection Behaviour of Phytophagous Insects*, Chapman & Hall, New York.
7. Bernays, E. A. and Simpson, S. J. (1982) Control of food intake. *Adv. Insect Physiol.*, **16**, 59–118.
8. Bernays, E. A., Howard, J. J., Champagne, D. and Estesen, B. J. (1991). Rutin, a phagostimulant for the polyphagous acridid *Schistocerca americana*. *Entomol. Exp. Appl.*, **60**, 19–28.
9. Blaakmeer, A., Stork, A., van Veldhuizen, A., van Beek, T. A., de Groot, Æ., van Loon, J. J. A. and Schoonhoven, L. M. (1994a) Isolation, identification, and synthesis of miriamides: new hostmarkers from eggs of *Pieris brassicae*. *J. Nat. Prod.*, **57**, 90–99.
10. Blaakmeer, A., Hagenbeek, D., van Beek, T. A., de Groot, Æ., Schoonhoven, L. M. and van Loon, J. J. A. (1994b) Plant response to eggs vs. host marking pheromone as factors inhibiting oviposition by *Pieris brassicae*. *J. Chem. Ecol.*, **20**, 1657–1665.
11. Bland, R. G. (1989) Antennal sensilla of Acrididae (Orthoptera) in relation to subfamily and food preference. *Ann. Entomol. Soc. Am.*, **82**, 368–384.
12. Blaney, W. M. and Simmonds, M. S. J. (1987) Control of mouthparts by the subesophageal ganglion, in *Arthropod Brain*, (ed. A. P. Gupta), John Wiley, New York, pp. 303–322.
13. Blaney, W. M. and Simmonds, M. S. J. (1990) A behavioural and electrophysiological study of the role of tarsal chemoreceptors in feeding by adults of *Spodoptera*, *Heliothis virescens* and *Helicoverpa armigera*. *J. Insect Physiol.*, **36**, 743–756.
14. Blaney, W. M., Schoonhoven, L. M., and Sim-

monds, M. S. J. (1986). Sensitivity variations in insect chemoreceptors: a review. *Experientia*, **42**, 13–19.
15. Blaney, W. M., Simmonds, M. S. J., Ley, S. V. and Katz, R. B. (1987) An electrophysiological and behavioural study of insect antifeedant properties of natural and synthetic drimane-related compounds. *Physiol. Entomol.*, **12**, 281–291.
16. Blau, P. A., Feeny, P. and Contardo, L. (1978) Allylglucosinolate and herbivorous caterpillars, a contrast in toxicity and tolerance. *Science*, **200**, 1296–1298.
17. Blom, F. (1978) Sensory activity and food intake: a study of input–output relationships in two phytophagous insects. *Neth. J. Zool.*, **28**, 277–340.
18. Bowers, M. D. (1983) The role of iridoid glycosides in host-plant specificity of checkerspot butterflies. *J. Chem. Ecol.*, **9**, 475–494.
19. Chapman, R. F. (1977) The role of the leaf surface in food selection by acridids and other insects. *Coll. Intern. CNRS*, **265**, 133–149.
20. Chapman, R. F. (1982) Chemoreception: the significance of receptor numbers. *Adv. Insect Physiol.*, **16**, 247–356.
21. Chapman, R. F. and Bernays, E. A. (1989) Insect behavior at the leaf surface and learning as aspects of host plant selection. *Experientia*, **45**, 215–222.
22. Chapman, R. F. and Fraser, J. (1989) The chemosensory system of the monophagous grasshopper, *Bootettix argentatus* Bruner (Orthoptera, Acrididae). *Int. J. Insect Morphol. Embryol.*, **18**, 111–118.
23. Chew, F. S. and Renwick, J. A. A. (1995) Chemical ecology of hostplant choice in *Pieris* butterflies, in *Chemical Ecology of Insects*, 2nd edn. (eds R. T. Cardé and W. J. Bell), Chapman & Hall, New York, pp. 214–238.
24. Chiang, H.-S. and Norris, D. M. (1983) Morphological and physiological parameters of soy bean resistance to Agromyzid bean flies. *Environ. Entomol.*, **12**, 260–265.
25. De Boer, G. (1993) Plasticity in food preference and diet-induced differential weighting of chemosensory information in larval *Manduca sexta. J. Insect Physiol.*, **39**, 17–24.
26. De Candolle, A. P. (1804) *Essai sur les Propriétés Médicales des Plantes, Comparées avec leurs Formes Extérieures et leur Classification Naturelle*, Didot Jeune, Paris.
27. Derridj, S., Gregoire, V., Boutin, J. P. and Fiala, V. (1989) Plant growth stages in the interspecific oviposition preference of the European corn borer and relations with chemicals present on the leaf surfaces. *Entomol. Exp. Appl.*, **53**, 267–276.
28. Derridj, S., Wu, B. R., Stammitti, L., Garrec, J. P. and Derrien, A. (1996) Chemicals on the leaf surface: information about the plant available to insects. *Entomol. Exp. Appl.*, **80**, 197–201.
29. Dethier, V. G. (1941) Chemical factors determining the choice of food by *Papilio* larvae. *Am. Nat.*, **75**, 61–73.
30. Dethier, V. G. (1973) Electrophysiological studies of gustation in lepidopterous larvae. II. Taste spectra in relation to food-plant discrimination. *J. Comp. Physiol.*, **82**, 103–133.
31. Dethier, V. G. (1976) *The Hungry Fly*, Harvard University Press, Cambridge, MA.
32. Dethier, V. G. (1982) Mechanisms of host plant recognition. *Entomol. Exp. Appl.*, **31**, 49–56.
33. Dethier, V. G. and Crnjar, R. M. (1982) Candidate codes in the gustatory system of caterpillars. *J. Gen. Physiol.*, **79**, 549–569.
34. Dethier, V. G. and Kuch, J. H. (1971) Electrophysiological studies of gustation in lepidopterous larvae. I. Comparative sensitivity to sugars, amino acids, and glycosides. *Z. Vergl. Physiol.*, **72**, 343–363.
35. Doss, R. P., Shanks, C. H., Chamberlain, J. D. and Garth, J. K. L. (1987) Role of leaf hairs in resistance of a clone of strawberry, *Fragaria chiloensis*, to feeding by adult black vine weevil, *Otiorhynchus sulcatus* (Coleoptera, Curculionidae). *Environ. Entomol.*, **16**, 764–766.
36. Dreyer, D. L. and Jones, K. C. (1981) Feeding deterrency of flavonoids and related phenolics towards *Schizaphis graminum* and *Myzus persicae*: aphid feeding deterrents in wheat. *Phytochemistry*, **20**, 2489–2493.
37. Du, Y.-J., van Loon, J. J. A. and Renwick, J. A. A. (1995) Contact chemoreception of oviposition stimulating glucosinolates and an oviposition deterrent cardenolide in two subspecies of *Pieris napi. Physiol. Entomol.*, **20**, 164–174.
38. Eigenbrode, S. D. and Espelie, K. E. (1995) Effects of plant epicuticular lipids on insect herbivores. *Annu. Rev. Entomol.*, **40**, 171–194.
39. Fabre, J. H. (1886) *Souvenirs Entomologiques*, vol. 3, Delagrave, Paris.
40. Feeny, P., Sachdev-Gupta, K., Rosenberry, L. and Carter, M. (1988) Luteolin 7-*O*-(6'-*O*-malonyl)-β-D-glucoside and *trans*-chlorogenic

acid: oviposition stimulants for the black swallowtail butterfly. *Phytochemistry*, **27**, 3439–3448.
41. Fraenkel, G. S. (1959) The raison d'être of secondary plant substances. *Science*, **129**, 1466–1470.
42. Frazier, J. L. (1992) How animals perceive secondary plant compounds, in *Herbivores. Their Interactions with Secondary Plant Metabolites*, 2nd edn, vol. 2, (eds G. A. Rosenthal and M. R. Berenbaum), Academic Press, New York, pp. 89–133.
43. Frazier J. L. and Chyb, S. (1995) Use of feeding inhibitors in insect control, in *Regulatory Mechanisms in Insect Feeding*, (eds R. F. Chapman and G. de Boer), Chapman & Hall, New York, pp. 364–381.
44. Glendinning, J. I. (1995) Electrophysiological support for two bitter transduction mechanisms within the same taste receptor of an insect. *Chem. Senses*, **20**, 697–698.
45. Gregory, P., Avé, D. A., Bouthyette, P. J. and Tingey, W. M. (1986) Insect-defensive chemistry of potato glandular trichomes, in *Insects and the Plant Surface*, (eds B. Juniper and T. R. E. Southwood), Edward Arnold, London, pp. 173–183.
46. Gupta, P. D. and Thorsteinson, A. J. (1960) Food plant relationships of the diamond-back moth (*Plutella maculipennis*). II. Sensory regulation of oviposition of the adult female. *Entomol. Exp. Appl.*, **3**, 305–314
47. Haley Sperling, J. L. and Mitchell, B. K. (1991) A comparative study of host recognition and the sense of taste in *Leptinotarsa*. *J. Exp. Biol.*, **157**, 439–459.
48. Harris, M. O. and Miller, J. R. (1984) Foliar form influences ovipositional behaviour of the onion fly. *Physiol. Entomol.*, **9**, 145–155.
49. Honda, K. (1990) Identification of host-plant chemicals stimulating oviposition by swallowtail butterfly, *Papilio protenor*. *J. Chem. Ecol.*, **16**, 325–337.
50. Honda, K., Tada, A., Hayashi, N., Abe, F. and Yamauchi, T. (1995) Alkaloidal oviposition stimulants for a danaid butterfly, *Ideopsis similis* L., from a host plant, *Tylophora tanakae* (Asclepiadaceae). *Experientia*, **51**, 753–756.
51. Hsiao, T. H. (1985) Feeding behavior, in *Comprehensive Insect Physiology, Biochemistry and Pharmacology*, vol. 9, (eds G. A. Kerkut and L. I. Gilbert), Pergamon Press, New York, pp. 497-512.
52. Hsiao, T. H. (1988) Host specificity, seasonality and bionomics of Leptinotarsa beetles, in *Biology of Chrysomelidae*, (eds P. Jolivet, E. Petitpierre and T. H. Hsiao), Kluwer, Dordrecht, pp. 581–599.
53. Huang, X. and Renwick, J. A. A. (1994) Relative activities of glucosinolates as oviposition stimulants for *Pieris rapae* and *P. napi oleracea*. *J. Chem. Ecol.*, **20**, 1025–1037.
54. Jeffree, C. E. (1986) The cuticle, epicuticular waxes and trichomes of plants, with reference to their structure, functions and evolution, in *Insects and the Plant Surface*, (eds B. Juniper and T. R. E. Southwood), Edward Arnold, London, pp. 23–64.
55. Jermy, T. (1958) Untersuchungen über Auffinden und Wahl der Nahrung beim Kartoffelkäfer (*Leptinotarsa decemlineata* Say). *Entomol. Exp. Appl.*, **1**, 179–208.
56. Jermy, T. (1966) Feeding inhibitors and food preference in chewing phytophagous insects. *Entomol. Exp. Appl.*, **9**, 1–12.
57. Jermy, T. (1994) Hypotheses on oligophagy: how far the case of the Colorado potato beetle supports them, in *Novel Aspects of the Biology of Chrysomelidae*, (eds P. H. Jolivet, M. L. Cox and E. Petitpierre), Kluwer, Dordrecht, pp. 127–139.
58. Jermy, T. and Szentesi, Á. (1978) The role of inhibitory stimuli in the choice of oviposition site by phytophagous insects. *Entomol. Exp. Appl.*, **24**, 458–471.
59. Jones, C. G. and Firn, R. D. (1979) Some allelochemics of *Pteridium aquilinum* and their involvement in resistance to *Pieris brassicae* (Lep., Pieridae). *Biochem. Syst. Ecol.*, **7**, 187–192.
60. Kennedy, J. S. and Booth, C. O. (1951) Host alternation in *Aphis fabae* Scop. I. Feeding preferences and fecundity in relation to the age and kind of leaves. *Ann. Appl. Biol.*, **38**, 25–65.
61. Kent, K. S. and Hildebrand, J. G. (1987) Cephalic sensory pathways in the central nervous system of *Manduca sexta* (Lepidoptera, Sphingidae). *Phil. Trans. Roy. Soc. Lond. B*, **315**, 3–33.
62. Khan, Z. R., Ward, J. T. and Norris, D. M. (1986) Role of trichomes in soybean resistance to cabbage looper. *Entomol. Exp. Appl.*, **79**, 928–935.
63. Klingauf, F., Nöcker-Wenzel, K. and Röttger, U. (1978) Die Rolle peripherer Pflanzenwachse für den Befall durch phytophage Insekten. *Z. Pflanzenkrh. Pflanzenschutz*, **85**, 228–237.

64. Kubo, I., Lee, Y.-W., Balogh-Nair, V., Nakanishi, K. and Chapya, A. (1976) Structure of ajugarins. *J. C. S. Chem. Comm.*, **1976**, 949–950.
65. Luo, L.-E., van Loon, J. J. A. and Schoonhoven, L. M. (1995) Behavioural and sensory responses to some neem compounds by *Pieris brassicae* larvae. *Physiol. Entomol.*, **20**, 134–140.
66. Lupton, F. G. H. (1967) The use of resistant varieties in crop protection. *World Rev. Pest-Contr.*, **6**, 47–58.
67. Ma, W.-C. (1972) Dynamics of feeding responses in *Pieris brassicae* Linn. as a function of chemosensory input, a behavioural, ultrastructural and electrophysiological study. *Meded. Landbouwhogeschool Wageningen*, **72-11**, 1–162.
68. Ma, W.-C. and Schoonhoven, L. M. (1973) Tarsal chemosensory hairs of the large white butterfly *Pieris brassicae* and their possible role in oviposition behaviour. *Entomol. Exp. Appl.*, **16**, 343–357.
69. McIver, S. B. (1985) Mechanoreception, in *Comprehensive Insect Physiology, Biochemistry and Pharmacology*, vol. 6, (eds G. A. Kerkut and L. I. Gilbert), Pergamon Press, New York, pp. 71–132.
70. Matsumoto, Y. (1969) Some plant chemicals influencing the insect behaviors, in *Proceedings of the 11th International Congress of Botany, Seattle (Abstracts)*, p. 143.
71. Matsumoto, Y. and Thorsteinson, A. J. (1968) Effect of organic sulfur compounds on oviposition in onion maggot *Hylemya antiqua* Meigen (Diptera, Anthomyiidae). *Appl. Entomol. Zool.*, **3**, 5–12.
72. Messchendorp, L., van Loon, J. J. A. and Gols, G. J. Z. (1996) Behavioural and sensory responses to drimane antifeedants in *Pieris brassicae* larvae. *Entomol. Exp. Appl.*, **79**, 195–202.
73. Metcalf, R. L. and Metcalf, E. R. (1992) *Plant Kairomones in Insect Ecology and Control*, Chapman & Hall, London.
74. Miller, J. R. and Strickler, K. L. (1984) Finding and accepting host plants, in *Chemical Ecology of Insects*, (eds W. J. Bell and R. T. Cardé), Chapman & Hall, New York, pp. 127–157.
75. Mitchell, B. K. (1974) Behavioural and electrophysiological investigations on the responses of larvae of the Colorado potato beetle (*Leptinotarsa decemlineata*) to amino acids. *Entomol. Exp. Appl.*, **17**, 255–264.
76. Mitchell, B. K. (1988) Adult leaf beetles as models for exploring the chemical basis of host-plant recognition. *J. Insect Physiol.*, **34**, 213–225.
77. Mitchell, B. K. and Gregory, P. (1979) Physiology of the maxillary sugar sensitive cell in the red turnip beetle, *Entomoscelis americana*. *J. Comp. Physiol.*, **132**, 167–178.
78. Mitchell, B. K. and Harrison, G. D. (1984) Characterization of galeal chemosensilla in the adult Colorado beetle, *Leptinotarsa decemlineata*. *Physiol. Entomol.*, **9**, 49–56.
79. Mitchell, B. K. and Itagaki, H. (1992) Interneurons of the subesophageal ganglion of *Sarcophaga bullata* responding to gustatory and mechanosensory stimuli. *J. Comp. Physiol. A*, **171**, 213–230.
80. Mitchell, B. K., Rolseth, B. M. and McCashin, B. G. (1990) Differential responses of galeal gustatory sensilla of the adult Colorado potato beetle, *Leptinotarsa decemlineata* (Say), to leaf saps from host and non-host plants. *Physiol. Entomol.*, **15**, 61–72.
81. Montgomery, M. E. and Arn, H. (1974) Feeding response of *Aphis pomi*, *Myzus persicae* and *Amphorophora agathonica* to phlorizin. *J. Insect Physiol.*, **20**, 413–421.
82. Morgan, E. D. and Mandava, N. B. (1990) *CRC Handbook of Natural Pesticides, vol. 6. Insect Attractants and Repellents*. CRC Press, Boca Raton, FL.
83. Mullin, C. A., Chyb, S., Eichenseer, H., Hollister, B. and Frazier, J. L. (1994) Neuroreceptor mechanisms in insect gustation, a pharmacological approach. *J. Insect Physiol.*, **40**, 913–931.
84. Nair, K. S. S. and McEwen, F. L. (1976) Host selection by the adult cabbage maggot, *Hylemya brassicae* (Diptera, Anthomyiidae): effect of glucosinolates and common nutrients on oviposition. *Can. Entomol.*, **108**, 1021–1030.
85. Nielsen, J. K., Larsen, L. M. and Sørensen, H. (1979) Host plant selection of the horseradish flea beetle, *Phyllotreta armoraciae* (Coleoptera, Chrysomelidae): identification of two flavonol glycosides stimulating feeding in combination with glucosinolates. *Entomol. Exp. Appl.*, **26**, 40–48.
86. Nishida, R. (1995) Oviposition stimulants of swallowtail butterflies, in *Swallowtail Butterflies. Their Ecology and Evolutionary Biology*, (eds J. M. Scriber, Y. Tsubaki and R. C. Lederhouse), Scientific Publications, Gainesville, FL, pp. 17–26.
87. Nishijima, Y. (1960) Host plant preference of

the soybean pod borer, *Gracholita glicinivorella* (Matsumura) (Lep., Encosmidae). *Entomol. Exp. Appl.*, **3**, 38–47.
88. Ozaki, M., Amakawa, T., Ozaki, K. and Tokunaga, F. (1993) Two types of sugar-binding protein in the labellum of the fly. *J. Gen. Physiol.*, **102**, 201–216.
89. Panda, N. and Khush, G. S. (1995) *Host Plant Resistance to Insects*, CAB International, London.
90. Pereyra, P. C. and Bowers, M. D. (1988) Iridoid glycosides as oviposition stimulants for the buckeye butterfly, *Junonia coenia* (Nympalidae). *J. Chem. Ecol.*, **14**, 917–928.
91. Peterson, S. C., Herrebout, W. M. and Kooi, R. E. (1990) Chemosensory basis of host colonization by small ermine moth larvae. *Proc. K. Ned. Akad. Wet.*, **93**, 287–294.
92. Peterson, S. C., Hanson, F. E. and Warthen, J. D. (1993) Deterrence coding by a larval *Manduca* chemosensory neurone mediating rejection of a non-host plant, *Canna generalis* L. *Physiol. Entomol.*, **18**, 285–295.
93. Quiras, C. F., Stevens, M. A., Rick, C. M. and Kok Yokomi, M. K. (1977) Resistance in tomato to the pink form of the potato aphid, *Macrosiphum euphorbiae* (Thomas): the role of anatomy, epidermal hairs and foliage composition. *J. Am. Soc. Hortic. Sci.*, **102**, 166–171.
94. Raffa, K. (1987) Influence of host plant on deterrence by azadirachtin of feeding by fall armyworm larvae (Lepidoptera, Noctuidae). *J. Econ. Entomol.*, **80**, 384–387.
95. Rees, C. J. C. (1969) Chemoreceptor specificity associated with choice of feeding site by the beetle, *Chrysolina brunsvicensis* on its foodplant, *Hypericum hirsutum*. *Entomol. Exp. Appl.*, **12**, 556–583.
96. Renwick, J. A. A. (1990) Oviposition stimulants and deterrents, in *CRC Handbook of Natural Pesticides, vol. 6. Insect Attractants and Repellents*, (eds E. D. Morgan and N. B. Mandava), CRC Press, Boca Raton, FL, pp. 151–160.
97. Renwick, J. A. A., Radke, C. D., Sachdev-Gupta, K. and Städler, E. (1992) Leaf surface chemicals stimulating oviposition by *Pieris rapae* on cabbage. *Chemoecology*, **3**, 33–38.
98. Robinson, S. J., Thurston, R. and Jones, G. A. (1980) Antixenosis of smooth leaf cotton *Gossypium* spp. to the ovipositional response of the tobacco budworm *Heliothis virescens*. *Crop Sci.*, **20**, 646–649.
99. Roessingh, P. and Städler, E. (1990) Foliar form, colour and surface characteristics influence oviposition behaviour in the cabbage root fly, *Delia radicum*. *Entomol. Exp. Appl.*, **57**, 93–100.
100. Roessingh, P., Städler, E., Schöni, R. and Feeny, P. (1991) Tarsal contact chemoreceptors of the black swallowtail butterfly *Papilio polyxenes*: responses to phytochemicals from host- and non-host plants. *Physiol. Entomol.*, **16**, 485–495.
101. Roessingh, P., Städler, E., Fenwick, G. R., Lewis, J. A., Nielsen, J. K. Hurter, J. and Ramp, T. (1992a) Oviposition and tarsal chemoreceptors of the cabbage root fly are stimulated by glucosinolates and host plant extracts. *Entomol. Exp. Appl.*, **65**, 267–282.
102. Roessingh, P., Städler, E., Hurter, J. and Ramp, T. (1992b) Oviposition stimulant for the cabbage root fly: important new cabbage leaf surface compound and specific tarsal receptors, in *Proceedings of the 8th International Symposium on Insect–Plant Relationships*, (eds S. B. J. Menken, J. H. Visser and P. Harrewijn), Kluwer, Dordrecht, pp. 141–142.
103. Roitberg, B. D. and Prokopy, R. J. (1987) Insects that mark host plants. *Bioscience*, **37**, 400–406.
104. Rothschild, M. and Schoonhoven, L. M. (1977) Assessment of egg load by *Pieris brassicae* (Lepidoptera, Pieridae). *Nature*, **266**, 352–355.
105. Sachdev-Gupta, K., Feeny, P. and Carter, M. (1993) Oviposition stimulants for the pipevine swallowtail, *Battus philenor*, from an *Aristolochia* host plant: synergism between inositols, aristolochic acids and a monogalactosyl diglyceride. *Chemoecology*, **4**, 19–28.
106. Schoonhoven, L. M. (1967) Chemoreception of mustard oil glucosides in larvae of *Pieris brassicae*. *Proc. K. Ned. Akad. Wet. C*, **70**, 556–568.
107. Schoonhoven, L. M. (1972) Secondary plant substances and insects. *Rec. Adv. Phytochem.*, **5**, 197–224.
108. Schoonhoven, L. M. (1981) Chemical mediators between plants and phytophagous insects, in *Semiochemicals: Their Role in Pest Control*, (eds D. A. Nordlund, R. L. Jones and W. J. Lewis), John Wiley, New York, pp. 31–50.
109. Schoonhoven, L. M. (1982) Biological aspects of antifeedants. *Entomol. Exp. Appl.*, **31**, 57–69.
110. Schoonhoven, L. M. (1987) What makes a caterpillar eat? The sensory code underlying feeding behaviour, in *Advances in Chemoreception and Behaviour*, (eds R. F. Chapman, E. A.

Bernays and J. G. Stoffolano), Springer-Verlag, New York, pp. 69–97.
111. Schoonhoven, L. M. (1990) Host-marking pheromones in Lepidoptera, with special reference to *Pieris* spp. *J. Chem. Ecol.*, **16**, 3043–3052.
112. Schoonhoven, L. M. (1991) 100 years of botanical instinct. *Symp. Biol. Hung.*, **39**, 3–14.
113. Schoonhoven, L. M. and Blom, F. (1988) Chemoreception and feeding behaviour in a caterpillar: towards a model of brain functioning in insects. *Entomol. Exp. Appl.*, **49**, 123–129.
114. Schoonhoven, L. M. and Derksen Koppers, I. (1976) Effects of some allelochemics on food uptake and survival of a polyphagous aphid, *Myzus persicae*. *Entomol. Exp. Appl.*, **19**, 52–56.
115. Schoonhoven, L. M. and Dethier, V. G. (1966) Sensory aspects of host-plant discrimination by lepidopterous larvae. *Arch. Néerl. Zool.*, **16**, 497–530.
116. Schoonhoven, L. M. and Luo, L. E. (1994) Multiple mode of action of the feeding deterrent, toosendanin, on the sense of taste in *Pieris brassicae* larvae. *J. Comp. Physiol. A* 175, 519–524.
117. Schoonhoven, L. M., Blaney, W. M. and Simmonds, M. S. J. (1992) Sensory coding of feeding deterrents in phytophagous insects, in *Insect–Plant Interactions*, vol. 4, (ed. E. A. Bernays), CRC Press, Boca Raton, FL, 59–79.
118. Shah, M. A. (1982) The influence of plant surfaces on the searching behaviour of coccinellid larvae. *Entomol. Exp. Appl.*, **31**, 377–380.
119. Shields, V. D. C. and Mitchell, B. K. (1995a) Sinigrin as a feeding deterrent in two crucifer feeding, polyphagous lepidopterous species and the effects of feeding stimulant mixtures on deterrency. *Phil. Trans. Roy. Soc. Lond. B*, **347**, 439–446.
120. Shields, V. D. C. and Mitchell, B. K. (1995b) Responses of maxillary styloconic receptors to stimulation by sinigrin, sucrose and inositol in two crucifer-feeding, polyphagous lepidopterous species. *Phil. Trans. Roy. Soc. Lond. B*, **347**, 447–457.
121. Shields, V. D. C. and Mitchell, B. K. (1995c) The effect of phagostimulant mixtures on deterrent receptor(s) in two crucifer-feeding lepidopterous species. *Phil. Trans. Roy. Soc. Lond. B*, **347**, 459–464.
122. Simmonds, M. S. J. and Blaney, W. M. (1991) Gustatory codes in lepidopterous larvae. *Symp. Biol. Hung.*, **39**, 17–27.
123. Simmonds, M. S. J. and Blaney, W. M. (1996) Azadirachtin: advances in understanding its activity as an antifeedant. *Entomol. Exp. Appl.*, **80**, 23–26.
124. Simmonds, M. S. J., Blaney, W. M., Mithen, R., Birch, A. N. and Fenwick, R. (1994) Behavioural and chemosensory responses of the turnip root fly (*Delia floralis*) to glucosinolates. *Entomol. Exp. Appl.*, **71**, 41–57.
125. Simmonds, M. S. J., Blaney, W. M., Ley, S. V., Anderson, J. C., Banteli, R., Denholm, A. A., Green, W., Grossman, R. B., Gutteridge, C., Jenneus, L., Smith, S. C., Toogood, P. L. and Wood, A. (1995) Behavioural and neurophysiological responses of *Spodoptera littoralis* to azadirachtin and a range of synthetic analogues. *Entomol. Exp. Appl.*, **77**, 69–80.
126. Singh, B. B., Hadley, H. H. and Bernard, R. L. (1971) Morphology of pubescence in soy beans in its relationship to plant vigor. *Crop Sci.*, **11**, 13–16.
127. Smith, B. D. (1966) Effect of the plant alkaloid sparteine on the distribution of the aphid *Acyrtosiphon spartii* (Koch). *Nature*, **212**, 213–214.
128. Smith, C. M. (1989) *Plant Resistance to Insects. A Fundamental Approach*, John Wiley, New York.
129. Southwood, T. R. E. (1986) Plant surfaces and insects – an overview, in *Insects and the Plant Surface*, (eds B. Juniper and T. R. E. Southwood), Edward Arnold, London, pp. 1–22.
130. Städler, E. (1978) Chemoreception of host plant chemicals by ovipositing females of *Delia* (*Hylemya*) *brassicae*. *Entomol. Exp. Appl.*, **24**, 711–720.
131. Städler, E. (1986) Oviposition and feeding stimuli in leaf surface waxes, in *Insects and the Plant Surface*, (eds B. Juniper and T. R. E. Southwood), Edward Arnold, London, pp. 105–121.
132. Städler, E. (1992) Behavioral responses of insects to plant secondary compounds, in *Herbivores: Their Interactions with Secondary Plant metabolites*, 2nd edn, vol. 2, (eds G. A. Rosenthal and M. R. Berenbaum), Academic Press, New York, pp. 45–88.
133. Städler, E. and Buser, H. R. (1984) Defense chemicals in leaf surface wax synergistically stimulate oviposition by a phytophagous insect. *Experientia*, **40**, 1157–1159.
134. Städler, E. and Roessingh, P. (1991) Perception of surface chemicals by feeding and ovipositing insects. *Symp. Biol. Hung.*, **39**, 71–86.

135. Städler, E., Ernst, B., Hurter, J. and Boller, E. (1994) Tarsal contact chemoreceptor for host marking pheromone of the cherry fruit fly, *Rhagoletis cerasi*: responses to natural and synthetic compounds. *Physiol. Entomol.*, **19**, 139–151.
136. Städler, E., Renwick, J. A. A., Radke, C. D. and Sachdev-Gupta, K. (1995) Tarsal contact chemoreceptor response to glucosinolates and cardenolides mediating oviposition in *Pieris rapae*. *Physiol. Entomol.*, **20**, 175–187.
137. Stoner, K. A. (1990) Glossy leaf wax and plant resistance to insects in *Brassica oleracea* under natural infestation. *Environ. Entomol.*, **19**, 730–739.
138. Stork, N. E. (1980) Role of waxblooms in preventing attachment to Brassicas by the mustard beetle, *Phaedon cochleariae*. *Entomol. Exp. Appl.*, **28**, 100–107.
139. Sutcliffe, J. F. and Mitchell, B. K. (1982) Characterization of galeal sugar and glucosinolate-sensitive cells in *Entomoscelis americana* adults. *J. Comp. Physiol. A*, **146**, 393–400.
140. Tanton, M. T. (1962) The effect of leaf toughness on the feeding of larvae of the mustard beetle, *Phaedon cochleariae* Fab. *Entomol. Exp. Appl.*, **5**, 74–78.
141. Van Drongelen, W. (1979) Contact chemoreception of host plant specific chemicals in larvae of various *Yponomeuta* species (Lepidoptera) *J. Comp. Physiol.*, **134**, 265–279.
142. Van Drongelen, W. and van Loon, J. J. A. (1980) Inheritance of gustatory sensitivity in F1 progeny of crosses between *Yponomeuta cagnagellus* and *Y. malinellus* (Lepidoptera). *Entomol. Exp. Appl.* **28**, 199–203.
143. Van Duyn, J. W., Turnipseed, S. G. and Maxwell, H. D. (1972) Resistance in soybeans to the Mexican bean beetle. II. Reactions of the beetle to resistant plants. *Crop Sci.*, **12**, 561–562.
144. Van Loon, J. J. A. (1990) Chemoreception of phenolic acids and flavonoids in larvae of two species of *Pieris*. *J. Comp. Physiol. A*, **166**, 889–899.
145. Van Loon, J. J. A. (1995) Unpublished results.
146. Van Loon, J. J. A. (1996) Chemosensory basis of feeding and oviposition behaviour in herbivorous insects, a glance at the periphery. *Entomol. Exp. Appl.*, **80**, 7–13.
147. Van Loon, J. J. A. and van Eeuwijk, F. A. (1989) Chemoreception of amino acids in larvae of two species of *Pieris*. *Physiol. Entomol.* **14**, 459–469.
148. Van Loon, J. J. A., Blaakmeer, A., Griepink, F. C., van Beek, T. A., Schoonhoven, L. M. and de Groot, Æ. (1992) Leaf surface compound from *Brassica oleracea* (Cruciferae) induces oviposition by *Pieris brassicae* (Lepidoptera, Pieridae). *Chemoecology*, **3**, 39–44.
149. Verschaffelt, E. (1910) The causes determining the selection of food in some herbivorous insects. *Proc. K. Ned. Akad. Wet.*, **13**, 536–542.
150. Wensler, R. J. D. (1962) Mode of host selection by an aphid. *Nature*, **195**, 830–831.
151. Yencho, C. G. and Tingey, W. M. (1994) Glandular trichomes of *Solanum berthaultii* alter host preference of the Colorado potato beetle, *Leptinotarsa decemlineata*. *Entomol. Exp. Appl.*, **70**, 217–225.

HOST-PLANT SELECTION: WHY INSECTS DO NOT BEHAVE NORMALLY

7.1	Geographical variation	195
7.2	Differences between populations in the same region	198
7.3	Differences between individuals	199
7.4	Environmental factors causing changes in host-plant preference	200
7.4.1	Seasonality	200
7.4.2	Temperature	202
7.5	Internal factors causing changes in host-plant preference	203
7.5.1	Developmental stage	203
7.5.2	Sex	204
7.6	Experience-induced changes in host-plant preference	204
7.6.1	Non-associative changes	204
7.6.2	Associative changes	210
7.7	Pre- and early-adult experience	214
7.8	Adaptive significance of experience-induced changes in host preference	215
7.9	Genetic variation in host-plant preference	216
7.9.1	Interspecific differences	216
7.9.2	Intraspecific differences	217
7.10	Conclusions	219
7.11	References	220

Variation is a basic characteristic of life. One obvious variability is that between species, families, orders, phyla and kingdoms, etc., but a surprisingly large residual component of variability exists within species. This applies not least to host-plant choices in herbivores. The host-plant range of an insect species is not a fixed and unchangeable property. Meticulous studies have shown that some individuals or even whole populations of an insect species may reject plants, although they belong to the 'normal' host-plant range of the species. Both host-plant range as well as preference ranking of acceptable plants are apparently often variable within and among populations and the assumption that herbivorous insects possess fixed host preferences is wrong. Differences in propensity to find or accept certain host plants may be genotypically determined or may result from previous experience. The latter category includes associative learning and other types of learning. From the multitude of observations described in the literature it appears that phenotypic variation in host-plant preference and use among individuals or populations of herbivorous insects is common. Genetic differentiation in host preference is most probably more common than the few reports on this phenomenon suggest [42, 44]. On the basis of the still limited information it is assumed that more often than not variation in host-selection behaviour or insect performance involves both genotypic and experiential factors. Thus each insect is an individual, which may deviate from the mean and which may possess its own set of food preferences and aversions. In this view 'aberrant' behaviour does not exist.

7.1 GEOGRAPHICAL VARIATION

Numerous cases are known in which insects in different parts of their distribution area show differences in host-plant preference. The leaf-mining moth *Phyllonorycter blancardella* is in North America restricted to members of the genus *Malus*, whereas in Europe, its native distribution area, it has a considerably broader host-plant range and thrives on plants belonging to at least seven other genera of the Rosaceae [79]. A reversed situation is met in the cicadellid *Graphocephala ennahi*. Whereas Nearctic indigenous populations feed exclusively on *Rhododendron*, this species shows polyphagous habits in Europe, where it became established in the beginning of this century. Here this insect can be found on various plant species belonging to as many as 13 different plant families. An explanation could be that after introduction into the new region this species has under the new ecological conditions started an expansion of its host-plant range, whereas the above-mentioned *P. blancardella* shows a host-range restriction in its newly occupied region that is possibly due to founding principles [114]. Why the two species show opposite responses remains obscure. Whatever their causes may be, these examples show that an insect's host range may appear to be changed after its introduction into a new region.

The occurrence of considerable variation in preference rankings of their host plants by different populations of the Colorado potato beetle is exemplified by the following observations. In southern parts of North America this insect feeds only on *Solanum rostratum* and *S. augustifolium* and can hardly survive on either *S. elaeagnifolium* or cultivated potato. Populations in Arizona, however, are uniquely adapted to *S. elaeagnifolium,* whereas beetles collected on cultivated potatoes in northern parts of the USA die on *S. elaeagnifolium* [52]. Populations in North Carolina flourish on *S. carolinense*, but populations from more northerly locations exhibit uniformly low survival on this host. A heritable variation in the ability to survive on *S. carolinense* exists both between and within beetle populations [49]. The host-range expansion of the Colorado potato beetle to include potato, as has happened in the recent past, must have involved major genetic changes [75].

Two more examples found among butterflies and grasshoppers are presented to illustrate the notion that interpopulational variations occur in all major insect taxa. The tiger swallowtail, *Papilio glaucus*, a truly polyphagous butterfly, covers most of the North American continent and feeds throughout its range on at least 13 plant families, but in any one geographical area this species is restricted to a subset of these host plants. Based on differences in food preferences and ability to utilize different host-plant species, several subspecies have been recognized [111, 113]. *Schistocerca shoshone* is a grasshopper that must be classified as highly polyphagous as a species, but some of its populations are clearly monophagous. The differences in feeding habits between populations are not simply the result of differences in the food available to the different populations but are probably based on genetic differences between the insects [127]. Many more cases can be cited, all showing that, even though a species may be polyphagous over its geographical range as a whole, larvae from local populations may be truly specialists [18].

Regional differences in host-plant preference often reflect adaptations to local conditions. Local factors, such as the presence of a competitor for food, may exert a selection pressure, resulting in host-plant specialization. This is well exemplified by the host preferences of two closely related weevil species, *Larinus sturnus* and *L. jaceae*, feeding in the heads of some thistles and related knapweeds. Wherever populations of the two insect species occur together, both species are represented by 'biotypes' that select different Cardueae species as breeding hosts (Table 7.1). Although the two species have almost identical food niches they avoid larval competition for food and space, possibly presenting a case of ecological character displacement [149]. The causes of these differences in host-plant preference are largely unknown, but it seems likely that genetic differences occur between geographically separated populations. Adult individuals of *L. sturnus* collected in different areas maintained in laboratory choice tests the differences in host preference reflecting their

Table 7.1 Regional host-plant differences in two related weevil species, *Larinus sturnus* and *Larinus jaceae*. n.p. = insect species not present. (Source: data from Zwölfer, 1970)

Geographical area	Larinus sturnus	Larinus jaceae
Switzerland, Jura	*Centaurea scabiosa*	*Carduus nutans*
Switzerland, Wallis	*Centaurea scabiosa*	n.p.
Germany, Pfelz	n.p.	*Centaurea scabiosa*
France, Alsace	*Carduus nutans*	n.p.

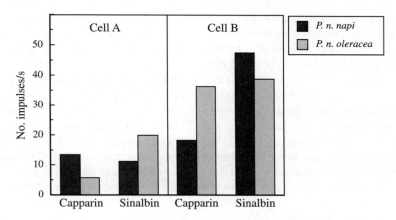

Figure 7.1 Sensitivity of chemoreceptor hairs on the tarsi of two *Pieris napi* strains for some glucosinolates. *P. n. napi* butterflies were collected in Europe, *P. n. oleracea* in North America. Nerve impulse frequencies are shown for two glucosinolate-sensitive cells (A and B) when stimulated with a solution of 100 μg/ml capparin or sinalbin. (Source: redrawn from Du et al., 1995.)

region of origin, suggesting a genetic basis for differences in host selection. The insects (and, don't forget, the thistle species as well) possibly constitute a mosaic of locally different or even sibling species that have evolved into the present pattern of insect–plant association [149, 150]. In other instances of geographical insect biotypes also the possibility of regional quality differences between plants cannot be excluded *a priori* as an important factor causing different host-plant choices.

As a proximate cause of differences in food-plant choices between populations it is conceivable that the sense organs in different populations are differently tuned and that behavioural differences can be traced back, at least to some extent, to different sensory inputs. Indeed, *Pieris napi* butterflies originating from different geographical regions (Europe and North America) show different sensory responses to various glucosinolates, compounds known to play a key role in host recognition (Fig. 7.1). These differences are probably related to differences in host-plant species between the two continents [143].

When the feeding activity of insects from different populations is analysed in greater detail, interpopulational differences in feeding behaviour, e.g. biting frequency, meal lengths and behavioural diversity in responses to plants, may be found. When the experimental protocol excludes epigenetic effects these differences can be attributed to genetic differences [130].

Besides the intraspecific variation in host-plant preference found in most insect species studied, the **limits of variation**, i.e. the conservative nature of insect–host relationships, is often an important feature. The Colorado potato beetle presents a case. Two North American and three European populations of this beetle were studied for similarities and dissimilarities between their behavioural responses to solanaceous host- and non-host-plant species. Despite the fact that populations in North America and in Europe have evolved separately for about 70 years, significant differences in food-related behaviour mostly appeared among populations of the same continent! This indicates that the ecological differences (alternative host plants, natural enemies) between the two continents, representing different selective pressures, have not caused basic behavioural changes so far [130]. This lack of major intercontinental differences is the more surprising because recently a new karyotypic (metacentric) race appeared in the USA, rapidly spread across the continent and hybridized with the original (acrocentric) race. The new race is absent in Europe [52].

7.2 DIFFERENCES BETWEEN POPULATIONS IN THE SAME REGION

Sometimes quite striking differences in host preference exist even between different insect populations occurring at short distances from each other. Adults of the nymphaline butterfly *Euphydryas editha* may oviposit in California on five different host species belonging to the Scrophulariaceae family. Some populations of this butterfly are strictly monophagous, whereas others are usually monophagous, but occasionally include a second host in their diet. A few populations exist in which the females oviposit on as many as four plant species, each belonging to a different genus. These interpopulation differences in breadth of host use are not primarily caused by differences in availability of host species, because

Table 7.2 (**A**) Egg masses of six different populations (A–F) of *Euphydryas editha* observed on six different host plant species in the field (percentages). All host plants are in the family Scrophulariaceae except for the genus *Plantago*, which is in the Plantaginaceae. 0 = plants of this genus are present in fair quantity, but no eggs have been found on them. A dash indicates absence or scarcity of the plant. (**B**) Oviposition of *E. editha* from six different populations (A–F) when offered a choice of six host plants in the laboratory (percentages). n.a. = plant not available for the test. (Source: modified from Singer, 1971)

Population	Plantago	Orthocarpus	Pedicularis	Collinsia	Castilleja	Penstemon
A						
A	98	2	0	–	–	–
B	17	83	–	–	–	–
C	–	–	94	6	0	–
D	0	0	–	100	0	0
E	–	–	0	100	0	0
F	–	–	–	–	100	0
B						
A	29	21	21	29	0	0
B	24	24	20	33	0	0
C	26	16	32	26	0	0
D	0	0	0	100	0	0
E	0	0	0	100	0	0
F	19	n.a.	6	6	50	19

they occur in habitats with apparently more or less identical arrays of potential host species (Table 7.2(A)).

Within the oligophagous populations, interestingly, some individuals are monophagous whereas others accept all potential host-plant species. The degree of host specialization in these populations appears to be a continuous variable [122]. Although under laboratory conditions host-plant preferences of this butterfly are less marked than in the field, the differences between populations as observed in the field are essentially maintained, again indicating a genetic basis for an extensive ecotypic variation in host use (Table 7.2(B)) [123]. As mentioned above, it is conceivable that different insect populations display different host-plant preferences, (partly) caused by interpopulational variation among the plants [41]. Indeed, concomitant to genetic variation in *E. editha* butterflies genetic variation has been found among their host-plant populations [124].

Thus conspecific plants occurring at different sites may be attacked by different insect species as a result of genetically determined differences in acceptability to these herbivores. Species with such regional differences, be it insects or plants, are obvious candidates for allopatric speciation.

7.3 DIFFERENCES BETWEEN INDIVIDUALS

Everyone investigating the feeding responses of insects to different plant species now and then encounters individuals that show 'aberrant' behaviour and do not seem to follow the preferences dictated by the 'normal' host range of the species. These idiosyncrasies occur too often to consider them as mere abnormalities. When ignoring the extent of variation in behavioural or physiological parameters, as biologists often tend to do under the influence of Platonic philosophical traditions, essential information is lost. Such 'tyranny of the Golden Mean' disregards some basic principles of life [6].

An illustration of 'deviant' behaviour was seen in an experiment with larvae of the privet hawkmoth (*Sphinx ligustri*), which normally feeds on privet and some related Oleaceae. Ten caterpillars were removed from their common food and, after one hour of starvation, placed each on an oleander (*Nerium oleander*) leaf, all taken from the same bush. As expected, the caterpillars did not feed at all during a 4-hour period, or at best took a few test bites only, with the exception of one individual. This insect soon started to eat and consumed $3.8\,cm^2$ of the leaf surface during the rest of the experiment. Apparently, this individual did not possess a behavioural barrier to this particular plant, which belongs to the Apocynaceae, a plant family characterized by a copious latex flow when damaged. The highly poisonous cardenolides in this plant did not seem to affect the caterpillar, because the insect's growth and development were not markedly affected by the unusual meal. Observations of this type are certainly not rare events [105] (Fig. 7.2).

Likewise, the abundance of records of

Figure 7.2 Feeding activity of five tobacco hornworm larvae (*Manduca sexta*), which were each confined to one leaf of a non-host plant (dandelion) for a period of 4 hours. All leaves were collected from the same plant. Three insects showed some nibbling but refused to eat while two insects started to feed immediately and consumed considerable amounts of leaf tissue.

Table 7.3 The plant families of the primary and secondary host plants of three genera of host-alternating aphids. (Source: H. Szelegiewicz, 1978, cited by Dixon, 1985)

Aphid genus	Primary host plants	Secondary host plants
Rhopalosiphum	Rosaceae	Gramineae
Pemphigus	Salicaceae	Compositae, Gramineae, Umbelliferae
Prociphilus	Caprifoliaceae, Oleaceae, Rosaceae	Pinaceae

oviposition 'mistakes' by insects ovipositing on plants outside their normal host range [85, 112] merit more attention than they usually receive. The significance of individuals with host choices deviating from normal is often underestimated because of an emphasis on the average amounting almost to a fetish. The presence of 'abnormal' individuals, however, probably reflects the built-in flexibility that enables a species to cope with changing conditions in its environment.

7.4 ENVIRONMENTAL FACTORS CAUSING CHANGES IN HOST-PLANT PREFERENCE

7.4.1 SEASONALITY

Host-plant preferences can vary with time among and within individuals in a population. Seasonal variation is pronounced in a number of aphid species, which show an obligatory shifting between unrelated host-plant taxa over successive generations (Fig. 7.3).

This habit is typical of about 10% of all extant aphid species, though many non-alternating species are thought to have been derived from alternating ancestors [71, 81, 115]. Briefly, the life cycle of host-alternating species is as follows. Winged females leave the primary host plant in spring and their offspring spend the summer months on the secondary host plants, fast-growing herbaceous species. The tremendous risks of not finding an appropriate food plant after leaving the primary host seem to be compensated by improved food quality once a secondary host plant is found. Aphid species feeding on herbs achieve higher growth and reproduction per unit of sap energy consumed (an average difference of 160%) than tree dwelling species (Fig. 7.4).

The better performance probably results from the higher amounts of amino-nitrogen available in the phloem sap of herbs [36]. The summer generations consist of wingless, rapidly maturing parthenogenetic individuals, which produce living young in quick succession. In early autumn winged forms, known as **sexuparae**, fly back to the primary host, on which eggs are laid.

The summer and winter hosts often belong to quite distinct plant families (Table 7.3).

The mealy plum aphid (*Hyalopterus pruni*), for instance, feeds during the warm season on common reed (*Phragmites communis*) and withdraws during the winter to some *Prunus* species. What makes the aphids migrate in late summer? It appears that seasonal cues, such as photoperiod, temperature and the physiological condition of the plant [78, 81], govern the production of alternative morphs. These cues are apparently also involved in the behavioural switch with respect to host-plant selection. The ultimate factor causing the persistence of host alternation in a species is the possibility of exploiting the complementary growth patterns of woody and herbaceous plants [71].

Seasonal changes in host preference occur in some other insects also. The cicad *Muellerianella fairmairei* is bivoltine in western Europe. It oviposits in spring on *Holcus lanatus* (Gramineae) but females of the second gener-

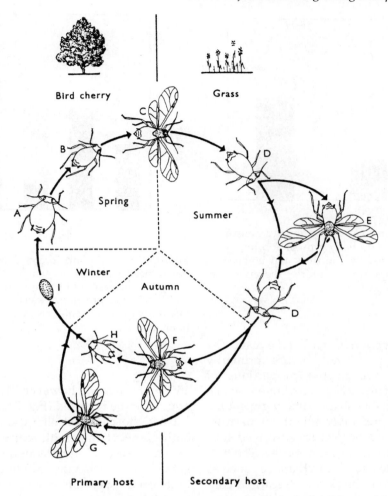

Figure 7.3 Life cycle of a heteroecious aphid species (bird-cherry–oat aphid, *Rhopalosiphum padi*). A = fundatrix; B = apterous fundatrigenia; C = emigrant; D = apterous exule; E = alate exule; F = gynopara; G = male; H = ovipara; I = egg. (Source: reproduced from Dixon, 1973, with permission.)

ation only accept *Juncus effusus* (Juncaceae) for egg laying [37]. Other examples are found in some bivoltine lepidopterans which may alternate between two entirely different host plants during successive generations. Thus, in the geometrid *Tephroclystis virgaureata*, caterpillars of the spring generation feed on some Compositae, e.g. *Solidago* and *Senecio*, whereas larvae of the summer generation occur on rosaceous plants, e.g. *Crataegus* and *Prunus* [69]. It would be interesting to know which factors govern the selection of oviposition sites by females of the two generations, but few attempts to analyse the causes of such changes in behaviour have been made. Seasonal factors may have changed potential host plants to such an extent that the insect switches from one plant species to another. Alternatively, the insect's innate preferences may have changed.

In the case of another lepidopteran, the pipevine swallowtail (*Battus philenor*), the seasonality of host-plant preference may be

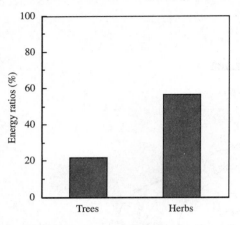

Figure 7.4 Percentages of the energy consumed that is converted into body growth and reproduction for aphids that live on trees and herbaceous plant species respectively. (Source: redrawn from Llewellyn, 1982.)

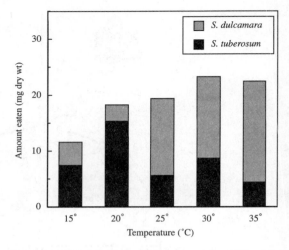

Figure 7.5 Food-plant preferences in Colorado potato beetles at different temperatures. Food consumption increases with temperature. Between 20° and 25°C food preference switches from *Solanum tuberosum* to *Solanum dulcamara*. (Source: data from Bongers, 1970.)

caused by changes on the part of the plant. The shift in host preference occurs primarily through a change in the proportions of ovipositing females using two different search modes based on leaf shape. At any given time some females preferably alight on narrow-leaved plants and neglect broad-leaved host plants, while others do the opposite. The butterflies appear to use a 'search image', such as is also known from birds, which learn to concentrate their foraging efforts on specific prey items [138]. Behavioural studies of this butterfly under experimental conditions have confirmed that most of the variability in search behaviour seen in nature is probably due to differences in adult experience [95].

7.4.2 TEMPERATURE

Food-plant preference sometimes changes with temperature. Colorado potato beetles offered a choice between potato and woody nightshade (*Solanum dulcamara*) normally show a predilection for potato. When choice experiments are performed at a temperature of 25°C or higher, however, the preference appears to be reversed (Fig. 7.5).

Does the chemical composition of one or both plants change with temperature, affecting their sensory impressions on the insect? Or is the insect's behaviour modified by changes in its central nervous system or chemoreceptors? We do not know.

Another example of the influence of temperature on feeding habits concerns the alfalfa aphid (*Terioaphis maculata*). Some alfalfa cultivars that are normally resistant to this insect become susceptible at a temperature of 10°C. Conceivably, reduced metabolic activity or lower mobility of the aphids at low temperatures might be contributing factors [101], but in this case also it is unknown whether temperature modifies the insect or the plant or both. Similar observations were recorded for some other aphids, heteropterans and Hessian flies (*Mayetiola destructor*), but the mechanisms involved remain obscure [84].

7.5 INTERNAL FACTORS CAUSING CHANGES IN HOST-PLANT PREFERENCE

Food plant preference of an individual insect does not always remain constant throughout its lifetime, but may change, for instance when nutritive needs change during its ontogeny.

7.5.1 DEVELOPMENTAL STAGE

In many herbivorous insect species the food-plant range becomes either smaller or wider during larval development. During their final instar many caterpillars accept more plant species than younger instars. Such diet broadening is quite striking in, for instance, larvae of the garden tiger moth (*Arctia caja*). The list of food plants of their younger instars is rather meagre, whereas later instar caterpillars reject hardly any plant species [80]. There are also many cases of the reverse reported in the literature – of developing larvae that are much more fastidious eaters than newly hatched larvae. This could be the result of 'preference induction', a kind of learning behaviour discussed in section 7.6.1.

Changes in food preference behaviour during development may be an expression of altered nutritional requirements. Indeed there is no reason to expect *a priori* that the insect's nutritive needs will remain constant during ontogeny. To investigate possible changes of feeding preferences over time, gypsy moth larvae (*Lymantria dispar*) were reared with continuous access to two types of an artificial diet. One diet contained a lower than optimal level of lipids; the other was low in protein content. Across early to late instars preference clearly shifted away from lipid-deficient, high-protein toward protein-deficient, high-lipid food. This change in food selection most probably reflects an adjustment to shifting nutritional demands [125]. The observation that larvae of this species in the wild often switch from plant to plant and may feed on several different plant species during development may, in view of the aforementioned experiment, be considered as food-quality optimization related to changes in nutritional requirements [3]. In migratory locusts (*Locusta migratoria*) also, the relative requirements for protein and carbohydrate change during somatic growth of the adults. In this insect the neural responsiveness of chemosensilla on the maxillary palps appeared to be consistent with alterations in protein and sugar ingestion, indicating the presence of feedback from nutritional demands to receptor sensitivity [119].

Some caterpillar species show a distinct switch to a different part of their host plant as they grow. The reasons for such switches in feeding behaviour are obscure. Nutritional factors or changes in predation risks with increasing size may exert selection pressure leading to a shift of place [96]. Seasonal changes in plant quality may also play a role, but it could equally well reflect changes in the ability to accept alternative feeding sites due to morphological changes resulting from increased body size, e.g. allowing feeding on tougher tissues.

A quite spectacular change in feeding habits during larval development is exhibited by a number of lycaenid caterpillars. At first these species are vegetarian but at some stage of their development the larvae stop eating and drop to the ground. They may then be adopted by ants and taken to the nest, where they are placed among the ant brood. The 'myrmecophilous' ('ant-liking') larvae now produce, from special glands, secretions that contain up to 20% sugars and sometimes also small amounts of amino acids. These substances appease ants and, in many species, reward them for protecting the caterpillars against predators and parasitoids. The caterpillars, once inside the brood chamber of the ant nest, devour the fluid tissues of the grubs of their hosts [5, 24]. The bizarre change from herbivory to predation of ant brood as a normal pattern in many Lycaenidae bears some analogy to the cannibalistic habit that, under certain conditions, is exhibited even by several highly specialized herbivorous insects [34, 132].

Large feeding differences may be found between the larval and adult stages of an insect. Larvae of the western corn rootworm (*Diabrotica virgifera virgifera*) are strictly monophagous, feeding only on corn roots, whereas the adult beetles are polyphagous [22]. As with many other biological traits, feeding behaviour may drastically change with the transition from larva to adult, due to altered nutritional requirements and environmental conditions.

7.5.2 SEX

In the above-cited experiments on dietary selection by gypsy moth larvae a significant difference was noted (in addition to that between early and late instar larvae) between the dietary preferences of male and female larvae. Male larvae ate a higher proportion from the protein-deficient, high-lipid diet than female larvae. Clearly this reflects a physiological adjustment to the fact that in this species only male moths possess wings and develop the capability to fly. Lipids need to be stored to serve as flight fuel. Females, on the other hand, require extra amounts of protein for egg development [125]. Hormonal differences between male and female larvae presumably control an accurate dietary balance of protein–lipid intake. In another, somewhat more natural type of feeding trial, tropical walking sticks (*Lamponius portoricensis*) were offered leaves of four of their most common food-plant species. Male and female insects ate different relative amounts of the various kinds of plants, thereby demonstrating, as in the gypsy moth larvae, foraging differences between the sexes [99].

Observations in the field have also revealed dietary divergence between the sexes in a number of grasshopper species. Whereas in *Oedaleus senegalensis* males at any stage and immature females predominantly fed on the leaves of millet plants, mature females showed a liking for the milky seed heads of this plant, thereby satisfying their increased protein demands during oogenesis [16]. In another field study on 14 grasshopper species it was also found that the composition of the diet and food preferences were significantly different between the two sexes. Therefore, strangely enough, male and female genotypes occupy different food niches [141].

7.6 EXPERIENCE-INDUCED CHANGES IN HOST-PLANT PREFERENCE

In contrast to vertebrates, in herbivorous insects behaviour in general and food-related behaviour in particular is primarily determined genetically. Nevertheless, in various instances different types of learning may change feeding or oviposition behaviour significantly. Two learning types can be distinguished: non-associative and associative changes. An associative change, i.e. **associative learning**, is the ability to associate a previously meaningless (ineffective) stimulus, the conditioned stimulus (CS), with a meaningful stimulus, the unconditioned stimulus (US), which causes either a positive or a negative response and coincides with the CS. Non-associative changes represent simpler types of learning in which a coupling of CS and US is lacking. From the experience-induced changes in host-plant preferences discussed below, habituation and induced food preferences are non-associative changes while food-aversion learning, dietary self-selection and experience-induced changes in oviposition behaviour represent associative changes [8, 129].

7.6.1 NON-ASSOCIATIVE CHANGES

(a) Habituation to deterrents

Habituation is a waning of the response to a repeatedly presented stimulus [136]. It is the simplest form of learning.

Since the specificity of plant acceptance by

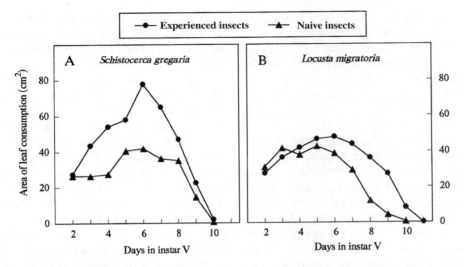

Figure 7.6 Quantities of deterrent-treated sorghum leaves consumed daily over the instar by experienced as compared to naive fifth instar nymphs of **(A)** the polyphagous desert locust (*Schistocerca gregaria*) and **(B)** the oligophagous migratory locust (*Locusta migratoria*). (Source: redrawn from Jermy et al., 1982.)

herbivorous insects is predominantly determined by the occurrence of secondary plant substances that inhibit feeding (section 6.6), habituation to deterrent stimuli has been studied in some detail. For example, freshly moulted fifth instar nymphs of two acridid species, *Schistocerca gregaria* and *Locusta migratoria*, were divided into two groups: one group was reared on untreated sorghum leaves ('naive group'), the other ('experienced group') received each day sorghum leaves treated with the deterrent nicotine hydrogen tartrate for 19 hours and untreated food for 5 hours. Each day ten naive and ten experienced nymphs were given deterrent-treated sorghum leaves in a no-choice situation and the amount consumed during a period of 19 hours was measured. Fig. 7.6(A) shows that experienced nymphs of the polyphagous *S. gregaria* habituated to the deterrent: they consumed much more from the deterrent leaf than naive nymphs. The nymphs of *L. migratoria*, a food specialist (Fig. 7.6(B)) also showed some habituation, albeit to a lesser degree [62].

Habituation occurred only to relatively weak stimuli, but it did not appear to chemicals that inhibited feeding for more than 12 hours [62, 128]. Insects do not habituate to non-host plants, even if they are just under the acceptance level [60]. The acceptance level of marginal host plants, however, may be increased after cross-habituation to certain feeding deterrents [53]. In the non-host plants of specialized insects it is most probably the presence of a whole complex of deterrent stimuli [59] that prevents habituation to the deterrent effect of these plants. This explanation is supported by the observation of *Spodoptera litura* larvae which, after repeated exposure, readily habituate to a single deterrent (azadirachtin) but not to a refined neem seed extract containing, besides azadirachtin, some other limonoids as well [14].

Is habituation a process taking place in the central nervous system or are changes in the peripheral chemoreceptor system responsible for the increased acceptability of a deterrent? The observation that incorporation of the deterrent compound salicin into the diet of *Manduca sexta* larvae reduced sensitivity of the deterrent receptor cell to this compound suggests participation of the chemoreceptors in the learning process. A concomitant effect was that such larvae readily accepted salicin-treated host plants that were normally rejected. These findings agree with the idea that, in addition to central learning processes, 'peripheral learning' takes place [104, 106]. On the other hand, a study with nymphs of *Schistocerca gregaria* on the neural basis of habituation to the deterrent nicotine hydrogen tartrate (NHT) did not reveal any peripheral changes, suggesting that in this insect only central learning is involved. In an elegant experiment small pieces of nylon tubing were fixed around the maxillary palps. These appendages house 30% of all mouthpart chemoreceptors. An experimental group of insects received NHT solution in the tubes for several hours daily, while in the control group the tubes were filled with distilled water. After four days the nymphs were tested as follows: the tubes were left in place, but were empty, and leaves treated with NHT were given as food to both groups. The experimental group consumed significantly more of the NHT-treated leaves as compared with the control group, despite the fact that direct perception of the chemical by the maxillary receptors was prevented by the tubes. These results clearly proved that habituation was mediated by the central nervous system: the information on NHT provided by the palpal receptors during the pretreatment was stored centrally and then compared with the information provided by the receptors on other mouthparts when the insects ate the NHT-treated leaves during the test [128].

(b) Induction of feeding preference (food imprinting)

In an early book on entomology Kirby and Spence [68] noted that, although insects may feed on various food plants, individuals of such generalist species, once they feed on a particular plant, often clearly show food specialization. 'It is worthy of remark, however, that when some of these have fed for a time on one plant they will die rather than eat another, which would have been perfectly acceptable to them if accustomed to it from the first.' This observation lucidly describes a phenomenon that in recent years has attracted much attention because it reflects some kind of learning in herbivorous insects, creatures that for a long time were considered to display only simple stimulus-response behaviour patterns. Kirby and Spence's observation fell into oblivion until it was rediscovered and given a firm experimental basis by Jermy *et al.* [61]. These authors reared larvae of the corn earworm, *Helicoverpa zea*, on an artificial diet until the end of the fourth instar. After moulting to the fifth instar the insects were divided into four groups and were then fed during the whole fifth instar on one of the following foods: artificial diet, *Pelargonium hortorum* (geranium), *Taraxacum officinale* (dandelion) or *Brassica oleracea* (cabbage). When the freshly moulted sixth instar larvae were given a choice of leaf discs from the three plants (Fig. 7.7), it appeared that in each group of larvae preference had increased for the plant species they had fed upon during the fifth instar as compared to the two other plants.

Those larvae that had no experience with any plant at all, since they were reared only on the artificial diet ('naive' larvae), showed a pattern of preference that differed from the other three groups (Fig. 7.8).

This effect of previous experience on food-plant choice has been termed 'induced preference' [61]. Its nature and function are largely unknown and it does not fit into the usual categories of learning [31]. Since the behavioural response shows some similarity to the phenomenon known as 'imprinting', that occurs in young vertebrate animals, induced feeding preference has also been labelled 'food imprinting' [129]. Insects reared on an artificial diet lacking any characteristic host-plant chemical seem to maintain the naivety of the newborn. They readily accept every host-plant

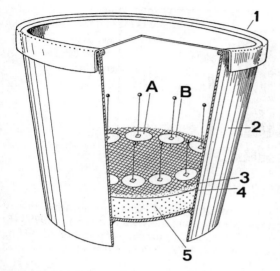

Figure 7.7 Arrangement for preference tests. A, B = leaf discs of two different plant species, mounted on pins; 1 = cover of Petri dish; 2 = paper food container cup; 3 = wire screen; 4 = moist filter paper; 5 = paraffin layer. (Source: reproduced from Jermy et al., 1968, with permission.)

species offered and even some non-hosts. Within 1–2 days on the plant food, however, their indifference is lost and an affinity for the plant experienced is established [103]. Food preferences in newly hatched larvae likewise develop with time. After 1–2 days feeding on the plants on which they were born larvae reject other foods (Fig. 7.9). Again, a comparison with imprinting is difficult to avoid.

The minimal duration of feeding needed for induction was demonstrated to be 4 hours with larvae of *Pieris brassicae* feeding on *Tropaeolum majus* [76]. The persistence of an induced preference varies considerably among insect species and depends also on the plant species on which the insect gained experience. For example, as shown in Fig. 7.8, induced preference for the plants in corn earworm larvae persisted through the moult from the fifth to the sixth instar. In another experiment with larvae of the same species an induced preference appeared to persist through two moults and the whole instar between these moults, during which the

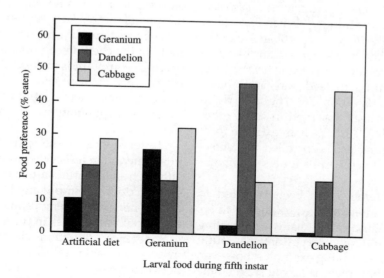

Figure 7.8 Food preference of sixth instar larvae of *Helicoverpa zea* reared on artificial diet until the end of fourth instar and then fed during the fifth instar on artificial diet, geranium, dandelion or cabbage. (Source: redrawn from Jermy et al., 1968.)

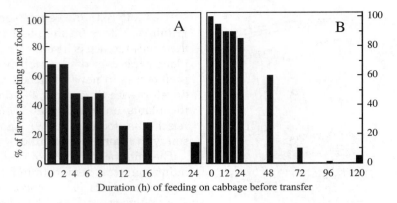

Figure 7.9 **(A)** Change in acceptance of nasturtium by first instar larvae of *Pieris rapae* as a function of duration of feeding experience on cabbage plants. (Source: redrawn from Renwick and Huang, 1995.) **(B)** Change in acceptance of artificial diet by first instar larvae of *P. brassicae* as a function of duration of feeding experience on cabbage plants. (Source: redrawn from Schoonhoven, 1977.) Note the difference in scale of the X-axes.

larvae were fed an artificial diet lacking any specific substance from their host plants [61].

Induced preference is in some cases very rigid. Darwin [26] cites the observation made by M. Michely: 'The caterpillars of *Bombyx hesperus* feed in a state of nature on the leaves of the *Café diable*, but, after having been reared on *Ailanthus*, they would not touch the *Café diable*, and actually died of hunger'. When *Pieris brassicae* larvae were reared on cabbage and the young fifth instar larvae were transferred to another host plant, *Tropaeolum majus*, the larvae also refused to feed and as a consequence died of starvation [76]. The same rigidity of induction was observed in some other lepidopterans [48, 110]. Such an extreme form of food imprinting has been called the 'starving-to-death-at-Lucullian-banquets' phenomenon [60].

Induced preference has been reported to occur in many insect species belonging to six different orders, indicating that we are dealing with a rather fundamental type of behavioural change. The most striking examples, however, are found in lepidopterous larvae [129] (Table 7.4).

That is not to say, however, that the phenomenon is universal among herbivores. It could not be found in several lepidopteran [60] as well as acridid [51] species.

A multiplicity of mechanisms has been suggested to produce induced food preferences and related changes [10], but the physiological processes involved are still largely beyond our grasp. As for the neural background of preference induction, it is not known what role the peripheral organs (receptors) or the central nervous system play. In some cases the sensitivity of the chemoreceptors increases for host specific compounds when the insects are reared on that plant (Fig. 7.10), which might cause stronger neural stimulation by plants experienced previously as compared to novel food plants.

The existence of the 'peripheral learning' phenomenon in insect chemosensory systems, although an important attribute of the nervous system, does not imply any lesser role for the central nervous system: primacy of preference behaviour still resides in the CNS.

Preference induction clearly proves that insects are able to discriminate not only between host- and non-host-plant species, but also between initially equally acceptable host-plant species. Thus, herbivorous insects are able to perceive a very detailed 'chemosensory

Table 7.4 List of insect species in which experience was found to change feeding preference (for references see Szentesi and Jermy, 1990, on which the table is based). (Source: reprinted with permission from E. A. Bernays, *Insect–Plant Interactions*, volume 2. Copyright CRC Press, Boca Raton, Florida.)

Orthoptera
　Schistocerca gregaria
Phasmodea
　Carausius morosus
　Bacillus rossius
Heteroptera
　Dysdercus koenigi
Homoptera
　Acyrthosiphon pisum
　Schizaphis graminum
Coleoptera
　Epilachna pustulosa
　Subcoccinella 24-punctata
　Haltica lythri
　Galerucella lineola
　Phratora vitellinae
　Leptinotarsa decemlineata
Lepidoptera
　Noctuidae
　　Helicoverpa armigera
　　H. zea
　　Spodoptera eridania
　Lymantriidae
　　Euproctis chrysorrhoea
　　Lymantria dispar
　Arctiidae
　　Hyphantrea cunea
　Sphingidae
　　Manduca sexta
　Saturniidae
　　Antheraea pernyi
　　A. polyphemus
　　Callosamia promethea
　　Hyalophora cecropia
　　Limenitis archippus
　　L. astyanax
　　L. hybrid rubidus
　Pieridae
　　Pieris brassicae
　　P. rapae
　Papilionidae
　　Papilio aegeus
　　P. glaucus
　　P. machaon
　Nymphalidae
　　Chlosyne lacinia
　　Polygonia interrogationis
　Pyraustidae
　　Loxostege sticticalis

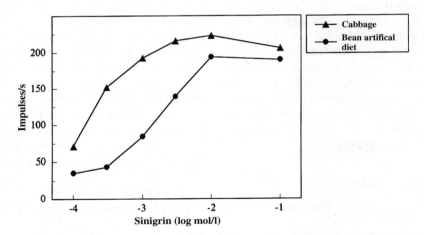

Figure 7.10 Neural responses (impulses/s) of the medial taste hair (sensillum styloconicum) on the maxillae of fifth instar larvae of *Spodoptera littoralis* when stimulated by a glucosinolate (sinigrin) at various concentrations. The insects were reared on cabbage leaves or on an artificial bean diet. Cabbage-reared caterpillars show a significantly higher sensitivity to the glucosinolate than those reared on the bean diet. (Source: redrawn from Schoonhoven *et al.*, 1987.)

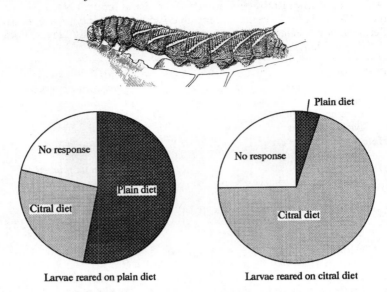

Figure 7.11 Orientation of fifth instar larvae of *Manduca sexta* reared on an artificial diet with or without citral. The two diets were presented in a choice test. Responses are shown as percentages of larvae moving towards either the citral diet or the plain diet. (Source: data from Saxena and Schoonhoven, 1978.)

profile' [27], or a 'chemical *Gestalt*' [70], of each host-plant species. This indicates a subtle complexity of stimuli involved in host recognition [29], based on both taste and olfaction. Experiments with caterpillars of the tobacco hornworm have shown that odours may be used as cues for recognizing the food on which preference induction occurred (Fig. 7.11).

The existence of the imprinting mechanism has to be taken into account when herbivorous insects whose introduction is being considered for the biological control of weeds are screened for their potential host-plant range in the region of introduction.

7.6.2 ASSOCIATIVE CHANGES

(a) Food aversion learning

Aversion learning in herbivorous insects is, according to Dethier's [30] definition, an acquired aversion for a plant that had induced temporary illness. This phenomenon, well-known in vertebrates, including man [91], was first studied in larvae of the arctiid moth, *Diacrisia virginica*. These caterpillars, while moving through short vegetation and taking bites from a number of plant individuals and species, forage as true polyphages. When offered leaves of *Petunia hybrida*, a plant not found in their natural habitat, they greedily consumed it. Ingestion for more than 24 hours, however, caused illness (regurgitation, decrease in activity, locomotor ataxia, mild convulsion and bloating in the thoracic region). When such larvae, after recovery from the illness, were placed in a field arena with other plant species, they avoided *Petunia*, in contrast to inexperienced (naive) larvae (Fig. 7.12) [32].

The survival value of this type of learning is self-evident, but it is still unknown how common aversion learning is in insects. So far it has been observed only in a few caterpillar species and in one or two acridids [7]. Possibly it is more common in polyphagous species than in specialists [33]. A further example of experience-induced food rejection was found

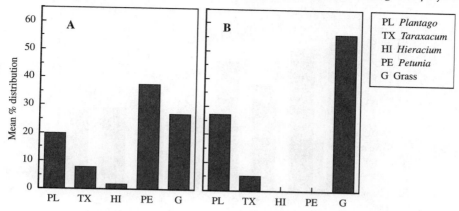

Figure 7.12 Mean percentage distribution of *Diacrisia virginica* caterpillars in field plots over a 2-day period. **(A)** Naive caterpillars. **(B)** Distribution after recovery of caterpillars fed exclusively on *Petunia* for 48 hours. (Source: data from Dethier, 1988.)

in oligophagous *Locusta migratoria* adults. At the first encounter this insect may first palpate and then take a bite before it rejects a non-host plant. At subsequent encounters with that particular plant species palpation only suffices to decide rejection [13]. In this case learning takes place by associating the information provided by the palpal chemoreceptors from the leaf surface with the information received when taking a bite from the distasteful ('punishment') leaf contents [12]. Such a learned response did not wane in another acridid, *Schistocerca americana*, even when the insect had been without food for over 2 hours [19].

(b) Dietary self-selection (mixed diet)

While most herbivorous insects are highly specific in host-plant choice and induction of feeding preference represents a further temporary or permanent restriction in food plant range (see 7.6.1b), in some species an opposite behaviour has been observed – a craving for a mixed diet. This is expressed by frequent shifts between different foods. This phenomenon was first described for the polyphagous larvae of *Malacosoma castrensis*. When caterpillars were confined to one of the food plant species on which they are found in nature, growth was retarded and mortality was high. When caged with a number of plant species, the caterpillars were seen to switch often between plants and their survival rate was much higher than on a single-plant diet. Likewise, *Arctia caja* caterpillars showed such voluntary food switches and, as a result, grew better [80]. The most frequent switching among food plants is found in grasshoppers and locusts. Several species are known to die, or at least to show reduced survival and fecundity rates, when restricted to a single food-plant species [4, 66]. Field observations showed that the grasshopper *Taeniopoda eques*, which lives in south-western parts of North America in habitats containing a great diversity of plants, may ingest up to eight different plant species even within one meal. Figure 7.13 shows the frequency of switches between different food items of an adult grasshopper in nature. The graph indicates that the probability of switches at the next meal declined with length of previous meal, yet even with apparently very acceptable food (as evidenced by longest feeding bouts), the likelihood of switching to a different plant species was as high as 25% [11].

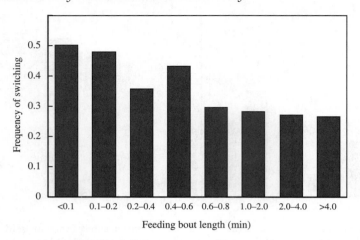

Figure 7.13 Frequency of switches (number of switches/number of feeding bouts) between different food items in relation to length of prior feeding bouts by an adult female of *Taeniopoda eques* in the field. (Source: redrawn from Bernays *et al.*, 1992.)

In grasshoppers three mechanisms causing dietary mixing have been recognized so far. The actual importance of each is most probably related to both the feeding behaviour of the species and the habitat in which the species occurs [9]. One of the possible mechanisms is food aversion learning. For example, when the polyphagous grasshopper *Schistocerca americana* is fed only spinach, it takes progressively shorter meals and finally rejects spinach completely. Since no changes in the chemoreceptor system were found, it was assumed that the insect associated the aversive consequences (malaise) of eating spinach with prior experience of its taste [72]. At the same time, alternative flavours became more attractive, suggesting that learned aversion may be accompanied by the novelty effect (**neophilia**). Learning predominates in polyphagous grasshoppers that live in simple habitats where few plant species are available and the insect may therefore be forced to feed for a longer time on a single plant species [9] (Table 7.5).

Positive associative learning is represented by the phenomenon of 'specific hunger'. Locusts which previously experienced a high protein diet that contained a specific odorous compound were significantly attracted to that odour when for some time deprived of protein [118].

A second possible mechanism involves the modulation of chemoreceptors. The sensitivity of the palpal receptors to amino acids or to sugars may change in response to haemolymph nutrient composition. Feeding on a diet low in protein results in a low concentration of amino acids in the haemolymph and this in turn results in an increase in the sensitivity of the chemoreceptors to amino acids in the food (section 4.4.3a). Consequently, the insect switches to a new diet rich in protein, i.e. containing more amino acids. The same mechanism operates with respect to carbohydrate content of the food: if the diet is low in carbohydrates, the sugar level of the haemolymph decreases, which in turn decreases in some unknown way the sensitivity threshold of the receptors to sugar [1]. Similar variations in neural activity corresponding to the pattern of protein and carbohydrate ingestion occur in caterpillars such as the polyphagous noctuid *Spodoptera littoralis* [117]. This chemoreceptor modulation mechanism occurs also in specialist species, such as the tobacco hornworm and the migratory locust (*Locusta migratoria*), which feeds on grass species (Table 7.5). Grasses are more dif-

Table 7.5 Predictions concerning mechanisms of dietary mixing in different situations. (Source: modified from Bernays and Bright, 1993) (Source: reprinted from *Comparative Biochemistry and Physiology*, volume 104A, E. A. Bernays and K. L. Bright: Mechanisms of dietary mixing in grasshoppers: a review, 125–131. Copyright 1992, with kind permission from Elsevier Science Ltd)

Learning predominates	*Chemosensory changes*	*Arousal with novelty phenomena*
Limited number of diverse dicots with characteristic chemical cues	Potential foods hard to distinguish – species of grass	Very large number of food items
Simple habitat with few food items	Plants have surface indicators of carbohydrate and protein (sugar, amino acids) and/or other nutrients	Complex habitat with numerous foods in close proximity
Fidelity to limited site (i.e. few hosts)	Monophagous species	Aposematic species
Infrequent locomotion	Large percentage of time in locomotion and rapid contact with many different foods	
Example: *Schistocerca* spp.	Example: *Locusta migratoria*	Example: *Taeniopoda eques*

ficult to distinguish from each other than dicots, because the latter group of plants are largely characterized by different secondary plant substances [9].

A third mechanism is arousal by novelty. The grasshopper *Taeniopoda eques*, for example, is a 'compulsive' switcher in which a newly detected odour, i.e. chemical novelty, provides an arousal stimulus leading to feeding. For this insect 'variety is the spice of life' [11]. This mechanism is important in grasshoppers that are very mobile and live in complex habitats with numerous plant species in close proximity [9] (Table 7.5).

The benefits of a mixed diet are also enjoyed by members of other insect orders, including the Hemiptera. The xylophagous leafhopper *Homalodisca coagulata* shows discrete shifts in host usage during its long life span, thereby increasing survival [17].

(c) Experience-induced changes in oviposition behaviour

Egg-laying females may gain experience when they oviposit on a specific substrate for the first time. This experience can influence subsequent choices of oviposition substrates. When apple maggot flies (*Rhagoletis pomonella*) were trained by letting them oviposit on apples and were then offered fruits of another host plant, *Crataegus mollis*, the latter were not accepted. Conversely, flies trained on *C. mollis* fruits rejected apple for oviposition [89]. This behaviour is similar to imprinting, but it may represent associative learning, since the flies associate the size and surface chemistry of the fruits they encountered first with the successful completion of egg-laying behaviour. The flies can discriminate not only different host species but even among different apple cultivars, since they develop a preference for the cultivar on which their first oviposition experience was gained [88].

Several butterflies use visual cues in oviposition site selection. For example, *Battus philenor* can discriminate between two *Aristolochia* host species visually by leaf shape. However, contact chemosensory experience with a host plant, even without an egg being laid, reinforces recognition of the specific leaf shape [86]. Females of *Heliconius* butterfly species, at the first encounter with a host plant, associate the shape of its leaves with their chemistry and afterwards search for the same leaf shape [43]. Laboratory experiments have shown that, by trial and error, females of *Pieris rapae* also learn to associate the appearance of the oviposition site with the presence of plants containing sinigrin, which stimulates oviposition [139]. When females of *Papilio machaon* have several times laid eggs on the same plant species dominant in a particular locality, they afterwards also approach non-host-plant species resembling the dominant host [148]. Not all lepidopterans, however, seem to learn during oviposition. The saturniid butterfly, *Hylesia lineata*, oviposits on four plant species in proportions predicted by their abundance, i.e. there appears to be no difference in innate preference and induced preference does not seem to occur, in this case suggesting a lack of discrimination among the four host plants [87].

When the host plants are recognized by leaf shape, the females supposedly use a 'search image' [93], although in most of these cases the use of additional chemosensory cues, notably plant volatiles, cannot be excluded.

Host finding and host recognition by females searching for oviposition sites is often a much more complex behaviour than food selection because, in flying insects, besides the whole process of navigation, visual as well as chemical stimuli (olfactory and contact) are operating. This, together with the fact that in nature flying insects may cover large distances between two successive egg depositions, makes an analysis of the role of experience in oviposition behaviour under natural conditions methodologically very difficult.

7.7 PRE- AND EARLY-ADULT EXPERIENCE

Some entomologists suggested long ago that larval experience gained on a host-plant species increases the preference for that plant as a feeding and/or oviposition substrate in the adult stage. The American entomologist A. D. Hopkins [50] was among the first to publish some evidence for this phenomenon and pointed to its relevance for evolutionary changes in host-plant preference. It is therefore called the 'Hopkins host-selection principle' [77]. Although several authors have since claimed to have observed this kind of memory transfer over metamorphosis, up to now no reliable experimental proof has been found. (This does not necessarily mean that Hopkins's inference was wrong. His experiments have not been repeated, because he used some insect species that are experimentally difficult to handle, e.g. cerambycids.) On the other hand, many authors who have sought a transfer of larval feeding experience to the adult stage could not find such effect [129].

Jaenike [55] proposed a 'neo-Hopkins host-selection principle', which states that the exposure of adult insects to a particular type of host will often increase their preference for that host. Such early adult experience has been demonstrated in *Drosophila* [137]: when fully developed larvae and/or newly formed puparia were thoroughly cleaned from the remainder of the larval culture medium by washing, the adults were less attracted to that food than adults developing from the unwashed larvae and/or puparia. Following this line of thought, Corbet [23] proposed the 'chemical legacy hypothesis', emphasizing that even if the puparia are washed, some chemical cues originating from the larval food might persist within the puparium and the emerging adult might contact them. Although it has not been proved experimentally, similar early-adult experience of the food plant may be present in insects that emerge as adults from the plant part where they developed as larvae, such as species developing in stalks and wood, in fruits and seeds, in mines, galls, etc.

Evidence that learning actually takes place in the immature insect and is transferred to the adult is scarce. In a recent study *Drosophila* larvae were trained to avoid electric shocks by a Pavlovian conditioning procedure using a specific odour. The odour avoidance behaviour was maintained through metamorphosis in spite of the extensive reorganization of the nervous system during metamorphosis [92, 140]. This experiment shows that memory transfer from the larval stage to the adult is in principle possible.

7.8 ADAPTIVE SIGNIFICANCE OF EXPERIENCE-INDUCED CHANGES IN HOST PREFERENCE

A common flaw of studies on experience-induced changes in host preference behaviour is that they are based, almost without exception, on laboratory experiments only. Therefore we do not know how far the results can be extrapolated to nature. Nevertheless, supposing that the above-discussed phenomena also have a role under natural conditions, it is logical to ask: What is the adaptive value of such changes?

Habituation to deterrents occurred in experiments using single compounds at concentrations that in a no-choice situation did not prevent feeding totally, while it did not occur with strong deterrents or with non-host plants. The adaptive significance of habituation to feeding-inhibitory stimuli could thus lie in the possibility of an insect overcoming behaviourally slight increases in the concentration of deterrent chemicals in its host plants due, for instance, to environmental factors. It could also allow insects to exploit marginally acceptable, initially deterrent plants, provided they are not toxic, in a situation where no better plants are available.

The adaptive advantage of **food aversion learning** is self-evident: it prevents the consumption of deleterious quantities of poisonous or nutritionally inadequate plants. Polyphagous insects are especially prone to ingest poisonous plants. Natural selection has therefore in these insects promoted the evolution of the capacity for aversion learning, whereas specialist feeders are prevented from feeding on physiologically unsuitable plants by their finickiness about host-plant choice. These species may therefore lack the capacity for aversion learning, but our knowledge on this point is scanty.

The adaptive advantage of **induced feeding preference** is less evident. It has been assumed that it reflects an adaptation of insects in which frequent changes of food type decreases the efficiency of food utilization [46, 47, 64, 107, 109, 110]. If forced to switch to a novel host several insect species (but not all – see, for instance, Pescador [87]) incurred a high physiological cost, evidenced by reduced growth and other fitness parameters (Table 7.6).

It seems that insects become physiologically or biochemically adapted to the host plant on

Table 7.6 Performance of *Colias philodice* larvae on two host plant species, *Medicago sativa* and *Melilotus alba*, and its modification by food plant switch. Fifth instar larvae were tested for various performance parameters when fed their rearing plant or an alternative food plant. Relative growth rate (RGR) = mg of dry matter of growth per mg of dry body weight per day. (Source: modified from Karowe, 1989)

Food plants (rearing/tested)	RGR	Instar duration (h)	Total food consumed (mg dry wt)	Pupal weight (mg dry wt)
sativa/sativa	0.26	104	125	29
sativa/alba	0.08	150	81	14
alba/alba	0.31	124	162	34
alba/sativa	0.11	141	89	16

which they have fed for some time and induced preference behaviour then prevents a change of food in a vegetation where various host plants may occur, for instance with intertwined shoots or leaves. While this hypothesis is intuitively attractive, the adaptive advantage of the 'starving-to-death-at-Lucullian-banquets' phenomenon found in various insects is difficult to comprehend [60]. Furthermore, a strong induced feeding preference, which forces insects to search for familiar food even in the presence of suitable alternative hosts, may slow down larval development and thus prolong exposure to predators and parasites. On the other hand, when a larva falls off its host plant, for instance, an increased sensitivity to host odour may help it to refind the host. Induction may also increase the intensity of normal food intake by heightening arousal and minimizing interruptions in feeding.

Induced oviposition preference is thought to enhance foraging efficiency within resource patches [89] and to reduce the likelihood that a female entering a patch of rare or unsuitable hosts will stay there and lose time in fruitless searching [90]. Furthermore, females depositing eggs on more abundant host-plant species would ensure a higher survival rate of their progeny [94]. However, in this case also one cannot help wondering what advantage might arise from behaviour that results in oviposition on one host-plant species in a habitat but prevents oviposition on several other suitable host species cooccurring with the host chosen originally [63, 94].

It should be considered that both induced feeding and induced oviposition preferences represent a temporary or permanent restriction of the innate host specificity of a herbivorous insect population or of an individual. It is most probable that not all induced changes of host preferences are adaptive. Cases like the 'starving-to-death-at-Lucullian-banquets' phenomenon or the avoidance of otherwise suitable oviposition substrates may indicate the limited ability of the insects' nervous system to switch from a behaviour just learned to another, innate, behaviour or to learn a new one. Thus, induction of preference might be a kind of (chemical or visual) 'tunnel vision' due to some basic inertia in the insect's nervous system [10, 129]. In conclusion, as yet no fully satisfactory answer can be given to the question of the adaptive advantage of experience-induced behaviour [60].

7.9 GENETIC VARIATION IN HOST-PLANT PREFERENCE

Variability in a population of organisms has a combined phenotypic and genotypic basis. Modern biometric genetics attempts to partition the variability into components derived respectively from variability in the genes and from variability in the environment. In this section we will focus on genetic variation in host preferences between species, populations and individuals. Studies that have attempted to unravel the genetic background of host-plant preference are relatively few and often incomplete, but their results are highly rewarding.

7.9.1 INTERSPECIFIC DIFFERENCES

Interspecific variation was studied in some closely related *Yponomeuta* species. The F_1 progeny of crosses between *Y. cagnagellus*, specialized to *Euonymus europaeus* (Celastraceae), and *Y. malinellus*, specialized to apple (Rosaceae), accepted the host plants of both parents in spite of the taxonomic distance between the two foods. There is some evidence that in the F_1 progeny chemoreceptor sensitivity to characteristic host chemicals of both parents is autosomally combined [142]. Frey et al. [40] demonstrated clear-cut differences between the antennograms of the apple and the hawthorn races of *Rhagoletis pomonella* and the closely related *R. mendax* to several host fruit odour components (see Fig. 5.20). The differences were heritable. Thus, antennal sensitivity to volatile plant substances presumably plays an important role in host shifts and speciation in these insects. Hybrids between the two species show significantly weaker antennal responses to odour compounds from the hosts of either parent. This presumably reflects reduced ability to locate host plants, which may be the cause of absence of gene flow in nature [39].

Females of a swallowtail butterfly, *Papilio zeliacon*, population from the western part of North America oviposit on two umbellifer species, *Lomatium grayi* and *Cymopterus terebinthus*, while females of the closely related *P. oregonius* population, at a distance of about 50 km from the above population, oviposit exclusively on *Artemisia dracunculus* (Compositae), although both the other plant species are also available. Laboratory crosses between the two species have shown that oviposition preference in these species is controlled significantly by one or more loci on the X chromosome and is modified by at least one locus on other chromosomes [133]. In several other hybridization experiments host preferences and performance traits were often found to be polygenically based, either autosomally or (partially) sex-linked (Table 7.7). This finding may be relevant for the observation that hybrids often demonstrate an expanded host range relative to either parent.

7.9.2 INTRASPECIFIC DIFFERENCES

Heritable intraspecific variation in host-plant preferences have been found in several insect species in laboratory assays and in plant breeding for insect resistance (see the question of biotypes in section 10.2.1). For example, females of laboratory-reared isofemale strains of the oligophagous swallowtail *Papilio zeliacon* and the monophagous *P. oregonius* showed variation among strains in the propensity to lay some eggs on the other species' host plants [133]. Females of *Helicoverpa virescens* from Mississippi show a greater oviposition preference for cotton than those collected in the US Virgin Islands. Crosses indicated that this difference is genetically determined and most probably resulted from a difference in the abundance of cotton between the two locations [102]. Some strains of the silkworm (*Bombyx mori*) that readily accept several other plants than mulberry to feed upon possess deterrent receptors with a strikingly reduced sensitivity to some secondary plant compounds that in normal silkworms strongly inhibit feeding (K. Asaoka, pers. comm.).

Unfortunately, no studies have yet documented genetically based individual differences in host responses using natural arrays of resources in the field [58]. As regards the adaptive and evolutionary significance of individual, i.e. intrapopulation, genetic variation in host-plant preference, Jaenike has pointed out that

> the ubiquity of such variation is reminiscent of electrophoretic variation, for which adaptive explanations have been sought, largely in vain, for the past 25 years. Thus, the idea that genetic variation for host selection is neutral and nonadaptive should be considered seriously, especially in the absence of evidence that such variation affects patterns of host use in the wild.
>
> Jaenike, 1990.

Certainly, very few if any biological features are always and under all conditions adaptive [45].

Studies on the genetic covariance between oviposition preference and larval performance have produced conflicting results [134, 145]. For example, covariance has been found in the agromyzid fly *Liriomyza sativae* [144]. However, the females used in the experiment were collected as larvae in the field, so an environmental effect cannot be ruled out [58]. For example, no measures were taken in these experiments to avoid early-adult induction of oviposition preference (see section 7.7). In another study employing the butterfly *Euphydryas editha*, ovipositing females showed different degrees of selectivity. Some very choosy females discriminated among individ-

Table 7.7 Host plant preferences and larval performance of interspecific hybrids and intraspecific crosses; A = autosomal genes involved; X = sex-linked genes involved; p = partial (in contrast to most other animals Lepidoptera (and birds) have a sex-determination system in which females are the heterogametic sex)

Species	Food-plant preference	Ovipositional preference	Performance	A/X	Remarks	Ref.
Interspecific hybrids						
Helicoverpa v. x *s.*	Dominance			A		2
Helicoverpa v. x *s.*			Dominance	A	Performance is polygenically determined	116
Yponomeuta c. x *m.*	Both parents			A	Information on chemoreceptor sensitivity of hybrids	142
Lymenitis a. x *a.*	Both parents				Information on preference induction in hybrids	48
Papilio o. x *z.*			Intermediate	A	Genes affecting performance differ from those affecting oviposition preference	135
Papilio o. x *z.*		P. dominance		X(A)	X chromosome has the largest effect, modifying effects by autosomal gene(s)	131
Papilio	Dominance				One parent species is monophagous, the other is polyphagous	126
Papilio g. x *c.*			x	A	Genes affecting performance differ from those affecting oviposition preference	113
Papilio g. x *c.*		x		X	Oviposition site preference probably based on relatively few loci	113
Procecidochares a. x A		Dominance		A	Oviposition site preference based on single-gene–two-allele system	54
Intraspecific crosses (different populations or strains)						
Oncopeltus fasciatus		Dominance			Oviposition site preference is polygenically inherited	73
Drosophila tripunctata		Dominance		A	Oviposition site preference is polygenically inherited	57
Drosophila tripunctata	P. dominance			A	Food preference is genetically independent of oviposition site preference	56
Helicoverpa virescens		Dominance		X	Oviposition site preference is inherited with paternal dominance	146

ual plants of *Pedicularis semibarbata* and it was found that their offspring grew less well on plant individuals that were avoided by these females than on those that were accepted. The less fussy females of this butterfly accepted all individuals of the host-plant species and their offspring grew equally well on all plants [83]. However, since all females were caught in the wild the correlation between preference and performance might also be due to migration among genetically differentiated populations specialized on different plant species [58].

Ovipositing females often do not recognize plant species that would be suitable for larval development (see Fig. 2.6). This is circumstantial evidence for the lack of an overall genetic covariance between oviposition preference and larval performance. The discrepancy may be caused both by ecological (environmental) and genetic factors [145]. Ecological factors that have been found to be operative include:

1. lack of synchrony between insect and plant phenology [120, 131];
2. constraints on foraging by the females, e.g. when habitats are avoided that contain optimal hosts [21, 120];
3. suboptimal hosts are much more abundant than optimal hosts [25];
4. nectar sources for the females occur at large distances from optimal larval hosts [82, 121].

As for the genetic factors, a lack of genetic variation for either host preference behaviour or digestive capability may hinder the evolution of correspondence [147]. Such incongruity is often quite evident in cases of introduced plants. North American *Pieris* species, for instance, readily oviposit on the crucifer *Thlaspi arvense*, although it is toxic to their larvae. This plant was recently introduced, and either there has not been enough time to evolve discrimination against it in the butterfly [20] or, alternatively, a lack of adequate genetic variation in host-choice behaviour prevented such an evolutionary change in the insect's behaviour.

7.10 CONCLUSIONS

This chapter aimed to demonstrate the existence of significant amounts of variation in feeding and oviposition behaviour below the species level for both host preference, or degree of specialization to particular host plants, and host suitability. The study of such variation contributes greatly to the understanding of resource use and other ecological processes, and of adaptation to stressed environments such as are caused by agricultural practices [28, 67].

Analysis of the variation or flexibility of an insect's behaviour is complicated by a similar variation on the part of the plant. Its nutritional quality varies over time, and its genetic constitution varies over place and time. Additionally, one can surmise that the nutritional requirements of insects change during ontogeny by virtue of shifting demands as a function of growth, reproduction and migration, processes that may affect an insect's food preference.

Some variations in food specialization within and among populations are also caused by differences in experience or maternal effects, while others reflect differences in genetic make-up. The fact that induced feeding preferences have been observed in all major herbivorous insect taxa suggests that it represents a basic phenomenon. At the same time, the fact that several species seem to lack this capacity entirely makes its function the more mysterious.

The finding that the memory of previous experiences is not solely located in the central nervous system, but may also be (partly) stored in chemoreceptor neurons provides evidence for the notion that all cells and organs are in some way or another modulated by previous influences. The memory located in neurons is complementary to memory shown by, for instance, the digestive system, where adaptation of detoxifying enzymes to different types of food occurs (section 4.4.5).

The various points discussed in this chapter have clearly shown that the relationship between an insect species and its host plants is not as absolute and fixed as might be thought at first sight. These relationships, rather, show considerable flexibility, which is on the one hand essential to maintain them and on the other a prerequisite for new evolutionary developments.

In conclusion, an insect that possesses the

average of all behavioural (and other biological) features does not exist. Each individual is a unique combination of inherited and acquired traits. The analysis of the resultant variation between individuals helps us to understand both mechanism and function, a prelude to understanding evolution.

7.11 REFERENCES

1. Abisgold, J. D. and Simpson, S. J. (1988) The effects of dietary protein levels and haemolymph composition on the sensitivity of the maxillary palp chemoreceptors of locusts. *J. Exp. Biol.*, **135**, 215–229.
2. Avison, T. I. (1988), cited in reference 116.
3. Barbosa, P., Martinet, P. and Waldvogel, M. (1986) Development, fecundity and survival of the herbivore *Lymantria dispar* and the number of plant species in its diet. *Ecol. Entomol.*, **11**, 1–6.
4. Barnes, O. L. (1965) Further tests of the effects of food plants on the migratory grasshopper. *J. Econ. Entomol.*, **58**, 475–479.
5. Baylis, M. and Pierce, N. (1993) The effects of ant mutualism on the foraging and diet of lycaenid caterpillars, in *Caterpillars. Ecological and Evolutionary Constraints on Foraging*, (eds N. E. Stamp and T. M. Casey), Chapman & Hall, New York, pp. 404–421.
6. Bennett, A. F. (1987) Interindividual variability: an underutilized resource, in *New Directions in Ecological Physiology*, (eds M. E. Feder, A. F. Bennett, W. W. Burggren and R. B. Huey), Cambridge University Press, Cambridge, pp. 147–169.
7. Bernays, E. A. (1992) Aversion learning and feeding, in *Insect Learning. Ecological and Evolutionary Perspectives*, (eds D. R. Papaj and A. C. Lewis), Chapman & Hall, New York, pp. 1–17.
8. Bernays, E. A. (1995) Effects of experience on feeding, in *Regulatory Mechanisms in Insect Feeding*, (eds R. F. Chapman and G. de Boer), Chapman & Hall, New York, pp. 297–306.
9. Bernays, E. A. and Bright, K. L. (1993) Mechanisms of dietary mixing in grasshoppers, a review. *Comp. Biochem. Physiol.*, **104A**, 125–131.
10. Bernays, E. A. and Weiss, M. R. (1996) Induced food preferences in caterpillars, the need to identify mechanisms. *Entomol. Exp. Appl.*, **78**, 1–8.
11. Bernays, E. A., Bright, K., Howard, J. J. and Raubenheimer, D. (1992) Variety is the spice of life, frequent switching between foods in the polyphagous grasshopper *Taeniopoda eques* Burmeister (Orthoptera, Acrididae). *Anim. Behav.*, **44**, 721–731.
12. Blaney, W. M. and Simmonds, M. S. J. (1985) Food selection by locusts: the role of learning in selection behaviour. *Entomol. Exp. Appl.*, **39**, 273–278.
13. Blaney, W. M. and Winstanley, C. (1982) Food selection behaviour in *Locusta migratoria*, in *Proceedings of the 5th International Symposium on Insect–Plant Relationships, Wageningen, 1982*, (eds J. H. Visser and A. K. Minks), Pudoc, Wageningen, pp. 365–366.
14. Bomford, M. K. and Isman, M. B. (1996) Desensitization of fifth instar *Spodoptera litura* (Lepidoptera, Noctuidae) to azadirachtin and neem. *Entomol. Exp. Appl.*, **81**, 307–313.
15. Bongers, W. (1970) Aspects of host-plant relationships of the Colorado beetle. *Meded. Landbouwhogeschool Wageningen*, **70-10**, 1–77.
16. Boys, H. A. (1978) Food selection by *Oedaleus senegalensis* (Acrididae, Orthoptera) in grassland and millet fields. *Entomol. Exp. Appl.*, **24**, 278–286.
17. Brodbeck, B. V., Andersen, P. C. and Mizell, R. F. (1995) Differential utilization of nutrients during development by the xylophagous leafhopper, *Homalodisca coagulata*. *Entomol. Exp. Appl.*, **75**, 279–289.
18. Cates, R. G. (1980) Feeding patterns of monophagous, oligophagous and polyphagous insect herbivores: the effect of resource abundance and plant chemistry. *Oecologia*, **46**, 22–31.
19. Chapman, R. F. and Sword, S. (1993) The importance of palpation in food selection by a polyphagous grasshopper (Orthoptera, Acrididae). *J. Insect Behav.*, **6**, 79–91.
20. Chew, F. S. (1977) Coevolution of pierid butterflies and their cruciferous foodplants. II. The distribution of eggs on potential foodplants. *Evolution*, **31**, 568–579.
21. Chew, F. S. (1981) Coexistence and local extinction in two pierid butterflies. *Am. Nat.*, **118**, 655–672.
22. Chyb, S., Eichenseer, H., Hollister, B., Mullin, C. A. and Frazier, J. L. (1995) Identification of sensilla involved in taste mediation in adult western corn rootworm (*Diabrotica virgifera virgifera* LeConte). *J. Chem. Ecol.*, **21**, 313–329.
23. Corbet, S. A. (1985) Insect chemosensory

responses, a chemical legacy hypothesis. *Ecol. Entomol.*, **10**, 143–153.
24. Cottrell, C. B. (1984) Aphytophagy in butterflies: its relationship to myrmecophily. *Zool. J. Linn. Soc.*, **79**, 1–57.
25. Courtney, S. P. (1982) Coevolution of pierid butterflies and their cruciferous foodplants. V. Habitat selection, community structure and speciation. *Oecologia*, 54, 101–107.
26. Darwin, C. (1875) *The Variation of Animals and Plants under Domestication*, 2nd edn, Methuen, London.
27. De Boer, G. and Hanson, F. E. (1984) Food plant selection and induction of feeding preference among host and non-host plants in larvae of the tobacco hornworm *Manduca sexta*. *Entomol. Exp. Appl.*, **35**, 177–193.
28. Denno, R. F. and McClure, M. S. (1983) *Variable Plants and Herbivores in Natural and Managed Systems*, Academic Press, San Diego, CA.
29. Dethier, V. G. (1973) Electrophysiological studies of gustation in lepidopterous larvae. II. Taste spectra in relation to foodplant discrimination. *J. Comp. Physiol.*, 82, 103–134.
30. Dethier, V. G. (1980) Food-aversion learning in two polyphagous caterpillars, *Diacrisia virginica* and *Estigmene congrua*. *Physiol. Entomol.*, 5, 321–325.
31. Dethier, V. G. (1982) Mechanism of host-plant recognition. *Entomol. Exp. Appl.*, **31**, 49–56.
32. Dethier, V. G. (1988) Induction and aversion-learning in polyphagous arctiid larvae (Lepidoptera) in an ecological setting. *Can. Entomol.*, **120**, 125–131.
33. Dethier, V. G. and Yost, M. T. (1979) Oligophagy and absence of food-aversion learning in tobacco hornworms *Manduca sexta*. *Physiol. Entomol.*, **4**, 125–130.
34. Dickinson, J. L. (1992) Egg cannibalism by larvae and adults of the milkweed leaf beetle (*Labidomera clivicollis*, Coleoptera, Chrysomelidae). *Ecol. Entomol.*, **17**, 209–218.
35. Dixon, A. F. G. (1973) *Biology of Aphids*, Edward Arnold, London.
36. Dixon, A. F. G. (1985) *Aphid Ecology*, Blackie, Glasgow.
37. Drosopoulos, S. (1977) Biosystematic studies on the *Muellerianella* complex (Dephacidae, Homoptera, Auchenorrhyncha). *Meded. Landbouwhogeschool Wageningen*, **77-14**, 1–133.
38. Du, Y.-J., van Loon, J. J. A. and Renwick, J. A. A. (1995) Contact chemoreception of oviposition-stimulating glucosinolates and an oviposition-deterrent cardenolide in two sub-species of *Pieris napi*. *Physiol. Entomol.*, **20**, 164–174.
39. Frey, J. E. and Bush, G. L. (1996) Impaired host odor perception in hybrids between the sibling species *Rhagoletis pomonella* and *R. mendax*. *Entomol. Exp. Appl.*, **80**, 163–165.
40. Frey, J. E., Bierbaum, T. J. and Bush, G. L. (1992) Differences among sibling species *Rhagoletis mendax* and *R. pomonella* (Diptera, Tephritidae) in their antennal sensitivity to host fruit compounds. *J. Chem. Ecol.*, **18**, 2011–2024.
41. Fritz, R. S. and Simms, E. L. (1992) *Plant Resistance to Herbivores and Pathogens. Ecology, Evolution and Genetics*, University of Chicago Press, Chicago, IL.
42. Futuyma, D. J. and Peterson, S. C. (1985) Genetic variation in the use of resources by insects. *Annu. Rev. Entomol.*, **30**, 217–238.
43. Gilbert, L. E. (1975) Ecological consequences of a coevolved mutualism between butterflies and plants, in *Coevolution of Animals and Plants*, (eds L. E. Gilbert and P. H. Raven), University of Texas Press, Austin, TX, pp. 210–240.
44. Gould, F. (1993) The spatial scale of genetic variation in insect populations, in *Evolution of Insect Pests. Patterns of Variation*, (eds K. C. Kim and B. A. McPheron), John Wiley, New York, pp. 67–85.
45. Gould, S. J. and Lewontin, R. C. (1979) The spandrels of San Marco and the Panglossian paradigm, a critique of the adaptationist programme. *Proc. Roy. Soc. Lond. B*, **205**, 581–598.
46. Grabstein, E. M. and Scriber, J. M. (1982a) The relationship between restriction of host plant consumption, and postingestive utilization of biomass and nitrogen in *Hyalophora cecropia*. *Entomol. Exp. Appl.*, **31**, 202–210.
47. Grabstein, E. M. and Scriber, J. M. (1982b) Host plant utilization by *Hyalophora cecropia* as affected by prior feeding experience. *Entomol. Exp. Appl.*, **32**, 262–268.
48. Hanson, F. E. (1976) Comparative studies on induction of food preferences in lepidopterous larvae. *Symp. Biol. Hung.*, **16**, 71–77.
49. Hare, J. D. and Kennedy, G. G. (1986) Genetic variation in plant–insect associations: survival of *Leptinotarsa decemlineata* populations on *Solanum carolinense*. *Evolution*, **40**, 1031–1043.
50. Hopkins, A. D. (1917) A discussion of C. C. Hewitt's paper on 'Insect Behavior'. *J. Econ. Entomol.*, **10**, 92–93.
51. Howard, G. and Bernays, E. A., cited in reference 127.

52. Hsiao, T. H. (1985) Ecophysiological and genetic aspects of geographic variations of the Colorado potato beetle, in *Proceedings of a Symposium on the Colorado potato beetle*, XVIIth International Congress of Entomology, *Research Bulletin 704, Massachusetts Agricultural Experimental Station*, Amherst, MA, pp. 63–77.
53. Huang, X. P. and Renwick, J. A. A. (1995) Cross habituation to feeding deterrents and acceptance of a marginal host plant by *Pieris rapae* larvae. *Entomol. Exp. Appl.*, **76**, 295–302.
54. Huettel, M. D. and Bush, G. L. (1972) The genetics of host selection and its bearing on sympatric speciation in *Procecidochares* (Diptera, Tephritidae). *Entomol. Exp. Appl.*, **15**, 465–480.
55. Jaenike, J. (1983) Induction of host preference in *Drosophila melanogaster*. *Oecologia*, **58**, 320–325.
56. Jaenike, J. (1985) Genetic and environmental determinants of food preference in *Drosophila tripunctata*. *Evolution*, **39**, 362–369.
57. Jaenike, J. (1987) Genetics of oviposition preference in *Drosophila tripunctata*. *Heredity*, **59**, 363–369.
58. Jaenike, J. (1990) Host specialization in phytophagous insects. *Annu. Rev. Ecol. Syst.*, **21**, 243–273.
59. Jermy, T. (1983) Multiplicity of insect antifeedants in plants, in *Natural Products for Innovative Pest Management*, (eds D. L. Whitehead and W. S. Bowers), Pergamon Press, Oxford, pp. 223–236.
60. Jermy, T. (1987) The role of experience in the host selection of phytophagous insects, in *Perspectives in Chemoreception and Behavior*, (eds R. F. Chapman, E. A. Bernays and J. G. Stoffolano), Springer-Verlag, New York, pp. 142–157.
61. Jermy, T., Hanson, F. E. and Dethier, V. G. (1968) Induction of specific food preference in lepidopterous larvae. *Entomol. Exp. Appl.*, **11**, 211–230.
62. Jermy, T., Bernays, E. A. and Szentesi, Á. (1982) The effect of repeated exposure to feeding deterrents on their acceptability to phytophagous insects, in *Proceedings of the 5th International Symposium on Insect–Plant Relationships, Wageningen, 1982*, (eds J. H. Visser and A. K. Minks), Pudoc, Wageningen, pp. 25–32.
63. Jones, R. E. and Ives, P. M. (1979) The adaptiveness of searching and host selection behaviour in *Pieris rapae* (L.). *Austr. J. Ecol.*, **4**, 75–86.
64. Karowe, D. N. (1989) Facultative monophagy as a consequence of prior feeding experience: behavioral and physiological specialization in *Colias philodice* larvae. *Oecologia*, **78**, 106–111.
65. Deleted in proof.
66. Kaufman, T. (1965) Biological studies of some Bavarian Acridoidea (Orthoptera) with special reference to their feeding habits. *Ann. Entomol. Soc. Am.*, **58**, 791–800.
67. Kim, K. C. and McPheron, B. A. (1993) *Evolution of Insect Pests. Patterns of Variation*, John Wiley, New York.
68. Kirby, W. and Spence, W. (1863) *An Introduction to Entomology*, 7th edn, Longman, Green, Longman, Roberts & Green, London.
69. Klos, R. (1901) Zur Lebensgeschichte von *Tephroclystia virgaureata*. *Verh. K. K. zool.-bot. Ges. Wien*, **51**, 785.
70. Kogan, M. (1977) The role of chemical factors in insect/plant relationships, in *Proceedings of the 15th International Congress of Entomology, Washington, DC*, Entomological Society of America, College Park, MD, pp. 211–227.
71. Kundu, R. and Dixon, A. F. G. (1995) Evolution of complex life cycles in aphids. *J. Anim. Ecol.*, **64**, 245–255.
72. Lee, J. C. and Bernays, E. A. (1988) Declining acceptability of food plant for a polyphagous grasshopper *Schistocerca americana*: the role of food aversion learning. *Physiol. Entomol.*, **13**, 291–301.
73. Leslie, J. F. and Dingle, H. (1983) A genetic basis of oviposition preference in the large milkweed bug, *Oncopeltus fasciatus*. *Entomol. Exp. Appl.*, **34**, 215–220.
74. Llewellyn, M. (1982) The energy economy of fluid-feeding herbivorous insects, in *Proceedings of the 5th International Symposium on Insect–Plant Relationships, Wageningen, 1982*, (eds J. H. Visser and A. K. Minks), Pudoc, Wageningen, pp. 243–251.
75. Lu, W. and Logan, P. (1994) Genetic variation in oviposition between and within populations of *Leptinotarsa decemlineata* (Coleoptera, Chrysomelidae). *Ann. Entomol. Soc. Am.*, **87**, 634–640.
76. Ma, W. C. (1972) Dynamics of feeding responses in *Pieris brassicae* Linn. as a function of chemosensory input, a behavioural, ultra-

structural and electrophysiological study. *Meded. Landbouwhogeschool Wageningen*, **72-11**, 1–162.

77. Mackenzie, A. (1992) The evolutionary significance of host-mediated conditioning. *Antenna*, **16**, 141–150.

78. Mackenzie, A. and Dixon, A. F. G. (1990) Host alternation in aphids: constraint versus optimization. *Am. Nat.*, **136**, 132–134.

79. Maier, C. T. (1985) Rosaceous hosts of *Phyllonorycter* species (Lepidoptera, Gracillariidae) in New England. *Ann. Entomol. Soc. Am.*, **78**, 826–830.

80. Merz, E. (1959) Pflanzen und Raupen. Über einige Prinzipien der Futterwahl bei Großschmetterlingsraupen. *Biol. Zentralbl.*, **78**, 152–188.

81. Moran, N. A. (1992) The evolution of aphid life cycles. *Annu. Rev. Entomol.*, **37**, 321–348.

82. Murphy, D. D., Menninger, M. S. and Ehrlich, P. R. (1984) Nectar source distribution as a determinant of oviposition host species in *Euphydrias chalcedona*. *Oecologia*, **62**, 269–271.

83. Ng, D. (1988) A novel level of interaction in plant–insect systems. *Nature*, **334**, 611–613.

84. Panda, N. and Khush, G. S. (1995) *Host Plant Resistance to Insects*, CAB International, Oxford.

85. Papaj, D. (1986a) An oviposition 'mistake' by *Battus philenor* L. (Papilionidae). *J. Lepidop. Soc.*, **40**, 348–349.

86. Papaj, D. (1986b) Conditioning of leaf shape discrimination by chemical cues in the butterfly *Battus philenor*. *Anim. Behav.*, **34**, 1281–1288.

87. Pescador, A. R. (1993) The effects of a multi-species sequential diet on the growth and survival of tropical polyphagous caterpillars. *Entomol. Exp. Appl.*, **67**, 15–24.

88. Prokopy, R. J. and Papaj, D. R. (1988) Learning of apple fruit biotypes by apple maggot flies. *J. Insect Behav.*, **1**, 67–74.

89. Prokopy, R. J., Averill, A. L., Cooley, S. S. and Roitberg, C. A. (1982) Associative learning of apple fruit biotypes by apple maggot flies. *Science*, **218**, 76–77.

90. Prokopy, R. J., Papaj, D. R., Cooley, S. S. and Kallet, C. (1986) On the nature of learning in oviposition site acceptance by apple maggot flies. *Anim. Behav.*, **34**, 98–107.

91. Provenza, F. D. and Cincotta, R. P. (1993) Foraging as a self-organizational learning process: accepting adaptability at the expense of predictability, in *Diet Selection*, (ed. R. N. Hughes), Blackwell, Oxford, pp. 78–101.

92. Punzo, F. (1988) Learning and localization of brain function in the tarantula spider, *Aphonopelma chalcodes* (Orthognatha, Theraphosidae). *Comp. Biochem. Physiol.*, **A89**, 465–470.

93. Rausher, M. D. (1978) Search image for leaf shape in a butterfly. *Science*, **200**, 1071–1073.

94. Rausher, M. D. (1980) Host abundance, juvenile survival, and oviposition preference in *Battus philenor*. *Evolution*, **34**, 342–355.

95. Rausher, M. D. (1985) Variability for host preference in insect populations: mechanistic and evolutionary models. *J. Insect Physiol.*, **31**, 873–889.

96. Reavy, D. and Lawton, J. H. (1991) Larval contribution to fitness in leaf-eating insects, in *Reproductive Behaviour of Insects, Individuals and Populations*, (eds W. J. Bailey and J. Ridsdill-Smith), Chapman & Hall, London, pp. 293–329.

97. Renwick, J. A. A. and Huang, X. P. (1995) Rejection of host plant by larvae of cabbage butterfly: diet-dependent sensitivity to an antifeedant. *J. Chem. Ecol.*, **21**, 465–475.

98. Sanderson, E. D. (1915) *Insect Pests of Farm, Garden and Orchard*, John Wiley, New York.

99. Sandlin, E. A. and Willig, M. R. (1993) Effects of age, sex, prior experience, and intraspecific food variation on diet composition of a tropical folivore (Phasmatodea, Phasmatidae). *Environ. Entomol.*, **22**, 625–633.

100. Saxena, K. N. and Schoonhoven, L. M. (1978) Induction of orientation and feeding preferences in *Manduca sexta* larvae for an artificial diet containing citral. *Entomol. Exp. Appl.*, **23**, 72–78.

101. Schalk, J. M., Kindler, S. D. and Manglitz, G. R. (1969) Temperature and the preference of the spotted alfalfa aphid for resistant and susceptible alfalfa plants (*Therioaphis maculata*, Hem., Hom., Aphididae). *J. Econ. Entomol.*, **62**, 1000–1003.

102. Schneider, J. C. and Roush, R. T. (1986) Genetic differences in oviposition preference between two populations of *Heliothis virescens*, in *Evolutionary Genetics of Invertebrate Behavior*, (ed. M. D. Huettel), Plenum Press, New York, pp. 163–171.

103. Schoonhoven, L. M. (1967) Loss of host plant specificity by *Manduca sexta* after rearing on

an artificial diet. *Entomol. Exp. Appl.*, **10**, 270–272.
104. Schoonhoven, L. M. (1969) Sensitivity changes in some insect chemoreceptors and their effect on food selection behaviour. *Proc. K. Ned. Akad. Wetensch.*, **C72**, 491–498.
105. Schoonhoven, L. M. (1977) On the individuality of insect feeding behaviour. *Proc. K. Ned. Akad. Wetensch. Amsterdam*, **C80**, 341–350.
106. Schoonhoven, L. M. (1987) What makes a caterpillar eat? The sensory code underlying feeding behavior, in *Perspectives in Chemoreception and Behavior*, (eds R. F. Chapman, E. A. Bernays and J. G. Stoffolano), Springer-Verlag, New York, pp. 67–97.
107. Schoonhoven, L. M. and Meerman, J. (1978) Metabolic cost of changes in diet and neutralization of allelochemics. *Entomol. Exp. Appl.*, **24**, 689–693.
108. Schoonhoven, L. M., Blaney, W. M. and Simmonds, M. S. J. (1987) Inconstancies of chemoreceptor sensitivities, in *Insects–Plants. Proceedings of the 6th International Symposium on Insect–Plant Relationships*, (eds V. Labeyrie, G. Fabres and D. Lachaise), Junk, Dordrecht, pp. 141–145.
109. Scriber, J. M. (1981) Sequential diets, metabolic cost, and growth of *Spodoptera eridania* feeding upon dill, lima bean and cabbage. *Oecologia*, **51**, 175–180.
110. Scriber, J. M. (1982) The behavioural and nutritional physiology of southern armyworm larvae as a function of plant species consumed in earlier instars. *Entomol. Exp. Appl.*, **31**, 359–369.
111. Scriber, J. M. (1984) Host-plant suitability, in *Chemical Ecology of Insects*, (eds W. J. Bell and R. T. Cardé), Chapman & Hall, New York, pp. 159–202.
112. Scriber, J. M., Giebink, B. L. and Snider, D. (1991) Reciprocal latitudinal clines in oviposition behavior of *Papilio glaucus* and *P. canadensis* across the great Lakes hybrid zone: possible sex-linkage of oviposition preferences. *Oecologia*, **87**, 360–368.
113. Scriber, J. M. and Lederhouse, R. C. (1992) The thermal environment as a resource dictating geographic patterns of feeding specialization of insect herbivores, in *Effects of Resource Distribution on Animal–Plant Interactions*, (eds M. D. Hunter, T. Ohgushi and P. Price), Academic Press, San Diego, pp. 430–466.
114. Sergel, R. (1987) On the occurrence and ecology of the *Rhododendron*-leafhopper *Graphocephala fennahi* Young 1977, in the western Palearctic region (Homoptera, Cicadellidae). *Anz. Schädlungskd. Pflanzenkrh. Pflanzenschutz*, **60**, 134–136.
115. Shaposhnikov, G. Ch. (1987) Evolution of aphids in relation to evolution of plants, in *Aphids: Their Biology, Natural Enemies, and Control*, World Crop Pests, vol. 2A, (eds A. K. Minks and P. Harrewijn), Elsevier, Amsterdam, pp. 409–414.
116. Sheck, A. L. and Gould, F. (1993) The genetic basis of host range in *Heliothis virescens*: larval survival and growth. *Entomol. Exp. Appl.*, **69**, 157–172.
117. Simmonds, M. S. J., Simpson, S. J. and Blaney, W. M. (1992) Dietary selection behaviour in *Spodoptera littoralis*: the effects of conditioning diet and conditioning period on neural responsiveness and selection behaviour. *J. Exp. Biol.*, **162**, 73–90.
118. Simpson, S. J. and White, P. (1990) Associative learning and locust feeding: evidence for a learned hunger for protein. *Anim. Behav.*, **40**, 506–513.
119. Simpson, C. L., Chyb, S. and Simpson, S. J. (1990) Changes in chemoreceptor sensitivity in relation to dietary selection by adult *Locusta migratoria*. *Entomol. Exp. Appl.*, **56**, 259–268.
120. Singer, M. C. (1971) Evolution of food plant preference in the butterfly *Euphydryas editha*. *Evolution*, **25**, 383–389.
121. Singer, M. C. (1982) Quantification of host preference by manipulation of oviposition behavior in the butterfly *Euphydryas editha*. *Oecologia*, **52**, 224–229.
122. Singer, M. C. (1983a) Determinants of multiple host use by a phytophagous insect population. *Evolution*, **37**, 389–403.
123. Singer, M. C. (1983b) Heritability of oviposition preference and its relationship to offspring performance within a single insect population. *Evolution*, **37**, 575–596.
124. Singer, M. C. and Parmesan, C. (1993) Sources of variation in patterns of plant–insect association. *Nature*, **361**, 251–253.
125. Stockoff, B. A. (1993) Ontogenetic change in dietary selection for protein and lipid by gypsy moth larvae. *J. Insect Physiol.*, **39**, 677–686.
126. Stride, G. O. and Straatman, R. (1962) The host plant relationship of an Australian swallowtail *Papilio augeus*, and its significance in the evo-

lution of host plant selection. *Proc. Linn. Soc. NSW,* **87**, 69–78.
127. Sword, G. A. and Chapman, R. F. (1994) Monophagy in a polyphagous grasshopper, *Schistocerca shoshone. Entomol. Exp. Appl.,* **73**, 255–264.
128. Szentesi, Á. and Bernays, E. A. (1984) A study of behavioural habituation to a feeding deterrent in nymphs of *Schistocerca gregaria. Physiol. Entomol.,* **9**, 329–340.
129. Szentesi, Á. and Jermy, T. (1990) The role of experience in host plant choice by phytophagous insects, in *Insect–Plant Interactions,* vol. 2, (ed. E. A. Bernays), CRC Press, Boca Raton, FL, pp. 39–74.
130. Szentesi, Á. and Jermy, T. (1993) A comparison of food-related behaviour between geographic populations of the Colorado potato beetle (Coleoptera, Chrysomelidae), on six solanaceous plants. *Entomol. Exp. Appl.,* **66**, 283–293.
131. Tabashnik, B. E. (1986) Evolution of host utilization in *Colias* butterflies, in *Evolutionary Genetics of Invertebrate Behavior,* (ed. M. D. Huettel), Plenum Press, New York, pp. 173–184.
132. Tarpley, M. D., Breden, F. and Chippendale, G. M. (1993) Genetic control of geographic variation for cannibalism in the southwestern corn borer, *Diatraea grandiosella. Entomol. Exp. Appl.,* **66**, 145–152.
133. Thompson, J. N. (1988) Evolutionary genetics of oviposition preference in swallowtail butterflies. *Evolution,* **42**, 1223–1234.
134. Thompson, J. N. and Pellmyr, D. (1991) Evolution of oviposition behavior and host preference in Lepidoptera. *Annu. Rev. Entomol.,* **36**, 65–89.
135. Thompson, J. N., Wehling, W. and Podolsky, R. (1990) Evolutionary genetics of host use in swallowtail butterflies. *Nature,* **344**, 148–150.
136. Thompson, R. F. and Spencer, W. A. (1966) Habituation, a model phenomenon for the study of neuronal substrates of behaviour. *Psychol. Rev.,* **73**, 16–43.
137. Thorpe, W. H. (1939) Further studies in preimaginal olfactory conditioning in insects. *Proc. Roy. Soc. Lond. B.,* **127**, 424–433.
138. Tinbergen, L. (1960) The natural control of insects in pine woods. I. Factors influencing the intensity of predation by songbirds. *Arch. Néerl. Zool.,* **13**, 265–336.

139. Traynier, R. M. M. (1986) Visual learning in assays of sinigrin solution as an oviposition releaser for the cabbage butterfly, *Pieris rapae. Entomol. Exp. Appl.,* **40**, 25–33.
140. Tully, T., Cambiazo, V. and Kruse, L. (1994) Memory through metamorphosis in normal and mutant *Drosophila. J. Neurosci.,* **14**, 68–74.
141. Ueckert, D. N. and Hansen, R. M. (1971) Dietary overlap of grasshoppers on sandhill rangeland in northeastern Colorado. *Oecologia,* **8**, 276–295.
142. Van Drongelen, W. and van Loon, J. J. A. (1980) Inheritance of gustatory sensitivity in F1 progeny of crosses between *Yponomeuta cagnagellus* and *Y. malinellus* (Lepidoptera). *Entomol. Exp. Appl.,* **28**, 199–203.
143. Van Loon, J. J. A. (1996) Chemosensory basis of feeding and oviposition behaviour in herbivorous insects: a glance at the periphery. *Entomol. Exp. Appl.,* **80**, 7–13.
144. Via, S. (1986) Genetic covariance between oviposition preference and larval performance in an insect herbivore. *Evolution,* **40**, 778–785.
145. Via, S. (1990) Ecological genetics and host adaptations in herbivorous insects: the experimental study of evolution in natural and agricultural systems. *Annu. Rev. Entomol.,* **35**, 421–446.
146. Waldvogel, M. and Gould, F. (1990) Variation in oviposition preference of *Heliothis virescens* in relation to macroevolutionary patterns of heliothine host range. *Evolution,* **44**, 1326–1337.
147. Wasserman, S. S. and Futuyma, D. J. (1981) Evolution of host plant utilization in laboratory populations of the southern cowpea weevil, *Callosobruchus maculatus* Fabricius (Coleoptera, Bruchidae). *Evolution,* **35**, 605–617.
148. Wiklund, C. (1982) Generalist versus specialist utilization of host plants among butterflies, in *Proceedings of the 5th International Symposium on Insect–Plant Relationships, Wageningen, 1982,* (eds J. H. Visser and A. K. Minks), Pudoc, Wageningen, pp. 181–191.
149. Zwölfer, H. (1970) Der 'Regionale Futterpflanzenwechsel' bei phytophagen Insekten als evolutionäres Problem. *Z. Angew. Entomol.,* **65**, 233–237.
150. Zwölfer, H. (1988) Evolutionary and ecological relationships of the insect fauna of thistles. *Annu. Rev. Entomol.,* **33**, 103–122.

THE ENDOCRINE SYSTEM OF HERBIVORES LISTENS TO HOST-PLANT SIGNALS

8

8.1	Development	227
8.1.1	Morphism	227
8.1.2	Diapause	230
8.2	Reproduction	231
8.2.1	Maturation	231
8.2.2	Mating behaviour	232
8.3	Conclusions	235
8.4	References	236

Few places on earth provide conditions for growth of plants and animals which remain constant throughout the year. Usually favourable seasons alternate with periods of low temperatures or drought. Plants tide themselves over these adverse periods by shedding their leaves, or the aboveground parts may die back completely. Annual species wait for the new growing season as dormant seeds in the soil. Insects may cease growth and reproduction and enter diapause, a state of regulated inactivity. But even during the growing season plants change in their nutritive value (Chapter 4) as well as in the amounts of secondary plant substances they harbour (Chapter 3). When a herbivorous insect strives to optimally exploit short-lived plants or particular developmental stages, an accurate synchronization of its life cycle with that of its host is of great adaptive value. Obviously, this is particularly true of food specialists and less important to generalist feeders. Synchronization is often quite well attained when both insects and plants respond to the same geophysical variables, notably changes in photoperiod and related factors such as temperature and rainfall. Sensitivity to these cues allows insects to anticipate future environmental changes and prepare for them through various inductive responses. A more refined synchronization of their phenology may be achieved when insects 'listen' to signals indicative of the developmental stage of their host plants. This ensures an accurate coupling of the insect's life cycle to that of its host, even when for some reason or other the host plant is slightly earlier or later than would be expected on the basis of, for instance, critical day length. Indeed, a great many insects monitor early indications of physiological changes in their host plant in order to set off the neuroendocrine mechanisms governing their development and reproduction at the appropriate time.

This chapter deals with plant factors, including olfactory, gustatory and tactile stimuli, that may be used by herbivores to synchronize their life cycles as closely as possible to those of their hosts and, as far as they are known, the underlying physiological mechanisms.

8.1 DEVELOPMENT

Developmental processes which may be governed by stimuli from the host plant include form determination (morphism) and induction of diapause.

8.1.1 MORPHISM

Aphids are well-known for their environmentally determined polymorphism (or polyphenism). They often show different gen-

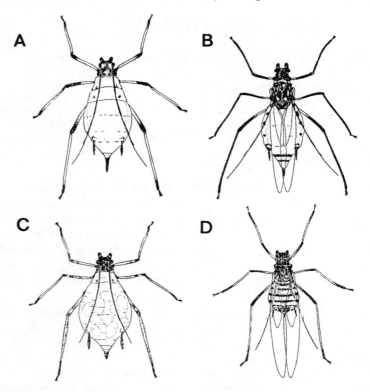

Figure 8.1 Polymorphism in the vetch aphid, *Megoura viciae*. **(A)** Apterous virginoparous female. **(B)** Alate virginoparous female. **(C)** Oviparous female. **(D)** Male. (Source: reproduced from Lees, 1961, with permission.)

erations of seasonal forms in relation with host alternation [10]. In many species the summer population can be either **apterous** (wingless) or **alate** (winged; see Fig. 8.1).

The two morphs are both 'virginoparous' (since they reproduce by parthenogenesis) but differ in their capacity to migrate to another host when the food plant becomes overexploited and its nutritional value starts to deteriorate. Food quality, with its gustatory and nutritional aspects, appears to be an important factor influencing wing formation. In an experiment with pea aphids (*Acyrthosiphon pisum*) it was shown that the age of the host plant affected the occurrence of wing development. Likewise, stimuli arising from crowded conditions stimulate wing development in the offspring of apterous females, and the effects of host-plant age and crowding are additive (Fig. 8.2).

The question that immediately arises concerns the nature of the factor or factors in the plant that induce the morphogenetic change to wing development. The answer to this question depends on which aphid–plant combination is studied, since many variations on the same physiological theme have been found. Often, the quality and amount of amino acids, sugars and/or water content are involved as agents responsible for the formation of alates. Generally, a deterioration in the nutritional value of host saps evokes an irreversible formation of wings, thereby allowing the insect to search for better hosts. It is noteworthy that

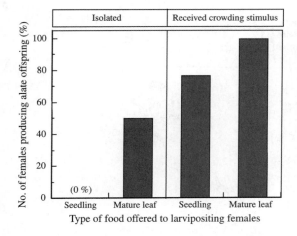

Figure 8.2 Role of host plant and crowding in production of winged forms by apterous pea aphids. All aphids were reared in isolation on broad bean (*Vicia faba*) seedlings and as adults were either subjected to a crowding stimulus or not. Larvipositing adults were kept on seedlings or on mature, fully expanded green leaves. (Source: data from Sutherland, 1967.)

wing development does not necessarily indicate a nutritional deficiency but rather a response to some plant factor (or factors) that precedes the change in food quality. Complete development and vigorous growth of alate adults occurs on plants that are still fully adequate food for normal growth and reproduction [10, 13].

Cultures of aphids on an artificial diet, i.e. a chemically defined liquid food, provide an elegant method of determining the influence of dietary components on morph determination. When green peach aphids (*Myzus persicae*) were grown on such a diet it appeared that omission of vitamin C from the standard composition caused an increase in the relative proportion of alate offspring from 28 to 62%. When, on the other hand, the amount of amino acids was halved, the frequency of alates produced was reduced [34]. Apparently the developmental switch responsible for wing formation is affected in a number of insects by these specific nutrients. The fact that the switch is not activated in all individuals may be due to the signal being more complex under natural conditions.

In several aphid species the production of sexual forms in late summer or autumn is also influenced by host-plant factors. Hitherto unknown changes in aging plants serve as signals to stimulate the production of sexuparae [13]. A striking example of a host effect on aphid morphs is found in *Eriosoma pyricola*, a species that lives on the roots of pear trees. During a limited period in late summer and early autumn alate sexuparous adults are produced, which emerge from the soil and then fly to elm trees, their primary host. The pear tree factor inducing the production of sexuparae is related to cessation of root growth, which, in its turn, is regulated by photoperiod and temperature conditions affecting the aboveground tree parts. Direct effects of these environmental variables on the aphids could be excluded, but the nature of the effective plant factor has not been identified [31].

What physiological mechanism links environmental triggers such as food quality to the developmental switch that governs the production of either alate or apterous adults? The juvenile hormone (JH) appears to play a key role in this process [8]. Larvae with low levels of JH produce winged females, whereas high JH levels, by mediating gene regulators, suppress wing formation. The question of how food quality affects the regulator of JH production, however, remains to be resolved.

Although the most dramatic examples of environmental polymorphism are found among aphids, the phenomenon also occurs in other insect taxa. For example, larvae of *Nemoria arizonaria*, a North American bivoltine moth species, show spectacular differences in appearance and behaviour between generations. Larvae of the spring generation feed on oak catkins and develop into highly cryptic mimics of catkins, showing a yellow rugose integument with reddish-brown stamen-like dots (Fig. 8.3(A)).

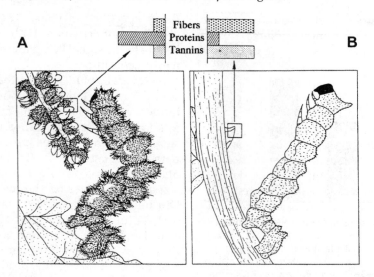

Figure 8.3 (A) Larvae of *Nemoria arizonaria* of the spring generation feeding on oak catkins and (B) larvae of the summer generation feeding on oak leaves develop into catkin morphs and twig morphs respectively. Catkins contain less fibre and polyphenols (tannins) but more unbound protein than leaves. (Source: reproduced from Dettner, 1989; after Greene, 1989, with permission.)

Larvae of the summer brood feed on oak leaves and have a greenish-grey, less rugose integument. Caterpillars of the latter form adopt the typical geometrid posture when disturbed (Fig. 8.3(B)). When raised on an artificial diet with tannin concentrations reflecting (high) natural food levels, caterpillars developed into twig morphs. Conversely, on tannin-free diets catkin morphs were produced. In this case also it is unclear how the plant stimulus, i.e. tannin levels, elicits the appropriate developmental response [7].

8.1.2 DIAPAUSE

Colorado potato beetles, after completion of metamorphosis, emerge from the soil and start feeding on the foliage of potato plants or a few related species. Under long-day photoperiods the beetles soon start to reproduce but under short photoperiods they withdraw again into the soil and enter a period of diapause. When fed old rather than young potato leaves the beetles show a clear tendency to enter dia-

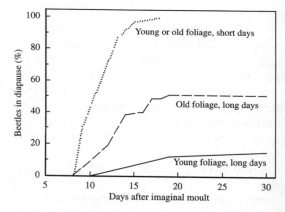

Figure 8.4 Effects of food quality (young *versus* old potato foliage) and day length (short *versus* long photoperiod) on diapause induction in adult Colorado potato beetles (*Leptinotarsa decemlineata*). (Source: redrawn from Danilevski, 1965.)

pause, even under long-day conditions (Fig. 8.4).

This response cannot be attributed to a nutritive deficiency because, when the corpora allata of active insects (which produce JH) are implanted into diapausing recipients, normal activity and reproduction is resumed, even when fed aging foliage [5]. The observation that wide differences in the incidence of diapause exist between field populations feeding on *Solanum dulcamara* (nightshade) and *S. tuberosum* (potato) likewise indicates that plant factors contribute to diapause induction. The adaptive significance of an earlier start of diapause in insects living on nightshade as compared to potato is probably related to the fact that insects on late-season nightshade are confronted with very high levels of glycoalkaloids, causing high mortality rates [9].

Several other instances of diapause induction in response to food quality have been reported in the literature [2, 21, 32] but the nature of the effective cues is in most cases unknown. Changes in moisture content and lipid levels have sometimes been found to contribute to diapause induction [2], but many other factors, including plant growth regulators, may also act as signals for timing neuroendocrine processes that regulate diapause.

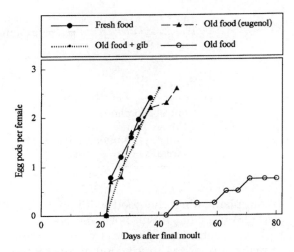

Figure 8.5 Egg-laying activity in desert locusts as a function of age and type of food. All insects were fed on cabbage leaves. Insects were fed old (senescent) leaves or fresh (full-grown normal) leaves. One group of insects was fed old food supplemented with 1 μg gibberelin A_2 per locust per day. Another group of insects was fed old food, but was treated externally with 1 μl of eugenol on the seventh day after the final moult. (Source: redrawn from Ellis *et al.*, 1965.)

8.2 REPRODUCTION

Herbivorous insects often oviposit on or near their larval food plants. For various reasons it may be a good strategy to postpone egg production and mating behaviour until the female insect is sure that food plants at the right developmental stage are available. Indeed in many species ripening of the gonads and mating behaviour are delayed in the absence of stimulating cues from host plants.

8.2.1 MATURATION

In natural populations the reproductive activity of desert locust (*Schistocerca gregaria*) adults can be delayed for periods of up to 9 months. Within a population, however, the onset of reproduction is remarkably well synchronized between individuals. Some environmental cue must trigger the maturation process, with the result that, after 1–2 weeks, coinciding with the beginning of the rainy season, all females are ready to oviposit. The signal stimulating egg development comes from the food. When the adult locusts eat bursting buds from some *Commiphora* species, such as *C. myrrha*, compounds that initiate sexual maturation are ingested. *Commiphora* buds have been known from ancient times to contain large amounts of essential oils. The supposition that some oil constituents, i.e. terpenoids such as pinene (**46**), eugenol (**22**) and limonene (**34**), induce sexual development could be verified in laboratory experiments. When young adults are fed senescent cabbage leaves only, sexual maturation is suppressed for extensive periods of time (Fig. 8.5).

When normal full-grown foliage is given, egg laying starts 3 weeks after the last moult. A single treatment of the insects with eugenol,

Table 8.1 Effect of presence of host plant after 3 days on oogenesis in *Plutella xylostella* females. (Source: modified from Hillyer and Thorsteinson, 1969)

Stimulus	No. of females	Mean no. eggs per ovariole
Control (no plant)	47	2.4
Cabbage plant	38	5.7
0.1 ml allyl isothiocyanate	32	4.3
0.2 ml allyl isothiocyanate	31	5.1

or supplementing the food with gibberelic acid (25) (a plant growth hormone present in high concentrations in the leaves of new growth) normal egg production occurs on the deficient (senescent) food [6] (Fig. 8.5). These compounds apparently serve as signals which in nature activate the neuroendocrine system, thereby stimulating sexual development.

Similar associations, although often less marked, have been found in other insect groups. The prereproductive period in female diamondback moths (*Plutella xylostella*) becomes longer when no host plant (cabbage) is present. Experiments employing host-plant specific volatiles indicated that the host influence on ovarian development can be simulated by a single host-plant odour constituent, allyl isothiocyanate [12] (Table 8.1).

In the bean weevil (*Acanthoscelides obtectus*) host odours are not effective and oogenesis is stimulated only after palpal contact with either bean seeds, leaves or pods. Beans are extensively palpated, which causes increased oogenesis within days. When the beans are varnished, however, or when the maxillary palps are ablated no egg ripening occurs, indicating that these palps have a crucial function in the perception of gustatory stimuli. Indeed bean seed washings elicit an oogenesis response in weevils, albeit to a lesser degree than beans themselves [23]. Although chemical cues, olfactory as well as gustatory, generally play a paramount role in stimulating sexual maturation, effects of physical factors cannot be excluded (Fig. 8.6).

In the literature some other instances of accelerated egg maturation due to kairomones from the host plant have been reported [30] (Table 8.2). In all but one case it concerns specialist feeders, a fact that is probably not fortuitous. Host-plant effects, as described above for food specialists but also for the polyphagous desert locust, may be considered as an 'anticipation' on the availability of sufficient food for the offspring. Some flexibility in timing of reproductive development contributes to a life history strategy that allows adaptation to unpredictable changes in host availability.

8.2.2 MATING BEHAVIOUR

Amateur entomologists rearing moths in captivity have known for a long time that often mating is stimulated by the presence of host plants. Riddiford and Williams [29] reported the first experimental proof of the influence of host plants on mating in the polyphemus moth (*Antheraea polyphemus*). Copulation in this insect occurred only in the presence of volatiles from oak leaves, their host plant. Since then several examples have been brought to light of female moths that begin pheromone release or 'calling' at a younger age, begin calling earlier in the night and spend more time calling when in the vicinity of their host plants [15, 18, 28]. In the small ermine moth, *Yponomeuta cagnagellus*, a long-lived species that reaches sexual maturity about one week after emergence, host-plant

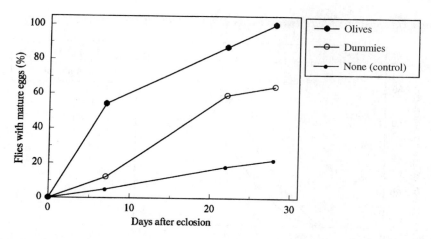

Figure 8.6 Ovarian development of female olive flies (*Dacus oleae*) in the presence or absence of olive fruits or dummies (orange-coloured wax domes). (Source: redrawn from Koveos and Tzanakakis, 1990.)

Table 8.2 Insects in which oogenesis is stimulated by kairomones from their host plants (G = generalist; S = specialist)

Insect species	Host range	Host plant	Reference
Orthoptera			
Acrididae			
Schistocerca gregaria	G	Various	6
Homoptera			
Aphalaridae			
Euphyllura phillyreae	S	Oleaceae	24
Lepidoptera			
Plutellidae			
Acrolepiopsis assectella	S	Leek	‡
Plutella xylostella	S	Brassicaceae	12, 22
Pyralidae			
Homoeosoma electellum	S	Sunflowers	19
Gelechiidae			
Scrobipalpa ocellatella	S	Beet	‡
Phtorimaea operculella	S	Potato	‡
Sitotroga cerealella	S	Grains	‡
Tortricidae			
Zeiraphera diniana	S	Larch	‡
Diptera			
Tephritidae			
Philophylla heraclei	S	Umbelliferae	‡
Dacus oleae	S	Olive fruits	‡, 14
Chloropidae			
Oscinella frit	S	Grasses	‡
Coleoptera			
Bruchidae			
Acanthoscelides obtectus	S	Bean seeds	‡, 23
Careydon serratus	S	Peanut (seeds)	3

‡ see Robert, 1986 for references

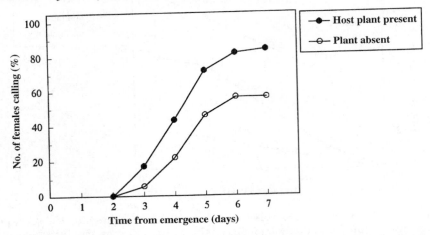

Figure 8.7 Calling behaviour in females of *Yponomeuta evonymellus* in the presence or absence of host plants. (Source: redrawn from Hendrikse and Vos-Bünnemeyer, 1987.)

volatiles act as releaser of calling behaviour. In the absence of host material calling is delayed or even permanently suppressed [11] (Fig. 8.7).

In this case the association between the odour of a suitable host and the age at which females initiate calling may contribute to the speciation process within the genus *Yponomeuta*, since it promotes reproductive isolation among sibling species that use different plant species as larval food [20]. Plant odours also affect male insects. In several nocturnal species males respond more strongly to female sex attractants when they occur mixed with green leaf volatiles [17]. Green leaf volatiles and some terpenoids can interact with pheromones at the receptor level, thereby modulating the sensitivity of the pheromone receptors [35].

Sometimes a very specific part of the host plant may accelerate female calling behaviour, as in the sunflower moth, *Homoeosoma electellum*, which oviposits preferentially in newly-opened flowers. Most females initiate calling for the first time during the first day after emergence, whereas in the absence of pollen calling behaviour may be delayed for periods of up to 2 weeks (Fig. 8.8).

Neonate larvae of this species require free pollen which, following anthesis, will due to biotic and abiotic influences rapidly decrease with time. The ability to rapidly initiate or delay calling behaviour (and concomitant ovarian development) in response to pollen availability permits the moth to cope with the unpredictability of food for its offspring [19].

As yet knowledge of which plant chemicals stimulate calling behaviour (and thus act as kairomones) is scarce. Conversely, new insights have recently been gained into the physiological mechanisms that control the production of pheromone and calling behaviour [27]. Females of the corn earworm (*Helicoverpa zea*), like several other species, locate a host prior to mating. Once they have found a suitable food source for its offspring, pheromone production and its release is started. Calling behaviour appears to be elicited by volatiles emanating from the silk of corn ears. Some of their purified components, for example phenylacetaldehyde, evoke the same response. Ethylene, a plant hormone widely involved in fruit ripening, is also effective (Fig. 8.9).

This compound might act as a common host

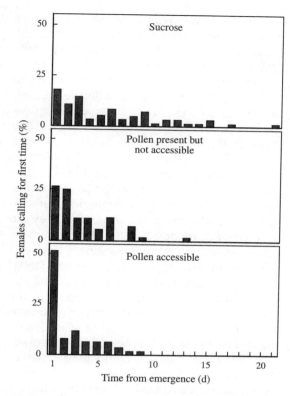

Figure 8.8 The age at which females of sunflower moths (*Homoeosoma electellum*) initiate calling behaviour for the first time as a function of age. Insects were held in the presence of sucrose only, or sucrose and sunflower pollen that was not directly accessible, or sucrose and accessible pollen. (Source: redrawn from McNeil and Delisle, 1989b.)

cue, because *H. zea* larvae feed on the fruiting parts of many different plants [28]. Treatment of moth antennae with silver nitrate (silver ions are known to inhibit ethylene responses in plants) suppressed pheromone release in the presence of ethylene, suggesting the involvement of ethylene receptors in the female olfactory system. When stimulated by host-plant odours corn earworm females release from their corpora cardiaca a pheromone biosynthesis activating neuropeptide (PBAN), which stimulates the production of pheromone in the abdominal glands [25, 27, 28] (Fig. 8.10).

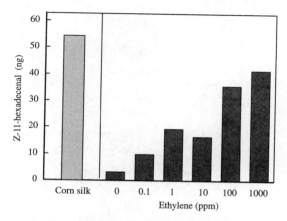

Figure 8.9 Pheromone production (ng per female) in *Helicoverpa zea* in the presence of (left) corn silk or (right) various concentrations of ethylene. (Source: redrawn from Raina et al., 1992.)

The examples given suffice to show that in a number of lepidopteran species virgin females first locate a suitable host before they start to attract males. Mating starts off egg production. In this way short-lived insects optimally exploit their limited time and energy resources.

8.3 CONCLUSIONS

During the past 50 years the role of plant chemicals in host-plant selection by ovipositing female insects and subsequently in food recognition by their hungry larvae has been extensively documented. The role of host plants in timing insect development has in comparison received very little attention. In this chapter some cases of insects adapting their life cycles to that of their host plants have been discussed briefly. Although few of this type of plant effect on insects are known, from a taxonomic point of view they cover a strikingly broad variety of insects, indicating that we have most probably encountered a fairly common phenomenon (see, for instance, Table 8.2). Not only different kinds of insect but also different developmental stages appear to be rous insects must use signals from host organisms with regulatory systems that are totally different from those regulating their own growth and development. Whereas insect parasitoids can employ host hormones identical to their own internal signal molecules as cues to synchronize their life cycles, and rabbit fleas use host hormones to adapt their reproductive cycles to that of their hosts, herbivores must rely on signals that are in no way related to physiologically familiar compounds. The plant signals that advertise the presence of a host or details about its physiological state are predominantly chemicals. Sometimes they are host-specific; in other cases they are more general compounds such as plant hormones.

The identity of most of the relevant chemicals largely remains to be elucidated, and the way in which insects detect and decipher the chemical signals also needs further investigation. The chain of events taking place inside the insect, likewise, leaves many open questions but at least there is compelling evidence that all physiological and behavioural responses are mediated by the neuroendocrine system.

A striking aspect of the described effects of plant signals is that, unlike responses to photoperiod, they rarely evoke an all-or-none reaction. In most cases development is only accelerated and at least part of the population will complete development, albeit with some delay, even in the continued absence of the plant signal.

We foresee that the topic of fine adjustments of insect life cycles to host-plant phenology will give an extra dimension, one of great richness, to the field of insect–plant relationships.

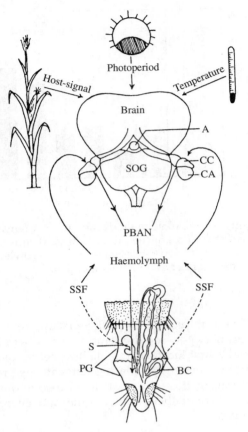

Figure 8.10 A schematic representation of the neurohormonal control of sex pheromone production in females of *Helicoverpa zea*. PBAN (pheromone biosynthesis activating neuropeptide) is produced in the suboesophageal ganglion (SOG) and transferred to the corpora cardiaca (CC). External factors, such as host-plant odour, photoperiod and temperature control its release from the corpora cardiaca (CC) into the haemolymph and it then stimulates pheromone production in the abdominal pheromone gland (PG). BC = bursa copulatrix. (Source: reproduced from Raina, 1988, with permission.)

affected, another indication of the importance of the capacity to tune in to subtle host signals.

In contrast to insect parasitoids or insect parasites living on vertebrate hosts, herbivo-

8.4 REFERENCES

1. Danilevski, A. S. (1965) *Photoperiodism and Seasonal Development of Insects*, Oliver & Boyd, Edinburgh.
2. Danks, H. V. (1987) *Insect Dormancy: an Ecological Perspective*, Biological Survey of Canada, Ottawa.

3. Delobel, A. (1989) Influence des gousses d'arachide (*Arachis hypogea*) et de l'alimentation imaginale sur l'ovogénèse, l'accouplement et la ponte chez la bruche *Careydon serratus*. *Entomol. Exp. Appl.*, **52**, 281–289.
4. Dettner, K. (1989) Chemische Ökologie. Ein interdisziplinäres Forschungsgebiet zwischen Biologie und Chemie. *Z. Umweltsch. Ökotox.*, **4**, 29–36.
5. De Wilde, J., Bongers, W. and Schooneveld, H. (1969) Effects of hostplant age on phytophagous insects. *Entomol. Exp. Appl.*, **12**, 714–720.
6. Ellis, P. E., Carlisle, D. B. and Osborne, D. J. (1965) Desert locusts: sexual maturation delayed by feeding on senescent vegetation. *Science*, **149**, 546–547.
7. Greene, E. (1989) A diet-induced developmental polymorphism in a caterpillar. *Science*, **243**, 643–646.
8. Hardie, J. and Lees, A. D. (1985) Endocrine control of polymorphism and polyphenism, in *Comprehensive Insect Physiology, Biochemistry and Pharmacology*, vol. 8, (eds G. A. Kerkut and L. I. Gilbert), Pergamon Press, Oxford, pp. 441–490.
9. Hare, J. D. (1983) Seasonal variation in plant–insect associations: utilization of *Solanum dulcamara* by *Leptinotarsa decemlineata*. *Ecology*, **64**, 345–361.
10. Harrewijn, P. (1978) The role of plant substances in polymorphism of the aphid *Myzus persicae*. *Entomol. Exp. Appl.*, **24**, 198–214.
11. Hendrikse, A. and Vos-Bünnemeyer, E. (1987) Role of the host-plant stimuli in sexual behaviour of the small ermine moths (*Yponomeuta*). *Ecol. Entomol.*, **12**, 363–371.
12. Hillyer, R. J. and Thorsteinson, A. J. (1969) The influence of the host plant or males on ovarian development or oviposition in the diamondback moth *Plutella maculipennis* (Curt.). *Can. J. Zool.*, **47**, 805–816.
13. Kawada, K. (1987) Polymorphism and morph determination, in *Aphids: Their Biology, Natural Enemies, and Control, World Crop Pests*, vol. 2A, (eds A. K. Minks and P. Harrewijn), Elsevier, Amsterdam, pp. 255–268.
14. Koveos, D. S. and Tzanakakis, M. E. (1990) Effect of the presence of olive fruit on ovarian maturation in the olive fruit fly, *Dacus oleae*, under laboratory conditions. *Entomol. Exp. Appl.*, **55**, 161–168.
15. Landolt, P. J., Heath, R. R., Millar, J. G., Davis-Hernandez, K. M., Dueben, B. D. and Ward, K. E. (1994) Effects of host plant, *Gossypium hirsutum* L., on sexual attraction of cabbage looper moths, *Trichoplusia ni* (Hübner) (Lepidoptera, Noctuidae). *J. Chem. Ecol.*, **20**, 2959–2974.
16. Lees, A. D. (1961) Clonal polymorphism in aphids. *Symp. Roy. Entomol. Soc. Lond.*, **1**, 68–79.
17. Light, D. M., Flath, R. A., Buttery, R. G., Zalom, F. G., Rice, R. C., Dickens, J. C. and Jang, E. B. (1993) Host-plant green-leaf volatiles synergize the synthetic sex pheromones of the corn earworm and codling moth (Lepidoptera). *Chemoecology*, **4**, 145–152.
18. McNeil, J. N. and Delisle, J. (1989a) Are host plants important in pheromone-mediated mating systems of Lepidoptera? *Experientia*, **45**, 236–240.
19. McNeil, J. N. and Delisle, J. (1989b) Host plant pollen influences calling behavior and ovarian development of the sunflower moth, *Homoeosoma electellum*. *Oecologia*, **80**, 201–205.
20. Menken, S. B. J., Herrebout, W. M. and Wiebes, J. T. (1992) Small ermine moths (*Yponomeuta*): their host relations and evolution. *Annu. Rev. Entomol.*, **37**, 41–66.
21. Numata, H. and Yamamoto, K. (1990) Feeding on seeds induces diapause in the cabbage bug, *Eurydema rugosa*. *Entomol. Exp. Appl.*, **57**, 281–284.
22. Pittendrigh, B. R. and Pivnick, K. A. (1993) Effects of a host plant, *Brassica juncea*, on calling behaviour and egg maturation in the diamondback moth, *Plutella xylostella*. *Entomol. Exp. Appl.*, **68**, 117–126.
23. Pouzat, J. (1978) Host plant chemosensory influence on oogenesis in the bean weevil, *Acanthoscelides obtectus* (Coleoptera, Bruchidae). *Entomol. Exp. Appl.*, **24**, 601–608.
24. Prophetou-Athanasiadou, D. A. (1993) Diapause termination and phenology of the olive psyllid, *Euphyllura phillyreae* on two host plants in coastal northern Greece. *Entomol. Exp. Appl.*, **67**, 193–197.
25. Rafaeli, A. (1994) Pheromonotropic stimulation of moth pheromone gland cultures in vitro. *Arch. Insect Biochem. Physiol.*, **25**, 271–286.
26. Raina, A. K. (1988) Selected factors influencing neurohormonal regulation of sex pheromone production in *Heliothis* species. *J. Chem. Ecol.*, **14**, 2063–2069.
27. Raina, A. K. (1993) Neuroendocrine control of sex pheromone biosynthesis in Lepidoptera. *Annu. Rev. Entomol.*, **38**, 329–349.

28. Raina, A. K., Kingan, T. G. and Mattoo, A. K. (1992) Chemical signals from host plant and sexual behavior in a moth. *Science*, **255**, 592–594.
29. Riddiford, L. M. and Williams, C. M. (1967) Volatile principle from oak leaves: role in sex life of the polyphemus moth. *Science*, **155**, 589–590.
30. Robert, P. C. (1986) Les relations plantes–insectes phytophages chez les femelles pondeuses: le rôle des stimulus chimiques et physiques. Une mise au point bibliographique. *Agronomie*, **6**, 127–142.
31. Sethi, S. L. and Swenson, K. G. (1967) Formation of sexuparae in the aphid *Eriosoma pyricola*, on pear roots. *Entomol. Exp. Appl.*, **10**, 97–102.
32. Steinberg, S., Podoler, H. and Applebaum, S. W. (1992) Diapause induction in the codling moth, *Cydia pomonella*: effect of larval diet. *Entomol. Exp. Appl.*, **62**, 269–275.
33. Sutherland, O. R. W. (1967) Role of host plant in production of winged forms by a green strain of pea aphid *Acyrthosiphon pisum* Harris. *Nature*, **216**, 387–388.
34. Sutherland, O. R. W. and Mittler, T. E. (1971) Influence of diet composition and crowding on wing production by the aphid *Myzus persicae*. *J. Insect Physiol.*, **17**, 321–328.
35. Van der Pers, J. N. C., Thomas, G. and den Otter, C. J. (1980) Interactions between plant odours and pheromone reception in small ermine moths (Lepidoptera, Yponomeutidae). *Chem. Senses*, **5**, 367–371.

ECOLOGY: LIVING APART TOGETHER 9

9.1	Host plants affecting herbivorous insect demography	240
9.1.1	Host-plant quality	240
9.1.2	Host-plant phenology	240
9.1.3	Vegetation structure	241
9.1.4	Why are so many herbivorous insect species 'rare'?	242
9.2	Herbivorous insects affecting plant demography	244
9.2.1	Seed production and plant recruitment	244
9.2.2	Herbivory and plant competitiveness	246
9.2.3	Plant tolerance and compensation for damage	246
9.3	The composition of insect–plant communities	247
9.3.1	General considerations	247
9.3.2	Species–area relationships	248
9.3.3	Plant structural diversity and architecture	249
9.3.4	Plant chemistry	250
9.3.5	Intrapopulational and intraindividual heterogeneity of host plants	250
9.3.6	Changes in herbivorous insect assemblages through time	251
9.4	Interactions in insect–plant communities	256
9.4.1	Competition and facilitation among herbivorous insects	256
9.4.2	Interactions through natural enemies	260
9.4.3	Plants affecting the third trophic level	262
9.4.4	Natural enemies affecting herbivorous insect damage	267
9.4.5	The role of microorganisms in insect–plant relationships	268
9.5	Herbivorous insects affecting plant communities	268
9.5.1	Early successional plant communities	268
9.5.2	Established (climax) plant communities	269
9.6	Energy and mineral flow in communities	270
9.7	Conclusions	272
9.8	References	272

This chapter deals with the ecological aspects of insect–plant relationships at the level of insect and plant populations and at the level of insect–plant communities. First, we discuss whether or not food plants affect the population dynamics (demography) of insect herbivores and to what extent, and whether or not insect herbivores influence the demography of host-plant populations and to what extent (sections 9.1 and 9.2). The term 'population' is defined as a group of interbreeding or potentially interbreeding individuals [80]. 'Population dynamics' means variations in population density, i.e. in the number of individuals per unit area in time. The population dynamics of both insects and plants are the result of many more biotic and abiotic factors than we can discuss here, so we use also the more neutral term 'demography'. Secondly, we deal with insect–plant communities (sections 9.3–9.6). According to Price's [94] definition, 'biological communities' consist of coexisting interacting populations. A plant population, the populations of herbivores exploiting that plant and the populations of natural enemies (including pathogens) of the herbivores compose an assemblage of populations that can be regarded as a community comprising three trophic levels. We survey the composition of insect–plant communities, the insect–insect and insect–plant interactions shaping them, as well as the energy and mineral flow through such communities.

9.1 HOST PLANTS AFFECTING HERBIVOROUS INSECT DEMOGRAPHY

9.1.1 HOST-PLANT QUALITY

That the quality of the host plant as a food for herbivorous insects may vary because of both genetic variation and environmental factors has been discussed in detail in Chapter 4. Here we recapitulate briefly the effect of host-plant quality on insect population dynamics, with the help of two models representing the multitude of interactions involved [51] (Fig. 9.1).

The 'phytocentric model' stresses the plant's resource acquisition and allocation, the outcome of which is the set of characteristics of the plant organs, such as the presence or absence of primary and secondary plant substances, physical attributes, etc. These are determined by intrinsic factors (genotype and ontogeny) and by extrinsic (abiotic and biotic environmental) effects. The 'exploiter model' indicates that modes of resource exploitation by the insect, i.e. the mode of feeding and food quality (the output of the 'phytocentric model') together result in the insect's specific performance attributes, such as feeding preference, food consumption, growth, development, survival, fecundity and oviposition preference. Consumption itself may be a factor influencing plant quality in a negative (induced resistance) or a positive (facilitation) sense (section 3.14). The insects' life history parameters, combined with specific performance attributes, determine insect fitness and population growth, which results in their population dynamics. The latter, of course, is also influenced by other extrinsic factors such as natural enemies and abiotic environmental factors as well as by the presence of other members of the community in which a certain insect–plant association is embedded. The insects, in turn, may also affect plant characteristics.

9.1.2 HOST-PLANT PHENOLOGY

Phenology of the host plant may be of crucial importance for the performance of herbivorous insects. Weather conditions disrupting the synchronization between insect and plant phenology may strongly influence insect population dynamics. For example, most foliage-feeding lepidopterans, such as the larvae of *Operophthera brumata*, are able to attack oaks (*Quercus robur*) only from just after bud break

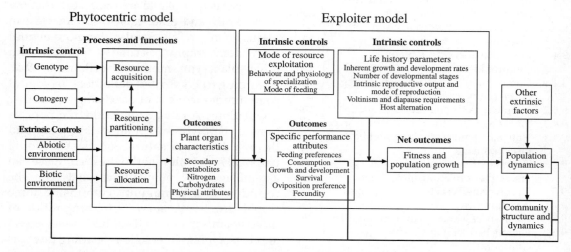

Figure 9.1 Interactions determining the performance of plants and their herbivorous insect exploiters. See text for explanation. (Source: redrawn from Jones and Coleman, 1991.)

Host plants affecting insect demography 241

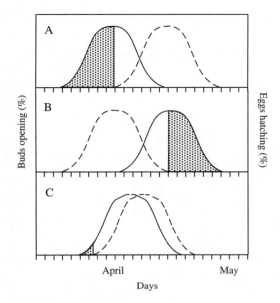

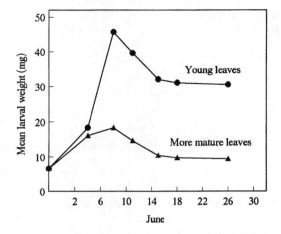

Figure 9.2 Timing of *Operophthera brumata* egg hatch (solid line) and oak bud break (broken line). Stippled area indicates starvation of larvae. (Source: redrawn from Feeny, 1976.)

Figure 9.3 The mean fresh weight of fourth-instar *Operophthera brumata* larvae reared on young and more mature oak leaves. The initial rise in weight, representing fourth- and fifth-instar larval feeding phase, is followed by a decline during the prepupal phase until pupation is complete. The difference in performance is due to the decrease in protein content and increase in tannin content of oak leaves with age. (Source: modified from Feeny, 1970.)

until the extension of the first set of leaves. Figure 9.2 shows that if bud opening precedes egg hatch (B) or egg hatch precedes bud opening (A) a large proportion of the larvae starve. However, the more precisely the timing of the two events coincides (C), the better the performance of the larvae (Fig. 9.3) and the stronger the defoliation [25, 26]. Clearly, individual trees which bud either very early or very late may remain unattacked by these insects ('phenological escape').

9.1.3 VEGETATION STRUCTURE

Most herbivorous insect species get to their host plants by active locomotion or passive dispersal so the probability of finding the host plant strongly depends on the structure of the vegetation in which the host occurs. The **resource concentration hypothesis** proposes that herbivores are more likely to find and remain on hosts that are growing in dense or nearly pure stands and that the most specialized species frequently attain higher relative densities in simple environments [105]. Field experiments comparing the abundance of specialized insects per host plant in various combinations of host and non-host plants often gave results similar to those depicted in Figure 9.4: i.e. in a monoculture the abundance of a specialist herbivore per individual host plant is much higher than in a polyculture of the same plant species [124].

Different vegetation structures may be accompanied by differences of microclimate, which in turn may affect the abundance of specialized herbivorous insects. For example, chrysomelid beetles attacking bean and squash were less abundant in a maize–bean–squash polyculture than in monocultures, partially because they avoided bean and squash plants shaded by maize plants [102].

Studies on wild plant species that form patches of various size intermingled with

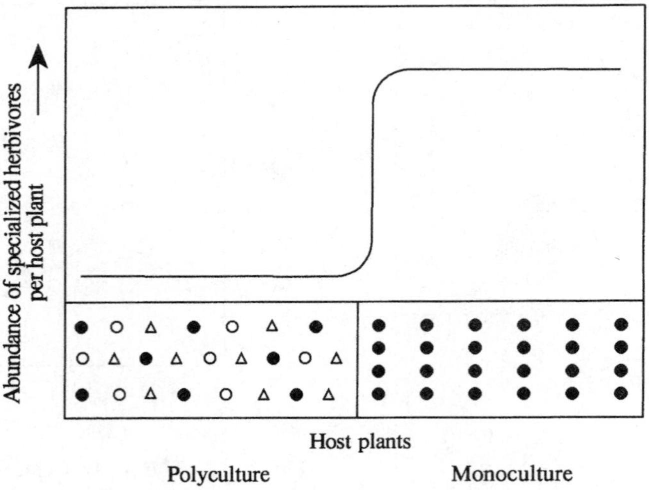

Figure 9.4 The effect of poly- *versus* monoculture on the abundance per host plant of specialized herbivores on the plant species represented by black dots. (Source: redrawn from Strong *et al.*, 1984.)

other plant species provided variable results. The density of cinnabar moth adults (*Tyria jacobaeae*) was greater in high-density plots of their host plant, the ragwort (*Senecio jacobaea*), than in low density plots [131]. In contrast, the butterfly *Anthocaris cardamines* showed an opposite behaviour: the egg-laying females preferred the edges of host-plant patches, so isolated host plants were more prone to be oviposited on than individual plants within patches [13].

As will be discussed in more detail in section 12.3.2, the **enemy hypothesis** predicts that there will be a greater abundance and diversity of entomophages in polycultures, first of all because of an increased availability of alternate host insects [105]. This results in reduced numbers of herbivorous insects in such vegetations.

Agricultural landscapes are characterized by habitat fragmentation, i.e. by the presence of isolated host-plant patches (fields) among non-host plants. Kruess and Tscharntke [64] studied the effect of habitat fragmentation on herbivores and their natural enemies by establishing 1.2 m² red clover (*Trifolium pratense*) patches isolated by a distance of 100–500 m from the nearest meadow with naturally occurring clover plants. They found that not only the immigration rate of herbivorous insects but even more the number of parasitoid species following the herbivores decreased with increasing distance from the nearest reservoir population. Thus, habitat fragmentation affects natural enemies more than the herbivores, thereby increasing the probability of pest outbreaks in isolated patches. This is an important aspect that must be considered in biological control of pest insects with parasitoids.

Clearly, the effect of vegetation structure on the relative abundance and the population dynamics of herbivorous species is highly unpredictable, because each species behaves differently [54].

9.1.4 WHY ARE SO MANY HERBIVOROUS INSECT SPECIES 'RARE'?

It is common knowledge among entomologist taxonomists that in almost all higher insect taxa (genera, families) many or even most

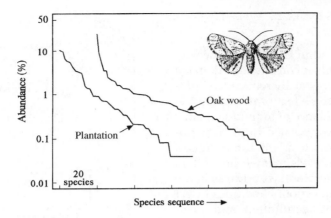

Figure 9.5 The relative abundance of moth species in an oak forest and in a conifer plantation in Northern Ireland. In both plantations there are a few abundant moth species while most species are rare or very rare. (Note the logarithmic scale of abundance. The sequence of species corresponds to their decreasing relative abundance.) (Source: redrawn from Magurran, 1988.)

species are 'rare', i.e. difficult to find or absent in localities and periods where and when they might be expected to be present given their life-history, host-plant availability, etc. [32]. For example, in Hungary the following ratios of 'rare species' within the most thoroughly studied herbivorous insect groups have been found: 48.5% in Chrysomelidae [56], 62.2% in Tephritidae [81] and 65.6% in Coccoidea [60]. Moreover, different herbivorous insect species associated with particular plant species may differ greatly in their average abundance: a few species are often abundant and/or produce outbreaks, while the abundance of most species is always low [78], as shown in Figure 9.5.

An explanation can be sought in the hypothesis that herbivores are seldom food-limited but appear most often to be enemy-limited [38]. This has been supported by, for instance, Root and Cappuccino [106], in a 6-year study of the herbivorous insect assemblage associated with a natural goldenrod (*Solidago altissima*) population: of the 138 species only seven were abundant and even these seldom reached densities at which they caused a decrease in the density of the remaining species. A further explanation could be that host-plant quality is not optimal for the insects (see Chapter 4 for a detailed discussion of this). According to Wratten [139] this fully explains why the density of herbivores is generally much lower than the available food resources would support. Extreme weather conditions could be a further partial explanation for the 'rarity' of many species and the cause of temporary local extinctions, such as observed in *Euphydrias* butterflies [23]. In a long series of studies Den Boer [18] often found local extinction in carabid beetle populations. His conclusion is most probably also valid for herbivorous species: 'Populations do not exist very long, but are continually becoming extinct and being founded again'. Even if one supposes that 'rarity' results from the combined action of the above mentioned factors, the intriguing question remains unanswered: Why are some insect species abundant and why are other (often closely related) species 'rare'? Yet, the 'rarity' of many herbivorous insect species is a fact that should be thoroughly considered before generalizations are made concerning ecology and evolution of insect–insect and insect–plant interactions [32].

9.2 HERBIVOROUS INSECTS AFFECTING PLANT DEMOGRAPHY

As has been discussed in Chapter 2.8, herbivorous insects may inflict considerable damage on plants and may drastically reduce seed production. However, the effect of insects on the reproductive performance of individual plants cannot simply be extrapolated to plant populations, and even less to plant communities, because at the latter levels many other factors may interfere [4]. Therefore, we address in the following only the question: How does insect damage affect plant recruitment and plant competitiveness?

9.2.1 SEED PRODUCTION AND PLANT RECRUITMENT

In natural communities plants usually produce a large number of propagules (seeds, spores, tubers, etc.) per unit area, but not every propagule develops, even in the absence of herbivores, into a mature plant during the next generation. On the contrary, in closed plant communities only a small fraction of the seeds will develop into productive plants, mainly due to intra- and interspecific plant competition. This means that with most plant species recruitment in the subsequent plant generation is microsite-limited. Therefore, in most cases the reduction in seed production by herbivores must reach a certain degree before it really affects the population dynamics of a plant species. As Figure 9.6 shows, quite a large loss of seeds by seed 'predation', i.e. by the destruction of seeds by animals (granivores), can be tolerated without affecting the number of adult plants in the next generation, especially in the vicinity of the parental plant.

However, the effect of granivory on plant population dynamics strongly depends on the species' life-history traits. As can be seen from the model in Figure 9.7, granivory has little effect on plants that have a long life cycle (perennials), that also propagate vegetatively, that produce many small seeds and that show

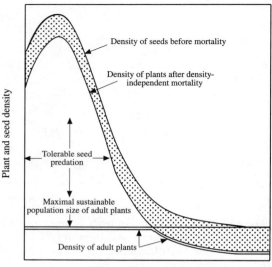

Figure 9.6 An idealized relationship between distance of seed dispersal and the density of plants and seeds, showing the level of predation that is tolerable without affecting the density of adult plants in the progeny. (Source: modified from Harper, 1977.)

masting (i.e. cyclic seed production) or form a persistent seed bank (i.e. their seeds remain dormant for several years in the soil) [73].

Thus, trees, shrubs and perennial herbs can tolerate considerable seed losses. Seed loss may, on the other hand, strongly affect annuals that propagate by seeds only, that produce few large seeds and that form only a transient seed bank, i.e. their seeds do not remain dormant in the soil for several years. The thistle *Cirsium canescens* is an example of the latter type. In an experiment this monocarpic plant of sand prairies in middle USA was protected from flower- and seed-feeding insects by insecticide treatments. This resulted in a three- to 37-fold increase in the number of progeny that reached maturity per parent plant as compared with the controls: water spray or no treatment (Fig. 9.8).

In this species the availability of microsites and the effect of vertebrate herbivores appeared to have less impact on the popula-

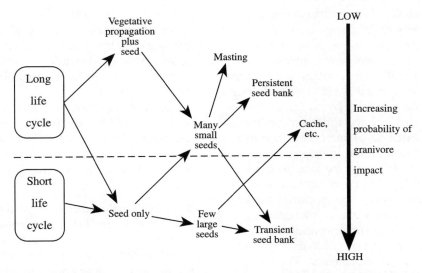

Figure 9.7 A conceptual model suggesting the interrelationship of plant life-history traits that may affect the impact of seed predation (granivory) on plant recruitment. (Source: modified from Louda, 1995.)

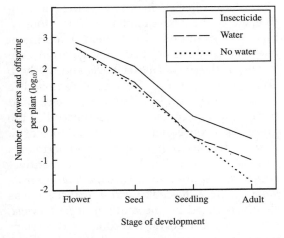

Figure 9.8 Response of *Cirsium canescens* individuals in three treatments (insecticide spray, water spray and no spray) over the stages of development in a 6-year experiment. Flower = total number of flowers initiated per plants; Seed = total number of viable seeds released per plant after flower and seed predation by insects; Seedling = number of seedlings established per plant; Adult = number of seedlings that matured to flower and set seed. (Source: redrawn from Louda and Masters, 1993.)

tion dynamics of this plant [74, 75]. In another experiment insecticide treatments also increased seed production in the shrubs *Haplopappus squarrosus* and *H. venetus*. However, plant recruitment increased only in *H. squarrosus* because recruitment of *H. venetus* was microsite-limited [71, 72].

Naturally, not only seed predation but also damage inflicted on vegetative plant parts may reduce seed production. The model in Figure 9.7 is also valid for such cases. For example, as has been mentioned in Chapter 2, leaf-feeding herbivorous insects that cause little defoliation on oak can still strongly reduce the number of acorns produced [14]. It is unknown, however, whether or not this affects the number of fruiting progeny of an individual oak tree. The tree, after all, may live for centuries and may produce millions of acorns during its lifetime, while only a minute fraction of them find space to develop into a mature tree during this time. Hence the effect of reduction in seed production on the population dynamics of long-lived trees is most probably negligible. Although acorn numbers may theoretically affect the spreading of oaks, for instance *via* their dispersal by jays, it is impossible even to guess how this would affect the spread of oak over centuries.

In stable plant communities each parent individual of annual and biennial plant species will be replaced by roughly one offspring individual. This is indicated by the fact that the composition of such plant communities, e.g. a mesophilic meadow, does not change drastically from one year to the next. Nevertheless, in most cases several orders of magnitude more seeds are produced than can develop into mature plants to replace each parent individual. In such communities reduction of seed production probably has little direct influence on recruitment. The situation is, however, very different for plant communities of early secondary plant succession, such as occur in disturbed habitats. Here insect exclusion experiments, such as the abovementioned one with *Cirsium canescens*, revealed considerable effects of herbivorous insects on plant demography [74, 75], because in such habitats microsites are much less limited than in established plant communities.

The experience gained from biological control of weeds by herbivorous insects may help to elucidate the effect of insects on plant distribution and abundance in general, since in a number of cases introduced weeds have been successfully controlled by introduced herbivorous insect species (see examples in Chapter 12). In this connection, however, the following should be considered.

1. Unsuccessful attempts at biological weed control with herbivorous insects outnumbers successful programmes [31, 117].
2. For control purposes the insects are introduced, very carefully, without their specific natural enemies (parasitoids, etc.), which means that their potential rate of reproduction can be much higher than in their original habitats.
3. If an introduced weed is successfully controlled by an insect species, this does not mean that the same plant is kept at low population levels in its area of origin by the same insect, because both the plant and the insect now occur under ecological conditions that differ fundamentally from the original habitat.
4. Biological weed control has been carried out mainly in disturbed communities, because introduced weeds mostly invade habitats, e.g. pastures, that differ widely from the undisturbed natural vegetation [40].

Thus the partly positive experiences of biological weed control do not necessarily prove the overall effect of those insect species on the population dynamics of their host plants in the original natural plant communities.

9.2.2 HERBIVORY AND PLANT COMPETITIVENESS

Herbivorous insects may reduce the competitiveness of their host plants, as in the following two examples. In a field experiment insecticides were applied to vegetation consisting of the grass *Holcus mollis* and the herb *Galium saxatile*. The treatment removed a grass-feeding aphid, *Holcaphis holci* and thereby increased the competitiveness of the grass. As a consequence, the abundance of *G. saxatile* decreased [90]. In field experiments carried out in Switzerland, the competitiveness of the spotted knapweed, *Centaurea maculosa*, with a fescue, *Festuca pratensis*, was clearly reduced by the moth *Agapeta zoegana*, whose larvae feed on the roots of the knapweed [84]. Thus, insect herbivores may markedly affect the balance between competing plant species in natural vegetations.

9.2.3 PLANT TOLERANCE AND COMPENSATION FOR DAMAGE

We have illustrated in section 2.9 how, under various conditions, plants can tolerate, compensate or even overcompensate herbivore damage. Here we deal briefly with the probable effects of these plant traits on plant population dynamics. As an example, the spindle tree (*Euonymus europaeus*) may be defoliated almost yearly by *Yponomeuta* spp. (Lepidoptera) and often suffers considerable bark damage from rabbits. Nevertheless, it is common in coastal dunes and inland hedge growth in Holland, demonstrating very strong regrowth capacity [132]. Experiments carried out on the morning glory (*Ipomoea purpurea*) also indicate that plant tolerance may largely counterbalance damage caused by herbivorous insects [119].

A thought-provoking, although still strongly debated idea was put forward by Owen [89] – that plants occasionally benefit from being eaten by animals. He suggested, for example, that the honeydew accumulating on the ground beneath aphid-infested plants increases the number of free-living nitrogen-fixing bacteria and hence makes more nitrogen available to the plants. Thus, especially on soils with a short supply of nitrogen, a plant may benefit more from increased nitrogen uptake than it loses to phloem-tapping aphids. According to a recent study [76] the microorganisms that develop rapidly on gypsy moth frass on the oak forest soil temporarily immobilize the nitrogen content of the frass by incorporating it into their body mass. By this they hinder for some months the recycling of the nitrogen present in the frass to the oaks, but they also prevent the loss of nitrogen from the ecosystem.

Thus the effect of insect feeding on plant demography appears extremely variable. As regards its overall importance, Crawley [15] concluded:

> Comparison of fencing and insecticide-exclusion experiments leaves little room for doubt that in most ecosystems, vertebrate herbivores have substantially greater impact on plant population dynamics and community structure than invertebrates. Even in forest communities, where the main canopy defoliators tend to be invertebrates, vertebrates may still exert a controlling influence on plant dynamics through seed and seedling predation.

(See also section 9.5.2.)

9.3 THE COMPOSITION OF INSECT–PLANT COMMUNITIES

9.3.1 GENERAL CONSIDERATIONS

The literature provides various definitions of the term 'community'. As mentioned before, we have chosen Price's [94] definition, according to which a community is composed of coexisting and interacting populations. Plant populations, herbivorous insect populations and populations of the insects' natural enemies (including pathogens) comprise an assemblage that can be regarded as an 'insect–plant community', since such populations interact in various ways. This 'subcommunity', however, represents only one of the compartments composing a 'full community', for example, a meadow. In the following we restrict our discussion to the question: How do plants determine the composition of herbivorous insect assemblages?

An extensive discussion is going on among ecologists on the question of how communities are organized. According to one extremist view communities are supraindividual organizations in which each member population has an irreplaceable function. Therefore the removal of any population would overthrow the stability or equilibrium of the community. The interrelations among the populations are so tight that the evolution of each community member influences the evolution of the other members [134]. In such a reasoning the assumption of interspecific competition, as an organizing force, plays a decisive role. At the other extreme it is supposed that the assemblage of herbivorous insects especially ('herbivorous insect communities') results from the individualistic colonization of the plants by insect species, in the absence of competition

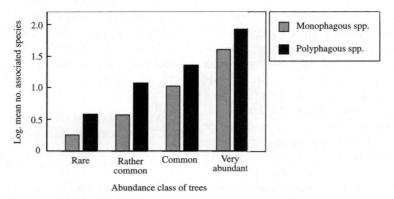

Figure 9.9 The mean number of herbivorous insect species associated with Hawaiian trees of different abundance classes. (Source: redrawn from Southwood, 1973.)

[93]. From studies on herbivorous insects attacking bracken, Lawton et al. [70] concluded that accidents of evolutionary history and large-scale biogeographical processes dominate local community structure; if species interactions are present, their effects are extremely feeble. In this view 'insect–plant communities' are little more than haphazard assemblages. This does not mean, of course, that there is no interaction among populations (see examples below). However, there is no evidence for the assumption that the populations of a community are totally interdependent like the cogwheels of a machine, so that the removal or addition of any population would disturb the whole community. In the following we treat some factors that are supposed or were found to determine the composition of herbivorous insect assemblages associated with plant populations or plant communities.

9.3.2 SPECIES–AREA RELATIONSHIPS

As noted in Chapter 2.4, interesting correlations have been found between the number of herbivorous insect species associated with a plant species and the size of the plant's area of distribution. Figure 9.9 shows this relationship in the case of Hawaiian tree species [121] and Figure 9.10 gives the same for British perennial herbs [68].

Two hypotheses provide an explanation of this relationship. According to the **encounter-frequency hypothesis** [120] the number of herbivorous insect species associated with a plant species depends on the probability of the insects' encountering a potentially suitable host-plant species. The more widespread a plant the greater the probability of encounter. The **habitat-heterogeneity hypothesis** [138] states that the more widespread a plant the more kinds of habitats will occur within its area; consequently, the plant supports more insect species because the insects' habitat requirements vary. As a matter of fact, the two hypotheses are complementary: the more widespread a plant the more likely it is to occur in different habitats, which in turn increases the probability of encounters.

As different sizes of area may also imply differences in habitat diversity, Simberloff [118] experimentally attempted to separate the effects of habitat diversity and area size. In pure mangrove (*Rhizophora mangle*) stands he surveyed the arboreal arthropod fauna, then reduced the size of some mangrove islands by cutting down a number of trees. In this way

The composition of insect–plant communities 249

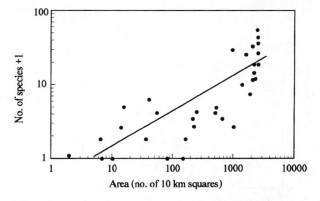

Figure 9.10 Number of herbivorous insect species associated with perennial herbs in Britain as a function of the plant's geographical range. (Source: modified from Lawton and Schröder, 1977.)

habitat diversity was not affected; nonetheless the species richness of arthropods decreased compared to the control islands whose sizes remained unchanged. Thus, area was the prime factor determining species richness.

9.3.3 PLANT STRUCTURAL DIVERSITY AND ARCHITECTURE

The richness of herbivorous insect assemblages is strongly influenced by the plants' structural diversity, i.e. by the distribution of plant structures in the vertical plane. Furthermore, it is also affected by the plants' architectural attributes, i.e. by the availability and distribution of plant parts or structures in space above the ground level [6]. Figure 9.11 demonstrates that the number of herbivorous insect species associated with British plants increases both with increasing area of distribution and with increasing structural complexity from monocotyledons to bushes and shrubs.

The effect of plant architecture on herbivorous insect diversity has been found even within one plant genus. As presented in Figure 9.12, architectural complexity of *Opuntia* species is correlated with the number of herbivorous insect species associated with them. Undoubtedly there is also an effect of size on

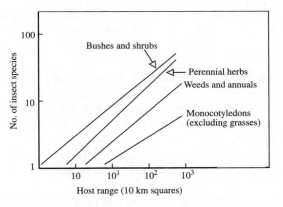

Figure 9.11 The species richness of herbivorous insects on plants increases with the complexity of plant structure and area size. Number of herbivorous insect species associated with British plant species as a function of host-plant area and architectural complexity. The number of associated insect species rises as architectural complexity increases. Host range is expressed in terms of 10 km map grid squares in which the plant occurs. Note logarithmic scale of Y-axis. (Source: reproduced from Strong *et al.*, 1984, with permission, based on data from Lawton and Schröder, 1977.)

the composition of a plant's insect fauna, but a partitioning between size and heterogeneity has rarely been made [83].

In summary, then, a greater variety of insect

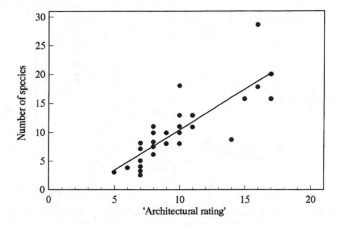

Figure 9.12 The number of herbivorous insect species associated with 28 North and South American *Opuntia* species as a function of the plants' architecture. 'Architectural rating' is the sum of the following variables scored from 1–4: (1) height of mature plant; (2) mean number of cladodes; (3) cladode size (cm^2); (4) development of woody stem; (5) cladode complexity: quality of cladode surface, presence and density of spines. (Source: redrawn from Strong *et al.*, 1984, based on data from Moran, 1980.)

species is to be found on larger plant species, with more complex architecture, that have a common and widespread distribution and perhaps live in the same area as related species.

9.3.4 PLANT CHEMISTRY

Considering the paramount importance of plant chemistry in the insects' host-plant selection (Chapters 5, 6, 7) and in the process of host-plant specialization (Chapter 10), it seems logical to suppose that plant chemistry decisively determines the composition of herbivorous insect assemblages. In this connection several hypotheses have been developed. Both the **biochemical barrier hypothesis** [124] and the **diverse defence hypothesis** [52] predict a negative correlation between insect species richness and chemical diversity and/or uniqueness, because colonization of such plants would be reduced. Conversely, the **common chemistry hypothesis** [20] and the **enemy escape hypothesis** [96] predict an opposite situation: plants with diverse chemistry are more likely to share chemicals with other plant taxa, facilitating host shifts, and herbivorous insects colonizing chemically unusual plants may be less exposed to entomophages that locate their hosts by plant odours.

The implicit contradiction among the above hypotheses itself indicates that the relation between plant chemistry and composition of herbivorous insect communities (guilds) is extremely multifaceted and largely unpredictable.

9.3.5 INTRAPOPULATIONAL AND INTRAINDIVIDUAL HETEROGENEITY OF HOST PLANTS

The variability of the plants' physical and chemical characteristics have been discussed in detail in Chapters 3 and 4. Here we deal with the effects of this heterogeneity on the assemblage of insect populations. The abundance of herbivorous insect populations may vary tremendously even among individuals of the same plant population. Price [93] proposed the **plant genotypic-heterogeneity hypothesis**, pointing out that the genetic diversity of

Table 9.1 Number of sawflies (four species) on willow clones in two different plots (I and II). The shoots developed from stems that grew the year before. (Source: reproduced from Price, 1983, with permission)

		Galls (no.)			
Plot	Clone	Euura sp. Stem	Euura sp. Petiole	Pontania sp. Leaf	Phyllocolpa sp. Leaf folds (no.)
I	1	0	0	4	0
	2	60	1	2	0
	3	15	2	0	3
	4	0	1	1	0
	5	1	0	3	0
	6	8	1	4	41
	7	3	0	1	0
II	8	5	3	235	9
	9	0	0	119	0
	10	9	0	114	0
	11	3	0	125	0
	12	3	0	50	0
	13	3	0	262	0
	14	15	1	100	0
	15	13	0	0	0
	16	36	0	2	0

a host-plant population may influence the species diversity of herbivorous insects associated with the plant. Table 9.1 shows the striking differences in presence and abundance of four sawfly species on 16 different willow clones growing in two plots.

The willow clones differ in many characters: phenology, chemistry, stem and bud colour, stem diameter, growth characteristics, etc. Habitat differences between the two plots were not excluded [93]. Genotypic intraindividual differences may also affect the presence and abundance of the insect populations on different parts of the same plant individual, as has been observed in, for instance, gall-making aphids on poplar trees (Fig. 9.13).

The phenotypic differences among individual plants caused by local ecological factors, may also influence the composition of herbivorous insect assemblages. For example, bracken fern supports partially different herbivorous insect species in shaded *versus* unshaded places in the same area [66, 67].

9.3.6 CHANGES IN HERBIVOROUS INSECT ASSEMBLAGES THROUGH TIME

(a) **Seasonal changes**

Especially in temperate regions, herbivorous insect populations associated with a plant species show drastic seasonal changes due to the insects' phenology. For example, the guilds formed by microlepidopteran populations on fruit trees in Hungary show two clearly distinguishable phenological stages (seasonal aspects) that are characterized by the presence and/or absence of the moth species' larvae. The spring stage covers the spring months from bud opening to blossom; the summer–autumn stage lasts from petal drop to the end of the growing season. It is noteworthy that from the 51 monitored species only 13

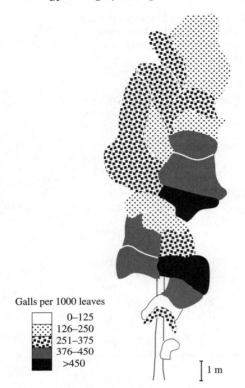

Figure 9.13 The distribution of about 53000 *Pemphigus betae* (Aphididae) galls over 20 branches of a 20.1 meter high *Populus angustifolia* tree. The size of each branch reflects total leaf area. (Source: redrawn from Whitham, 1983.)

were present in both stages and six occurred only in the summer–autumn stage [98] (Table 9.2).

(b) Successional changes

Secondary plant succession, i.e. the recolonization and further development of the plant community on a site that became bare by some disturbance, is naturally also followed by colonization by herbivorous insects. Studies carried out in Britain have shown a rapid change in the vegetation during the first 5 years. Plant species diversity reached a maximum after 7–11 years, followed by a gradual decrease [6]. Data from another study shown in Figure 9.14 indicate that species diversity of herbivorous insects and predators initially increased rapidly but after reaching a plateau it did not decrease despite a reduction in plant species diversity.

The number of leaf miner species even increased with time [34]. Unfortunately, the processes determining the successional changes in herbivorous insect assemblages related to plant succession are largely unknown.

(c) Colonization of introduced plants by herbivorous insects

Humans have unconsciously made an interesting experiment by transferring exotic plants to new geographic areas, which might shed some light on the adaptation rate of a local entomofauna to 'neophytes' (i.e. plants that were settled in new geographic regions after 1500 AD). According to the available data the colonization curve (also called 'recruitment curve') is asymptotic, i.e. the number of insects adopting an introduced plant species as host increases rapidly in the beginning but later gradually slows down, although colonization probably never stops [124]. The following hypotheses have been proposed as explanations for the asymptotic nature of the colonization curve.

The **niche saturation hypothesis** [77] (or ecological-saturation hypothesis) states that colonization is limited by the number of available niches, because the establishment of subsequent colonizers becomes increasingly difficult, as a result of interspecific competition. However, the rarity of interspecific competition among herbivorous insects, the occurrence of 'vacant niches' and the possibility of multiple occupation of the same niche (see 9.4.1) negate this hypothesis.

According to the **enemy-free space hypothesis** [65] (or enemy escape hypothesis) the natural enemies associated with the introduced plant limit colonization by herbivores. Although such a role of natural enemies can not be excluded, so far no case studies on

Table 9.2 Seasonal changes in the species composition of microlepidopterous guilds (larvae) harboured by six fruit tree species in Hungary. The spring stage comprises data from bud opening to blossom, the summer stage from petal dropping to the end of the growing season. Data were collected during 1954–1974. (Source: modified from Reichart, 1977)

	Spring stage						Summer–autumn stage					
	Pear	Apple	Apricot	Peach	Plum	Cherry	Pear	Apple	Apricot	Peach	Plum	Cherry
Yponomeutidae												
Scythropia crataegella					+							
Yponomeuta padellus					+							
Yponomeuta malinellus		+										
Swammerdamia pyrella	+	+	+									
Argyresthia pruniella		+										
Argyresthia conjugella								+				
Ypsolophidae												
Ypsolophus persicella			+	+								
Ypsolophus asperella			+									
Glyphipterigidae												
Choreutis pariana	+	+		+	+	+	+	+	+		+	
Oecophoridae												
Diurnea fagella		+	+									
Dasystoma salicella	+	+	+	+	+							
Coleophoridae												
Coleophora serratella	+	+	+		+							
Coleophora paripennella		+										
Coleophora bernoulliella	+	+	+			+						
Agonoxenidae												
Blastodacna atra	+	+										
Gelechiidae												
Anarsia lineatella		+	+	+	+			+	+	+		
Anacampis obscurella				+	+							
Gelechia scotinella			+		+							
Recurvaria leucatella	+	+	+	+	+	+						
Recurvaria nanella	+	+	+	+	+	+						
Tortricidae												
Sparganothis pilleriana		+		+								
Pandemis dumetana		+		+	+	+						
Pandemis heparana	+	+	+	+	+	+	+	+	+	+	+	+
Pandemis cerasana	+	+	+	+	+	+	+	+	+	+	+	
Argyrothenia ljungiana		+	+		+	+	+	+	+	+		
Choristoneura hebenstreitella		+			+		+	+				
Archips rosana		+		+	+	+	+	+	+	+		
Archips crataegana		+	+			+						
Archips xylosteana		+	+			+						
Archips podana		+	+	+	+	+	+	+	+	+	+	

Table 9.2 (Continued)

	Spring stage						Summer–autumn stage					
	Pear	Apple	Apricot	Peach	Plum	Cherry	Pear	Apple	Apricot	Peach	Plum	Cherry
Clepsis spectrana	+											
Adoxophyes orana		+	+	+	+	+	+	+	+	+	+	+
Ptycholoma lecheana		+	+		+	+						
Neosphaleroptera nubilana		+	+		+	+						
Aleimma loeflingiana			+			+						
Croesia holmiana		+				+						
Acleris rhombana	+				+	+						
Acleris ferrugana	+	+	+	+	+	+						
Hedia pruniana					+	+						
Hedia dimidioalba	+	+	+	+	+	+	+	+				
Ancyllis achatana		+	+		+		+	+				
Ancyllis tineana				+								
Spilonota ocellana	+	+	+	+	+	+	+	+				
Enarmonia formosana									+	+	+	+
Cydia molesta	+	+		+			+	+	+	+	+	+
Cydia lobarzewskii											+	
Cydia funebrana									+		+	
Cydia pomonella							+	+	+	+		
Cydia pyrivora							+					
Phyticidae												
Trachycera advenella		+										
Acrobasis obtusella	+											

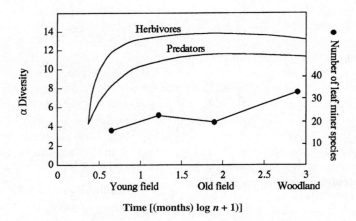

Figure 9.14 Species diversity of herbivorous and predatory insects (excluding Diptera) along a secondary succession gradient. Leaf miners are expressed as number of species found on herbs. (Source: redrawn from Brown and Southwood, 1987, based on a figure in Godfray, 1985.)

introduced plants are available to test this hypothesis.

The **pool exhaustion hypothesis** [69] provides the simplest explanation. It proposes that the pool of resident potential (pre-adapted) colonist herbivorous insect species is quickly exhausted, because it consists of (1) a number of polyphagous insect species that easily adopt the new plant, (2) a few specialists living on plants that are closely related to the introduced plant; these are potential colonizers of the new plant, while (3) the majority of local herbivorous insect species is specialized on plants not related to the newcomer. The latter are unlikely ever to adapt to the neophyte. Especially narrowly specialized mining and galling insects appear to be handicapped in colonizing new plants [124]. This hypothesis is supported by many examples. A Swiss survey of insects on goldenrod (*Solidago altissima*), a North American weed that was introduced into Europe more than 100 years ago, showed 18 herbivores feeding on it. Except for two insect species these were all opportunistic, unspecialized ectophages and not closely attuned to the growth cycle of *S. altissima*. In contrast, out of the 314 insect species found on this goldenrod in North America 58 species (25%) are specialist feeders. The endemic European goldenrod, *Solidago virgaurea*, houses 88 insect species, 23 (28%) of which are specialists [49]. Likewise, a solanaceous weed, *Solanum mauritanum*, introduced more than a century ago from South America into South Africa, remains relatively unscathed by local herbivores, whereas several endemic *Solanum* species often show a high incidence of damage [88]. A similar relationship has been found when the numbers of herbivorous arthropods associated with three tree species in their original habitat and in places of introduction were compared [122] (Fig. 9.15).

A broader study, involving many introduced plant species, confirms that underexploitation of neophytes is a common phenomenon. In an extensive survey of insects

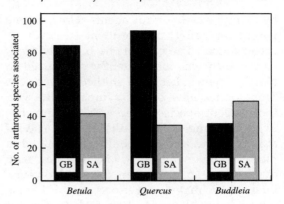

Figure 9.15 Three tree species, *Betula pendula*, *Quercus robur* and *Buddleia* spp., are found both in Britain (GB) and in South Africa (SA). *Betula* and *Quercus* are native to Britain but were introduced to South Africa, while for *Buddleia* the converse is true. Clearly the species richness of herbivorous arthropods associated with the trees is smaller where the trees were introduced compared to where they are native. (Source: redrawn from Southwood *et al.*, 1982.)

on exotic plants in Central Europe only 166 herbivorous insect species were observed on 100 different plant species. This contrasts sharply with the fact that endemic plants are known to be regularly attacked by at least five to ten insect species per plant species [58]. Probably, herbivores need some time to 'discover' (and adapt to) potential new resources. The insect needs to adapt its recognition system, as well as its physiological mechanisms. The importance of digestive adaptation, even among comparatively generalist insects, can be inferred from the observation that Californian grasshoppers were found to grow better and to show greater survival on native than on introduced grass species [50].

(d) Establishment of introduced herbivorous insects

Herbivorous insect species have been introduced to many parts of the world unintentionally as pests and purposefully as biological weed control agents. How do the newcomers affect the resident herbivorous insect fauna? The case of the butterfly *Pieris napi*, which was inadvertently introduced to North America, has been studied extensively. This species now occurs sympatrically with the resident species, *P. oleracea*. Their niches overlap strongly: the adults of both species may be found in the same field, mating takes place in the same area and females may oviposit on the same plant species. However, interspecific mating never occurs and there is no evidence of competition [11]. In Central Europe the apple tree harbours some 14 resident scale insect species, including *Quadraspidiotus ostreaeformis*, *Q. marani*, *Q. pyri*, and *Epidiaspis leperii*. Nevertheless, the introduced San José scale (*Q. perniciosus*), which infests the same parts of the apple tree, has easily established without driving off any of the resident scale insect species [61]. Such cases negate the competitive exclusion principle but support the assumption of 'vacant niches', or the possibility of multiple occupation of niches (see 9.4.1). However, no investigations have been carried out to check whether all dimensions of the niche occupied by an introduced insect are fully identical to the niche dimensions of one or more closely related resident species whose life history seems identical with that of the newcomer. Establishment of herbivorous insects introduced for biological control of weeds has failed in several cases, but there is no indication that this was due to the presence of other, either introduced or native, herbivorous species [35]. In conclusion, introduced herbivorous insects do not seem to markedly affect resident herbivore assemblages.

9.4 INTERACTIONS IN INSECT–PLANT COMMUNITIES

Virtually no insect species lives alone on its host plant. It shares it with other herbivorous species, with indwelling parasitoids and predators, with fungal mutualists and with

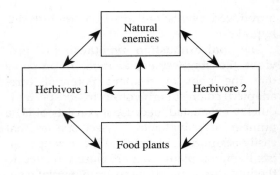

Figure 9.16 The pathways of direct, food-plant mediated, and enemy-mediated interactions among herbivores as well as between food plants and natural enemies. (Source: modified from Damman, 1993.)

pathogens. They all may affect positively or negatively the target species, thus forming intricate networks of interactions. A herbivorous species inhabiting a plant population interacts in various ways with the populations of natural enemies. The scheme depicted in Figure 9.16 indicates possible directions of various interactions [16].

The direct responses of the plants to herbivore attack, i.e. the phenomenon of induced resistance, have been discussed in detail in section 3.14. Below we deal with the remaining routes of interaction.

9.4.1 COMPETITION AND FACILITATION AMONG HERBIVOROUS INSECTS

It is self-evident that direct interactions between herbivorous insect populations are most likely to occur between those exploiting the same plant species, especially the same plant parts (section 2.3). Different herbivorous insect species feeding on a plant population form **guilds**. According to Root's [104, 105] terminology a guild is defined as 'a group of species that exploit the same class of environmental resources in a similar way'. Guilds can be regarded as elementary communities. Herbivorous insects feeding on, for example, collards (*Brassica oleracea*) compose the following guilds:

1. **pit-feeding guild**: flea beetles, weevils eating small patches on the leaves;
2. **strip-feeding guild**: caterpillars chewing the leaves from the side;
3. **sap-feeding guild**: aphids, leaf hoppers, tingids.

Populations or individuals of a guild influence each other more the less they are separated in space and time [43]. Separation in space may result, e.g. from the intrapopulational and intraindividual heterogeneity of the host plant (section 9.3.5) [30]. Temporal separation is also important; as indicated by Figure 2.12, eight insect species associated with the stinging nettle show little overlap of population peaks in time, which strongly reduces the probability of direct interactions. Of course, the possibility of indirect effects, i.e. *via* host-plant quality, remains. Thus, early season damage to mountain birch (*Betula pubescens*) foliage significantly impaired the performance of sawfly larvae (*Dineura virididorsata*), which is a typical late-season feeder. This example demonstrates that niche segregation between species of herbivores is not *a priori* enough to prevent interspecific interactions [86].

Theoretical ecologists (e.g. MacArthur [77]) assume that community structure is mainly determined by the number of niches a certain resource provides. Each niche can be occupied only by one species, because if two or more species require the same niche, then the strongest will outcompete the others. This is the classic **competitive exclusion principle** [33]. If we take the host plant as the primary niche dimension of herbivorous insects, then it would be logical to assume that the structure of herbivorous insect guilds associated with a plant species would be determined primarily by the interspecific competition of guild members. A survey of available data [124] has shown, however, that in 24 case studies no competition could be demonstrated, while in 18 other studies **amensalism**, i.e. strongly asymmetrical competition, was found when the plant was exploited by one insect species depriving the other species of food. This is the case during outbreaks when individual plants are defoliated or destroyed. However, even in these cases competition does not always occur. For example, from 21 experiments carried out on collards with two flea beetles (*Phyllotreta cruciferae, Ph. striolata*) and with the larvae of *Pieris rapae*, only in two instances were significant competitive effects found [53]. **Symmetrical competition** was observed in eight cases only. Interestingly, competition was most frequently observed among species of Hemiptera [124]. A more recent review [19] arrived at a different conclusion. This report examined 193 case studies (among them 19 laboratory or glasshouse experiments) on interactions between pairs of herbivorous insect species. It was found that interspecific competition occurred in 76% of interactions, was often asymmetrical and was frequent in most guilds (sap feeders, wood and stem borers, seed and fruit feeders), but not in free-living mandibulate folivores. In 18% of cases the interactions had no detectable effect on the populations involved, while in 6% facilitation has been observed.

As regards the overall conclusiveness of such largely contradictory literature surveys, the following should be considered. First, the spectrum of case studies may be biased by an understandable desire by researchers to select systems in which interspecific competition is strongly suspected; cases of apparent harmonious coexistence may not be chosen in the first place or may go unreported [123]. Second, for the same reason only abundant herbivorous insect species are chosen for such studies, since the probability of competition between rare species is extremely low and studying such cases would cause insurmountable methodological difficulties. However, as noted before (section 9.1.4), in natural communities most herbivorous insect species are most of the time present at low or very low population

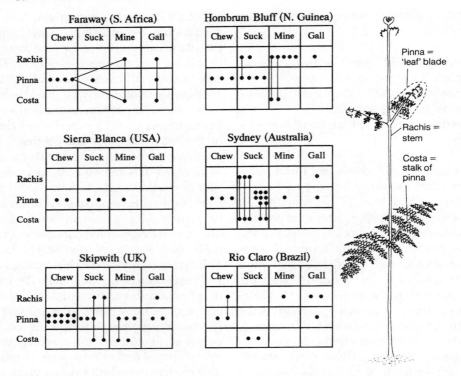

Figure 9.17 Feeding sites and feeding types of herbivorous arthropods attacking bracken fern (*Pteridium aquilinum*) in six different parts of the world. Each dot refers to one arthropod species; feeding sites of insects exploiting more than one plant part are joined by lines. (Source: redrawn from Lawton *et al.*, 1993.)

densities as compared to the abundance of their host plants. Therefore, the results of case studies in which competition among herbivorous insect species could be investigated experimentally or by observation are not representative of the majority of species and communities. Consequently, we still cannot judge the precise incidence of interspecific competition across a range of insect herbivore communities [123], so the overall importance of competition as an organizing force in structuring insect herbivore communities is still an open question. This contrasts sharply with other groups of organisms, including plants, carnivores and detritivores.

As mentioned before, one reason that competition among herbivores is relatively rare is that the plant world offers many unoccupied niches. An impression of the number of vacant niches (or, more precisely, resources) may be obtained by comparing the occurrence of specialized herbivorous insects on the same plant species in different geographical regions. Figure 9.17 shows that bracken fern (*Pteridium aquilinum*) is exploited to very different degrees in six geographical regions.

This indicates that (1) there are several unused resource dimensions ('vacant niches') on bracken in, for example, a North American habitat compared to bracken in England, and (2) the same resource can be exploited by several insect species, e.g. ten species in England chew the pinna. Lawton *et al.* [70] concluded that the colonization of bracken by herbivorous insects over evolutionary time has been largely a stochastic process that has

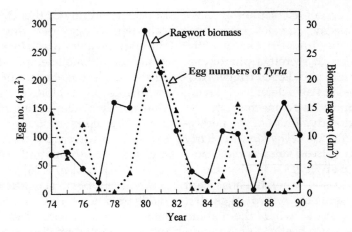

Figure 9.18 Fluctuation in the numbers of *Tyria jacobaeae* eggs per 4 m² and in the biomass of *Senecio jacobaea* estimated by percentage of ground cover per dm². (Source: redrawn from van der Meijden *et al.*, 1991.)

not been constrained by interspecific interactions between the herbivores. In other words, 'niche segregation' may have evolved by chance without any channelling forces [47]. The failure to occupy all available niche space argues against a powerful role for interspecific competition.

Obviously, MacArthur's [77] idea that tight 'species packing' on resources determines community structure does not hold for herbivorous insects. Price [93] rightly emphasized that

> in evolutionary time, in contrast to the tight species packing invoked by theoreticians dealing with competition, species will not be packed at all, being more like sardines in an ocean than sardines in a tin, with much ecological space between species. (. . .) In ecological space and time the conclusion is similar – local occupation of habitats is very patchy, many patches remain uncolonized, and the probability of encounter between potential competitors is low.

Intraspecific competition for food may affect the density of herbivorous insect populations, which, in turn, may influence interspecific interactions. Intraspecific competition is obvious in cases when local food sources become limited or are exhausted, e.g. when larvae of a folivorous species defoliate the host plant before completing their development. Such population outbreaks are very spectacular with forest pest insects, but are less pronounced in herbaceous vegetations when small patches or isolated individuals of a plant population are defoliated. For example, in a 17-year study the cinnabar moth (*Tyria jacobaeae*) showed periodic crashes in numbers (Fig. 9.18), because the larvae exhausted locally their food resource, the ragwort (*Senecio jacobaea*), by complete defoliation before the larvae were fully grown.

Starvation of the larvae and increased parasitism resulted in periodic depressions of the moth population. The plant population recovered between the outbreaks of the moth [133]. Intraspecific competition may seriously reduce insect fitness. It is therefore not surprising that behavioural mechanisms, such as territoriality and spacing, have evolved that decrease the probability of competition. Partitioning of food by territorial behaviour have been found, for example, in females of bruchids, butterflies and tephritid flies that mark their oviposition sites by substances (termed 'marking' or 'epideictic pheromones') that inhibit conspecific females from egg-laying at the same site [101, 103]. Larvae of the moth *Ephestia kuehniella* secrete a pheromone from their mandibular glands that triggers dispersal of the larvae and also deters egg-laying females [12]. A sophisticated mechanism regulates the population dynamics of the herbivorous ladybird beetle, *Epilachna niponica*. Early in the season the long-lived females avoid host plants containing many eggs and disperse by flight. Later in the season, when the host plant deteriorates, the females resorb eggs but oviposition may resume when the habitat recovers [87]. The bark beetle, *Dendroctonus pseudotsugae*, deters conspecifics by an antiaggregation pheromone, which is released by each sex at the sonic stimulus of the other sex [110]. Larvae of some microlepidopterous species that are solitary feeders in cobwebs prevent the settlement of conspecific and heterospecific larvae on the same leaf by specific knocking sounds and aggressive behaviour [111].

In contrast to the above, the feeding instars of several herbivorous species aggregate on their host plants. The adaptive advantage of such behaviour may be sought in the relative protection it provides against predators [62] or increased resource exploitation. For example, massive attack by bark beetles accelerates the breakdown of the tree's resistance [114] and aggregation of Japanese beetles (*Popillia japonica*) on tree tops in response to damage-induced plant volatiles increases mating behaviour [92]. Both territoriality and aggregation may occur in subsequent ontogenetic phases. For instance, the egg-laying females of *Pieris brassicae* show dispersal activity in that they avoid plants already carrying conspecific eggs [107], whereas their young-instar larvae aggregate when feeding and moulting.

Facilitation may occur among members of a guild if the presence of one herbivorous insect population provides conditions that promote another species in exploiting the common host plant. For example, females of the cecidomyid fly, *Dasyneura papaveris*, can deposit eggs into the poppy-head only through the holes made by egg-laying females of the curculionid beetle *Ceutorhynchus maculaalba*. Thus, in the absence of this beetle the fly is unable to exploit its host plant [125]. Damage by a stem-boring moth attacking the reed *Phragmites australis* converts resistant shoots into susceptible ones, thereby facilitating the growth of gall midge populations (*Giraudiella inclusa*) [128].

Another type of facilitation is using a shelter made by other herbivores. For instance, many late-season tortricids and pyralids on oak oviposit preferentially on or near leaf rolls built by leaf-rolling caterpillars feeding earlier in the season [10].

9.4.2 INTERACTIONS THROUGH NATURAL ENEMIES

Entomophages may strongly affect relative abundance of herbivorous insects. In this connection Lawton [65] proposed the **enemy-free space hypothesis** and conjectured that competition of herbivorous insects for such space is more likely to occur than competition for food. This can be demonstrated by the simple hypothetical model [45] presented in Figure 9.19.

Of the two herbivores, H1 is attacked both by a specialist and a generalist parasitoid while H2 is decimated only by the common

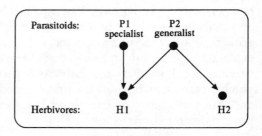

Figure 9.19 A scheme presenting the possible interactions between two parasitoid and two herbivorous insect species. (Source: modified from Holt and Lawton, 1993.)

generalist. When the numbers of H2 increase, the numbers of the generalist parasitoid increase too; this will reduce the abundance of H1. Thus 'apparent competition' [44] between the two herbivores occurs. A mathematical analysis has shown that, if H2 is sufficiently productive, this could result in the exclusion of both H1 and its specialized parasitoid from the community. However, as Holt and Lawton [45] stress: 'Many of the detailed theoretical arguments (...) lack direct empirical support'.

Nevertheless, there is quite a lot of circumstantial evidence supporting the postulates of 'enemy-free space' and 'apparent competition'. For instance, the lack of enemy-free space has often been blamed for the non-establishment of herbivorous insect species introduced for biological control of weeds. Among others, the cochineal species *Dactylopius opuntiae*, introduced to Africa, successfully controlled *Opuntia* species, except near the coast and at higher elevations where lower summer temperatures prevented a rapid multiplication of the herbivore. Here the cochineal suffered considerable predation by two indigenous coccinellid predators, which are also present in the warmer regions but there cannot keep up with the herbivore [36]. Faeth [24] reported on another type of enemy-mediated interaction. He simulated folivore damage by hole-punching one edge of oak leaves, then attached these structurally damaged leaves to oak leaves with mines of a *Cameraria* moth species. Undamaged leaves attached to leaves with mines served as a control. The mortality of the leafminer was significantly higher in the presence of structurally damaged leaves, indicating that both parasitoids and predators were attracted to the mines by the presence of such manipulated leaves.

Ants that attend honeydew-producing herbivorous insects may also mediate interactions between herbivores, since they may remove non-attended herbivore populations [7, 8]. However, ants may cause also positive interactions between two herbivores. For instance, the larvae of the beetle *Odontata dorsalis* mining in the leaves of locust trees near colonies of an ant-attended membracid (*Vanduza arquata*) were less often attacked by predators and parasitoids than the larvae mining far from the membracids [29].

The relative importance of various interactions between herbivores has been analysed in 79 field studies. Figure 9.20 demonstrates that resource- (plant-)mediated interactions were

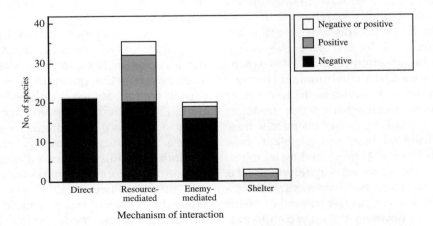

Figure 9.20 The relative importance of the mechanisms implicated in 79 case studies that showed clear evidence of interactions between herbivorous insect populations. (Source: redrawn from Damman, 1993.)

most numerous, followed by direct and then by enemy-mediated interactions, the 'factor shelter' being the least common.

The number of negative outcomes was highest in the case of direct interactions followed by the resource- and enemy-mediated ones, although the differences were small. Positive interactions were mostly found with resource-mediated interactions [16]. Yet, it is difficult to tell what role the above-discussed interactions play in various natural communities. Damman [16] has pointed out: 'At the moment, good experimental field studies are sufficiently scarce that only the most pronounced patterns can be detected.'

9.4.3 PLANTS AFFECTING THE THIRD TROPHIC LEVEL

Various characteristics of plants may affect herbivorous insects' natural enemies, including insect pathogens.

(a) Plant chemistry

Plant chemistry mediates interactions among members of all three trophic levels. The action of invertebrate predators, parasitoids or insect pathogens may be enhanced or retarded directly by plant chemicals or structure and indirectly by plant chemicals ingested by the herbivores or by quality variations in the food ingested by the herbivores. For example, terpenoids produced by the pine tree act as **allomones**, because they repel and deter many herbivorous insects; as **kairomones**, because they attract the bark beetles to the tree; finally as **synomones**, because bark beetle predators are also attracted by these chemicals from which both the predator and the plant benefits. Cucurbitacins (**14**) produced by members of the Cucurbitaceae act as deterrents (i.e. allomones) to many herbivores, such as *Epilachna tredecimnotata*. To the cucumber beetle (*Diabrotica* sp.), however, the same compounds are phagostimulants (i.e. kairomones). These insects store those compounds in their body, where they serve as deterrents (i.e. allomones) against the beetle's predators. In the latter case the plant's resistance compounds fail to protect their producer against this adapted herbivore. Trichomes of wild *Nicotiana* spp. contain alkaloids that are toxic to parasitoids of tobacco herbivores, so the alkaloids negatively affect both the plant and the parasitoid (i.e. here they are antimones) because the herbivorous insects attacking the plant are not restricted by the parasitoids [137]. A direct effect of host-plant allelochemicals ingested by herbivorous insects on the survival of parasitoids has been demonstrated by experiments with *Manduca sexta* larvae fed with tobacco cultivars containing different amounts of nicotine. From larvae feeding on low-nicotine cultivars, significantly more *Cotesia congregata* (Hymenoptera) adults emerged than from larvae fed upon high-nicotine cultivars [1].

Striking effects of host-plant species on the parasitization of the polyphagous African cotton bollworm (*Helicoverpa armigera*) were observed when this insect was exposed to one of its natural enemies on several different food plants. When chickpea plants with feeding caterpillars were exposed to ovipositing *Microplitis demolitor* wasps, no parasitization at all occurred, whereas under the same conditions parasitization levels of up to 75% were recorded on cotton and soybean plants [85] (Fig. 9.21).

Likewise, oviposition activity by an aphid parasitoid varied markedly with the plant species on which the aphids were feeding, demonstrating the importance of host-plant cues in the acceptance of aphids by the parasitoid [3]. Another influence of the host plant on its herbivore parasitoids concerns the latter's reproduction strategy. The plant's nutritional condition may influence the sex ratio of hymenopterous parasitoids, as in the case of the ichneumonid *Diadegma insulare*. When the females of this parasitoid oviposited on diamondback moth larvae placed on strongly fertilized plants, the ratio of females in the progeny was much higher than on

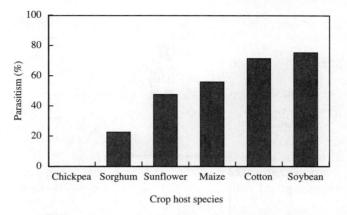

Figure 9.21 Effect of host plant on parasitism of *Helicoverpa armigera* by *Microplitis demolitor* on six crop hosts in a glasshouse experiment. (Source: data from Murray and Rynne, 1994.)

poorly-fertilized plants. This finding suggests that the parasitoid females used plant cues about the quality of the host larvae for sex allocation decision, i.e. whether to fertilize the eggs before deposition or not [28].

Not only may the availability of nutrients affecting host-plant quality influence a herbivore's natural enemies, but also other environmental stress factors acting on the plant are undoubtedly translated into enhanced or reduced efficacy of third-trophic-level organisms. Additionally, there is ample evidence that phytopathogenic infections may render the plant either better, or suboptimal or even unsuitable food for its insect herbivores. Exposure of the herbivore to diseased host plants may in turn have striking effects on its parasitoids and predators [39].

Sequestration of host-plant allelochemicals by herbivorous insects is widespread. They generally derive protection from these substances to natural enemies and pathogens [109]. Different herbivores may store different compounds out of the plant's chemical arsenal, as exemplified by four different insects on oleander, which accumulate different combinations of the range of compounds found in it [108] (Fig. 9.22).

Sequestration of plant allelochemicals by parasitoids *via* their host insects also occurs. Larvae of the tachinid fly *Zenilla adamsoni* contained cardenolides when they developed in *Danaus plexippus* larvae fed upon *Asclepias curassavica* [100]. Adult braconid and chalcidoid parasitoids emerging from *Bruchidius villosus* that developed in the seeds of *Laburnum anagyroides*, contained 1.3–3 μg per gram (fresh weight) of quinolizidine alkaloids originating from the plant [126]. It is not known, however, how the sequestered allelochemicals influence the parasitoids' fitness.

When the larvae of *Pieris brassicae* were reared on Brussels sprout, Swedish turnip, rape or nasturtium the overall performance of the larvae was highest on Brussels sprout and Swedish turnip, intermediate on rape and lowest on nasturtium. The differences were most likely due to differences in the allelochemical content of the leaves. However, the overall performance of the parasitoid, *Cotesia glomerata*, was best in nasturtium-fed larvae, although the between-plant differences in parasitoid performance were much less pronounced than in the caterpillars. Parasitization probably reduced consumption by the caterpillars, and a tendency was found for larger clutch sizes to be associated with larger combined weight of the caterpillar–parasitoid

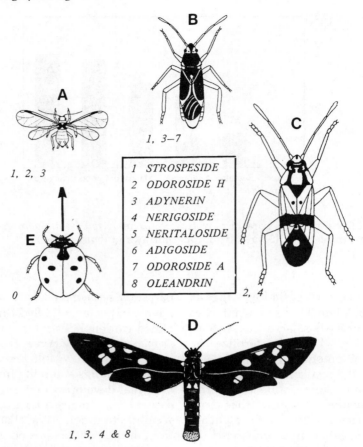

Figure 9.22 Four aposematic insect species feeding on *Nerium oleander* sequester and store cardenolides produced by their host plant. Different species have developed marked specializations. Some species store a few, others several of the cardenolides available. (**A**) *Aphis nerii*. (**B**) *Caenocoris nerii*. (**C**) *Spilostethus pandurus*. (**D**) *Syntomeida epilais*. (**E**) *Coccinella septempunctata*, a predator of *A. nerii*, which does not store cardenolides, although the related *C. undecimpunctata* does. (Source: reproduced from Rothschild *et al.*, 1972, with permission.)

complex. Thus, in this tritrophic system a greater number of parasitoids, i.e. an increased future extrinsic protection of the plant, could be achieved only by an increased cost (damage) to the plant. In this case an evolutionary increase of extrinsic protection is strongly constrained [55].

All tritrophic relationships discussed so far are based on constitutive chemicals produced by the herbivore's host plant. At present a rapidly growing body of literature attests to the fact that plants, in response to herbivore damage, produce volatiles that attract natural enemies of the herbivore. Hence the attacked plants, by emitting synomones, actively foster the presence of parasitoids and predators [21, 129] (more details can be found in sections 3.8 and 3.14). There is even some proof that plants may release different odours depending on the insect species feeding on them. These differences have been found to be perceived by parasitoids, which can learn to associate certain plant bouquets with the presence of a particular insect host species [130, 135]. From these findings a picture emerges of intricate multitrophic interactions and communication networks that extend far beyond 'simple' insect–plant relationships.

Several studies have clearly demonstrated that secondary plant substances may also influence the virulence of insect pathogens. For example, high concentrations of polyphenolic tannins in their host tree leaves protect the larvae of the gypsy moth (*Lymantria dispar*) from nuclear polyhedrosis virus infection. Larvae feeding on *Quercus* species and on *Acer rubrum*, which are rich in tannins, showed a 15–50 times lower mortality from the virus compared to larvae consuming leaves of *Populus* or *Pinus* species, which are low in tannins. On the other hand, gypsy moth larvae grew best on low-tannin leaves, i.e. high food quality was positively correlated with increased mortality due to the virus. Consequently, host-tree populations growing on poor soils or otherwise stressed are the loci of chronic moth outbreaks, although such trees are suboptimal food for the moth larvae [112].

Many plant allelochemicals show bactericidal activity. Nevertheless, the effect of a certain plant compound on the interaction between a herbivorous insect and, for instance, *Bacillus thuringiensis* (*B. t.*), depends on the sensitivity of the herbivore to that plant compound, on the sensitivity of the herbivore to *B. t.* and on the synergistic or antagonistic interaction between that plant compound and *B. t.* [99]. For example, larvae of *Manduca sexta*, which are oligophagous on solanaceous plants, tolerate high nicotine concentrations in their diet. The larvae are very sensitive to infestation with *Bacillus thuringiensis*. However, when fed a diet containing more than 0.4% nicotine, survival of the larvae is enhanced drastically (Fig. 9.23), indicating a strong effect of nicotine on the pathogen [63].

In summary, there is considerable evidence that plant factors significantly influence the effectiveness of natural populations of a herbivore's natural enemies. As our understanding of these tritrophic interactions increases, it will be possible to employ this knowledge in planning more long-term strategies to control insect pests in agricultural systems [57, 116].

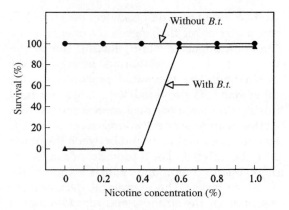

Figure 9.23 Survival to pupation of *Manduca sexta* larvae reared on six concentrations of nicotine incorporated into a synthetic diet with or without *Bacillus thuringiensis*. (Source: modified from Krischik *et al.*, 1988.)

(b) Extrafloral nectaries and honeydews

Extrafloral nectaries occur widely among angiosperms and have been found in at least 93 different families (they occur also in some ferns, but are absent in gymnosperms) [37, 59]. Their exudates contain mainly sugars and amino acids. These organs probably play an important role in providing food for natural enemies of herbivorous insects. The fact that as many as 60 species of Ichneumonidae fed at the extrafloral nectaries of faba bean, *Vicia faba* [9], forms in this context a telling observation. All amino acids required by insects for growth and reproduction rarely occur together in extrafloral nectaries, suggesting that most insects cannot totally depend on such nectars and need to search for other supplementary food, e.g. insects. This means that predatory insects such as coccinellid beetles, chrysopids and the braconid *Macrocentrus ancylivorus* that were found to feed on peach leaf nectaries [37] can partly cover their needs for energy, but a 'protein-hunger' remains, forcing the predators to prey on herbivorous insects on the peach tree. The parasitoid braconid, on the other hand, is provided with adequate food during its search for host insects on the tree. Cotton plants carry extrafloral nectaries on the leaf midribs and bracts and between the bracts and the calyces. They provide nectar 45–50 days before blooming, which exposes the floral nectaries. The extrafloral nectaries are essential for the survival and reproduction of several hymenopterous parasitoids attacking insect pests of cotton. Many arthropod predators are much more abundant on cultivars with extrafloral nectaries than on nectariless cotton. The number of pest insects is also reduced on nectariless cultivars, although often to a lesser degree than the number of predators [113].

Honeydews, which are excretions from homopteran insects deposited on the host-plant surface, contain mostly fructose, glucose, sucrose and also some trehalose and melezitose. Amino acids usually also occur, but rarely are all ten essential amino acids present in one honeydew. Honeydews are important food for a variety of entomophagous species, such as chrysopids, coccinellids, cantharids, tachinid flies, syrphids and many hymenopterous parasitoids. It is generally recognized that honeydew producers receive some degree of protection from natural enemies by attending ants. On the other hand, the honeydews often do not cover the protein needs of the ants, especially that of the brood, and ants therefore occasionally consume also the honeydew producers [37].

(c) **Plant morphology and architecture**

This may also affect third-trophic-level organisms. For example, plant surface structure can strongly influence entomophage efficacy:

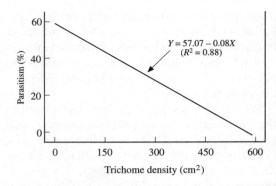

Figure 9.24 Percentage parasitization of *Helicoverpa zea* eggs by *Trichogramma pretiosum* on three cotton genotypes varying in trichome density. (Source: reproduced from Schuster and Calderon, 1986, with permission, based on data from Treacy *et al.*, 1984.)

hirsute cotton genotypes support fewer entomophagous insects than glabrous genotypes. Figure 9.24 shows that the percent parasitism of *Helicoverpa zea* eggs by *Trichogramma pretiosum* was the higher the lower trichome density was. However, predation by some bugs (*Geocoris* spp.) tends to be greater on more hirsute cotton genotypes [113].

As regards the effect of plant architecture, according to a survey of available data the number of parasitoid species per herbivorous insect species increases significantly with increasing complexity of plant architecture [42] (Fig. 9.25). This is supposedly in some way related to the numerical increase of herbivorous species with increasing plant size and complexity (see Fig. 9.11).

(d) **Plant galls**

Plant galls have been assumed to provide improved nutrition for the galling insect as well as protection from hypothermal stresses and from natural enemies [97]. Indeed, in tenthredinid sawflies the mean number of parasitoid species per host species was 16 on exposed feeders, as opposed to only four on shoot gallers [95]. By contrast, a survey of

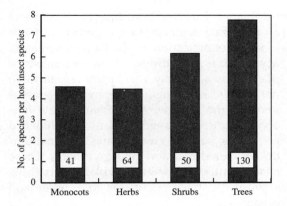

Figure 9.25 Number of parasitoid species per host insect species as a function of complexity of host-plant architecture for 285 British herbivores. Figures in bars refer to numbers of herbivores per plant category. (Source: redrawn from Hawkins and Lawton, 1987.)

species richness for parasitoids of British herbivorous insects found significantly higher rates of parasitism in gall-forming species than in external feeders [42]. A similar relation occurred between two chloropid flies: mortality rates caused by parasitoids were much higher in *Lipara lucens*, which forms gall chambers on reed (*Phragmites australis*), than in *L. pullitarsis*, the larvae of which develop between the young ensheathed reed leaves [17].

From the examples given it is clear that the difference in degree of parasitization between gall-forming and exposed feeding herbivorous species is very variable. Apparently it is not the (only) driving force of the galling habit. It is noteworthy that in the case of plant galls, as in other feeding modes, the question of the advantage of one life-form (galling) over the other (non-galling) is meaningless, as both forms exist and often coexist. Evidently they are equally adapted to survive in their own niche. Therefore, it would be a forlorn undertaking to find out which of such very different life-histories is more advantageous than the other.

(e) The structure of vegetation

Vegetation structure has also been found to influence the contribution of the third trophic level in the regulation of herbivorous insects. Presumably, different natural enemies may respond differently to the diversity of crop plant communities. Diversification (polyculture) may increase the efficacy of generalist entomophages by increasing the availability of alternative hosts (preys), but specialist enemies may be more effective in less diverse crop communities because concentration of host plants may increase the attraction of most specialist entomophages [115]. For example, larch sawfly larvae were less parasitized by tachinid flies when pine trees were near to larches. Pine odours most probably masked the odour of larch, which attracts the tachinids [82].

9.4.4 NATURAL ENEMIES AFFECTING HERBIVOROUS INSECT DAMAGE

It is common knowledge that entomophages reduce the damaging potential of pest insects by reducing their abundance. It can be shown, however, that the reduction of damage depends also on which ontogenetic stage of the pest is decimated by the entomophage. For example, let us assume that in a field there occur 1000 eggs per square metre of the cutworm, *Agrotis segetum*. An egg-parasitoid, *Trichogramma* sp., destroys 99% of the eggs, leaving ten caterpillars per square metre, which develop without further losses to adults. The 10 adults produce a total of 1000 eggs, by which initial abundance is restored (Fig. 9.26).

In another case the 1000 eggs develop into 1000 caterpillars, but the fully grown caterpillars are parasitized by the tachinid fly *Gonia ornata*, which causes 99% mortality, so that again ten adults emerge as before (Fig. 9.26). Thus, the population density measured in number of adults or eggs was the same in both

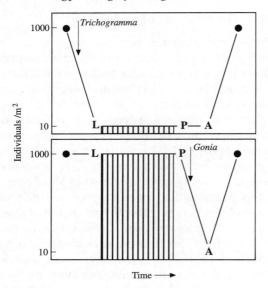

Figure 9.26 A hypothetical scheme of the changes in population density (individuals per m^2) of the four ontogenetic stages of the noctuid moth *Agrotis segetum* as a function of parasitization by *Trichogramma evanescens* (Hym.) or by *Gonia ornata* (Dipt.). ● = egg; L = larva; P = pupa; A = adult. (Source: redrawn from Jermy, 1957.)

cases; however, in the first case ten and in the second 1000 caterpillars per square metre damaged the plants [46].

9.4.5 THE ROLE OF MICROORGANISMS IN INSECT–PLANT RELATIONSHIPS

So far studies of insect–plant interactions have primarily focused on the relationships between macroorganisms, whereas until recently the participation of microorganisms in multitrophic relationships received little attention. Yet these unseen fellow creatures play a pivotal role as external or internal symbionts, as pathogens or as mutualists of plants, herbivorous arthropods or carnivorous arthropods. Fungi and bacteria, because of their short generation time and high mutability, represent important selective forces in ecosystems and may modulate and/or direct multitrophic level interactions. Insects can depend on microbial symbionts for digestion of plant food or detoxification of plant allelochemicals. Mycorrhizal and nitrogen-fixing mutualists of plants may adversely affect plant resistance to herbivores, and fungal endophytes in grasses may do the reverse. At the same time herbivory induces a range of plant physiological responses, such as growth rates and reallocation of stored and new photosynthates. These changes affect below-ground plant mutualists, saprophytic microorganisms and other soil organisms.

It is to be expected that unsolved riddles, for instance the question why some insect species adapt to a certain plant while others do not, will become intelligible when a holistic approach, including the scope and magnitude of microbial mediation of plant–insect interactions is taken. Some recent literature studies [2, 22, 41] have already shown a great diversity of microbial effects on insect–plant relationships, which must profoundly affect all terrestrial systems.

9.5 HERBIVOROUS INSECTS AFFECTING PLANT COMMUNITIES

As has been mentioned in section 9.1.3, herbivorous insects may occasionally affect the demography of their host plants. Therefore the assumption is obvious that insects may influence both plant succession and primary production in established (climax) plant communities.

9.5.1 EARLY SUCCESSIONAL PLANT COMMUNITIES

An early successional plant community was studied in England, applying both foliar and soil insecticide treatments to exclude herbivorous insects. The exclusion of folivorous insects resulted in a large increase of perennial grass growth and a reduction in plant species richness, because the grasses were able to outcompete many forbs. On the other hand, when

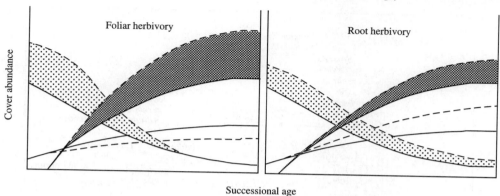

Figure 9.27 A model to illustrate the relative effect of above- and below-ground insect herbivores on different life-history groupings of plants during early succession. Solid lines = control; broken lines = application of insecticides; unshaded = perennial forbs; light shaded = annual forbs; dark shaded = perennial grasses. (Source: redrawn from Brown and Gange, 1992.)

root-feeding insects were excluded, this greatly increased perennial forb colonization and consequently, plant species richness [5] (Fig. 9.27).

It should be stated, however, that the responses to such insect exclusions may also be due to other causes than reduced herbivory. For example, soil-applied insecticides may exclude collembola that graze on mycorrhizal fungi, so the efficiency of water and nutrient uptake by the plants may increase [27]. Since insecticide treatments may also be repellent to small mammals that graze on various plants, it remains unclear which mechanism is responsible for producing the response of the plant community [15].

9.5.2 ESTABLISHED (CLIMAX) PLANT COMMUNITIES

Forests support an enormous number of herbivorous insect species, but only a few of them occasionally cause defoliation or even destruction of forest trees. Nevertheless, herbivorous insects are generally regarded as forest pests. From an ecological point of view, however, it is interesting to note that a closer look at primary production and the nutrient cycle in forests revealed that herbivorous insects function much like cybernetic regulators of primary production. In other words, they tend to ensure optimal output of plant production in the long term and their activity appears to vary inversely with the vigour and productivity of the forest as an ecosystem. This is achieved by the following.

1. Trees are able to compensate for partial defoliation of less than 40–50% with practically no loss in production. Light defoliation has occasionally been found even to stimulate vegetative growth.
2. Severe defoliation due to insect outbreaks occurs mostly in middle-aged to mature forests or in clusters of trees that are close to the end of their productivity cycle. Weakened, old and suppressed trees often die after defoliation, thus providing more nutrients, light, moisture, etc. for the remaining more vigorous trees.
3. Defoliation causes increased litter fall of insect excrement, exuviae, dead bodies, leaf parts, etc. enhancing the activity of soil microorganisms that increase nutrient flux among the remaining plants.
4. In severely defoliated forests the under-

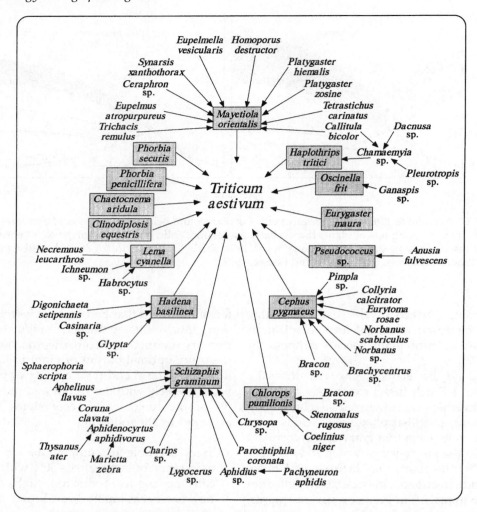

Figure 9.28 The most common herbivorous insect species (grey boxes) and their parasitoids, hyperparasitoids, and a predator (*Chamaemyia* sp.), associated with winter wheat in Hungary. (Source: redrawn from Jermy and Szelényi, 1958.)

story vegetation produces enough biomass to compensate for the production lost in the overstory trees.

Thus the ecosystem as a whole is only mildly affected by insect outbreaks. There are good reasons for supposing that in natural climax plant communities insect–plant interactions are in the long run mutualistic, despite temporary parasitic relations [79].

9.6 ENERGY AND MINERAL FLOW IN COMMUNITIES

The organisms forming a community are interconnected in an intricate web of trophic relations. Figure 9.28 illustrates how complex a food web can be even in a simple case like the herbivorous insect populations and their natural enemies associated with wheat. It should be noted that in the figure only the

Energy and mineral flow in communities 271

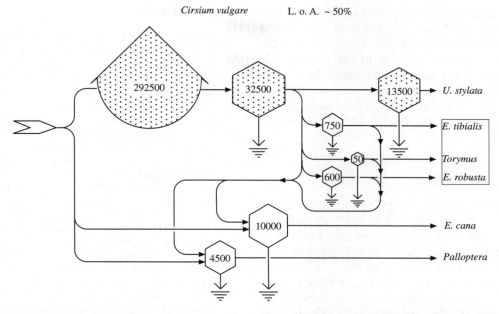

Figure 9.29 Amount of energy (in joules) temporarily stored in and direction of energy flow through populations of a tritrophic food web formed in the flower head of *Cirsium vulgare*. The data were calculated for 100 flower heads. The 'tank symbol' (upper left corner) represents the non-trophic, protective tissues of the gall induced by the tephritid fly *Urophora stylata*. The two stippled hexagons represent the larvae and the adults of *U. stylata* respectively. The open hexagons represent the further consumers. Lines and arrows show pathways of energy flow. The arrow at the left side symbolizes the input of energy (assimilates) into the system. The output consists of adult herbivores and parasitoids (right side of the chart) as well as heat produced by respiration ('heat sink' symbols). L. o. A. = energy loss by achenes. (Source: redrawn from Zwölfer, 1994.)

main herbivorous insect species are presented, together with their most common hymenopterous parasitoids, a predatory fly (*Chamaemyia* sp.), as well as some hyperparasitoids [48].

The flow of energy along the food web ensures the existence and activity of the communities. Quantified energy flow diagrams have been constructed for various natural communities. It is remarkable how little stored energy (i.e. biomass) the populations of the grazing food chain (i.e. the herbivores) represent and how little energy flows through this chain compared to the energy stored in the plant biomass and the energy loss by plant respiration. The energy turnover of the detritus food chain is greater by one order of magnitude than that of the grazing food chain [91].

Zwölfer [140] calculated energy storage and energy flow through different compartments of a tritrophic food web, consisting of the flower heads of *Cirsium vulgare* infested by the gall making tephritid fly, *Urophora stylata*, and its parasitoids (Fig. 9.29).

The ratio between the amount of energy stored in the non-trophic, protective gall tissues ('tank' symbol) and the mature fly larvae (left stippled hexagon), is about 9:1. Roughly 42% of the energy present in the fly larvae reaches the adult fly (right stippled hexagon), while part of it enters the hymenopterous parasitoids *Eurytoma tibialis*, *E. robusta*, and *Torymus* sp. The energy flow chart is complicated by the presence of a herbivorous moth species, *Eucosma cana* and a saprophagous fly species, *Palloptera* sp., the larvae of which eat parts of the flower head but may also consume any insect larva they encounter. The chart clearly indicates the numerous paths of energy flow through the populations of the second and third trophic levels.

9.7 CONCLUSIONS

The ecology of insect–plant interactions is a rapidly developing and vast domain of biology. Pressed for space, we could only deal here with some selected questions. We have shown that host plants may affect the demography of herbivorous insects by their quality as insect food, by their phenology and by the structure of the vegetation in which they occur. The rarity of most herbivorous insect species remains an unresolved enigma.

Herbivorous insects may affect host-plant demography in various ways. The result depends on the type of life history of the plant species, on the plant community to which the species belongs and on several other environmental factors. Herbivore exclusion experiments suggest, however, that vertebrate herbivores have a substantially greater impact on plant demography and plant community structure than insects.

The composition of herbivorous insect assemblages associated with a plant species largely depends on the geographical area the plant species occupies as well as on the plant's architecture and chemistry. Introduced plant species harbour fewer herbivorous insects than indigenous plants. The interactions among the members of insect–plant communities cover a broad spectrum from competition to facilitation and to lack of effect. The manifold trophic relations between the populations of the second and third trophic levels complicate the flow of energy along food webs in insect–plant communities.

9.8 REFERENCES

1. Barbosa, P. (1988) Natural enemies and herbivore–plant interactions: influence of plant allelochemicals and host specificity, in *Novel Aspects of Insect–Plant Interactions*, (eds P. Barbosa and D. K. Letourneau), John Wiley, New York, pp. 201–229.
2. Barbosa, P., Krischik, V. A. and Jones, C. G. (1991) *Microbial Mediation of Plant–Herbivore Interactions*, John Wiley, New York.
3. Braimah, H. and van Emden, H. F. (1994) The role of the plant in host acceptance by the parasitoid *Aphidius rhopalosiphi* (Hymenoptera, Braconidae). *Bull. Entomol. Res.*, **84**, 303–306.
4. Brown, B. J. and Allen, T. F. H. (1989) The importance of scale in evaluating herbivory impacts. *Oikos*, **54**, 189–194.
5. Brown, V. K. and Gange, A. C. (1992) Secondary plant succession: how is it modified by insect herbivory. *Vegetatio*, **101**, 3–13.
6. Brown, V. K. and Southwood, T. R. E. (1987) Secondary succession: patterns and strategies, in *Colonization, Succession and Stability*, (eds A. J. Gray, M. J. Crawley and P. J. Edwards), Blackwell Scientific Publications, Oxford, pp. 315–337.
7. Buckley, R. (1987a) Ant–plant–homopteran interactions. *Adv. Ecol. Res.*, **16**, 53–85.
8. Buckley, R. (1987b) Interactions involving plants, Homoptera, and ants. *Annu. Rev. Ecol. Syst.*, **18**, 111–135.
9. Bugg, R. L., Ellis, R. T. and Carlson, R. W. (1989) Ichneumonidae (Hymenoptera) using extrafloral nectar of faba bean (*Vicia faba* L., Fabaceae) in Massachusetts. *Biol. Agric. Hortic.*, **6**, 107–114.
10. Carroll, M. R., Wooster, M. T., Kearby, W. H. and Allen, D. C. (1979) Biological observations on three oak leaftiers, *Psilocoris querciella*, *P. reflexella*, and *P. cryptolechella* in Massachusetts and Missouri. *Ann. Entomol. Soc. Am.*, **72**, 441–447.
11. Chew, F. S. (1981) Coexistence and local extinction in two pierid butterflies. *Am. Nat.*, **118**, 655–672.
12. Corbet, S. A. (1973) Oviposition pheromone in larval mandibular glands of *Ephestia kuehniella*. *Nature*, **243**, 537–538.
13. Courtney, S. P. and Courtney, S. (1982) The 'edge-effect' in butterfly oviposition: causality in *Anthocaris cardamines* and related species. *Ecol. Entomol.*, **7**, 131–137.

14. Crawley, M. J. (1985) Reduction of oak fecundity by low-density herbivore populations. *Nature*, **314**, 163–164.
15. Crawley, M. J. (1989) The relative importance of vertebrate and invertebrate herbivores in plant population dynamics, in *Insect–Plant Interactions*, vol. 1, (ed. E. A. Bernays), CRC Press, Boca Raton, FL, pp. 45–71.
16. Damman, H. (1993) Patterns of interaction among herbivore species, in *Caterpillars. Ecological and Evolutionary Constraints on Foraging*, (eds N. E. Stamp and T. M. Casey), Chapman & Hall, New York, pp. 132–169.
17. De Bruyn, L. (1992) Plant resistance versus parasitoid attack in the evolution of the gall-forming fly *Lipara lucens*, in *Proceedings of the 8th International Symposium on Insect–Plant Relationships*, (eds S. B. J. Menken, J. H. Visser and P. Harrewijn), Kluwer, Dordrecht, pp. 339–340.
18. Den Boer, P. J. (1981) On the survival of populations in a heterogeneous and variable environment. *Oecologia*, **50**, 39–53.
19. Denno, R. F., McClure, M. S. and Ott, J. R. (1995) Interspecific interactions in phytophagous insects: Competition reexamined and resurrected. *Annu. Rev. Entomol.*, **40**, 297–331.
20. Dethier, V. G. (1941) Chemical factors determining the choice of food plants by *Papilio* larvae. *Am. Nat.*, **75**, 61–73.
21. Dicke, M. (1994) Local and systemic production of volatile herbivore-induced terpenoids: their role in plant–carnivore mutualism. *J. Plant Physiol.*, **143**, 456–472.
22. Dicke, M. (1996) The role of microorganisms in tritrophic interactions in systems of plants, herbivores and carnivores, in *Microbial Diversity in Time and Space*, (eds R. R. Colwell and U. Simidu), Plenum Press, New York, pp. 71–84.
23. Ehrlich, P. R., Murphy, D. D., Singer, M. C., Sherwood, C. B. White, R. R. and Brown, I. L. (1980) Extinction, reduction, stability and increase; the responses of checkerspot butterfly (*Euphydryas*) populations to the California drought. *Oecologia*, **46**, 101–105.
24. Faeth, S. H. (1990) Structural damage to oak leaves alters natural enemy attack on a leaf miner. *Entomol. Exp. Appl.*, **57**, 57–63.
25. Feeny, P. (1970) Seasonal changes in oak leaf tannins and nutrients as a cause of spring feeding by winter moth caterpillars. *Ecology*, **51**, 565–581.
26. Feeny, P. (1976) Plant apparency and chemical defense. *Rec. Adv. Phytochem.*, **10**, 1–40.
27. Fitter, A., cited in reference 15.
28. Fox, L. R., Kester, K. M. and Eisenbach, J. (1996) Direct and indirect responses of parasitoids to plants: sex ratio, plant quality and herbivore diet breadth. *Entomol. Exp. Appl.*, **80**, 289–292.
29. Fritz, R. S. (1983) Ant protection of a host plant's defoliator: consequence of an ant-membracid mutualism. *Ecology*, **64**, 789–797.
30. Fritz, R. S., Sacchi, C. F. and Price, P. W. (1986) Competition versus host plant phenotype in species composition: willow sawflies. *Ecology*, **67**, 1608–1618.
31. Gassmann, A. (1995) Europe as a source of biological control agents of exotic invasive weeds: status and implications. *Mitt. Schweiz. Entomol. Ges.*, **68**, 313–322.
32. Gaston, K. J. (1994) *Rarity*, Chapman & Hall, London.
33. Gause, G. F. (1934) *The Struggle for Existence*, Williams & Wilkins, Baltimore, MD.
34. Godfray, H. C. J. (1985) The absolute abundance of leaf miners on plants of different successional stages. *Oikos*, **45**, 17–25.
35. Goeden, R. D. and Louda, S. M. (1976) Biotic interference with insects imported for weed control. *Annu. Rev. Entomol.*, **21**, 325–342.
36. Greathead, D. J. (1971) A review of biological control in the Ethiopian region. *Commonw. Inst. Biol. Contr. Tech. Commun.*, **5**.
37. Hagen, K. S. (1986) Ecosystem analysis: plant cultivars (HPR), entomophagous species and food supplements, in *Interactions of Plant Resistance and Parasitoids and Predators of Insects*, (eds D. J. Boethel and R. D. Eikenbary), Ellis Horwood, Chichester, pp. 151–197.
38. Hairston, N. G., Smith, F. E. and Slobodkin, L. B. (1960) Community structure, population control and competition. *Am. Nat.*, **94**, 421–425.
39. Hammond, A. M. and Hardy, T. N. (1988) Quality of diseased plants as hosts for insects, in *Plant Stress–Insect Interactions*, (ed. E. A. Heinrichs), John Wiley, New York, pp. 381–432.
40. Harper, J. L. (1977) *Population Biology of Plants*, Academic Press, New York.
41. Hatcher, P. E. (1995) Three-way interactions between plant pathogenic fungi, herbivorous insects and their host plants. *Biol. Rev.*, **70**, 639–694.

42. Hawkins, B. A. and Lawton, J. H. (1987) Species richness for parasitoids of British phytophagous insects. *Nature*, **326**, 788–790.
43. Hawkins, C. P. and MacMahon, J. A. (1989) Guilds: the multiple meanings of a concept. *Annu. Rev. Entomol.*, **34**, 423–452.
44. Holt, R. D. (1977) Predation, apparent competition and the structure of prey communities. *Theor. Popul. Biol.*, **12**, 197–199.
45. Holt, R. D. and Lawton, J. H. (1993) Apparent competition and enemy-free space in insect host-parasitoid communities. *Am. Nat.*, **142**, 623–645.
46. Jermy, T. (1957) (Der Pflanzenschutz aus produktionsbiologischem Gesichtspunkt betrachtet.) *Ann. Inst. Prot. Plant. Hung.*, **7**, 23–33.
47. Jermy, T. (1985) Is there competition between phytophagous insects? *Z. Zool. Syst. Evol.-Forschung*, **23**, 275–285.
48. Jermy, T. and Szelényi, G. (1958) (Die Zoozönose des Winterweizens.) *Állatt. Közl., Budapest*, **46**, 229–241.
49. Jobin, A., Schaffner, U. and Nentwig, W. (1996) The structure of the phytophagous insect fauna on the introduced weed *Solidago altissima* in Switzerland. *Entomol. Exp. Appl.*, **79**, 33–42.
50. Joern, A. (1989) Insect herbivory in the transition to California annual grasslands: did grasshoppers deliver the coup de grass?, in *Grassland Structure and Function: California Annual Grassland*, (eds H. A. Mooney and L. F. Huennecke), Kluwer, Dordrecht, pp. 117–134.
51. Jones, C. G. and Coleman, J. S. (1991) Plant stress and insect herbivory: toward an integrated perspective, in *Response of Plants to Multiple Stresses*, (eds H. A. Mooney, W. E. Winner and E. J. Pell), Academic Press, New York, pp. 249–280.
52. Jones, C. G. and Lawton, J. H. (1991) Plant chemistry and insect species richness of British umbellifers. *J. Anim. Ecol.*, **60**, 767–777.
53. Kareiva, P. (1982) Exclusion experiments and the competitive release of insects feeding on collards. *Ecology*, **63**, 696–704.
54. Kareiva, P. (1983) Influence of vegetation structure on herbivore populations: resource concentration and herbivore movement, in *Variable Plants and Herbivores in Natural and Managed Systems*, (eds R. F. Denno and M. S. McClure), Academic Press, New York, pp. 259–298.
55. Karowe, D. N. and Schoonhoven, L. M. (1992) Interactions among three trophic levels: the influence of host plant on performance of *Pieris brassicae* and its parasitoid, *Cotesia glomerata*. *Entomol. Exp. Appl.*, **62**, 241–251.
56. Kaszab, Z. (1962) *Levélbogarak – Chrysomelidae, Fauna Hungariae. No. 63*, Akadémiai Kiadó, Budapest.
57. Kim, K. C. and McPheron, B. A. (1993) *Evolution of Insect Pests. Patterns of Variation*, John Wiley, New York.
58. Klausnitzer, B. (1983) Bemerkungen über die Ursachen und die Entstehung der Monophagie bei Insekten, in *Verhandlungen des X. Internationalen Symposiums über Entomofaunistik Mitteleuropas, Budapest*, pp. 5–12.
59. Koptur, S. (1992) Extrafloral nectary-mediated interactions between insects and plants, in *Insect–Plant Interactions*, vol. 4, (ed. E. A. Bernays), CRC Press, Boca Raton, FL, pp. 81–129.
60. Kosztarab, M. and Kozár, F. (1978) *Pajzstetvek – Coccoidea, Fauna Hungariae. No. 131*, Akadémiai Kiadó, Budapest.
61. Kozár, F. (1987) The probability of interspecific competitive situations in scale insects (Homoptera, Coccoidea). *Oecologia*, **73**, 99–104.
62. Krebs, J. R. and Davis, N. B. (1981) *An Introduction to Behavioural Ecology*, Blackwell, Oxford.
63. Krischik, V. A., Barbosa, P. and Reichelderfer, C. F. (1988) Three trophic level interactions: allelochemicals, *Manduca sexta* (L.), and *Bacillus thuringiensis* var. *kurstaki* Berliner. *Environ. Entomol.*, **17**, 476–482.
64. Kruess, A. and Tscharntke, T. (1994) Habitat fragmentation, species loss, and biological control. *Science*, **264**, 1581–1584.
65. Lawton, J. H. (1978) Host-plant influences on insect diversity: the effects of space and time. *Symp. Roy. Entomol. Soc. Lond.*, **9**, 105–125.
66. Lawton, J. H. (1982) Vacant niches and unsaturated communities, a comparison of bracken herbivores at sites on two continents. *J. Anim. Ecol.*, **51**, 573–595.
67. Lawton, J. H. (1984) Non-competitive populations, non-convergent communities, and vacant niches: the herbivores on bracken, in *Ecological Communities: Conceptual Issues and the Evidence*, (eds D. R. Strong, D. Simberloff and L. G. Abele), Princeton University Press, Princeton, NJ, pp. 67–101.
68. Lawton, J. H. and Schröder, D. (1977) Effect of plant type, size of geographical range and tax-

onomic isolation on number of insect species associated with British plants. *Nature*, **265**, 137–140.
69. Lawton, J. H. and Strong, D. R. (1981) Community patterns and competition in folivorous insects. *Am. Nat.*, **118**, 317–338.
70. Lawton, J. H., Lewinsohn, T. M. and Compton, S. G. (1993) Patterns of diversity for insect herbivores on bracken, in *Species Diversity in Ecological Communities*, (eds R. E. Ricklefs and D. Schluter), University of Chicago Press, Chicago, IL, pp. 178–184.
71. Louda, S. M. (1982) Limitation of the recruitment of the shrub *Haplopappus squarrosus* (Asteraceae) by flower- and seed-feeding insects. *J. Ecol.*, **70**, 43–53.
72. Louda, S. M. (1983) Seed predation and seedling mortality on the recruitment of a shrub, *Haplopappus venetus* (Asteraceae), along a climatic gradient. *Ecology*, **64**, 511–521.
73. Louda, S. M. (1995) The effect of seed predation on plant regeneration: evidence from Pacific Basin Mediterranean scrub communities, in *Ecology and Biogeography of Mediterranean Ecosystems in Chile, California and Australia*, (eds M. T. K. Arroyo, P. H. Zedler and M. D. Fox), Springer-Verlag, New York, pp. 311–344.
74. Louda, S. M. and Masters, R. A. (1993) Biological control of weeds in Great Plains rangelands. *Great Plains Res.*, **3**, 215–247.
75. Louda, S. M., Potvin, M. A. and Collinge, S. K. (1992) Predispersal seed predation in limitation of native thistle, in *Proceedings of the 8th International Symposium on Insect–Plant Relationships*, (eds S. B. J. Menken, J. H. Visser and P. Harrewijn), Kluwer, Dordrecht, pp. 30–32.
76. Lovett, G. M. and Ruesink, A. E. (1995) Carbon and nitrogen mineralization from decomposing gypsy moth frass. *Oecologia*, **104**, 133–138.
77. MacArthur, R. N. (1972) *Geographical Ecology. Patterns in the Distribution of Species*, Harper & Row, New York.
78. Magurran, W. E. (1988) *Ecological Diversity and its Measurement*, Princeton University Press, Princeton, NJ.
79. Mattson, W. J. and Addy, M. N. (1975) Phytophagous insects as regulators of forest production. *Science*, **190**, 515–522.
80. Mayr, E. (1963) *Animal Species and Evolution*, Harvard University Press, Cambridge, MA.
81. Mihályi, F. (1960) *Fúrólegyek – Trypetidae, Fauna Hungariae. No. 56*, Akadémiai Kiadó, Budapest.
82. Monteith, L. G. (1960) Influence of plants other than the food plants of their host on host-finding by tachinid parasites. *Can. Entomol.*, **92**, 641–652.
83. Moran, V. C. (1980) Interactions between phytophagous insects and their *Opuntia* hosts. *Ecol. Entomol.*, **5**, 153–164.
84. Müller-Schärer, H. (1991) The impact of root herbivory as a function of plant density and competition: survival, growth and fecundity of *Centaurea maculosa* in field plots. *J. Appl. Ecol.*, **28**, 759–776.
85. Murray, D. A. H. and Rynne, K. P. (1994) Effect of host plant on parasitism of *Helicoverpa armigera* (Lep., Noctuidae) by *Microplitis demolitor* (Hym., Braconidae). *Entomophaga*, **39**, 251–255.
86. Neuvonen, S., Hanhimäki, S., Suomela, J. and Haukioja, E. (1988) Early season damage to birch foliage affects the performance of a late season herbivore. *J. Appl. Entomol.*, **105**, 182–189.
87. Ohgushi, T. (1992) Resource limitation on insect herbivore populations, in *Effects of Resource Distribution on Animal–Plant Interactions*, (eds M. D. Hunter, T. Ohgushi and P. Price), Academic Press, New York, pp. 199–241.
88. Olckers, T. and Hulley, P. E. (1989) Insect herbivore diversity on the exotic weed *Solanum mauritianum* Scop. and three other species in the eastern Cape Province. *J. Entomol. Soc. S. Africa*, **52**, 81–93.
89. Owen, D. F. (1980) How plants may benefit from animals that eat them. *Oikos*, **35**, 230–235.
90 Packham, J., cited in reference 15.
91. Pianka, E. R. (1988) *Evolutionary Ecology*, 4th edn, Harper & Row, New York.
92. Potter, D. A., Loughrin, J. H., Rowe, W. J. and Hamilton-Kemp, T. R. (1996) Why do Japanese beetles defoliate trees from the top down? *Entomol. Exp. Appl.*, **80**, 209–212.
93. Price, P. W. (1983) Hypotheses on organization and evolution in herbivorous insect communities, in *Variable Plants and Herbivores in Natural and Managed Systems*, (eds R. F. Denno and M. S. McClure), Academic Press, New York, pp. 559–596.
94. Price, P. W. (1984) *Insect Ecology*, 2nd edn, John Wiley, New York.
95. Price, P. W. and Pschorn-Walcher, H. (1988)

95. Are galling insects better protected against parasitoids than exposed feeders?: A test using tenthredinoid sawflies. *Ecol. Entomol.*, **13**, 195–205.
96. Price, P. W., Bouton, C. E., Gross, P., McPheron, B. A., Thompson, J. N. and Weiss, A. E. (1980) Interactions among three trophic levels: influence of plants on interactions between insect herbivores and natural enemies. *Annu. Rev. Ecol. Syst.*, **11**, 41–65.
97. Price, P. W., Fernandes, G. W. and Waring, G. L. (1987) Adaptive nature of insect galls. *Environ. Entomol.*, **16**, 15–24.
98. Reichart, G. (1977) Data to the knowledge of moth communities occurring on fruit trees and shrubs in Hungary. *Acta Phytopathol. Hung.*, **12**, 359–373.
99. Reichelderfer, C. F. (1991) Interactions among allelochemicals, some Lepidoptera, and *Bacillus thuringiensis* Berliner, in *Microbial Mediation of Plant–Herbivore Interactions*, (eds P. Barbosa, V. A. Krischik and O. G. Jones), John Wiley, New York, pp. 507–524.
100. Reichstein, T., von Euw, J., Parsons, J. A. and Rothschild, M. (1968) Heart poisons in the monarch butterfly. *Science*, **161**, 861–866.
101. Renwick, J. A. A. and Chew, F. S. (1994) Oviposition behavior in Lepidoptera. *Annu. Rev. Entomol.*, **39**, 377–400.
102. Risch, S. (1981) Insect herbivore abundance in tropical monocultures and polycultures: an experimental test of two hypotheses. *Ecology*, **62**, 1325–1340.
103. Roitberg, B. D. and Mangel, M. (1988) On the evolutionary ecology of marking pheromones. *Evol. Ecol.*, **2**, 289–315.
104. Root, R. B. (1967) The niche exploitation pattern of the blue-gray gnatcatcher. *Ecol. Monogr.*, **37**, 317–350.
105. Root, R. B. (1973) Organization of plant arthropod association in simple and diverse habitats: the fauna of collards (*Brassica oleracea*). *Ecol. Monogr.*, **43**, 95–124.
106. Root, R. B. and Cappuccino, N. (1992) Patterns in population change and the organization of the insect community associated with goldenrod. *Ecol. Monogr.*, **62**, 393–420.
107. Rothschild, M. and Schoonhoven, L. M. (1977) Assessment of egg load by *Pieris brassicae* (Lepidoptera, Pieridae). *Nature*, **266**, 352–355.
108. Rothschild, M., von Euw, J. and Reichstein, T. (1972) Some problems connected with warningly coloured insects and toxic defense mechanisms. *Mitt. Basler Afrika Bibliogr.*, **4–6**, 135–158.
109. Rowell-Rahier, M. and Pasteels, J. M. (1992) Third trophic level influences of plant allelochemicals, in *Herbivores. Their Interactions with Secondary Plant Metabolites*, 2nd edn, vol. 2, (eds G. A. Rosenthal and M. R. Berenbaum), Academic Press, San Diego, CA, pp. 243–277.
110. Rudinsky, J. A., Ryker, L. C., Michael, R. R., Libbey, L. M. and Morgan, M. E. (1976) Sound production in Scolytidae: female sonic stimulus of male pheromone release in two *Dendroctonus* beetles. *J. Insect Physiol.*, **22**, 1675–1681.
111. Russ, K. (1971) Der Einfluss von Territorialverhaltensweisen von Mikrolepidopteren auf die Populationsdichte. *Acta Phytopathol. Acad. Sci. Hung.*, **6**, 147–152.
112. Schultz, J. C. and Keating, S. T. (1991) Host-plant-mediated interactions between the gypsy moth and a baculovirus, in *Microbial Mediation of Plant–Herbivore Interactions*, (eds P. Barbosa, V. A. Krischik and O. G. Jones), John Wiley, New York, pp. 489–506.
113. Schuster, M. F. and Calderon, M. (1986) Interactions of host plant resistant genotypes and beneficial insects in cotton ecosystems, in *Interactions of Plant Resistance and Parasitoids and Predators of Insects*, (eds D. J. Boethel and R. D. Eikenbary), Ellis Horwood, Chichester, pp. 84–97.
114. Schwerdtfeger, F. (1970) *Die Waldinsekten*, Paul Parey, Hamburg.
115. Sheehan, W. (1986) Response by specialist and generalist natural enemies to agroecosystem diversification, a selective review. *Environ. Entomol.*, **15**, 456–461.
116. Shepard, M. and Dahlman, D. L. (1988) Plant-induced stresses as factors in natural enemy efficacy, in *Plant Stress–Insect Interactions*, (ed. E. A. Heinrichs), John Wiley, New York, pp. 363–379.
117. Sheppard, A. W. (1992) Predicting biological weed control. *Trends Ecol. Evol.*, **7**, 290–291.
118. Simberloff, D. S. (1976) Experimental zoogeography of islands: effects of island size. *Ecology*, **57**, 629–648.
119. Simms, E. L. (1992) The evolution of plant resistance and correlated characters, in *Proceedings of the 8th International Symposium on Insect–Plant Relationships*, (eds S. B. J. Menken, J. H. Visser and P. Harrewijn), Kluwer, Dordrecht, pp. 15–25.

120. Southwood, T. R. E. (1961) The evolution of insect host-tree relationship – a new approach, in *Verhandlungen XI Internationaler Kongress für Entomologie, Wien*, Organisationskomitee des XI Internationale Kongresses für Entomologie, Vienna, vol. I, pp. 651–654.
121. Southwood, T. R. E. (1973) The insect/plant relationship – an evolutionary perspective. *Symp. Roy. Entomol. Soc. Lond.*, **6**, 3–30.
122. Southwood, T. R. E., Moran, V. C. and Kennedy, C. E. J. (1982) The richness, abundance and biomass of arthropod communities on trees. *J. Anim. Ecol.*, **51**, 635–649.
123. Stewart, A. J. A. (1996) Interspecific competition reinstated as an important force structuring insect herbivore communities. *Trends Ecol. Evol.*, **11**, 233–234.
124. Strong, D. R., Lawton, J. H. and Southwood, T. R. E. (1984) *Insects on Plants. Community Patterns and Mechanisms*, Blackwell, Oxford.
125. Szelényi, G. (1939) Die Schädlinge des Ölmohns in Ungarn, in *Verhandlungen des 7 Internationalen Kongress für Entomologie, Berlin*, vol. 4, pp. 2625–2639.
126. Szentesi, Á. and Wink, M. (1991) Fate of quinolizidine alkaloids through three trophic levels: *Laburnum anagyroides* (Leguminosae) and associated organisms. *J. Chem. Ecol.*, **17**, 1557–1573.
127. Treacy, M. F., Benedict, J. H. and Segers, J. C. (1984) Effects of smooth, hirsute and pilose cottons on the functional responses of *Trichogramma pretiosum* and *Chrysopa rufilabris*, in *Proceedings of the Beltwide Cotton Producers Research Conference, Atlanta, GA*, pp. 372–373.
128. Tscharntke, T. (1989) Attack by a stem-boring moth increases susceptibility of *Phragmites australis* to gall-making by a midge: mechanisms and effects on midge population dynamics. *Oikos*, **55**, 93–100.
129. Turlings, T. C. J. (1994) The active role of plants in the foraging success of entomophagous insects. *Norw. J. Agric. Sci.*, **16**, 211–219.
130. Turlings, T. C. J., Wäckers, F. L., Vet, L. E. M., Lewis, W. J. and Tumlinson, J. H. (1993) Learning of host-finding cues by hymenopterous parasitoids, in *Insect Learning. Ecological and Evolutionary Perspectives*, (eds D. R. Papaj and A. C. Lewis), Chapman & Hall, New York, pp. 51–78.
131. Van der Meijden, E. (1979) Herbivore exploitation of a fugitive plant species: local survival and extinction of the cinnabar moth and ragwort in a heterogenous environment. *Oecologia*, **42**, 307–323.
132. Van der Meijden, E., Wijn, M. and Verkaar, H. J. (1988) Defence and regrowth: alternative plant strategies in the struggle against herbivores. *Oikos*, **51**, 355–363.
133. Van der Meijden, E., van Wijk, C. A. M. and Kooi, R. E. (1991) Population dynamics of the cinnabar moth (*Tyria jacobaeae*): oscillations due to food limitation and local extinction risks. *Neth. J. Zool.*, **41**, 158–173.
134. Van Valen, L. (1973) A new evolutionary law. *Evol. Theory*, **1**, 1–30.
135. Vet, L. E. M. and Dicke, M. (1992) Ecology of infochemical use by natural enemies in a tritrophic context. *Annu. Rev. Entomol.*, **37**, 141–172.
136. Whitham, T. G. (1983) Host manipulation of parasites: within-plant variation as a defense against rapidly evolving pests, in *Variable Plants and Herbivores in Natural and Managed Systems*, (eds R. F. Denno and M. S. McClure), Academic Press, New York, pp. 15–41.
137. Whitman, D. W. (1988) Allelochemical interactions among plants, herbivores, and their predators, in *Novel Aspects of Insect–Plant Interactions*, (eds P. Barbosa and D. K. Letourneau), John Wiley, New York, pp. 11–64.
138. Williams, C. B. (1964) *Patterns in the Balance of Nature*, Academic Press, New York.
139. Wratten, S. (1992) Population regulation in insect herbivores – top-down or bottom-up? *NZ J. Ecol.*, **16**, 145–147.
140. Zwölfer, H. (1994) Structure and biomass transfer in food webs: stability, fluctuations, and network control, in *Flux Control in Biological Systems*, (ed. E. D. Schulze), Academic Press, New York, pp. 365–419.

EVOLUTION: WHO DRIVES WHOM? 10

10.1	Paleontological aspects	279
10.1.1	Chronology	279
10.1.2	How did insect diversity evolve?	283
10.2	Speciation in herbivorous insects	284
10.2.1	Infraspecific categories	284
10.2.2	Types of speciation	284
10.2.3	Reproductive isolation	285
10.2.4	Rates of speciation	287
10.2.5	Reciprocal speciation	289
10.3	Why does host-plant specificity prevail?	289
10.3.1	Adaptive significance of host-plant specialization	289
10.3.2	Phyletic relations	292
10.4	Evolutionary driving forces	295
10.4.1	Selection pressures exerted by plants on herbivorous insects	295
10.4.2	Selection pressures exerted by herbivorous insects on plants	296
10.4.3	Reciprocity of selection in insect–plant interactions	301
10.5	Theories	302
10.5.1	Coevolution	302
10.5.2	Diffuse coevolution, community coevolution	303
10.5.3	The geographic mosaic theory of coevolution	304
10.5.4	Sequential evolution	305
10.6	Conclusions	307
10.7	References	308

The great evolutionary geneticist, Dobzhansky [31], formulated splendidly the importance of evolutionary considerations in interpreting biological phenomena: 'Nothing in biology makes sense except in the light of evolution'. Since herbivorous insect species and the species of green plants that they consume make up roughly one-half of all presently known macroscopic species (see Fig. 2.1), it is not surprising that these two large and extremely diverse groups of organisms stimulated studies on the evolution of their interactions. Unfortunately, the processes of evolution cannot be studied experimentally, therefore most assumptions about how the extant relationships between organisms came into being are based on circumstantial evidence only. The interpretation of such evidence, however, varies greatly depending on the authors' scientific viewpoints, but it varies also in time as a function of the emergence of new knowledge, especially in genetics and paleobiology. As a consequence, the literature dealing with the evolution of insect–plant relationships abounds with partly contradictory or disputable assumptions. In the following we discuss the main aspects of the presumed evolutionary processes in question and we set forth a critical review of the various hypotheses.

10.1 PALEONTOLOGICAL ASPECTS

10.1.1 CHRONOLOGY

The face of planet Earth has emphatically changed during past aeons (Fig. 10.1).

The paleontological history of the Plant Kingdom shows several periods of increased diversification in the Devonian, the Mid-Carboniferous, the Upper Cretaceous and the Tertiary. The diversification involved an increase both in number of plant taxa and in complexity of plant structure [27, 83] (Fig. 10.2).

Fossil orthopterans are known from the Upper Carboniferous, but the type of food they consumed at that time is unknown. Herbivorous insects presumably evolved various

Figure 10.1 Reconstructed view of Late Carboniferous vegetation. Pteridophytic trees reached their fullest expression in the warm swamps of this period. Tree ferns of great height, rising to over 50 m, grew with an understory of bushy and herbaceous ferns and horsetails. Seed plants, such as the conifer *Cordaites* (upper right corner) grew to about 30 m with a trunk diameter of 1 m. It had long, strap-like leaves up to 1 m in length and 15 cm across. (Source: reproduced from Mägdefrau, 1959, with permission.)

modes of exploiting living plants in the following sequence: sucking is most probably the oldest mode of feeding on plants, followed by chewing, while mining and gall-making were established much later [118, 129].

Nowadays an overwhelming majority of the host plants of herbivorous insects belong to the phylum of flowering plants (Angiospermae). The time relationship between the diversification of the main herbivorous insect groups and the dramatic diversification of the angiosperms towards the end of the Early Cretaceous has been the subject of much discussion, because it may help to determine whether plants enhanced the evolution of insects or whether the opposite interaction was also important. The extensive spread of several extant herbivorous insect orders, especially Lepidoptera, Coleoptera (Chrysomelidae, Curculionidae), Diptera (Agromyzidae, Cecidomyidae) and Hymenoptera (Cynipidae) occurred after the appearance and radiation of angiosperms [129]. This would suggest that the appearance and evolution of the flowering plants accelerated the evolution of these groups. Other paleontological data show, however, that the familial radiation (increase of the number of families within the

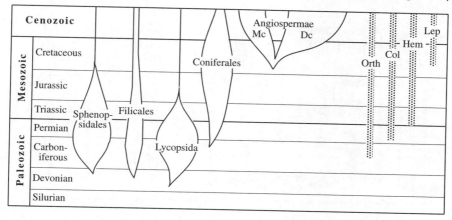

Figure 10.2 The great proliferation of angiosperms started towards the end of the Mesozoic, and within a relatively short span of time during the Mid-Cretaceous they became established worldwide as the dominant terrestrial plant group. The divergence of monocotyledons (Mc) and dicotyledons (Dc) may date right back to the origins of angiosperms. Bars represent the first occurrences of selected insect taxa germane to insect–plant relationships. Orth = Orthoptera; Col = Coleoptera; Hem = Hemiptera; Lep = Lepidoptera.

orders) in several modern insects began 245 million years ago, i.e. about 100 million years earlier than the appearance and rise to dominance of the angiosperms, which occurred 144–66 million years ago [86]. The spindle diagrams of Figure 10.3 indicate that the familial radiation of Coleoptera, Lepidoptera, Diptera and Hymenoptera occurred in the Upper Mesozoic, i.e. coinciding with the appearance and ascendancy of the flowering plants which, thus, may have accelerated the radiation of these groups of insects.

No such effect, however, is apparent in Orthoptera, Homoptera and Heteroptera. Analysing the number of all insect families through time, Labandeira and Sepkoski [86] even came to the conclusion that the appearance and ascendancy of the angiosperms coincided with a slow down rather than an acceleration of insect familial diversification (Fig. 10.4).

We have to emphasize, however, that familial diversification is not necessarily identical with species diversification. Thus, from the presently available information there is no general coincidence in time between the evolution of higher plants and insect taxa. This lack of correlation is, of course, quite relevant for any discussion on the evolution of insect–plant interactions.

The chemical evolution of the Plant Kingdom must have been of decisive importance to the evolution of insect–plant relations, a notion well expressed in the title of Schultz's paper [116]: 'Many factors influence the evolution of herbivorous diets, but plant chemistry is central'. The assumption that plants produce allelochemicals under selection pressure from herbivorous insects or even that the *raison d'être* of these chemicals is to defend plants against insect attacks [46] has been generally accepted. However, contrary to this widely held opinion Jones and Firn [76], reviewing the relevant literature, concluded that

the general metabolic traits that confer diversity [of allelochemicals] may have been selected for very early in the evolution of plants, perhaps before the evolution of terrestrial plants (...) the sole requirement for plant defence (...) may be the continual

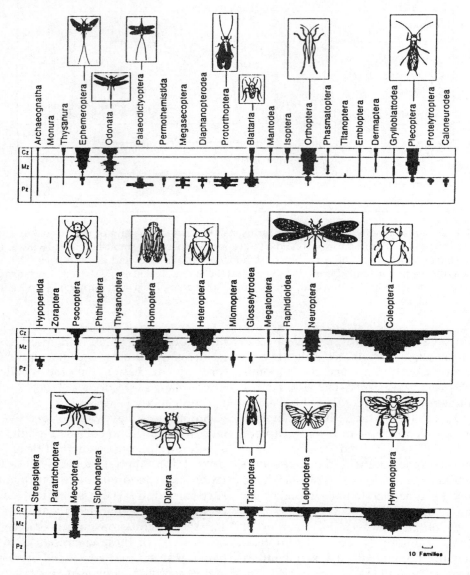

Figure 10.3 Spindle diagrams displaying diversities in fossil families within insect orders in stratigraphic stages of the Phanerozoic. A scale bar is shown in the lower right corner. Pz = Paleozoic; Mz = Mesozoic; Cz = Cenozoic. Angiosperms appeared approximately two-thirds of the way up the band for Mesozoic, i.e. above the 'M' in 'Mz'. (Source: reproduced with permission from C. C. Labandeira and J. J. Sepkoski: Insect diversity in the fossil record. *Science* **261**: 310–315. Copyright 1993 American Association for the Advancement of Science.)

production of a diversity of secondary metabolites. The evolution of plant defence may therefore have proceeded independent of consumer adaptation, once these fundamental traits were in place.

The question of whether or not the production of secondary plant substances has been modified as a result of the selection pressure exerted by the insects on the plants will be discussed in section 10.4.2.

Paleontological aspects

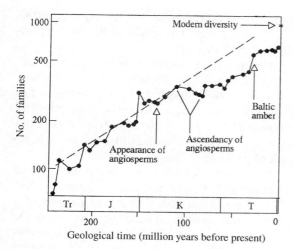

Figure 10.4 Insect familial diversity from the Triassic to the recent, plotted on semilogarithmic coordinates. The dashed line is interpretative, illustrating possible exponential diversification beginning in the Triassic and possibly continuing into Early Cretaceous. Tr = Triassic; J = Jurassic; K = Cretaceous; T = Tertiary. (Source: reproduced with permission from C. C. Labandeira and J. J. Sepkoski: Insect diversity in the fossil record. *Science* **261**: 310–315. Copyright 1993 American Association for the Advancement of Science.)

10.1.2 HOW DID INSECT DIVERSITY EVOLVE?

Several hypotheses have been propounded in order to explain the evolution of the striking diversity of Insecta and in particular of herbivorous insects.

According to the **ecological saturation hypothesis** [93] there has been always a roughly constant number of niches that could be occupied by insect species or higher insect taxa. A new insect taxon could get established only if another was excluded by competition and became extinct. However, competition among extant herbivorous insect species is rare or lacking (section 9.4.1), so it is unlikely that interspecific competition for suitable host plants played any substantial role in the evolutionary past in determining the number of herbivorous insect species associated with a certain plant taxon.

The proponents of the **expanding resource hypothesis** [139] argue that the resources provided by the plants to the insects increased both in quantity and in the ratio of niches per resource. In this view, an increase in plant structural and architectural diversity opens new possibilities for an increase in insect diversity. This opinion is well supported by the fact that more complex plants harbour more herbivorous species (section 9.3.3). Nevertheless, this hypothesis also implies that herbivorous insect diversity, in general, is primarily determined by the Plant Kingdom.

These views, however, are challenged by paleontological findings. As evinced by Figure 10.3, the insect orders encompassing herbivorous groups greatly differ in speed and extent of familial diversification. This suggests that diversification depends primarily on some intrinsic trends rather than merely on environmental (ecological) conditions. Similar variation in velocity of evolution can be found among the extant insect families within particular insect orders. That is to say, each order includes families that have not shown any diversification for a long time in evolutionary terms ('primitive' families), while others are in a state of rapid diversification. Even at the genus and species level it was found [22] that animal species introduced to oceanic islands show great differences in propensity to split off (**speciose** *versus* **non-speciose** lineages). Taken together, this strongly indicates, firstly, that the genomes of various herbivorous insect taxa greatly differ in the extent of variation, which is a precondition for fast diversification, and, secondly, that the propensity for genetic variation is largely independent of both phylogenetic status and environmental factors, including the taxonomic and structural diversity of the flora. Thus, autonomous processes (and chance?), independently of selection pressure exerted by the environment, seem to play an important role in the diversification of Insecta. We may term this postulate as the **intrinsic trend of diversification hypothesis** (see also section 10.5.4). Interestingly, some paleobotanists [136], working from the genetic background of speciation in plants, likewise suppose that the evolution of the Plant Kingdom occurred in an autonomous way, i.e. not as a consequence of competition and/or selection.

10.2 SPECIATION IN HERBIVOROUS INSECTS

Darwin, in his book *On the Origin of Species*, called the formation of species the 'mystery of mysteries'. The first decisive steps towards unveiling this mystery were made as long ago as the late 1930s and early 1940s by T. Dobzhansky, H. J. Muller and E. Mayr. Although their fundamental statements are widely accepted, considerable dispute remains about the details of speciation, i.e. about the processes that result in the emergence of new (daughter) species from an existing (parental) species.

In the following discussion the term 'biological species' is defined as a group of interbreeding or potentially interbreeding populations that are reproductively isolated from other groups of populations [95].

10.2.1 INFRASPECIFIC CATEGORIES

Different populations of the same herbivorous insect species may differ in host-plant preference and/or performance on particular hosts, or in other traits of their relation to plants. Such populations may be regarded as taxonomic units below the species level (therefore **infraspecific**). The general term **biotype** is often used to describe such differing populations. This term covers a number of categories reflecting different mechanisms that underlie biotype formation. Diehl and Bush [30] recognize the following categories:

Polymorphic populations are defined as 'the occurrence together in the same habitat of two or more discontinuous forms, or "phases", of species in such proportions that the rarest of them cannot be maintained merely by recurrent mutation' [43]. For example, several Hessian fly (*Mayetiola*

destructor) biotypes have been selected artificially that survive differently on various wheat varieties [54].

Geographical race (some semispecies, subspecies) is a geographically separated biotype differing, for instance, in host preference. For example, in North America the butterfly *Papilio glaucus* occurs in two geographic subspecies that differ, among other things, in host-plant preference [117].

Host race (biological race) is a population of a species that is partially reproductively isolated from other conspecific populations as a direct consequence of adaptation to a specific host [30]. For example, the time differences in the emergence of apple maggot (*Rhagoletis pomonella*) adults developing in three different host species may largely prevent mating between the three host races [19, 39] (Fig. 10.5).

Sibling species are morphologically nearly indistinguishable and can be recognized only by biochemical, cytological or behavioural traits. They may or may not hybridize in the laboratory. For example, the membracid species group *Enchenopa binotata* was long considered as one polyphagous species. However, a detailed experimental analysis clearly proved the existence of nine sympatric sibling species specialized on different plant species [142].

The above infraspecific categories may all represent pre-steps of speciation in herbivorous insects. The term 'ecotype' is regarded as a synonym of 'race' and 'subspecies'. The term 'strain' is quite loosely used in the literature; it mainly designates insect populations manipulated by man, e.g. a laboratory strain [7].

10.2.2 TYPES OF SPECIATION

The most widely used classification of speciation events is based upon the presence or absence of spatial (geographical) isolation between populations, i.e. on whether exchange of genes (gene flow) between populations is lacking or is possible to some degree.

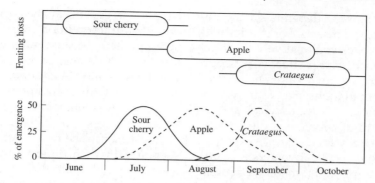

Figure 10.5 Adult emergence pattern of three host races of *Rhagoletis pomonella* in Wisconsin and fruiting time of the hosts. The last switch from *Crataegus* to apple occurred after apple was introduced into the USA 150 years ago. (Source: redrawn from Bush, 1975b.)

Two main types, two extremes of a continuum, are distinguished [18].

Allopatric speciation occurs when two or more subpopulations of a species are geographically isolated and develop into daughter species. In such cases gene flow is nil.

Sympatric speciation takes place when a new species develops within the parental species area of distribution, i.e. within the 'cruising range' of individuals of the parental species. In pure allopatry gene flow between populations is hindered right from the beginning by an extrinsic barrier, while in pure sympatry gene flow may continue to some extent for a number of generations after the populations became separated. Whether or not sympatric speciation exists is a much discussed issue. However, several cases of host races and sibling species among herbivorous insects strongly support the advocates of this concept [37, 38].

The role of so-called 'founder populations' [94] has been supposed to be important in speciation. These are very small populations isolated by any barrier (section 10.2.3) from the parental population, for instance a few individuals of a species introduced to an island. However, a large-scale experiment with *Drosophila pseudoobscura* did not support the claim that founder events are likely to result in speciation [53]. Levin [89] even suggested that **metapopulations**, rather than small, isolated populations, are the arenas of local speciation, since they are more prone to persist and to undergo genomic reorganization than small populations. (A metapopulation is a collection of local populations loosely connected by migration and isolated from the remainder of the species [90]. For example, populations of a specialized herbivorous insect species living on scattered patches of its host plant within the cruising ranges of the adults constitute a metapopulation.)

10.2.3 REPRODUCTIVE ISOLATION

The emergence of a daughter species from a parental species is possible only if some barrier prevents or restricts the gene flow between two populations. Below we discuss only some types of barrier that were also found in herbivorous insects. They cause either premating reproductive isolation (spatial and behavioural barriers, allochrony in life history) or postmating isolation (hybrid incompatibility).

(a) Spatial barriers

The most obvious barrier of gene flow is geographic isolation of populations by a mountain range, sea, river, desert, etc. The efficiency of such barriers largely depends on the dispersal capacity of the insects. For species with a sedentary life style even relatively short distances may be sufficient for effective isolation. This is indicated by genetic differences found between such populations. For example, the chrysomelid species *Oreina cacaliae* and *O. globosa*, which are oligophagous on *Petasites*, *Senecio* and *Adenostyles* species (Asteraceae) show considerable genetic divergence, as estimated by electrophoretic methods, among populations separated by only 40–250 km in Switzerland and Germany [112]. Surprisingly, even distances of 10 to a few hundred metres may suffice for spatial isolation. Non-dispersive monophagous insects, such as scale insects, leafminers and gall midges living on trees were found to represent genetically highly different subpopulations (termed **demes**) on individual trees. Insects transferred from one tree to another tree of the same species performed poorly compared to conspecific insects transferred within the same tree. This is explainable by the fact that such insects may produce hundreds of generations on the same tree; thus the phytochemical and microhabitat differences among individual trees, acting as selective forces, may result in genetically different demes [100].

(b) Behavioural barriers

Differences in feeding preference and/or in oviposition preference of herbivorous insects can result in effective isolation and, consequently, most probably offer an opportunity for sympatric speciation. In such cases gene flow can be totally absent between coexisting insect populations narrowly specialized to different host-plant species. For example, apple maggot (*Rhagoletis pomonella*) adults are attracted to specific chemicals occurring in apples, while a closely related (sibling) species, *R. mendax*, is attracted by different chemicals to the blueberry fruit [47]. Although the two species can easily be hybridized in the laboratory, genetic analysis has shown that under natural conditions there is no gene flow between them, because mating only occurs on their respective host plants [20, 37, 38]. Likewise, the aphid *Aulacorthum solani* s. str., which is polyphagous but avoids *Pulmonaria officinalis*, does not hybridize in nature with *A. solani langei*, which lives monophagously on *P. officinalis*. The two subspecies are totally separated by their different host-plant preferences, even though their host plants grow intermingled and often in physical contact with each other. Another subspecies, however, *A. s. aegopodii*, feeding monophagously on *Aegopodium podagraria*, may occasionally hybridize with *A. solani* s. str. [103] (Fig. 10.6).

Further types of behavioural barrier between populations are represented, for example, by differences in the composition of sex pheromones between two European corn borer (*Ostrinia nubilalis*) populations in North America [82] or by differences in acoustic mate recognition signals in planthoppers (Homoptera) [23].

(c) Allochrony in life history

Allochrony (the opposite of synchrony) in various parts of the insects' life histories, especially in mating periods, may cause reproductive isolation. For instance, the North American membracid species-complex, *Enchenopa binotata*, contains nine sympatric species specialized on coexisting host trees. Their allochronic life histories on different hosts are the primary factor in initiating and maintaining reproductive isolation. This has led to asynchronic mating periods and ultimately to speciation [142]. As shown in Figure 10.5, the emergence pattern of three apple maggot races overlaps only partially, which probably strongly reduces mating among the races [19]. The question arises, however, whether allochrony was the cause or the result of speciation.

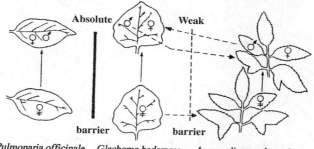

Figure 10.6 Premating isolating barriers of different strength caused by differentially strong preference for host plants. (Source: redrawn from Müller, 1985.)

(d) Hybrid incompatibility

Different populations of the same species may differ genetically to such an extent that the zygotes of hybrids are non-viable. For example, the European populations of the cherry fruit fly (*Rhagoletis cerasi*) are divided into at least two geographic races (Fig. 10.7), which show unidirectional incompatibility: crosses between males of the southern complex (circles) and females of the north-east populations (triangles) produce low hatch rates of eggs, while the opposite crosses yield normal levels of fertility.

This unilateral incompatibility may be due to either genetic or cytoplasmic factors. A third cause may be related to the absence or presence of microbial symbionts [16]. According to Thompson [131], symbionts in concert with environmental factors often play an important role in speciation. The interaction between a given symbiont and its host may be antagonistic in one environment but commensal or even mutualistic in other environments. In this way, differential selection exerted by different environments on the symbiont–host interactions may magnify the differences among different insect populations and thereby lead to speciation. An example of symbiont-caused incompatibility is found in the alfalfa weevil (*Hypera postica*). In experiments with three American and one European population, the crosses between populations harbouring a rickettsia and populations free of *Rickettsia* proved incompatibility [64].

10.2.4 RATES OF SPECIATION

The number of new (daughter) species emerging per unit of time is determined primarily by the occurrence of genetic variation through time in the parental species and secondarily by the forces of natural selection and drift. Evidence for a primary role of genetic variation is provided by studies on young oceanic islands like the Big Island of the Hawaiian Archipelago, which began to emerge from the Pacific Ocean less than 400 000 years ago. As mentioned above, here the rate of speciation in various species following colonization by an ancestral immigrant varied greatly. This can be explained by differences in the propensity of genomes for genetic disorganization and reorganization (speciose *versus* non-speciose lineages) [22]. Why some gene-complexes are so stable while others evolve rapidly (**genetic revolution**), remains as yet an unsolved problem [94].

The rate of speciation depends also on generation span, because meiosis causes the

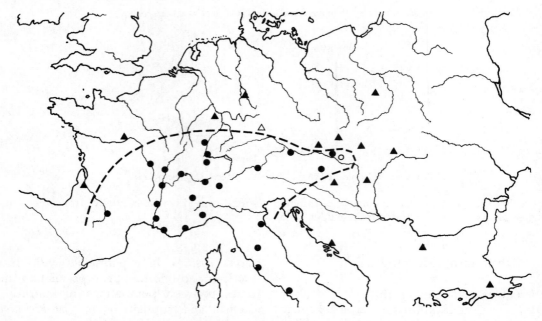

Figure 10.7 The distribution of the southern (●) and northern races (▲) of *Rhagoletis cerasi* which show incompatibility. Open symbols (○,△) indicate transitional populations. (Source: redrawn from Boller *et al.*, 1976.)

majority of mutations. Herbivorous insects show great variability in this respect. For example, the development of the North American periodic cicada (*Magicicada septendecim*) lasts 17 years [98] while the diamondback moth (*Plutella maculipennis*) has up to 28 generations per year in tropical environments [62]. Clearly, supposing the same mutation rate, species with short generation times – such as most insects – are potentially more prone to fast speciation than those with long generation times, e.g. perennial plants. Thus, a herbivorous insect species' adaptation to evolutionary changes of its host tree is potentially much faster than the evolutionary response of the tree to attack by the insect, supposing equally strong selection pressures exerted by the partners on each other.

There is also an ecological aspect to speciation. According to the niche concept, each new species must find an unoccupied niche in order to get established. However, the aforementioned studies on Hawaiian insect species indicate that the problem of finding an unoccupied niche is a secondary prerequisite of the evolution of a new species, the primary prerequisite being genetic preadaptation (see also the question of 'vacant niches' in section 9.4.1). On the other hand, if an insect genome contains genes that enable adaptation to a new plant species, then sudden changes in the habitat may provide an opportunity for changes in host-plant preferences. This is exemplified by the butterfly *Euphydryas editha* which adapted genetically to changes in human land use practices in North America. Within a period of only 8 years the proportion of females preferring the introduced weed *Plantago lanceolata* (Plantaginaceae) over their ancestral host plant *Collinsia parviflora* (Scrophulariaceae) rose from about 7% to more than 50% [123] (Fig. 10.8). Such a change

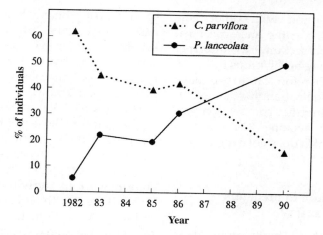

Figure 10.8 Changes over time in results of oviposition preference trials done with *Euphydryas editha* females in the field. Broken line = preference for the original host, *Collinsia parviflora*; solid line = preference for the introduced weed, *Plantago lanceolata*. (Source: redrawn from Singer *et al.*, 1993.)

in host preferences may be the prelude to speciation although, based on a mitochondrial RNA survey of 24 *E. editha* populations, Radtkey and Singer [108] denied this, since mating in this species is independent of host plant.

10.2.5 RECIPROCAL SPECIATION

Reciprocal speciation as a consequence of interactions between organisms is called **diversifying coevolution** [133]. Ehrlich and Raven [33] implicitly supposed that the interaction between herbivorous insects and plants might result in speciation of both partners (see the discussion in section 10.5.1). Reviewing the vast literature on coevolution and on relevant case studies, Thompson [133] concluded, however, that only some pollinators and plants as well as hosts and maternally inherited intracellular symbionts represent kinds of interaction that best fit the conditions of diversifying coevolution. At present there is no convincing evidence for reciprocal speciation as a result of interactions between plants and herbivorous insects. We treat this question in section 10.4.3.

10.3 WHY DOES HOST-PLANT SPECIFICITY PREVAIL?

In order to discuss the evolutionary processes that probably resulted in host-plant specialization, two aspects need special attention: i.e. the adaptive significance of specialization and the various phyletic relations between insects and their host plants.

10.3.1 ADAPTIVE SIGNIFICANCE OF HOST-PLANT SPECIALIZATION

Since specialists far outnumber generalists in most groups of herbivorous insects (Chapter 2), it has been conjectured that food specialization has selective advantage over generalist feeding habits [15] (see Jermy [73] and Menken [96] for other references). Implicitly or explicitly, principally the following advantages were regarded as explanations and/or evolutionary driving forces of host-plant specialization:

(a) Escape from interspecific competition

Ehrlich and Raven [33] supposed that switches to other host plants may have resulted in escape from competition with other herbivorous species utilizing the ancestral host. However, as shown in section 9.4.1, interspecific competition among herbivorous insects is rare or lacking and therefore can hardly have propelled host specialization in the evolutionary past. Assuming that competition was important is merely evoking 'the ghost of competition past', which does not explain anything [26].

(b) Reduced exposure to predators (enemy escape) [56, 124]

It was supposed that if a herbivorous insect population that lives on both plants A and B is exposed to more intensive predation on plant A than on B, then this selection pressure would result in the insect's increased preference (specialization) for plant B *versus* plant A. Experimental evidence [8, 9, 15] indicates that polyphagous herbivorous insects are more prone to attack by generalist predators (such as vespids and ants) than specialist herbivores. This, however, might be the result rather than the cause of host-plant specialization [72, 73]. It should also be considered that generalist predators and parasitoids typically have a very flexible searching behaviour. Their ability to quickly change search patterns is much better developed than the herbivores' potential to switch to new host-plant species (Fox [44] and references therein). Furthermore, selection for preferring plant B *versus* plant A is possible only if there is genetic variation in the insect for separately recognizing the two plant species. It is, however, largely unknown whether such genetic variation exists. For instance, the phyletic relation between *Yponomeuta* species and their host plants (see section 10.3.2), which represents a type of relationship common in various groups of insects, strongly argues against the enemy escape hypothesis. In other words, most *Yponomeuta* species are specialized on taxonomically very different plant species, which often occur in the same habitat. Therefore, these insect species most probably had in the evolutionary past, and still have, common natural enemies. It is hard to conceive how predators or parasitoids might have pushed the *Yponomeuta* species to specialize just on those plant species that they prefer now.

Thompson [133] has also questioned the role of predation in host-plant specialization. He has proposed instead that, with small specialists such as herbivorous insects 'it is the process of adaptation to an **intimate association with an individual host** throughout development, with all its attendant problems, including predation, that appears to be the reason why extreme specialization is common across all parasitic taxa' (our emphasis). However, this assumption hardly applies to herbivorous insects in general, since monophagy and broad polyphagy may occur even among congeneric species (section 10.3.2).

(c) Increased efficiency of detoxifying plant allelochemicals

This has been proposed as a further advantage [140], because specialists that are adapted physiologically to detoxify the allelochemicals specific to their hosts would utilize their food more efficiently. Comparative studies, however, have failed to demonstrate a correlation between narrow specialization and increased efficiency of food utilization [45], or else the results were contradictory [50, 51, 58].

It has also been assumed that deterrent allelochemicals betoken toxicity of plants [6], i.e. that plants, by developing toxicants, have affected insect specialization, because insects responded by avoiding such plants on the basis of the deterrent properties of the toxicants. Host specialization and even speciation in insects [105] is then the result of relationships with plants that were unsuccessful in becoming resistant. Although it has been shown that the botanical occurrence of deterrent chemicals basically determines host-plant specificity in insects [13, 28, 68], no general correlation between deterrence and toxicity was found [10, 12, 14].

(d) Genetically based trade-offs in offspring performance

According to Jaenike [66], the hypothesis of genetically based trade-offs in offspring performance is the most appealing explanation of host specialization. In this view, an evolutionary increase in offspring performance on one plant species entails a reduction in adaptation to other potential hosts. Joshi and Thompson [77] suggested that, once an insect population has adapted to two hosts, a negative additive genetic correlation between larval performance on the two hosts may constitute a selective force that favours specialization on to one host or the other. This may lead, through some kind of reproductive isolation, to host-race formation. Alternatively, an amelioration of the trade-off in larval performance on the two hosts might occur that would result in a more generalist phenotype of the population. What role such trade-offs actually play in host specialization is still an open question.

(e) Increased efficiency in host finding

This is also considered as a possible advantage [49]. However, as discussed in Chapter 5, specialist herbivorous insects are often unable to locate their host from a distance of more than several decimetres, especially in herbaceous vegetation. Most insects distinguish between host and non-host after contact only. Thus, specialization is rather a disadvantage in host finding.

In particular cases, host-plant specificity may facilitate mate finding [87], may enhance chemical defence of the insect by the sequestration of specific allelochemicals from the host [32], may enable development of specific means for efficient attachment to the plant [79] or may enhance crypsis [115]. These and similar aspects, however, are restricted to a few groups of insects only.

(f) Conclusion

In conclusion, none of the above advantages satisfactorily explains the strong dominance of specialists. However, it is conceivable that the various mechanisms operate in combination, leading to a system of selection pressures that is extremely difficult to disentangle. From our present knowledge it is an arduous task to find out which of the mechanisms mentioned play the dominant role in a particular case. Alternatively, it could even be that the initial assumption, i.e. that specialist feeding habits are advantageous, is erroneous [4]. The great numerical dominance of specialists over generalists may simply reflect (1) the relative rates of speciation and extinction between specialists and generalists [138], (2) the evolutionary irreversibility of specialization [40, 96] or (3) some constraints on the evolution of the insects' nervous system [13, 71, 73, 74] resulting predominantly in the emergence of new specialists from specialists. It also must be considered that, according to crossing experiments with *Papilio* butterflies, changes at relatively few genetic loci could have large effects on the host-preference hierarchy of these butterflies [133]. Studies on function and genetics of insect chemoreceptors even suggest that a single mutation could change monophagy to polyphagy and *vice versa* [29]. This problem is part of the debate on whether or not specialization is an evolutionary dead end. While several authors [40, 101] regard specialization as a dead end, others (Thompson [133] and references therein) negate this view and suppose that even the most specialized insects might enter a new evolutionary process. As a matter of fact, there is no clear phylogenetic evidence to answer this question adequately.

No doubt, polyphagous habits also confer an advantage on certain insect groups. The relatively high incidence of polyphagy among (ancient) Orthoptera, for instance, may be related to their relatively slow growth. Polyphagous grasshopper populations demon-

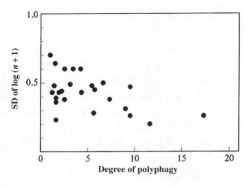

Figure 10.9 Relationship between degree of polyphagy and variability in population size of coexisting grasshopper species in Nebraska. Population variability was calculated from 25 years of sampling and the degree of polyphagy from the number of plant species reported to be fed on by each grasshopper species. (Source: redrawn from Joern and Games, 1990.)

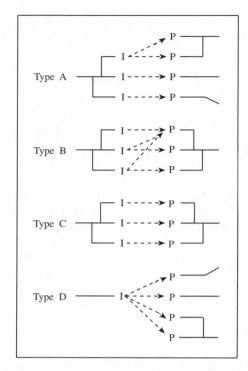

Figure 10.10 Types of cladogram between closely related insect species or a single insect species and the host plants. I = herbivorous insect species; P = host plant; broken lines with arrows = trophic relations. (Source: redrawn from Jermy, 1984.)

strate smaller fluctuations in population density than specialist species [75] (Fig. 10.9). The polyphages are less susceptible to short periods of suboptimal food quality of a particular host than specialists, who can not switch to other host species.

10.3.2 PHYLETIC RELATIONS

When the phyletic relations of the host plants of closely related insect species are considered, four types can be distinguished, as shown in the theoretical cladograms of Figure 10.10.

Type A: Closely related insect species live oligophagously or monophagously on distantly related plant species (incongruent cladograms). For example, the European *Yponomeuta* species are narrowly specialized on host species of four plant families (Fig. 10.11), which belong to three different plant orders [96, 97].

Thus, speciation in this genus was accompanied by at least two shifts between plant families. This type of relation is common among Hemiptera, Coleoptera, Lepidoptera, Diptera, and Hymenoptera. Host races and sibling species that often can be distinguished primarily by their host-plant preferences, also belong to this type.

Type B: Closely related insect species live oligophagously (or partly monophagously) on closely related plant species (partly congruent cladograms). This type is also common in the same insect orders where Type A occurs. Typical examples are *Pieris brassicae*, *P. napi* and *P. rapae*, which prefer roughly the same species of Cruciferae [130]. An extreme case is found in the chrysomelid species *Crioceris (Crioceris) asparagi*, *C. 5-punctata*, *C. 12-punctata*, and *C. 14-punctata*, all living monophagously on *Asparagus officinalis* (Fig. 10.12).

Type C: Closely related insect species live monophagously on closely related plant

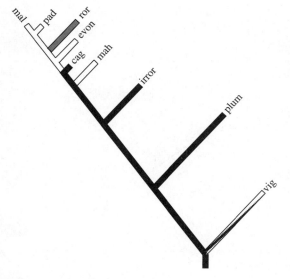

Figure 10.11 Phylogenetic tree of the nine West European *Yponomeuta* species based on allozyme data and the botanical status of their host plants. *Yponomeuta* species: cag = *cagnagellus*; evon = *evonymellus*; irror = *irrorellus*; mah = *mahalebellus*; mal = *malinellus*; pad = *padellus*; plum = *plumbellus*; ror = *rorellus*; vig = *vigintipunctatus*. Host plant affiliations: black = Celastraceae; white = Rosaceae; shaded = Salicaceae; black and white = *Y. vigintipunctatus* feeds on Crassulaceae, but its sister species, *Y. yamagawanus*, feeds on *Euonymus* (Celastraceae). (Source: redrawn from Menken et al., 1992.)

species. This type is rare. Figure 10.13 illustrates such a case [34].

The insects and their host plants form strongly congruent cladograms, although there are a few exceptions. Similar examples are: species of Adelgidae (Homoptera) on species of Pinaceae [128]; *Larinus* spp. (Coleoptera) on *Centaurea* spp. [143]; the herbivorous pollinator species of Agaonidae (Hymenoptera) on *Ficus* spp. [141]; or *Tegeticula* spp. (Lepidoptera) on *Yucca* spp. [106].

Type D: A polyphagous insect species feeds on plant species that belong to different plant families. This type is less common than A and B, because polyphagous species are outnumbered by oligophagous ones in most higher insect taxa. Species whose subsequent ontogenetic stages obligatorily feed in sequence on distantly related plant species, or insect species that alternate between distantly related host plants during subsequent generations (e.g. holocyclic aphids), also belong to this type.

The above types may appear together among congeneric species. For example, the two subgenera of *Crioceris* show quite different host specialization: all species of *Crioceris* s. str. are monophagous while the species of *Lilioceris* are oligophagous on several genera of Liliaceae (Fig. 10.12). Two scale insects belonging to the genus *Chionaspis* (Coccidae) live oligophagously on species of Pinaceae and Fagaceae respectively, whereas a third one is extremely polyphagous, infesting members of 17 different plant families (Fig. 10.14).

Types B and C suggest phylogenetic conservatism; in other words, speciation in herbivorous insects is often accompanied by shifts

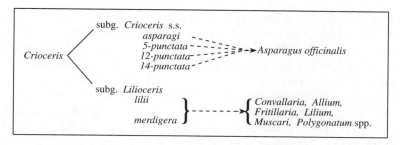

Figure 10.12 Host-plant relations of the genus *Crioceris* (Chrysomelidae). (Source: data from Kaszab, 1962.)

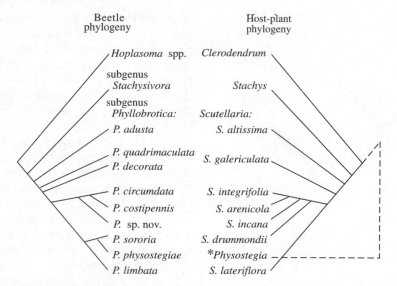

Figure 10.13 Cladograms of the chrysomelid genus *Phyllobrotica* (with the sister genus, *Hoplasoma*) and of its host plants. Each insect taxon is placed opposite its host. Beetle species with unknown hosts and plant species that are not hosts to *Phyllobrotica* lineage are excluded. (Source: redrawn from Farrell and Mitter, 1990.)

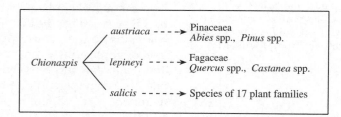

Figure 10.14 Host-plant relations of three *Chionaspis* spp. (Coccoidea). (Source: data from Kosztarab and Kozár, 1988.)

between closely related plant taxa. Ehrlich and Raven [33], in their extensive survey on butterflies associated with various plant families, emphasized that chemical traits characterizing higher plant taxa are responsible for this association. This agrees with the finding that lycaenid butterflies are associated mostly with species of Fabales, Santanales and Rosidae, i.e. with families characterized by the accumulation of phenolics, while they do not feed on members of plant families known to produce deterrent (toxic) compounds, such as Aristolochiaceae, Brassicaceae, Asclepiadaceae, etc. [42]. Farrell and Mitter [35] analysed 25 herbivorous insect groups for which cladograms of insect species and host-plant records were available. Rating the frequency of shifts between plant families they found that in 13 insect groups 16% or less of the speciation events were accompanied by such shifts. These data suggest that Types B and C, representing phylogenetic conservatism, are more common than Type A.

A genetic analysis of two chrysomelid species (*Ophraella*) concerning their feeding responses and performance on host plants of their congeners indicated that constraints on genetic variation guide host shifts and are responsible for phylogenetic conservatism [52]. In some cases speciation was found to be associated with the botanical occurrence of a specific plant compound. For example, Menken [96] suggested that the switch of *Yponomeuta* species from Celastraceae to Rosaceae could happen because dulcitol, which serves as a phagostimulant to the larvae, occurs in Celastraceae as well as in the new host genus *Prunus* (Rosaceae). Some authors regard such cases also as phylogenetic conservatism [35, 96].

Unfortunately, very little is known about the details of the plants' chemical profile as perceived by an insect. It is clear, however, that 'chemical similarity' does not mean the same to a phytochemist and to an insect. For example, to a phytochemist potato, which contains steroidal alkaloids, is quite dissimilar from *Hyoscyamus niger*, which contains tropane alkaloids instead. Yet the Colorado potato beetle readily accepts both plant species but does not accept *Solanum demissum*, because it contains demissine, an alkaloid with a molecular structure that differs only slightly from the potato alkaloids [114].

10.4 EVOLUTIONARY DRIVING FORCES

The following three questions arise in connection with any evolutionary explanation of extant insect–plant relationships: Do plants exert selection pressure on the insects? Do insects exert selection pressure on the plants? Are the selective interactions between insects and plants reciprocal?

10.4.1 SELECTION PRESSURES EXERTED BY PLANTS ON HERBIVOROUS INSECTS

It was shown above (section 10.3.1) that none of the single factors (or 'advantages') that were thought to have selected the insects for host specialization could have a decisive significance in the evolution of extant insect–plant relations. In the following we discuss whether or not plants have 'pushed' the insects towards food specialization.

Plant abundance has been hypothesized to promote the adaptation of insects to new plants [49]. This assumption is valid in the sense that, if within an insect population a new genome emerges that is tuned to a plant other than its parental host plant, then the probability of it becoming established is the higher the more abundant, i.e. the easier to meet, the new host plant is. However, the mere presence of a plant species does not evoke switches of insects to that plant if no insect population harbours adequate heritable trait(s) by which it is able to recognize and use that plant as a host (see also the 'pool exhaustion hypothesis' in section 9.3.6). If, however, such trait(s) do occur in an insect species, then the adaptation to a new host can develop very rapidly, as was found in the case of *Euphydryas editha* (Fig. 10.8).

Plant resistance, i.e. plant traits (mainly chemical) that signal the insect to recognize a plant as a non-host and thus prevent feeding and/or oviposition, must have a strong selective effect on insects. In other words, a newly emerged insect genotype (mutant) with novel feeding or oviposition preferences can become established only if it finds a plant whose (chemical) traits cause a sensory message that fits a hypothetical standard in the insect brain, like a key fits a lock [113]. Otherwise the new genotype will perish. As a consequence, the immense chemical and structural diversity of the plant world represents a decisive selection pressure on herbivorous insects (see section 10.5.4 for further discussion).

10.4.2 SELECTION PRESSURES EXERTED BY HERBIVOROUS INSECTS ON PLANTS

It is a general characteristic of any plant species that it is fed on and/or oviposited on by only a limited number of herbivorous insect species while it is avoided by most other insects (Chapter 2). This limitation of use by insects is called **plant resistance** or **defence**. Many authors use these terms as synonyms. As is also pointed out in Chapter 3, there is, however, an important conceptual difference between them. According to Rausher's definitions: '**resistance** is the degree to which a plant (or plant part) avoids damage by herbivores', and 'a **resistance trait** is any plant character that influences the amount of damage a plant suffers', while '**defence** or **defensive trait** is any resistance trait that has evolved or is maintained in a plant population because of selection pressure exerted by herbivores' [109]. This means that the term 'defence' should be used only in those cases where the evolutionary origin of resistance can be traced back to infestation by herbivores [104]. Before answering the question of how far plant resistance to insects can be regarded as defence, we have to deal with three hypotheses on plant defence or resistance.

The **plant apparency hypothesis** proposes that the type of defence depends on how easily a plant can be found by herbivores [41]. 'Apparent plants' are large, perennial, highly abundant plant species with a long vegetation time. They are easy to find. Such plants possess relatively large amounts of **quantitative defence compounds**, such as tannins or silica. On the other hand, herbaceous plants that are ephemeral in space and time are relatively hard for specialist herbivores to find. They are protected by small amounts of **qualitative defence compounds**, such as glucosinolates, alkaloids, etc. [40, 41]. The distinction between the two types of defence is by no means clear-cut. Thus trees, which are apparent plants, may also contain chemicals that can be regarded as providing qualitative defence [5]. Furthermore, the quantities of secondary plant substances vary widely and do not exhibit classes that can be neatly divided. The plant apparency theory has been largely replaced by the resource availability hypothesis.

The **resource availability hypothesis** [25] proposes that the availability of resources to the plant mainly determines both the amount and type of plant defence. For example, plants adapted to favourable habitats, such as light gaps in tropical forests, can grow fast enough to survive high levels of damage resulting from low levels of defence. On the other hand, limited resources favour plants with slow growth, which in turn favours high levels of defence.

Both the above hypotheses have been integrated into one model. The following is a summary of interesting considerations by Feeny [41] (who uses the terms 'defence' and 'resistance' as synonyms). The variation in the life histories of organisms can be predicted in terms of two environmental variables: adversity and disturbance [24, 60, 125]. Since the resource availability hypothesis refers primarily to the adversity axis, while the apparency hypothesis refers to the disturbance axis, it has been suggested to integrate the two hypotheses into the **habitat template model** [24, 126, 127] (Fig. 10.15).

According to this model, ephemeral herbaceous plants, such as **ruderals** [60], which inhabit disturbed habitats, are predicted to have low levels of resistance (qualitative, mobile), because of high resource availability (which means high productivity) and low apparency. At the other extreme, plants occupying stable but adverse habitats (**stress tolerators**, Grime [60]) show high levels of resistance (quantitative, immobile) because of low availability of resources and high apparency. Between these two extremes are situated stable, undisturbed habitats that enable high productivity. These habitats are occupied by plants called **competitors** [60], which are characterized by medium levels of

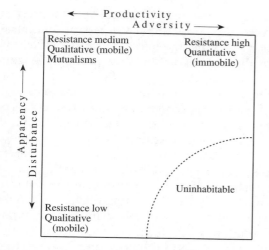

Figure 10.15 The habitat template model. See text for details. (Source: redrawn from Feeny, 1991.)

resistance (qualitative, mobile). Interactions among species are predicted to be best developed in such habitats [59, 126]. Under such conditions plants may have evolved mutualistic associations with natural enemies of herbivores. Finally, habitats that are highly adverse and disturbed are essentially uninhabitable [60]. A weak point of the habitat template model is the implicit assumption that quantitative resistance is high while qualitative resistance is low. Unfortunately, no experimental or observational data support such a distinction. A ruderal plant such as poison hemlock (*Conium maculatum*), possessing qualitative resistance (alkaloids), can be at least as resistant to generalist herbivores as oak trees containing tannins (quantitative resistance). This case also shows that, although models as discussed have a heuristic value, it may be impossible to devise a model incorporating all the ecological factors that influence plant resistance [41].

Now we come to the question: How far can plant resistance to herbivorous insects be regarded as **defence** or, in other words, did plant resistance evolve as a result of selection pressure exerted by herbivorous insects?

As mentioned before (section 9.2), damage inflicted by herbivorous insects does not necessarily reduce plant fitness; rather, it may occasionally even increase fitness. Thus it would be premature to assume that attacks by herbivorous insects always signify a selection pressure that results in an increase of resistance traits in the plants. The following considerations further support this conclusion.

1. Selection of a plant population by a herbivorous insect population would be successful only if the insect selectively reduced the fitness of those individual host plants that were most susceptible to that insect. By this means the frequency of genes that confer some degree of resistance to the insect would increase in the plant population. Selective preference of herbivorous insects for different host-plant genotypes does indeed occur. For example, egg-laying females of four sawfly species preferred different willow genotypes (clones; see Table 9.1) that are characterized by morphological traits [48, 107]. However, in natural communities the great majority of herbivorous insect species occur at low or very low population densities (section 9.1.4), even when their host plants are abundant. In most cases, therefore, they damage only a few individuals of the host-plant population. In addition, several specialist herbivorous insects find their host plants by chance only, especially in herbaceous vegetation (Chapter 5), and do not test several individual host plants before they begin to feed or oviposit. This means that they are unable to choose, for example, the most suitable (acceptable) individuals (genotypes) from the host-plant population. Furthermore, no overall genetic covariance has been found between oviposition site selection by the female and performance of the larvae [66, 132]. Therefore, the female may lay the eggs on individual host plants independently of whether the plants are optimal or suboptimal for larval development, i.e. whether the individual plant is susceptible or to some degree resistant or tolerant to the larva. As a consequence, most herbivorous insects do not really pick out

plant individuals belonging to a certain genome but rather randomly attack a few individuals, largely irrespective of their heritable traits. Herbivorous insect populations thus resemble a very ignorant plant breeder who tries to produce a new cultivar by randomly destroying a small part of his plant stand.

Moreover, insect abundance is extremely variable in space and time while other selective forces, such as plant–plant interactions (competition, allelopathy), pathogens, climatic and meteorological factors, as well as soil quality, exert a much more persistent selection pressure on the plants. All these selective forces may interact in various ways with, and sometimes counteract, the weak and occasional selection exerted by some insects. Thus, damage caused by an insect population does not necessarily result in selecting a plant population with resistance to that insect.

2. The maintenance of insect resistance traits in plants, such as the production of allelochemicals, is thought to entail a cost because it is energy-consuming and the materials used to synthesize the chemicals are taken from the metabolic pool that otherwise would most probably have gone into growth and reproduction, i.e. the plant's fitness. Therefore, selection would result in an optimal level of resistance, for example at a certain concentration of an allelochemical, which is determined by the trade-off between the costs and benefits of resistance. The benefit results, among others, from the reduction of damage caused by insects. As shown in Figure 10.16, fitness is maximal at an intermediate concentration of the resistance-providing chemical.

Thus, if a herbivorous insect were to exert a selection pressure on a plant for increased resistance, the concentration of resistance compound(s) will increase only to a certain value (R) [109]. This might produce an equilibrium in which the insect would still use the plant as a food, although to a lesser extent causing the insect performing suboptimally, while the plant would suffer less damage but at the same time would not be forced to further increase the production of costly allelochemical(s).

Reviewing the available experimental data, Simms [122] concluded that in many cases resistance to herbivory does involve costs, although some other cases suggest that these costs may be minimal. Considering in addition the cases in which overcompensation of herbivore damage was found (section 9.2.3) Rausher [109] suggested that 'there is no justi-

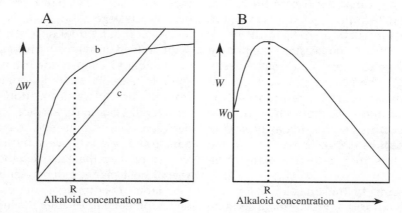

Figure 10.16 Cost – benefit model of selection on alkaloid concentration. **(A)** cost (c) and benefit (b) as a function of alkaloid concentration. ΔW = change in fitness. **(B)** Relationship between fitness (W) and alkaloid concentration; W_0 = base fitness. (Source: redrawn from Rausher, 1992.)

fication for assuming automatically that resistance necessarily benefits a plant by reducing herbivory'. Of course, a resistance trait in a plant may provide multiple benefits. Alkaloids in lupins, for instance, cause resistance not only against insects but also against molluscs and vertebrates. In addition, they inhibit pathogen infection, limit growth of neighbouring plants by allelopathic effects and function in nitrogen transport and storage. Furthermore, since resistance and tolerance traits are negatively correlated, selection for tolerance would reduce resistance and would result in a commensal relationship between herbivorous insects and plants. This seems to be the case in more insect–plant relations than was previously supposed [121].

Thus, even if an abundant insect population selects a plant population for its mild degree of resistance to that insect, most probably a dynamic equilibrium develops that is characterized by a certain level of resistance, which still allows the herbivore to operate, although probably at a somewhat reduced level. A similar conclusion was reached by quite different reasoning by Van der Meijden [135]. He considered the fact that specialist and generalist herbivorous insects may respond in opposite ways to different concentrations of plant resistance compounds. Specialists prefer individual plants containing high concentrations of such compounds, which often stimulate feeding and/or oviposition. Generalists, on the other hand, are deterred by high concentrations of the same compounds and thus preferably attack individual plants containing low concentrations of such compounds. Assuming that the intrapopulational variation in concentration of allelochemicals is determined genetically, it follows that specialist insects select the host-plant population for low, while generalists select for high concentrations of allelochemicals. These contrasting selection pressures exerted by specialist and generalist insects may partly explain why all plants have not evolved high levels of resistance and why there is considerable variation in the concentration of allelochemicals in natural plant populations.

3. Plant breeders have found that a plant lineage resistant to one herbivore is not necessarily resistant to another herbivore [51, 57]. Moreover, wheat varieties selected for resistance to a particular genotype of the Hessian fly (*Mayetiola destructor*) appear to be susceptible to another genotype of the same insect species [54]! Cucurbitacins in *Cucurbita moschata* inhibit feeding by *Epilachna tredecimnotata* but stimulate feeding by *Acalymna vittata*. Acyanogenic morphs of *Trifolium repens* are preferred by molluscs, while the fungus *Uromyces trifolii* damages mostly the cyanogenic morphs, and some weevil species behave variably [91]. Thus, if an insect population selects a plant population for increased production of an allelochemical that reduces the plant's palatability to that insect, this can lead to an increased susceptibility to other insects, pathogens, etc. On the other hand, in some cases the same secondary plant substance may provide resistance against both insects and pathogens, as has been found with the compound DIMBOA (**17**) in maize [80]. In conclusion, it is highly unpredictable how a resistance trait that appears incidentally in the plant as a result of selection by an insect will affect other organisms using the same plant as a resource.

4. There is ample evidence that several secondary plant substances have one or more primary functions other than resistance to herbivores. They may serve as resistance factors to pathogens, promote attraction of pollinators and dispersal agents, act as mediators in allelopathy, reduce transpiration, intercept UV radiation, protect against oxidative damage, regulate growth and development, promote pollen germination, act as storage molecules for nutrients and energy, stimulate nodule formation by nitrogen-fixing symbionts and stimulate mycorrhizal growth [55]. For example, the non-protein amino acid L-canavanine renders the seeds of some leguminous plants resistant to seed predators, but it also serves as

a nitrogen source for the germinating seedling [110]. On the other hand, recent experimental results disproved earlier assumptions [25] that nicotine serves as a nitrogen source in *Nicotiana sylvestris*. Nor does it protect *Datura stramonium* against UV radiation; it was therefore concluded that this compound is a 'defence' chemical [2, 3]. Yet, in our view nicotine is a resistance trait to which only a few insects were able to adapt; however, this does not mean that earlier insect attacks have caused the tendency to produce nicotine in this plant.

Clines of biological traits are often assumed to indicate a change of selection pressure with distance. An interesting phytochemical case is the observation that the percentage of plant species containing alkaloids decreases with latitude (Fig. 10.17).

This has been explained as being caused by higher grazing pressure exerted by all kinds of animals in the tropics throughout the year [88]. Other explanations could be, however, that resistance to pathogens, allelopathic interactions between plants and the abovementioned other functions of allelochemicals also require higher concentrations in warm climates. Also, at higher temperatures the production costs of allelochemicals could be lower, or the occurrence of alkaloids could even reflect plant responses to the latitudinal decrease in UVB radiation [21]. Therefore, in this case also what role herbivorous insects played or play in the increased production of secondary compounds at lower latitudes is an open question.

5. Brown and Lawton [17] have proposed that herbivorous insects may exert selection on leaf size and shape. They argue that these plant traits are involved in host-plant recognition and selection, can modify the efficiency with which the insects exploit leaves, may act as physical barriers to insects, also influence the number of species attacking plants and sometimes alter insect abundance. Indeed, there are data proving such effects but there is no evidence for the assumption that these plant traits resulted from 'selection past' by herbivorous insects. For example, narrow-leafed tree species support only about one-quarter the number of insect species found on tree species with broader leaves [102]. Yet to suppose that the narrow leaves of certain tree species have evolved under the selection pressure of some ancient herbivorous insects is mere speculation. In fact, it could be that broader leaves simply provide more niches for herbivorous insects. Brown and Lawton [17] themselves rightly emphasize: 'there are as yet

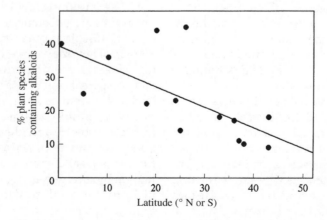

Figure 10.17 The relationship between percentage of alkaloid-bearing plants in floras of 14 countries and latitude. (Source: redrawn from Levin, 1976.)

no direct demonstrations that herbivore responses to leaf size and shape translate into effects on plant fitness, or plant performance'. This means that the role of herbivorous insects in the evolution of leaf size and shape, if any, remains an open question at best.

From the above points the following conclusions can be drawn.

1. In nature most herbivorous insect species inflict random damage on plants most of the time and reduce the fitness of only a small fraction of the host-plant population, largely independently of the genetic constitution (resistance traits) of individual plants.
2. In rare cases of strong selection pressure by insects, the resistance traits of plants may be raised to a level, determined by cost–benefit relations, that results in an equilibrium between insect damage and the cost of the resistance trait.
3. If an insect selects a plant for tolerance to the insect's attack, the outcome is another kind of equilibrium between the two organisms.
4. If two or more insect species exert selection pressure on the same plant species, this may lead to conflicting results, since resistance to one insect species may mean susceptibility to another.
5. Furthermore, because of the modular organization of plants (see Chapter 2), damage inflicted to one module does not necessarily represent selection pressure on the other modules and even less so on the plant as a whole.

In conclusion, the great majority of herbivorous insect species are unable to exert strong selection pressure on host-plant populations, unlike the reverse – the decisive selection pressures of plants on herbivorous insects.

Thinking in an evolutionary time scale, the objection to the above reasoning could be made that even a very weak selection pressure exerted by a herbivorous insect species on a plant species, the effect of which cannot now be measured, would be effective if it acted over millennia. This would be true if all other conditions, especially selection pressures by numerous other biotic and abiotic factors affecting both the insect and the plant, remained constant during that time. This is, of course, highly unlikely. It appears, therefore, more appropriate to stick to the observable and measurable facts.

10.4.3 RECIPROCITY OF SELECTION IN INSECT–PLANT INTERACTIONS

Plant pathologists have found that in several crop plant species the resistance of cultivars to a pathogen and the virulence of the latter form a **gene-for-gene** (GFG) interaction. In other words, for each gene for resistance in the host, there is a corresponding gene for avirulence in the parasite [81]. The best known examples are the cereal rusts (*Puccinia* spp.) and their hosts, wheat and oats.

It would be logical to suppose that similar GFG interactions may occur between herbivorous insects and their host plants. The case of the resistance of wheat cultivars to the Hessian fly has indeed been regarded as an example of a GFG relation. In wheat 20 genes for resistance to the fly were found and 11 fly biotypes were isolated that differed in their ability to attack wheat varieties with various resistance genes. Diehl and Bush [30], however, pointed out that in this case the GFG relation may be an artefact of the experimental method, since the fly biotypes were the result of artificial selection over several generations. Such a selection can hardly take place under natural conditions. Other cases of interactions between insects and plants, such as those between rice and the rice brown planthopper (*Nilaparvata lugens*) or wheat and the greenbug (*Schizaphis graminum*), turned out to be polygenically determined, which excludes simple GFG relations. Furthermore, oviposition preference and larval performance are controlled by different genes [66, 132, 133]. Because of such essential life-history differences between insects and pathogens, GFG interactions may be much more common with pathogens than with herbivorous insects [134]. The polygenic inheritance of factors determining most insect–plant interactions causes almost insurmountable methodological difficulties when natural populations are studied. No wonder that reciprocal selection has not so far been demonstrated in natural insect–plant interactions. As a matter of fact, the low efficiency of the selection pressure exerted by herbivorous insects on plants precludes the possibility of reciprocal selection.

10.5 THEORIES

As has been mentioned before, the huge number of herbivorous insect and green plant species, and especially the extreme variations in interactions between these organisms, provide a unique wealth of data for theoretical considerations on the evolutionary processes that may have led to extant insect–plant relationships. Below we survey some current theories.

10.5.1 COEVOLUTION

In a seminal paper Ehrlich and Raven [33] propounded an imaginative theory of 'coevolution' between plants and herbivorous insects. They inferred the following.

1. Plants 'through occasional mutations and recombinations, produced a series of chemical compounds not directly related to their basic metabolic pathways'.
2. 'Some of these compounds, by chance, serve to reduce or destroy the palatability of the plant in which they are produced. Such a plant, protected from the attack of phytophagous animals, would in a sense have entered a new adaptive zone. Evolutionary radiation of the plants might follow...'
3. 'If a new recombinant or mutant appeared in a population of insects that enabled individuals to feed on some previously protected plant group, selection could carry the line into a new adaptive zone. Here it would be free to diversify largely in the absence of competition from other phytophagous animals.'
4. Finally, Ehrlich and Raven [33] stated: 'Probably our most important overall conclusion is that the importance of **reciprocal selective responses** between ecologically linked organisms has been vastly underrated in considerations of the organic diversity. Indeed, the plant herbivore "interface" may be the major zone of interaction responsible for generating terrestrial organic diversity.' (Our emphasis).

The followers of Ehrlich and Raven's theory also regard the last point as the essence of coevolution and the phyletic relations of Types B and C (Fig. 10.10) are regarded as a result of reciprocal evolutionary interactions [35]. Thompson [133] also uses the term 'coevolution' to mean 'reciprocal evolutionary change in interacting species'. The validity of the above four points, however, is questionable, for the following reasons:

1. Taking points 1 and 2, since the same secondary plant compound may serve as a deterrent for one insect but may be ineffective against or even attractive to another, the appearance of such chemicals does not necessarily relieve the plant of herbivorous insect attacks (section 10.4.2).
2. The assumption of plant radiation (point 2) seems to be supported by the 'escalation–diversification hypothesis' [36], according to which plant taxa possessing 'escalated defence mechanisms' such as secretory canals (latex, resin) have diversified more extensively than those lacking such defences. However, there is no proof whatsoever of the implicit assumption that plants have evolved such 'defences' under selection pressure from herbivores and that such plants could diversify more effectively, because they were 'defended'

against insects. Furthermore, many insect species have adapted to live exclusively on canal-bearing plants.
3. Taking point 3, the processes of speciation in herbivorous insects that are known at present (section 10.1) do not support such an assumption, since competition among herbivorous insects has been found to be rare or often totally absent (Chapter 9).
4. Since the assumptions in points 1–3 are questionable, point 4 does not hold either, the more so because reciprocal speciation between interacting organisms has been found heretofore only for a number of pollinators and plants and for microbial symbionts and their hosts [133] (section 10.2.5).

Studies to test the coevolution theory have so far failed [133]. However, if the validity of this theory could be proved, it would explain only the evolution of some phyletic relations of Types B and C but by no means those of Types A and D (Fig. 10.10). In conclusion, the 'coevolution theory', although intuitively an elegant model, must be rejected on the basis of our present insights.

10.5.2 DIFFUSE COEVOLUTION, COMMUNITY COEVOLUTION

The theory of **diffuse coevolution** proposes that, instead of the pairwise reciprocal evolutionary interactions supposed by the classic coevolutionary theory,

> coevolution must be considered in a community context and not simply as an isolated two species interaction. The community affecting any one plant species includes several types of herbivores, alternative host plants, potentially competing plants and species in higher trophic levels that may respond directly to the herbivores or to the plants, changes induced by one member of the community may affect many species.
>
> Fox, 1988

This sounds a logical assumption, but even the much simpler pairwise coevolution between a plant species and a herbivorous insect species under natural conditions has not been proved in any single case so far. The same is even more true for cases of natural insect–plant interactions in a community context.

Similarly, the theory of **community coevolution** proposes that interactions between the members of a community are so tight that the evolution of any member affects the evolution of all other members [137]. A common flaw of both theories is the assumption that any evolutionary change of any community member ripples more or less through the whole community. This, again, has never been proved. On the contrary, an analysis of the consequences of species introduction into a community has shown that in the large majority of cases the appearance of a new species caused no structural or functional changes in the local community [120], which indicates that a new species can hardly affect the evolution of all or even the majority of other community members. The very low abundance and patchy distribution of most herbivorous insect species as compared to the abundance and distribution of their host plants, the rarity or absence of competition between herbivorous insects and the existence of 'vacant niches' (section 9.4.1) also contradict the assumption of tight evolutionary interactions within a community consisting of herbivorous insects.

It should also be considered that the interactions between species are studied in extant communities, so the results can hardly answer the question of how, when and why these relationships evolved. Actually, no evidence exists for the implicit assumption that members of extant communities have evolved together [67]. On the contrary, (1) many switchovers to new host plants must have occurred without previous evolutionary interactions between the insect and its new host plant (section 10.5.4) and (2) climatic changes during the recent geological past (ice ages), i.e. after the evolution of most recent insect and plant species, may have drastically changed the structure and distribution of plant and animal communities in most temperate zones. The member species must have reacted differently to such effects, as can be seen in recent changes of insect distribution in Central Europe presumably due to global warming [85].

10.5.3 THE GEOGRAPHIC MOSAIC THEORY OF COEVOLUTION

Thompson [133] summarized the essence of his theory as follows:

1. Interspecific interactions commonly differ in outcome among populations. These differences result from the combined effects of differences in the physical environment, the local genetic and demographic structure of populations, and the community context in which the interaction occurs.
2. As a result of these differences in outcomes, an interaction may coevolve in some populations, affect the evolution of only one of the participants in other populations, and have no effect on evolution in yet other local populations. Still other populations may fall outside the current geographic range of the interacting species.
3. In addition, populations differ in the extent to which they show extreme specialization to one or more species. Some populations may specialize on and sometimes coevolve locally with only one other species, other populations may specialize on and perhaps coevolve with different species, and yet others may coevolve simultaneously with multiple species.
4. These interpopulational differences in outcome and specialization create a geographic mosaic in interactions, which is the raw material for the dynamics of coevolution. This mosaic also creates the possibility that the overall evolution of a species is a result of coevolution with several species, even though individual populations are specialized to only one or two of the species.
5. Gene flow among populations and extinction of some demes reshape the geographic mosaic of coevolution as the adaptations and patterns of specialization developed locally spread to other populations or are lost. Characters evolved in some local populations will temporarily have the greatest effects on the overall direction of coevolution between a pair or group of species, whereas those evolved in other populations will contribute little.
6. The result is a continually shifting geographic pattern of coevolution between any two or more species. Much of the dynamics of the coevolutionary process need not result eventually in an escalating series of adaptations and counteradaptations that become fixed traits within species.

The geographic mosaic theory therefore suggests that the coevolutionary process is much more dynamic than is apparent from the study of individual populations or the distribution of characters found in phylogenetic trees.

Thompson, 1994

Thompson [133] defines coevolution as 'the reciprocal evolutionary change in interacting species' and implicitly assumes that all interacting species, at least in some locations, necessarily coevolve. The above points of the theory are supported by a wealth of examples taken from many case studies on interactions, especially between hosts and parasites, hosts and pathogens, flowering plants and pollinators. However, no cases are presented in which presumably reciprocal evolutionary interactions between (non-pollinator) herbivorous insects and their host plants, at least in local populations, would have been proved, apparently because such cases are not known. Nevertheless, Thompson seems to assume that herbivorous insects and host plants coevolve as well. We have shown above (section 10.4.3) that insect–host-plant interactions do not generally represent reciprocal selection pressures and therefore cannot result in reciprocal evolutionary changes. Thus, for the same reasons as Ehrlich and Raven's classic coevolution theory, the 'geographic mosaic theory of coevolution' does not satisfactorily explain how the relationships between herbivorous insects and their host plants came into being.

10.5.4 SEQUENTIAL EVOLUTION

The theory of **sequential evolution** [69, 70, 72, 73], also called **sequential colonization** [99], proposes that the evolution of herbivorous insects *follows* the evolution of plants without, however, significantly affecting plant evolution. The stage for the extensive radiation of herbivorous insects was set by the evolution of plant chemical diversity, which may have begun even before the appearance of terrestrial plants [76].

The sequential or asymmetrical nature of the evolution of insect–plant relationships is indicated by the following.

1. As stated in section 10.4, reciprocal selective interactions between plants and herbivorous insects has not been proved so far. Plants exert strong selection pressure on insects, whereas the latter exert selective pressure on the plants only in rare cases and even then only weakly.
2. As a consequence, speciation in herbivorous insects may be mediated by plants but speciation in plants has not been proved to occur as a consequence of interaction with herbivorous insects (section 10.2).
3. In several cases an evolutionary time lag has been found between the emergence of plant species or groups of species and their colonization by herbivorous insect species. For example, phylogenetic relations of Type A (Fig. 10.10) suggest frequent switchovers that have occurred often much later than the divergence of the plant taxa concerned. Several case studies support this observation [1, 63, 111, 119]. Ferns, which are regarded as an old plant group, are attacked by both ancient (e.g. sawflies) and advanced (e.g. Lepidoptera) herbivorous insect groups [61]. As yet no studies have proved that switchover of an insect species to a new host plant has changed the traits of that plant.

As for the processes involved in the evolution of insect–plant relations, the sequential evolution theory emphasizes the primary importance of the evolution of the plant recognition mechanism in insects. As has been shown before (Chapter 6), insects' ability to recognize their host plants, primarily by chemoreception, is the proximate cause of host-plant specificity (for references, see Jermy [73]). Since 'mutational changes at the receptor level or at the integrative level in the central nervous system could drastically alter food-plant preference' and 'a single mutation could change monophagy to polyphagy and *vice versa*' [29], it is logical to assume that heritable changes in the insects' plant recognition mechanism are the prime events in the evolution of insect–plant relationships [13, 70, 74, 96]. All characteristics of the evolution of host-plant specialization (such as changes from stenophagy to euryphagy and the reverse, permanent host fidelity, rapid changes in host-plant preference and switches to closely or to distantly related new host-plant species) depend on various kinds of genetic and developmental constraints determining the possibility and direction of changes within the insect's genome. Selection comes after the genetic changes [65] in the insect's plant recognition system because the establishment of a new genome that codes for new plant preferences will be successful only if it is able to tolerate the many ecological and physiological constraints to which it will be subjected.

The occurrence of such mutations and the establishment of new genotypes is most probably highly stochastic. The whole process is channelled primarily by genetic constraints and secondarily by selection. Both are characterized by chance events. It is, therefore, not surprising that the phyletic relations between insects and their host plants are extremely stochastic and unpredictable, as indicated by the examples given in section 10.3.2. Presumably the following genetic changes may result in the four main types of phyletic relations depicted in Figure 10.10.

1. Genetic changes (mutations, recombinations) alter the insect's plant recognition process so that the new insect genotype recognizes a plant as host whose chemical traits differ markedly from the traits of its parental host plant, i.e. the new host will most probably be distantly related to the parental host (**jumping host shift**). The result is a relation characterized by cladograms of Type A (Fig. 10.11). Such a change in the insect's chemoreceptors (or in the central coding of the incoming information) does not necessarily involve the perception of a series of plant compounds but may occasionally be due to the perception as an attractant of a single compound that may occur in taxonomically (and phytochemically) very different plant taxa, as has been mentioned above (section 10.3.2) in the case of *Yponomeuta* species. Such a shift seems possible without extensive modification in the overall neural perception of the switching species [96].
2. Genetic changes result in a new insect genotype that recognizes a plant as a host that is chemically only slightly different from the parental host plant, i.e. which is most probably closely related to the latter (**sliding host shift**). This leads to relationships of Type B (Fig. 10.10).
3. Sliding host shifts combined with monophagy produce relationships of Type C (Fig. 10.13). Of course, such relationships are no proof of simultaneous (reciprocal) radiation of plants and insects, even if the partner species appear to be of the same paleontological age [34]. They may just as well have resulted from sequential colonization with a time lag of up to a few hundred thousand years, time differences that can hardly be demonstrated in fossil records.
4. Polyphagous insects (Type D) supposedly originate from polyphagous or from oligophagous ancestors as a result of genetic changes that enable an insect to accept a wide range of chemically very different plant species as hosts. Nevertheless, polyphages are able to distinguish between their host-plant species. This has been clearly proved by experiments on induced food preference. Here 'experienced' polyphagous insects clearly preferred, i.e. recognized, the plant species they fed upon before to other host-plant species of which they had no experience but which were otherwise equally acceptable to 'naive' individual insects (section 7.6.1).

The theories of coevolution, diffuse coevolution and community coevolution, as well as the geographic mosaic theory of coevolution and, more generally, all assumptions concerning the evolution of insect–plant relationships based on evolutionary ecological premises, implicitly or explicitly suppose the following sequence of evolutionary steps.

1. Some selective forces, such as the occurrence of deterrent or toxic plant compounds, or the activity of competitor herbivores or natural enemies, force an insect (genotype) to specialize to one or a few plant species.
2. The insect becomes adapted physiologically to the given plant species.
3. The insect succeeds in recognizing those plant species as hosts.

This top-down sequence ignores the fact that host-plant specificity is based on the insects' plant recognition mechanism. Thus, the sequence of evolutionary steps must rather be of a bottom-up type, as follows.

1. Genetic changes alter the insect's plant recognition mechanism.
2. Selection imposed by ecological factors (especially the presence or absence of one or more plant species that are recognized as hosts by the new insect genotype, presence or absence of competitors and/or natural enemies) enables the establishment of the new genotype or wipes it out.
3. If it becomes established, further selection by nutritional factors enhances or prevents the physiological adaptation of the new genotype to the new host.

In conclusion, the evolutionary interactions between plants and herbivorous insects are asymmetrical. Plant evolution provided a profusion of niches: the chemically and structurally diverse plant species that enabled the radiation of herbivorous insects. Conversely, the insects exercised little if any influence on plant evolution. The evolution of the insects' nervous system, which determines the process of plant recognition, is the primary and highly autonomous process in the evolution of insect–plant interactions. (It is curious that some authors [35] dealing with obvious cases of sequential evolution (colonization), i.e. with phyletic relations where no reciprocal interactions have been proved, still use the term 'coevolution'.)

The theory of sequential evolution (TSE) has been criticized, especially by proponents of Ehrlich and Raven's [33] coevolution theory. A detailed criticism that comprises the main objections of other authors has been formulated by Thompson [133], and can be summarized as follows.

1. TSE denies any evolutionary influence of insects on plants.
2. TSE equates low insect population density with weak selection.
3. Sequential evolution is otherwise known as phylogenetic tracking.

These points are refuted as follows. The first objection is not exactly true, since TSE suggests that in rare cases insects may select, for example for higher concentrations of plant allelochemicals, although the evolutionary significance of such changes is unknown (see section 10.4.2). The second point is refuted by arguments given in section 10.4.2. The third statement reflects a misconception, since TSE relates not only to congruent phyletic relations between insects and their host plants but to all known kinds of such relations. Consequently, none of Thompson's objections invalidate the theory of sequential evolution.

10.6 CONCLUSIONS

In this chapter we dealt with the evolutionary interactions between herbivorous insects and plants. Speciation in herbivorous insects may result from differences in host-plant preferences between insect populations. This is indicated by the occurrence of infraspecific categories, also called biotypes, such as polymorphic populations, geographical races, host races and sibling species, which are nearly indistinguishable by morphological characteristics. However, they show basic differences in host-plant preference. Biotypes may represent pre-steps in the process of speciation in herbivorous insects.

Restriction or prevention of gene exchange between populations, i.e. reproductive isolation by some barrier, is a prerequisite of speciation. Spatial separation is the most obvious barrier to gene flow. For populations of insect species with low dispersal capacity, even short distances may represent effective barriers. Differences in feeding and/or oviposition preferences may also reduce gene flow considerably, even in sympatry. There is no evidence for reciprocal speciation between plants and herbivorous insects based on a gene-for-gene selection.

None of the several assumptions concerning the adaptive advantage of host-plant specificity in herbivorous insects can unequivocally explain why specialists far outnumber generalists in all main groups of insects. An acceptable explanation may be sought in some constraints on the evolution of the insects' nervous system, since plant recognition is a prerequisite for maintenance of host-plant specialization. Thus the preponderance of food specialists among herbivorous insects remains an enigma of insect–plant biology.

As for the evolutionary driving forces that resulted in extant insect–plant relations, selection pressures inflicted by plants on herbivorous insects account for various evolutionary changes in the insects. There is no unequivocal evidence, however, for evolutionary changes in plants that might have resulted from selection by herbivorous insects.

We have discussed the pros and cons of the theories of coevolution, diffuse coevolution, community coevolution, the geographic mosaic theory of coevolution and the theory of sequential evolution. At the present state of knowledge the last theory most plausibly explains how the extant insect–plant relationships came about.

10.7 REFERENCES

1. Abe, Y. (1991) Host race formation in the gall wasp *Andricus mukaigawae*. *Entomol. Exp. Appl.*, **58**, 15–20.
2. Baldwin, I. T. and Huh, S. (1994) Primary function for a chemical defense – nicotine does not protect *Datura stramonium* L. from UV damage. *Oecologia*, **97**, 243–247.
3. Baldwin, I. T. and Ohnmeis, T. E. (1994) Swords into plowshares? *Nicotiana silvestris* does not use nicotine as a nitrogen source under nitrogen-limited growth. *Oecologia*, **98**, 385–392.
4. Barbosa, P. (1988) Some thoughts on 'the evolution of host range'. *Ecology*, **69**, 912–915.
5. Barbosa, P. and Krischik, V. A. (1987) Influence of alkaloids on the feeding preference of eastern deciduous forest trees by the gypsy moth *Lymantria dispar*. *Am. Nat.*, **130**, 53–69.
6. Berenbaum, M. (1986) Post-ingestive effects of phytochemicals on insects: on Paracelsus and plant products, in *Insect–Plant Interactions*, (eds J. R. Miller and T. A. Miller), Springer-Verlag, New York, pp. 121–153.
7. Berlocher, S. H. (1979) Biochemical approaches to strain, race, and species discrimination, in *Genetics in Relation to Insect Management*, (eds M. A. Hoy and J. J. McKelvey), Rockefeller Foundation, New York, pp. 137–144.
8. Bernays, E. A. (1989) Host range in phytophagous insects: the potential role of generalist predators. *Evol. Ecol.*, **3**, 299–311.
9. Bernays, E. A. (1991a) Evolution of host-plant specificity. *Symp. Biol. Hung.*, **39**, 313–316.
10. Bernays, E. A. (1991b) Relationship between deterrence and toxicity of plant secondary compounds for the grasshopper *Schistocerca americana*. *J. Chem. Ecol.*, **17**, 2519–2526.
11. Bernays, E. A. and Chapman, R. F. (1977) Deterrent chemicals as a basis of oligophagy in *Locusta migratoria*. *Ecol. Entomol.*, **2**, 1–18.
12. Bernays, E. A. and Chapman, R. F. (1987) The evolution of deterrent responses in plant-feeding insects, in *Perspectives in Chemoreception and Behavior*, (eds R. F. Chapman, E. A. Bernays and J. G. Stoffolano), Springer-Verlag, New York, pp. 159–173.
13. Bernays, E. A. and Chapman, R. F. (1994) *Host-Plant Selection by Phytophagous Insects*, Chapman & Hall, New York.
14. Bernays, E. A. and Cornelius, M. (1992) Relationship between deterrence and toxicity of plant secondary compounds for the alfalfa weevil *Hypera brunneipennis*. *Entomol. Exp. Appl.*, **64**, 289–292.
15. Bernays, E. A. and Graham, M. (1988) On the evolution of host selection in phytophagous arthropods. *Ecology*, **69**, 886–892.
16. Boller, E. F., Russ, K., Vallo, V. and Bush, G. L. (1976) Incompatible races of European cherry fruit fly, *Rhagoletis cerasi* (Diptera, Tephritidae): their origin and potential use in biological control. *Entomol. Exp. Appl.*, **20**, 237–247.
17. Brown, V. K. and Lawton, J. H. (1991) Herbivory and the evolution of leaf size and shape. *Phil. Trans. Roy. Soc. Lond. B*, **333**, 265–272.
18. Bush, G. L. (1975a) Modes of speciation. *Annu. Rev. Ecol. Syst.*, **6**, 339–364.
19. Bush, G. L. (1975b) Sympatric speciation in phytophagous parasitic insects, in *Evolutionary Strategies of Parasitic Insects and Mites*, (ed. P. W. Price), Plenum Press, New York, pp. 187–206.
20. Bush, G. L. (1987) Evolutionary behavior genetics, in *Evolutionary Genetics of Invertebrate*

Behavior, (ed. M. D. Huettel), Plenum Press, New York, pp. 1–5.
21. Caldwell, M. M. and Robberecht, R. (1980) A steep latitudinal gradient of solar ultraviolet B radiation in the arctic-alpine life zone. *Ecology*, **61**, 600–611.
22. Carson, H. L. (1987) Colonization and speciation, in *Colonization, Succession and Stability*, (eds A. J. Gray, M. J. Crawley and P. J. Edwards), Blackwell Scientific Publications, Oxford, pp. 187–206.
23. Claridge, M. F. (1995) Species concepts and speciation in insect herbivores: planthopper case studies. *Boll. Zool.*, **62**, 55–58.
24. Coley, P. D. (1987) Interspecific variation in plant anti-herbivore properties: the role of habitat quality and rate of disturbance. *N. Physiol.*, **106**(Suppl.), 251–263.
25. Coley, P. D., Bryant, J. P. and Chapin, T. (1985) Resource availability and plant antiherbivore defense. *Science*, **230**, 895–899.
26. Connell, J. H. (1980) Diversity and coevolution of competitors, or the ghost of competition past. *Oikos*, **35**, 131–138.
27. Crane, P. R., Friis, E. M. and Pedersen, K. R. (1995) The origin and early diversification of angiosperms. *Nature*, **374**, 27–33.
28. Dethier, V. G. (1980) Evolution of receptor sensitivity to secondary plant substances with special reference to deterrents. *Am. Nat.*, **115**, 45–66.
29. Dethier, V. G. (1987) Analyzing proximate causes of behavior, in *Evolutionary Genetics of Invertebrate Behavior*, (ed. M. D. Huettel), Plenum Press, New York, pp. 319–328.
30. Diehl, S. R. and Bush, G. L. (1984) An evolutionary and applied perspective of insect biotypes. *Annu. Rev. Entomol.*, **29**, 471–501.
31. Dobzhansky, Th. (1973) Nothing in biology makes sense except in the light of evolution. *Am. Biol. Teacher*, **35**, 125–129.
32. Duffey, S. S. (1980) Sequestration of plant natural products by insects. *Annu. Rev. Entomol.*, **25**, 447–477.
33. Ehrlich, P. R. and Raven, P. H. (1964) Butterflies and plants, a study in coevolution. *Evolution*, **18**, 586–608.
34. Farrell, B. D. and Mitter, C. (1990) Phylogenesis of insect/plant interactions: have *Phyllobrotica* leaf beetles (Chrysomelidae) and Lamiales diversified in parallel? *Evolution*, **44**, 1389–1403.
35. Farrell, B. D. and Mitter, C. (1993) Phylogenetic determinants of insect/plant community diversity, in *Species Diversity in Ecological Systems. Historical and Geographical Perspectives*, (eds R. E. Ricklefs and D. Schluter), University of Chicago Press, Chicago, IL, pp. 253–266.
36. Farrell, B. D., Dussourd, D. E. and Mitter, C. (1991) Escalation of plant defense: do latex and resin canals spur plant diversification? *Am. Nat.*, **138**, 881–900.
37. Feder, J. L., Chilcote, C. A. and Bush, G. L. (1990a) The geographic pattern of genetic differentiation between host associated populations of *Rhagoletis pomonella* (Diptera, Tephritidae) in the eastern United States and Canada. *Evolution*, **44**, 570–594.
38. Feder, J. L., Chilcote, C. A. and Bush, G. L. (1990b) Regional, local and microgeographical allele frequency variation between apple and hawthorn populations of *Rhagoletis pomonella* in western Michigan. *Evolution*, **44**, 595–608.
39. Feder, J. L., Hunt, T. A. and Bush, L. (1993) The effects of climate, host plant phenology and host fidelity on the genetics of apple and hawthorn infesting races of *Rhagoletis pomonella*. *Entomol. Exp. Appl.*, **69**, 117–135.
40. Feeny, P. (1975) Biochemical coevolution between plants and their insect herbivores, in *Coevolution of Animals and Plants*, (eds L. E. Gilbert and P. H. Raven), University of Texas Press, Austin, TX, pp. 3–19.
41. Feeny, P. (1991) Theories of plant chemical defense, a brief historical survey. *Symp. Biol. Hung.*, **39**, 163–175.
42. Fiedler, K. (1996) Host-plant relationships of lycaenid butterflies: large-scale patterns, interactions with plant chemistry, and mutualism with ants. *Entomol. Exp. Appl.*, **80**, 259–267.
43. Ford, E. B. (1963) *Genetic Polymorphism*, MIT Press, Cambridge, MA.
44. Fox, L. R. (1988) Diffuse coevolution within complex communities. *Ecology*, **69**, 906–907.
45. Fox, L. R. and Morrow, P. A. (1981) Specialization: species property or local phenomenon? *Science*, **211**, 887–893.
46. Fraenkel, G. S. (1959) The raison d'être of secondary plant substances. *Science*, **129**, 1466–1470.
47. Frey, J. E. and Bush, G. L. (1990) *Rhagoletis* sibling species and host races differ in host odor recognition. *Entomol. Exp. Appl.*, **57**, 123–131.
48. Fritz, R. S. and Price, P. W. (1988) Genetic variation among plants and insect community

structure: willows and sawflies. *Ecology*, **69**, 845–856.
49. Futuyma, D. J. (1983) Selection factors in the evolution of host choice by phytophagous insects, in *Herbivorous Insects: Host Seeking Behaviour and Mechanisms*, (ed. S. Ahmad), Academic Press, New York, pp. 227–244.
50. Futuyma, D. J. (1991) Evolution of host specificity in herbivorous insects: genetic, ecological, and phylogenetic aspects, in *Plant–Animal Interactions. Evolutionary Ecology in Tropical and Temperate Regions*, (eds P. W. Price, T. M. Lewinsohn, G. W. Fernandes and W. W. Benson) John Wiley, New York, pp. 431–454.
51. Futuyma, D. J. and Peterson, S. C. (1985) Genetic variation in the use of resources by insects. *Annu. Rev. Entomol.*, **30**, 217–238.
52. Futuyma, D. J., Walsh, J. S., Morton, T., Funk, D. J. and Keese, M. C. (1994) Genetic variation in a phylogenetic context: responses of two specialized leaf beetles (Coleoptera, Chrysomelidae) to host plants of their congeners. *J. Evol. Biol.*, **7**, 127–146.
53. Galina, A., Moya, A. and Ayala, F. J. (1993) Founder-flush speciation in *Drosophila pseudoobscura*, a large-scale experiment. *Evolution*, **47**, 432–444.
54. Gallun, R. L., Starks, K. J. and Guthrie, W. D. (1975) Plant resistance to insects attacking cereals. *Annu. Rev. Entomol.*, **20**, 337–357.
55. Gershenzon, J. (1994) The cost of plant chemical defense against herbivory: a biochemical perspective, in *Insect–Plant Interactions*, vol. 5, (ed. E. A. Bernays), CRC Press, Boca Raton, FL, pp. 105–173.
56. Gilbert, L. E. and Singer, M. C. (1975) Butterfly ecology. *Annu. Rev. Ecol. Syst.*, **6**, 365–397.
57. Gould, F. (1983) Genetics of plant–herbivore systems: interactions between applied and basic study, in *Variable Plants and Herbivores in Natural and Managed Systems*, (eds R. F. Denno and M. S. McClure), Academic Press, New York, pp. 599–653.
58. Gould, F. (1984) Mixed function oxidases and herbivore polyphagy: the devil's advocate position. *Ecol. Entomol.*, **9**, 29–34.
59. Greenslade, P. J. M. (1983) Adversity selection and the habitat templet. *Am. Nat.*, **122**, 352–365.
60. Grime, J. P. (1977) Evidence for the existence of three primary strategies in plants and its relevance to ecological and evolutionary theory. *Am. Nat.*, **111**, 1169–1194.
61. Hendrix, S. D. (1980) An evolutionary and ecological perspective of the insect fauna of ferns. *Am. Nat.*, **115**, 171–196.
62. Ho, T. H. (1965) The life history and control of the diamondback moth in Malaya. *Mins. Agric. Co-op. Bull.*, **118**, 1–26.
63. Hsiao, T. H. (1989) Host plant affinity in relation to phylogeny of *Leptinotarsa* beetles. *Entomography*, **6**, 413–422.
64. Hsiao, T. H. and Hsiao, C. (1985) Hybridization and cytoplasmic incompatibility among alfalfa weevil strains. *Entomol. Exp. Appl.*, **37**, 155–159.
65. Jacob, F. (1981) *Le Jeu des Possibles*, Fayard, Paris.
66. Jaenike, J. (1990) Host specialization in phytophagous insects. *Annu. Rev. Ecol. Syst.*, **21**, 243–273.
67. Janzen, D. H. (1980) When is it coevolution? *Evolution*, **34**, 611–612.
68. Jermy, T. (1966) Feeding inhibitors and food preference in chewing phytophagous insects. *Entomol. Exp. Appl.*, **9**, 1–12.
69. Jermy, T. (1976) Insect–host-plant relationships – coevolution or sequential evolution? *Symp. Biol. Hung.*, **16**, 109–113.
70. Jermy, T. (1984) Evolution of insect/host plant relationships. *Am. Nat.*, **124**, 609–630.
71. Jermy, T. (1988) Can predation lead to narrow food specialization in phytophagous insects? *Ecology*, **69**, 902–904.
72. Jermy, T. (1991) Evolutionary interpretations of insect–plant relationships – a closer look. *Symp. Biol. Hung.*, **39**, 301–311.
73. Jermy, T. (1993) Evolution of insect–plant relationships – a devil's advocate approach. *Entomol. Exp. Appl.*, **66**, 3–12.
74. Jermy, T., Lábos, E. and Molnár, I. (1990) Stenophagy of phytophagous insects – a result of constraints on the evolution of the nervous system, in *Organizational Constraints on the Dynamics of Evolution*, (eds J. Maynard Smith and G. Vida), Manchester University Press, Manchester, pp. 157–166.
75. Joern, A. and Games, S. B. (1990) Population dynamics and regulation in grasshoppers, in *Biology of Grasshoppers*, (eds R. F. Chapman and A. Joern), John Wiley, New York, 415–482.
76. Jones, C. G. and Firn, R. D. (1991) On the evolution of plant secondary chemical diversity. *Phil. Trans. Roy. Soc. Lond. B*, **333**, 273–280.
77. Joshi, A. and Thompson, J. N. (1995) Trade-offs and the evolution of host specialization. *Evol. Ecol.*, **9**, 82–92.
78. Kaszab, Z. (1962) *Levélbogarak – Chrysomelidae*,

Fauna Hungariae No. 63, Akadémiai Kiadó, Budapest.
79. Kennedy, C. E. J. (1986) Attachment may be a basis for specialization in oak aphids. *Ecol. Entomol.*, **11**, 291–300.
80. Kennedy, G. G. and Barbour, J. D. (1992) Resistance variation in natural and managed systems, in *Plant Resistance to Herbivores and Pathogens*, (eds R. S. Fritz and E. L. Simms), University of Chicago Press, Chicago, IL, pp. 13–41.
81. Kerr, A. (1987) The impact of molecular genetics on plant pathology. *Annu. Rev. Phytopathol.*, **25**, 87–110.
82. Klun, J. A. and Maini, S. (1979) Genetic basis of an insect chemical communication system: the European cornborer. *Environ. Entomol.*, **8**, 423–526.
83. Knoll, A. H., Niklas, K. J. and Tiffney, B. H. (1979) Phanerozoic land-plant diversity in North America. *Science*, **206**, 1400–1402.
84. Kosztarab, M. and Kozár, F. (1988) *Scale Insects of Central Europe*, Akadémiai Kiadó, Budapest.
85. Kozár, F. (1992) Recent changes in the distribution of insects and the global warming, in *Proceedings of the 4th European Congress of Entomologists and XIII. Internationale Symposium für die Entomofaunistik Mitteleuropas, Gödöllö, Hungary*, vol. I, pp. 406–413.
86. Labandeira, C. C. and Sepkoski, J. J. (1993) Insect diversity in the fossil record. *Science*, **261**, 310–315.
87. Labeyrie, V. (1976) The importance of the coevolutive point of view in the investigation of the reproductive relations between insects and host-plants. *Symp. Biol. Hung.*, **16**, 133–136.
88. Levin, D. A. (1976) Alkaloid bearing plants: an ecogeographic perspective. *Am. Nat.*, **110**, 261–284.
89. Levin, D. A. (1995) Metapopulations: an arena for local speciation. *J. Evol. Biol.*, **8**, 635–644.
90. Levins, R. A. (1970) Extinction. *Am. Math. Soc.*, **2**, 77–107.
91. Linhart, Y. B. (1991) Disease, parasitism and herbivory: multidimensional challenges in plant evolution. *Trends Ecol. Evol.*, **6**, 392–396.
92. Mägdefrau, K. (1959) *Vegetationsbilder der Vorzeit*, 3rd edn, G. Fischer, Jena.
93. May, R. M. (1981) *Theoretical Ecology*, Blackwell, Oxford.
94. Mayr, E. (1954) Change of genetic environment and evolution, in *Evolution as a Process*, (eds J. Huxley, A. C. Hardy and E. B. Ford), Allen & Unwin, London, pp. 157–180.
95. Mayr, E. (1963) *Animal Species and Evolution*, Harvard University Press, Cambridge, MA.
96. Menken, S. B. J. (1996) Pattern and process in the evolution of insect–plant associations: *Yponomeuta* as an example. *Entomol. Exp. Appl.*, **80**, 297–305.
97. Menken, S. B. J., Herrebout, W. M. and Wiebes, J. T. (1992) Small ermine moths (*Yponomeuta*): their host relations and evolution. *Annu. Rev. Entomol.*, **37**, 41–88.
98. Metcalf, C. L. and Flint, W. P. (1939) *Destructive and Useful Insects. Their Habits and Control*, McGraw-Hill, New York.
99. Mitter, C. and Brooks, D. R. (1983) Phylogenetic aspects of coevolution, in *Coevolution*, (eds D. J. Futuyma and M. Slatkin), Sinauer, Sunderland, MA, pp. 65–98.
100. Mopper, S. (1996) Adaptive genetic structure in phytophagous insect populations. *Trends Ecol. Evol.*, **11**, 235–238.
101. Moran, N. (1988) The evolution of host plant alteration in aphids: evidence for specialization as a dead end. *Am. Nat.*, **132**, 681–706.
102. Moran, V. C. and Southwood, T. R. E. (1982) The guild composition of arthropod communities in trees. *J. Anim. Ecol.*, **51**, 289–272.
103. Müller, F. P. (1985) Genetic and evolutionary aspects of host choice in phytophagous insects, especially aphids. *Biol. Zbl.*, **104**, 225–237.
104. Owen, D. F. (1990) The language of attack and defense. *Oikos*, **57**, 133–146.
105. Pluthero, F. G. and Singh, R. J. (1984) Insect behavioral response to toxins: practical and evolutionary considerations. *Can. Entomol.*, **116**, 57–68.
106. Powell, J. A. (1992) Interrelationships of yuccas and yucca moths. *Trends Ecol. Evol.*, **7**, 10–15.
107. Price, P. W. (1983) Hypotheses on organization and evolution in herbivorous insect communities, in *Variable Plants and Herbivores in Natural and Managed Systems*, (eds R. F. Denno and M. S. McClure), Academic Press, New York, pp. 559–596.
108. Radtkey, R. R. and Singer, M. C. (1995) Repeated reversal of host-preference evolution in a specialist insect herbivore. *Evolution*, **49**, 351–359.
109. Rausher, M. D. (1992) Natural selection and the evolution of plant-insect interactions, in *Insect Chemical Ecology*, (eds B. D. Roitberg and M. B. Isman), Chapman & Hall, New York, pp. 20–88.
110. Rosenthal, G. A., Berge, M. A., Ozinkas, A. J.

and Hughes, C. G. (1988) Ability of L-canavanine to support nitrogen metabolism in jack bean, *Canavalia ensiformis* (L.) DC. *Agric. Food Chem.*, **36**, 1159–1163.
111. Roskam, J. C. and Zandee, M. (1992) Phylogenetic analysis of quantitative and qualitative characters of gall-inducing midges and the historical relation to their hosts, *Dasineura* (Diptera, Cecidomyiidae) on Rosaceae and monocots, in *Ordination in the Study of Morphology, Evolution and Systematics of Insects: Applications and Quantitative Genetic Rationals*, (eds J. T. Sorensen and R. Foottit), Elsevier, Amsterdam, pp. 325–347.
112. Rowell-Rahier, M. and Pasteels, J. M. (1992) Genetic relationships between *Oreina* species with different defensive strategies, in *Proceedings of the 8th International Symposium on Insect–Plant Relationships*, (eds S. B. J. Menken, J. H. Visser and P. Harrewijn), Kluwer, Dordrecht, pp. 341–342.
113. Schoonhoven, L. M. (1987) What makes a caterpillar eat? The sensory code underlying feeding behavior, in *Perspectives in Chemoreception and Behavior*, (eds R. F. Chapman, E. A. Bernays and J. G. Stoffolano), Springer-Verlag, New York, pp. 69–97.
114. Schreiber, K. (1958) Über einige Inhaltsstoffe der Solanaceen und ihre Bedeutung für die Kartoffelkäferresistenz. *Entomol. Exp. Appl.*, **1**, 28–37.
115. Schultz, J. C. (1983) Habitat selection and foraging tactics of caterpillars in heterogenous trees, in *Variable Plants and Herbivores in Natural and Managed Systems*, (eds R. F. Denno and M. S. McClure), Academic Press, New York, pp. 61–90.
116. Schultz, J. C. (1988) Many factors influence the evolution of herbivore diets, but plant chemistry is central. *Ecology*, **69**, 896–897.
117. Scriber, J. M. (1983) Evolution of feeding specialization, physiological efficiency, and host races in selected Papilionidae and Saturniidae, in *Variable Plants and Herbivores in Natural and Managed Systems*, (eds R. F. Denno and M. S. McClure), Academic Press, New York, pp. 373–412.
118. Shear, W. A. (1991) The early development of terrestrial ecosystems. *Nature*, **351**, 283–289.
119. Shields, O. and Reveal, J. L. (1988) Sequential evolution of *Euphilotes* (Lycaenidae, Scolitantidini) on their plant host *Eriogonum* (Polygonaceae, Eriogonidae). *Biol. J. Linn. Soc.*, **33**, 51–93.
120. Simberloff, D. S. (1981) Community effects of introduced species, in *Biotic Crisis in Ecological and Evolutionary Time*, (ed. M. H. Nitecki), Academic Press, New York, pp. 53–81.
121. Simms, E. L. (1992a) The evolution of plant resistance and correlated characters, in *Proceedings of the 8th International Symposium on Insect–Plant Relationships*, (eds S. B. J. Menken, J. H. Visser and P. Harrewijn), Kluwer, Dordrecht, pp. 15–25.
122. Simms, E. L. (1992b) Costs of plant resistance to herbivory, in *Plant Resistance to Herbivores and Pathogens*, (eds R. S. Fritz and E. L. Simms), University of Chicago Press, Chicago, IL, pp. 392–425.
123. Singer, M. C., Thomas, C. D. and Parmesan, C. (1993) Rapid human-induced evolution of insect-host associations. *Nature*, **366**, 681–683.
124. Smiley, J. (1978) Plant chemistry and the evolution of host specificity: new evidence from *Heliconius* and *Passiflora*. *Science*, **201**, 745–747.
125. Southwood, T. R. E. (1977) Habitat, the templet for ecological strategies? *J. Anim. Ecol.*, **46**, 337–365.
126. Southwood, T. R. E. (1988) Tactics, strategies and templets. *Oikos*, **52**, 3–18.
127. Southwood, T. R. E., Brown, V. K. and Reader, P. M. (1986) Leaf palatability, life expectancy and herbivore damage. *Oecologia*, **70**, 544–548.
128. Steffan, A. W. (1964) Problems of evolution and speciation in Adelgidae (Homoptera, Aphidoidea). *Can. Entomol.*, **96**, 155–157.
129. Strong, D. R., Lawton, J. H. and Southwood, T. R. E. (1984) *Insects on Plants. Community Patterns and Mechanisms*, Blackwell, Oxford.
130. Terofal, F. (1965) Zum Problem der Wirtsspezifität bei Pieriden (Lep.). *Mitt. Münch. Entomol. Ges.*, **55**, 1–76.
131. Thompson, J. N. (1987) Symbiont induced speciation. *Biol. J. Linn. Soc.*, **32**, 385–393.
132. Thompson, J. N. (1988) Evolutionary ecology of the relationship between oviposition preference and performance of offspring in phytophagous insects. *Entomol. Exp. Appl.*, **47**, 3–14.
133. Thompson, J. N. (1994) *The Coevolutionary Process*, University of Chicago Press, Chicago, IL.
134. Thompson, J. N. and Burdon, J. J. (1992) Gene-for-gene coevolution between plants and parasites. *Nature*, **360**, 121–125.
135. Van der Meijden, E. (1996) Plant defence, an evolutionary dilemma: contrasting effects of

(specialist and generalist) herbivores and natural enemies. *Entomol. Exp. Appl.*, **80**, 307–310.
136. Van Steenis, C. G. G. J. (1976) Autonomous evolution in plants. Differences in plant and animal evolution. *Garden's Bull.*, **29**, 103–126.
137. Van Valen, L. (1973) A new evolutionary theory. *Evol. Theory*, **1**, 1–30.
138. Vrba, E. S. (1980) Evolution, species and fossils: how does life evolve? *S. Afr. J. Sci.*, **76**, 61–84.
139. Whittaker, R. H. (1977) Evolution of species diversity in land communities. *Evol. Biol.*, **10**, 1–67.
140. Whittaker, J. B. and Feeny, P. P. (1971) Allelochemicals: chemical interactions between species. *Science*, **171**, 757–770.
141. Wiebes, J. T. (1979) Co-evolution of figs and their insect pollinators. *Annu. Rev. Ecol. Syst.*, **10**, 1–12.
142. Wood, T. K. (1993) Diversity in the New World Membracidae. *Annu. Rev. Entomol.*, **38**, 409–435.
143. Zwölfer, H., Frick, K. E. and Andres, L. A. (1971) A study of the host plant relationships of European members of the genus *Larinus* (Col., Curculionidae). *Commonw. Inst. Biol. Contr. Tech. Commun.*, **14**, 97–143.

INSECTS AND FLOWERS: THE BEAUTY OF MUTUALISM

11

11.1	Mutualism	317
11.2	Flower constancy	319
11.2.1	Flower recognition	320
11.2.2	Flower handling	323
11.3	Pollination energetics	324
11.3.1	Distance	325
11.3.2	Accessibility	326
11.3.3	Temperature	326
11.3.4	Food source evaluation	327
11.3.5	Reward strategy	328
11.3.6	Signalling nectar status	330
11.4	Pollinator movement within multiple-flower inflorescences	331
11.5	Competition	331
11.6	Evolution	333
11.7	Nature conservation	337
11.8	Economy	339
11.9	Conclusions	339
11.10	References	339

When C. K. Sprengel, rector of a Lutheran school in Germany, got depressed from his duties, his doctor advised him to study Nature and, to facilitate his recovery, taught his patient some elementary botany. Sprengel then gained an in-depth knowledge of flower morphology, nectar secretion and its function and published in 1793 a treatise [77] under the imaginative title *Das entdeckte Geheimniss der Natur im Bau und in der Befruchtung der Blumen* (*The Secret of Nature revealed in the Structure and Fertilization of Flowers*; Fig. 11.1).

In this landmark book he demonstrated that, although most angiosperm flowers are hermaphroditic, they usually require pollinating insects in order to set seed, and concluded that 'nature does not seem to allow any flower to be fertilized by its own pollen'. Because contemporary botanists considered Sprengel as a non-professional, the work was ignored and remained in oblivion until, after a dormancy of more than 60 years, it came to the attention of Charles Darwin. In his famous book *The Origin of Species* [16] Darwin expresses his approval of Sprengel's conclusions and reports additional experiments confirming the role of insects in pollination. Since then the rich variety of floral forms and shapes, and the ingenious associations with morphological and behavioural traits of insects that enhance the accuracy and efficiency of pollen transfer have stimulated many biologists to study the striking adaptations underlying these intimate relationships. Recent insights into ecological factors, such as cost–benefit analysis, and evolutionary patterns have added to the fascination of the subject.

Approximately two-thirds of all flowering plants are pollinated by insects. This service is not given *gratis*. In return for pollen transfer plants provide food for their partners in the form of nectar and pollen. These are desirable nutriments: nectar may contain 50% sugars and pollen 15–30% proteins and other essential elements. Because the two parties can survive barely or not at all without each other, this is an exemplary case of **mutualism**. Associations from which both partners benefit are widespread, but that between angiosperms and insect pollinators is probably the most spectacular and large-scale example of mutualism in the living world.

Although insects and flowers form an example *par excellence* of mutualism as a prin-

Figure 11.1 Title page of C. K. Sprengel's classic book, which describes the role of insects in pollination.

ciple, the degree of mutualism varies among species, and the interdependence of flowers and pollinators covers a broad spectrum. At the one end of it the partners are highly specialized and the interaction is a question of life and death. Figs, for instance, can only be pollinated by specialized fig wasps, a specialization reaching the extreme, because each fig species is pollinated by its own species of wasp. Female wasps pollinate and lay eggs within the flowers. The offspring develop within the seeds, eclose as adults and, still inside the fruit, mate. The females then fly off to lay their eggs in another fig inflorescence and then die within that fig. The interdependence is absolute. At the other end of the spectrum the relationship is antagonistic [81]. For example, *Ophrys* orchids can be regarded as 'sexual parasites' of their pollinators (see below), which can do very well without these flowers. Other insect–flower relationships are situated somewhere between these two extremes.

11.1 MUTUALISM

Many plant species that have conspicuous, coloured and scented flowers require insect pollination to optimize seed production [8] (Table 11.1).

Birdsfoot trefoil (*Lotus corniculatus*), for instance, produces practically no seeds in the absence of pollinators. Just one single honeybee visit results in the production of several seeds per flower, but to achieve maximum pollination as many as 12–25 visits are necessary [58]. The flowers of yellow bog saxifrage (*Saxifraga hirculus*) also require to be visited many times to ensure optimal seed setting. These flowers produce after about 200 visits from pollinators, during which roughly 350 pollen grains are deposited on their stigmas, an average of 30 seeds per flower [60]. Thus, flowers usually have to be visited more than once to maximize and to optimize seed setting. Different visitors bring pollen from different fathers and the risks of pollinators bringing incompatible pollen are compensated. In agricultural and horticultural crops fertilization and seed production is often suboptimal because there are insufficient numbers of natural pollinators. In that case yields can be improved considerably by moving honeybee colonies into the crop area [20] (Fig. 11.2).

From the insects' point of view pollen and nectar constitute important food sources. Apoidea (bees and bumblebees) even receive all their nourishment from these two flower products and they are well equipped to collect relatively large quantities of them. Bees are

Table 11.1 Effects of excluding insect visitors (primarily bumblebees) on the seed production of four ericaceous plant species in a bog ecosystem. The percentage of fruits producing seeds was compared on shoots that were enclosed in mesh bags and unenclosed shoots. (Source: modified from Reader, 1975)

Plant species	Seed production (%)	
	Enclosed	Unenclosed
Wild rosemary (*Andromeda glaucophylla*)	0.7	33.6
Swamp laurel (*Kalmia polifolia*)	0	55.6
Labrador tea (*Ledum groenlandicum*)	1.0	96.2
Large cranberry (*Vaccinium macrocarpon*)	4.0	55.7

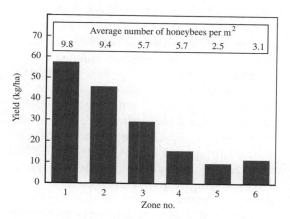

Figure 11.2 Seed yields of red clover (*Trifolium pratense*) decrease as distance from honeybee colonies on the edge of the field increases. Seed production was measured in six zones, each 122 m wide and parallel to the field edge. Zone 1: 0–122 m from the bee hives; Zone 6: 610–732 m. Figures represent number of honeybees observed in the various zones per unit of time. (Source: data from Braun *et al.*, 1953.)

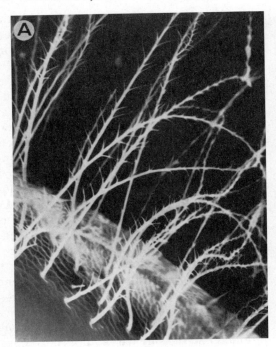 

Figure 11.3 (A) Plumose hairs covering the body surface of honeybees have teeth and hooks, which assist in collecting pollen. Magnification 160×. (B) Hairy coat of bumblebee head with some pollen grains sticking to the spiny hairs. Magnification 740×. (Source: reproduced from Barth, F. G.; *Insects and Flowers*. Copyright © 1985 by Princeton University Press.)

covered with a dense coat of feathery hairs, which, through the presence of small hooks, effectively catch and hold pollen grains when the insect touches the anthers of a flower. The stickiness of pollen from insect-pollinated flowers facilitates their adherence to an insect vector (Fig. 11.3). (Pollen from wind-pollinated plant species lacks the oily 'pollenkit' cover and is not sticky.) In addition to its pollen-carrying capacity the insulating fur allows its owners to maintain high body temperatures and thus to be active at low air temperatures [34]. During flight the bee rakes with its legs all the pollen from its hairy body surface and collects it in pollen baskets situated on the outer side of the tibia of both hind legs [74] (Fig. 11.4).

With this device a honeybee worker may carry a pollen load of as much as 10–20 mg home to the hive. A honeybee colony, consisting of approximately 10 000–50 000 insects, consumes approximately 20 kg of pollen and 60 kg of honey per year. The pollen provides the protein for growth and reproduction. To rear one honeybee about 125 mg is required, an amount equalling the body weight of the adult [72]. Nectar, in composition bearing some resemblance to phloem sap, contains anything from 10–70% sugars by weight. Other compounds, such as lipids, free amino acids, minerals and aromatic substances, are also present, albeit in small quantities. Whereas sugars are a most valuable reward for their energy content, the occurrence of amino acids in nectars is attractive to those pollinators that lack alternative resources [1].

Although plant and pollinator depend fully on each other, there is at the same time, as in any mutualism, an intrinsic conflict between the parties, in that each is under selection for

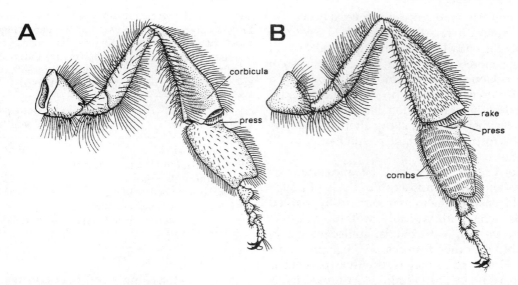

Figure 11.4 Hindleg of a worker honeybee. **(A)** Outer surface showing the pollen basket (corbicula), consisting of a bare concavity fringed with stiff hairs. The 'press' forces the pollen into the basket. **(B)** Inner surface with 'combs' and 'rakes', which manipulate pollen into the press before it is pushed into the lower end of the basket. (Source: redrawn from Snodgrass, 1956.)

increased exploitation of the other. Plants need to receive as many conspecific pollen grains on their stigmas as possible and the reciprocal transfer of their pollen to flower stigmas of other conspecifics. The ideal vector would, during each visit, contact anthers and stigmas, move rapidly among plants and search exclusively for conspecific flowers, even when other flowering plants abound. To force its pollinator to visit many flowers, selection will favour the secretion of a sufficient amount of nectar to be attractive to bees and reward them for taking the trouble to make the visit, but not so much that pollinators need to visit only a few flowers per trip to imbibe a full nectar load and go home. From the plant's perspective a harried, hungry and yet plant-species-constant pollinator is ideal. Insects, on the other hand, according to optimal foraging theory, will try to collect as much food as possible while minimizing energy and time expenditure. This means that flowers with copious nectar flow will be preferred and that it may be more efficient to visit flowers of different species during a foraging trip. This conflict between the interests of plants and their pollinators must have been a major force in shaping present-day plant–pollinator relationships [19, 43].

11.2 FLOWER CONSTANCY

Individual honeybees often restrict their visits to flowers of a single species and neglect rewarding flowers of alternative plant species. The tendency to specialize has been referred to as **flower constancy**. The phenomenon of flower constancy is of crucial importance to pollination ecology and evolution and thus deserves special attention. This type of specialization implicates learning processes based on flower recognition from a distance and on acquiring the skill to collect pollen and nectar from flowers of different architecture. The advantage of flower constancy behaviour has been premised on a limited ability to learn or remember how to recognize several flowers at once or how to handle different food resources most efficiently [86]. Accordingly, a bee would forage more efficiently if its sensory system and behaviour were temporarily fixed in a particular way. So far, no experimental support for this assumption has been obtained [23]. Neither is a satisfactory alternative explanation for this important phenomenon available.

Flower constancy can be measured by examining the composition of loads of pollen on the basis of their characteristically marked walls, which are typically sculptured, punctured, crossed with bonds, spined or recognizable by other features of the grain exine [57]. Flower constancy is usually expressed as the percentage of individuals with pure loads at the end of a foraging trip [15]. Many bee species show high degrees of flower constancy, as shown in Table 11.2.

They are even more constant than the figures in this table suggest, because the definition of a pure load is a strict one. 'Mixed' pollen loads often contain only very small amounts of heterospecific pollen. For instance, of the 19% mixed loads of *Apis mellifera* in Table 11.2, every one was 95–99% pure [30].

The duration of a period of flower constancy may vary considerably. Often bees keep to one flower species only during a single trip. Other individuals show longer periods of fidelity and visit the same kind of flowers for several hours or days. Different workers of a honeybee colony may show constancy to different flower species, and different colonies as a whole may also be specialized on different flowers. Absolute flower constancy would be counterproductive and prevent insects from discovering more rewarding resources. Thus solitary bees constantly check other flower species to assess whether more rewarding species are available, and consequently show lower degrees of flower constancy than, for example, honeybees. In the latter case efficiency is increased by 'scouts', which constantly monitor, sample and pool information about the best food sources available and, by employing their highly developed communication system, 'instruct' the 'recruits' on which sources to visit.

From both pollen load analysis and direct observations in the field it appears that not all kinds of pollinators show the same degree of faithfulness to one flower species. Although social bees are superior to other groups, a tendency to visit successive flowers of the same species has also been observed outside the Apoidea, for instance, in some butterflies, and to a lesser extent hoverflies (Syrphidae) [28, 51, 89].

Table 11.2 Flower constancy of some Apoidea. Flower constancy is expressed as the percentage of individuals with pollen from one plant species only. The figures do not include the (many) individuals which carry only a small fraction of different pollen species. (Source: modified from Grant, 1950)

Genus	Pure pollen load (%)
Apis (honeybee)	81
Megachile (leafcutter bee, solitary)	65
Bombus (bumblebee)	55
Halictus (sweet bee, solitary)	81
Andrena (mining bee, solitary)	54
Anthophora (mining bee, solitary)	20

11.2.1 FLOWER RECOGNITION

Bees can rapidly associate several flower characteristics with food reward. Odours and colours especially are easily remembered when offered in combination with a reward such as sugar solution. A floral odour can be learned in one trial with a reliability of 93–100%, but conditioning to colours takes four to six trials (Fig. 11.5). Odours of pure chemicals are usually more difficult to remember than floral scents (Fig. 11.6).

Shapes and patterns appear to be still more

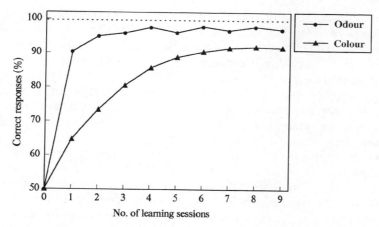

Figure 11.5 Learning curves for odour and colour in honeybees show that a typical floral scent is more rapidly learned than an average colour and that the accuracy of odour memory is higher. (Source: redrawn from Gould, 1993.)

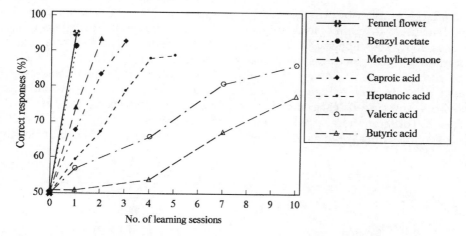

Figure 11.6 Odour learning curves in honeybees for a flower fragrance (fennel) and various pure chemicals. (Source: redrawn from Kriston, 1971.)

difficult to learn and require 10–30 trials to reach approximately the level of accuracy equivalent to that of colour memory [26]. Interestingly, more complex shapes can be learned faster than simple ones, due to an innate preference for shapes with high figure intensities, i.e. figures with high ratios between the contour length and the area enclosed. Attractiveness to visual cues can be further increased by the presence of nectar guides, adding once more to a flower's visual complexity.

A high degree of flower constancy requires not only the ability to quickly learn a rewarding flower species but also the capacity to identify conspecific flowers rapidly on the basis of characteristics sufficiently specific to minimize the chance of error. The rich menu of volatiles produced by flowers plays a prominent role in that specificity. Common constituents of their aromas are the monoterpenes and sesquiterpenes, but volatile aromatic phenols and simple alcohols, ketones and esters may also be present. Generally, floral volatiles are bouquets of at least a few and often many components, although the blend is often dominated by one or a few main components [17, 45, 80]. Thus chromatographic analyses showed the scent of sunflowers to be a mixture of as many as 144 constituents. Not less than 28 of them are relevant for constituting 'sunflower odour' as perceived by honeybees, indicating a finely tuned olfactory system in these highly adapted insects [62].

Given the large number of possible combinations of flower odour components and the fact that bees show the capacity to discriminate thousands of odour mixtures, they must be able to recognize many flowers by their scents only. Experimental evidence showed that honeybees can indeed learn and distinguish at least 700 different floral aromas. Floral scents, however, not only serve as identifiers, assisting the harried bee to recognize the flower species, but also enable pollinators after landing to forage efficiently. For that reason flowers may show spatial patterning of fragrance emission within a flower, both in the kind and amount of volatiles produced, forming an odorous nectar guide (Fig. 11.7).

Such an odour trail, in concert with tactile and gustatory stimuli, helps an experienced insect to rapidly find the pollen or nectaries [17, 18]. The observation that pollen produces characteristic volatiles, often quite different from the scents of flowers, is seen as evidence that pollen is deliberately provided as a reward to insect pollinators. This is supported by the fact that bees can discriminate between plant species on the basis of pollen odour.

Colour is another component of floral advertisement and one of the most important signals by which a pollinator locates, recognizes and discriminates between flowers at a greater distance. The selective spectral reflection of flowers and the colour vision systems of pollinators have developed together in a mutual relationship. To optimally exploit the flower constancy behaviour of Hymenoptera and to prevent 'mistakes' by its pollinators, a plant should have floral colours as different as

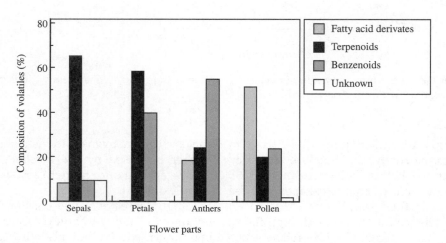

Figure 11.7 Percentage of four classes of volatile compound produced by different parts of the flowers of *Rosa rugosa*. (Source: data from Dobson *et al.*, 1990.)

Plate 1 Examples of flowers turning colour with age. Changes may involve different flower parts, including the corona, corolla throat, nectar guide, specialized petal, filaments, ovary or whole flower. **(A)** *Androsace lanuginosa* (Primulaceae): eye changes from yellow through orange to red. **(B)** *Myosotis* sp. (Boraginaceae): corona changes from yellow to white. **(C)** *Lantana camara* (Verbenaceae): whole flower changes from yellow to red. **(D)** *Lupinus nanus* (Leguminosae): banner petal spot changes from white to purple. **(E)** *Raphiolepis umbellata* (Rosaceae): filaments change from white to red. **(F)** *Ribes odoratum* (Saxifragaceae): petals change from yellow to red. (Source: reproduced, with permission, from *Nature*; M. R. Weiss, Floral colour changes as cues for pollinators, vol. 354: 227–229. Copyright 1991 Macmillan Magazines Limited.)

possible from sympatric flowers. Flower colours of different angiosperms do show sharp steps in their spectra at precisely those wavelengths where the pollinators are most sensitive to spectral differences. This indicates an evolutionary tuning of flower colours to the sensory system of bee pollinators or, alternatively, the result of a coevolutionary process [9].

11.2.2 FLOWER HANDLING

Naive bees, after landing on a flower, show an innate probing response but they must learn how to exploit flowers of increased complexity efficiently. Since learning involves investment of time and energy it befits a bee, once a flower species has been successfully probed, to continue to forage from it. Learning how to manipulate complex flower types is no easy task [48]. Different types of flower, with their nectaries often hidden at very specific places, require different handling techniques and bees have to learn such things as where exactly to alight, where exactly the nectaries are located and how to reach them as fast as possible (Fig. 11.8).

Food finding is relatively simple on the flat-topped inflorescences of umbellifers where bees, while rapidly moving around, collect pollen from the tiny flowers by pressing their bodies to the surface. More advanced procedures are needed on more complex flowers, such as *Chelone glabra* (Scrophulariaceae), where the petals must be pried apart to obtain access to the nectar, or monkshood (*Aconitum*

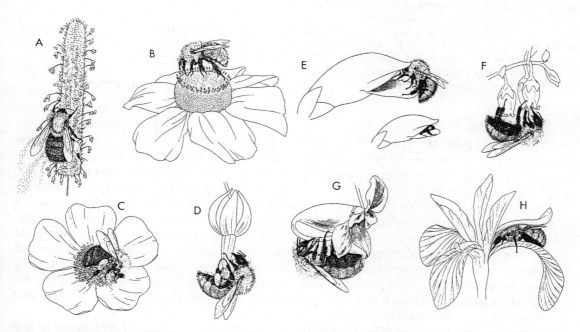

Figure 11.8 Handling of different kinds of flower by bumblebees collecting nectar or pollen. **(A)** Walking up inflorescence of grass (*Phleum* sp.) collecting pollen. **(B)** Collecting nectar and possibly pollen from a composite flower. **(C)** Grasping and vibrating groups of anthers in *Rosa* sp. during pollen collection. **(D)** Holding *Solanum dulcamara* blossom with legs and mandibles while shaking pollen from the tubular anthers by vibrating the flower. **(E)** Entering *Chelone* blossoms. **(F)** Collecting nectar from *Vaccinium* blossoms. **(G)** 'Robbing' nectar *via* a hole bitten in the spur of *Impatiens* sp. **(H)** Iris blossom being visited for nectar. (Source: reproduced from Heinrich, 1976, with permission.)

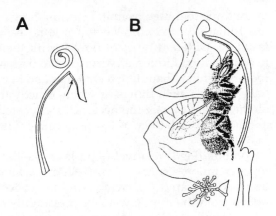

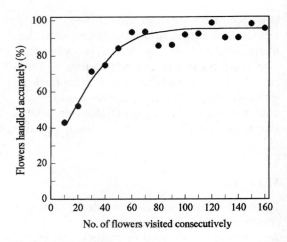

Figure 11.9 Monkshood flowers contain two vertical nectar petals, partly shaped into a tube, with nectaries located at the very end. **(A)** Nectar petal of *Aconitum vulparia* (arrow indicates tube entrance). **(B)** Flower of *Aconitum variegatum*, with worker bumblebee inserting its tongue into nectar petal. The bee, after entering at the bottom, must pass over the anthers to be able to probe into the tips of the two nectar petals. Monkshood species occur only in parts of the world where bumblebees occur. (Source: **(A)** reproduced from Knoll, 1956, with permission; **(B)** reproduced from Laverty and Plowright, 1988, with permission.)

Figure 11.10 Improvement in handling success of naive bumblebees after one to 150 contacts with *Impatiens biflora* flowers. (Source: redrawn from Heinrich, 1979a.)

spp.) flowers, where the greedy bee, after entering at the bottom, must pass over the anthers to reach the nectaries concealed deeply inside (Fig. 11.9).

Flower handling techniques learned on one plant species presumably interfere with previously learned techniques for other plant species because of limited neural capacity. Bumblebees with no experience in handling flowers with complex morphology and limited access to the nectar, such as jewelweed (*Impatiens biflora*), often could not find the rich nectar content. It took about 60–100 flower encounters before they had fully developed the skill to extract the nectar [33] (Fig. 11.10). The strategy of flower constancy must surely increase foraging efficiency, because the insect, once it knows where the nectaries of a particular type of flower are located and how to reach them with the least effort, saves energy and time.

11.3 POLLINATION ENERGETICS

Mutualistic relationships between two groups of organisms benefit both partners, as manifested by increased fitness. According to the optimization theory organisms try to maximize their survival chances and reproductive success by balancing costs against benefits for each activity or function. Application of cost–benefit analyses to insect–flower mutualism has proved useful in understanding the degree of mutual dependency. The optimal foraging theory holds that foraging strategies may involve decisions that maximize the net rate of food intake (i.e. net caloric gain per unit of time).

Pollination energetics can be studied in two ways. The first is a conceptual analysis based on models, which account for behavioural and physiological mechanisms that underlie foraging. The second approach aims to test in the field the predictions of the models on the energy balance of foragers. Factors that may be relevant for a cost–benefit analysis of a foraging insect are manifold, but include as a minimum the distance to the food source, its accessibility, the amount and quality of the food, and the ambient temperature.

11.3.1 DISTANCE

Honeybee foraging normally extends over a vast area around a colony's nest. A detailed study of one colony in a deciduous forest showed that the most common forage patch distance was 600–800 m, but many individuals flew out several kilometres from the nest. Because 95% of the colony's foraging activity occurred within a radius of 6 km, the food source area of this colony could be set at more than 100 km² [72]. Bumblebees, likewise, often forage at distances of several kilometres from their nests.

Food collection requires an enormous expenditure of energy. A foraging bumblebee weighing 0.5 g spends as much as 600 J per hour, which is equivalent to the energy bound in 40 mg of glucose! Flight activity accounts for by far the greatest share of energy consumption. Hovering in front of a flower, as larger insects often do to extract nectar, is particularly costly. To economize on energy expenditure bees will only travel to distant food sources if the reward makes the trip profitable. The higher the sugar concentration at an experimental feeding station the further honeybees will forage on it. The decision to collect food at a distant source takes into account not only flying energy to get there, but also loss of travel time. Therefore, the relationship between distance and minimum food concentration to make the trip worthwhile is not linear but takes an exponential form (Fig. 11.11).

Flowers at 3 km from the colony should provide at least 3.4 times more nectar than flowers near the hive to make foraging on them attractive. In spite of an appreciable energy consumption bees are highly efficient flyers. The travel costs of a return flight to a

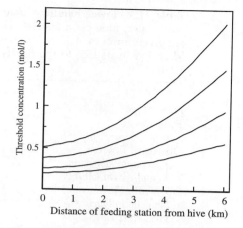

Figure 11.11 Honeybees finding a sugar solution will recruit other bees in the hive only when the feeding station presents a sugar solution above a threshold concentration. The threshold increases with distance. Threshold concentrations vary with environmental variables such as the presence of alternative food sources and weather conditions. The responses on four different days are represented by four different curves, reflecting different environmental conditions. (Source: data from Boch, 1956.)

food source located 4.5 km from a bee's nest are the equivalent of 10% of the nectar yield.

In nature bees are confronted with variability in nectar content among flowers. Do they notice it and if so, does it bother them? In an elegant experiment by Real and his coworkers insects were allowed to forage on a patch of artificial flowers of two distinct colours. All blue-coloured flowers contained the same reward. The yellow-coloured flowers contained variable amounts of sugar water, but on average they had the same amount as the blue ones. Bumblebees and paper wasps preferred the blue flowers with the lower variance in reward. When, however, the mean sugar content of the high-variance flowers was raised, the insects preferred to forage on this more risky type. Apparently the bees' foraging strategy includes a certain degree of risk avoidance, which can be offset by increased gain [67]. Thus, distance is one parameter on

Table 11.3 Flower handling time and foraging success of bumblebee workers on nine plant species with different floral complexity. Figures are given for bees visiting a flower species for the first time (1) or after 54 earlier visits (55) to the flower species. (Source: data from Laverty, 1994)

Flower type Plant species	Handling time (s)		Foraging success (%)
	1	55	1
Open-cup flowers			
Apocynum sibiricum	5.5	0.4	100
A. androsaemifolium	14.1	0.3	100
Open-tube flowers			
Prunella vulgaris	13.9	0.1	100
Vicia cracca	18.1	0.2	100
Impatiens capensis	20.4	1.7	70
Closed-tube flowers			
Gentiana andrewsii	44.4	6.5	45
Chelone glabra	196.6	8.1	40
Monkshood flowers			
Aconitum henryi	134.7	13.6	35
A. napellus	153.5	3	29

which foraging decisions are based, predictability is another, the caloric worth of nectar rewards is a third.

11.3.2 ACCESSIBILITY

Flower morphology affects the time needed to find and collect the nectar or pollen. Shallow, open-cup flowers require little handling skill, since the nectar is accessible from any position on the flower. Complex flowers, such as monkshood (*Aconitum* spp.; Fig. 11.9), demand more complex handling methods from their visitors because locating the reward is more difficult. It also takes more time. The flower compensates the pollinator for increased investment in time and effort by providing a rich nectar reward. The number of trials and the time needed for naive bumblebees to learn flower handling increases with floral complexity [48] (Table 11.3). The combination of copious rewards with a floral morphology requiring high learning capabilities of its pollinators promotes flower constancy.

11.3.3 TEMPERATURE

Bumblebees can be seen collecting food at temperatures near freezing or, in the Arctic, even below 0°C. Honeybees become active, depending on the season, between 10° and 16°C. Bees can forage at cool temperatures because they are **endothermic** and fly with a minimum thorax temperature of 30°C. They maintain a high body temperature by the heat produced from their flight metabolism and, when not in flight, by shivering their flight muscles with the wings uncoupled [34, 85]. Foraging at low temperatures is, however, expensive in energy. Food-collecting bumblebees at 5°C spend two or three times more energy than at 26°C to keep their thorax temperature at 30°C or higher (Fig. 11.12).

An elevated temperature is a prerequisite for normal functioning of the flight muscles. To maintain high body temperatures bees possess an unusually high activity of the enzyme fructose-1,6-diphosphatase, which enables heat generation by ATP hydrolysis. The

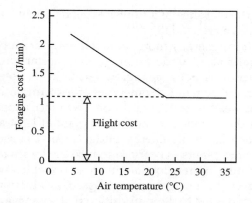

Figure 11.12 Calculated costs in relation to air temperatures for a bumblebee that regulates its thorax temperature at 30°C and spends half its time in flight and half on handling flowers. All costs above the dashed line are contributed to thermoregulation. (Source: redrawn from Heinrich, 1979b.)

activity of this enzyme is in some bumblebees about 40 times that in the honeybee, allowing them to forage at substantially lower temperatures than honeybees.

Investment in extra heat production under cold weather conditions must of course be compensated by high energy intake. Bumblebees therefore can be seen to forage in cool weather on rhododendron blossoms, which yield profitable amounts of nectar, while neglecting wild cherry (*Prunus avium*) and lambkill (*Kalmia angustifolia*) blooms, since these flowers produce too little nectar to break even in energy terms under these conditions.

Once the energy requirements of a forager at different temperatures are known, as well as the fuel needed to fly per unit of time, the extractable energy content per flower and the average distance between flowers, predictions can be made as to which flower species will be visited at various temperatures and which will not. B. Heinrich [34], in a fascinating book entitled *Bumblebee Economics*, shows on the basis of ingenious experiments that bumblebees employ a thermal strategy that accounts for many variables and thereby ensures maintenance of a positive energy balance.

Flowers that open early in the morning are visited mainly by large insects, which can regulate their body temperature. As temperatures rise small pollinators become active. In the early morning the blossoms of *Arctostaphylos otayensis* (Ericaceae) were found each to contain 6.3 J of sugar. Bumblebees when foraging at 2°C need about 3.4 J/min in order to make an energy gain on these flowers even while they are foraging at near-frosty temperatures. At noon each flower contained only 1.3 J of sugar. By that time bumblebees had lost interest and the flowers were visited predominantly by small-insect species [35].

To make flower visitation profitable at cool weather conditions nectar production should be relatively profuse or the flowers should grow closer together than under high temperature conditions so that they can be visited in rapid succession. The tendency for spring flowers to grow in clumps may be a strategy on the part of the plant relevant to pollination success. Also, the fact that plants growing further north secrete more nectar than conspecifics at lower latitudes and similar trends on elevation gradients suggests an adaptation to greater energy needs of their pollinators [35].

Although our knowledge about the relationship between nectar provision and the energy requirements of pollinators is far from complete, there is sufficient evidence to conclude that supply and demand are finely tuned to each other.

11.3.4 FOOD SOURCE EVALUATION

Honeybees returning from a foraging trip communicate details about location and quality of the food source to other members of the colony by means of the famous bee dance. Detailed analysis of this ritualized act and correlation of its subtle modifications with various manipulations of food sources allowed Karl von Frisch [22] and coworkers to determine which factors the bees use to calculate the profitability of their foraging activity.

They found that a whole range of food source characteristics are taken into account in the nature and duration of the dance, such as distance, nectar quality (i.e. sugar concentration), viscosity, ease of obtaining the nectar, uniformity of flow, odour, time of day and weather conditions. In addition to the direct costs, i.e. flight and handling energy and time, bees making foraging decisions probably also reckon indirect costs such as risk of predation and body wear and tear. It is unclear how time, for instance handling time, is measured, but some evidence suggests that bees measuring foraging gains and costs integrate their time budget in some way or other into their energy budget, and hence estimate time in terms of energy units [84]. Thus studies on pollination energetics include factors such as:

1. pollinator foraging behaviour (distance to foraging area, interplant flight distance, departure decisions, movement patterns, speed);
2. reward type and quantity (pollen, nectar composition, caloric value and spatiotemporal distribution);
3. flower handling costs (pollinator energy and time expenditure).

Figure 11.13 summarizes these and other factors affecting cost–benefit analyses.

11.3.5 REWARD STRATEGY

To promote outcrossing plants need visitors, such as insects and some vertebrates, which are rewarded for their service. Flowers must provide sufficient nectar to attract foragers but they must limit this reward so that pollinators will go on and visit other plants of the same species [43]. Nectar secretion per flower or per plant is carefully optimized and adapted to the season, time of day (Fig. 11.14) and the kind of pollinators the plant prefers to employ. A 100 mg bumblebee after landing on a flower may expend about 0.3 J/min, while a 3 g sphinx moth imbibing nectar while hovering at the entrance of a flower expends about 140 times as much [35].

It is generally thought (albeit without much quantitative underpinning) that nectar production is not a negligible factor in terms of carbon demand. (Sometimes, for example in hot weather, nectar secretion may place a load on the water economy of the plant at least as much as on its energy budget.) As a consequence plants may try to reduce the losses by limiting nectar flow as much as possible and direct the energy saved to seed production. This has led to the supposition that some plant individuals may attempt to cheat on their conspecifics and save energy by secreting little or no nectar at all. Bees that are conditioned to a particular flower species will continue to visit many more without receiving a single reward. This kind of deception is an example of **automimicry** [14]. The interactions between bees, nectar-producing plants and cheating automimics can be understood in more formal terms by employing the game theory [54], as has been done in a study of the nectar flow of individual flowers of *Cerinthe major* (Boraginaceae). About 25% produced copious amounts but the remaining flowers secreted only small quantities. The observed ratio between high secretors and low secretors closely fitted the value predicted on the basis of flower type, amounts of nectar produced and mean discrimination and handling times for a particular forager [24]. Other reports on substantial differences in nectar productivity between plants also indicate that plants can play various strategies and vary the proportion of cheating flowers they make.

Amino acids are found in floral nectars of primitive angiosperms, albeit in relatively small amounts. In some other plant taxa they occur at significant quantities. Their concentrations vary from 0.37–4.69 µmol/ml in herbaceous species. The discovery of a correlation between pollinator type and amino acid concentration in nectars led to the idea that

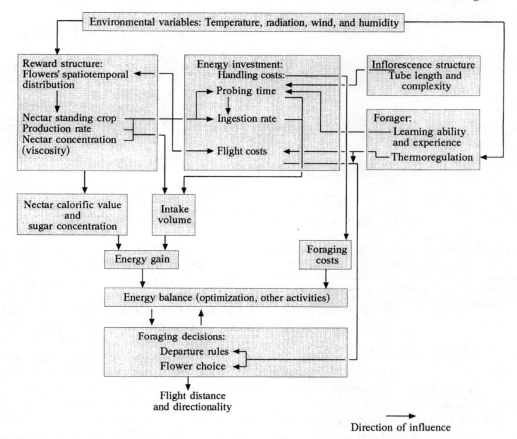

Figure 11.13 Relationships between the components of foraging behaviour and pollination energetics. (Source: redrawn from Dafni, 1992.)

their presence reflects an adaptation to pollinators that have no alternative resources, such as butterflies and moths. Thus tubular flowers adapted to pollination by lepidopterans contain higher levels than flowers that flies, for instance, feed on [2]. Experimental work on butterflies showed that *Pieris rapae* females prefer nectars containing amino acids over sugar-only nectars [1].

Some plant species have flowers with two kinds of stamen, some with reproductive anthers, producing normal pollen, and some with 'reward anthers' (Fig. 11.15).

The latter type is often more conspicuous and brightly coloured to attract potential pollinators. They may produce limited quantities of highly nutritious but sterile pollen. In other cases only some kind of milky juice is secreted. The reward anthers clearly serve to mimic normal anthers and attract pollinators by deceit. When manoeuvring to forage on them the insect automatically takes care of pollination with the fertile pollen. Presumably the development of reward anthers is advantageous to the plant in terms of production costs, but this has still to be proved.

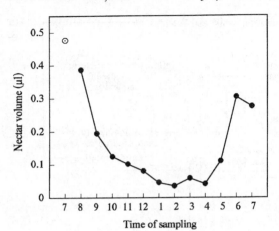

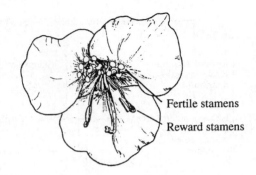

Figure 11.15 Flower of *Commelina tuberosa* with two stamen types. (Source: reproduced from Hess, 1983, with permission.)

Figure 11.14 Mean volumes of nectar secreted per hour by *Cerinthe major* blossoms during one day in the absence of insects. The first reading represents the overnight accumulation of nectar. (Source: redrawn from Gilbert *et al.*, 1991.)

11.3.6 SIGNALLING NECTAR STATUS

Foraging efficiency would be raised if bees, rather than moving randomly between flowers, avoided unrewarding flowers on one hand and recognized rich food sources on the other. Such behaviour has been seen in higher hymenopterans, which assess while still airborne the reward state of a blossom by the smell of 'footprints', volatile pheromones left by previous visitors. Honeybees and bumblebees collecting nectar label their empties by leaving odour traces. The scent marks are of short duration, i.e. in the order of minutes, and avoidance of visited flowers by conspecifics is easily observed [25, 70]. Bees can also deposit a message indicating that the food source is worthwhile to visit. This pheromone is also secreted by the tarsi and was found in bumblebees to consist of a complex mixture of alkanes and alkenes [71]. By recognizing the combination of pheromones left behind by themselves and other bees they can more easily select the least harvested, most productive flowers.

On the plant side something can be gained too by advertising the developmental state of its flowers. Pollination efficiency would increase if the plant signalled to its pollinators which flowers had already been pollinated by previous visitors. Many plants give precisely such a warning by changing flower colour, scent production and even geometric outline [78]. Thus the orange flowers of *Lotus scoparius* (Papilionaceae) turn yellow after pollination, whereas the blossoms of some other species change in ultraviolet reflection following pollination [40]. Flowers also change with age. Floral colours of members from at least 77 diverse angiosperm families undergo dramatic, often localized, changes in senescing blossoms [88] (Plate 1).

Sexual viability and nectar secretion of postchange flowers is low, because they lack pollen and appear non-receptive. Why do plants then keep flowers that have lost their reproductive capacity? It seems likely that retention of older flowers increases a plant's attractiveness to pollinators from a distance. At close range, however, the bees easily learn to discriminate floral colour phases and avoid postchange flowers [87, 89]. Thus, by changing their colour in response to pollination or concomitant with aging, flowers continue to serve the plant by attracting pollinators even after their time is over.

Alterations in the production of fragrances

following pollination have been little studied [78] but some cases involving different families have been reported in the literature [69, 82].

11.4 POLLINATOR MOVEMENT WITHIN MULTIPLE-FLOWER INFLORESCENCES

To facilitate flower recognition and thus increase the profit from insect visits many plant species have their individual blooms clustered into an inflorescence, thereby making a far more conspicuous display than single flowers might achieve. Vertically elongated inflorescences such as those of foxglove (Fig. 11.16), monkshood, willow herb and lupin are of special interest, since this spatial arrangement adds an extra dimension to the pollination economy of monoecious plants.

Within vertical inflorescences bees and flies follow a foraging route that typically starts near the bottom and runs upward. The lower flowers provide more nectar than the upper ones. In agreement with the optimal foraging theory the insects start where the largest nectar source is, i.e. at the base, and since the distance to an upper flower is small, lower nectar quantities are now acceptable. Moreover, sugar concentration is here often somewhat higher than in the lower flowers. The upward direction of pollinator movements suits the plant very well, since the flowers of these species (as in many other plants) are protandrous, i.e. the anthers mature some days before the stigmas. Each day a new flower opens at the top of the inflorescence, replacing a senescent flower at the bottom. The older (lower) flowers are functionally female, with receptive stigmas, whereas the top flowers are functionally males, with mature anthers but still immature stigmas. The foraging behaviour of starting at the bottom and visiting the pollen-containing upper flowers only before leaving the inflorescence obviously promotes cross-pollination and minimizes the chance of self-pollination. The plant's blooming strategy is nicely adapted to pollinator behaviour [34, 53].

11.5 COMPETITION

Since in insect–flower relationships some basic conditions of existence affecting both partners are involved, i.e. food availability (to pollinators) and reproductive success (on the plant side), competition for available resources is likely to occur on both sides. Since plant reproductive success is frequently limited by pollinator activity [8] species will compete for effective pollen carriers, whereas insects will be under selective pressure to exploit their food sources more efficiently than competing species.

Plants can compete for pollinators by producing more flowers. At the same time this increases the risks of geitonogamy (pollination between flowers on the same plant) and sets a limit to pollen export. There are several solutions to this dilemma [43, 75]. Plant species can escape competition by utilization of different pollinator species or guilds, e.g. by differences in floral morphology or by flowering at different times. Adaptation to different pollinator

Figure 11.16 The vertical inflorescence of foxglove (*Digitalis lutea*).

species, as exemplified by high rewards early in the morning so that bumblebees are attracted, or developing long corolla tubes so that only long-tongued insects can reach the nectaries [65], is undoubtedly a widespread phenomenon. Since flowering time is under genetic control it has been suggested that plant species that have a large pollinator overlap avoid competition by blooming at different times. Obviously this resource partitioning and character displacement is mutually beneficial to plants and pollinators. The timing strategy has been observed in some relatively simple plant communities. Thus different plant species in meadow communities in the Rocky Mountains show a regular (i.e. non-random) temporal segregation of blooming periods, thereby reducing competition for bumblebee pollinators [63]. Likewise, most insect-pollinated bog plants use the same species of bumblebee and bloom at different times. Those species that depend wholly or to a large extent on bumblebee pollination show a sharper separation in blooming periods than species that are less dependent on bumblebees [34].

In early spring few plants flower and relatively high numbers of pollinators compete for food. Advancing blooming time may therefore, in addition to the advantage of a longer seed growth period, be advantageous to a plant through higher pollination success. The risks of freezing, however, may act in the opposite direction.

In late spring and early summer there is a proliferation of blooms. This is nicely reflected in the seasonal changes in the threshold concentration of sucrose solutions that elicit recruitment dances by foraging honeybees. In early summer when nectar flow is abundant only high sucrose levels elicit recruitment. In midsummer natural food sources are not as rich any more and at the same time have to be shared with many other insects. Now food competition among pollinators is intense. As a result recruitment dances are elicited even by sugar concentrations 16 times lower than in late spring (Fig. 11.17). Apparently, bees adjust their acceptance level by force of circumstances and compare food source quality to generally available food.

In springtime, some plant species, such as willows (*Salix* spp.) and dandelion (*Taraxacum officinalis*), produce almost endless supplies of nectar and pollen. Since many insects feed on such copious food sources, it has been suggested that these plants use a 'dumping' strategy to attract many insects. As a result, seed

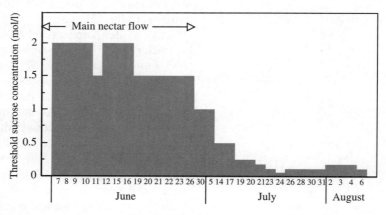

Figure 11.17 Threshold sugar concentrations required to elicit recruitment behaviour in honeybee foragers during early and midsummer. (Source: data from Lindauer, 1948.)

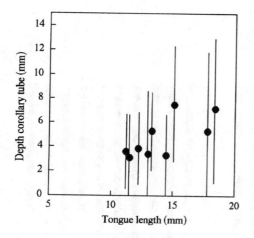

Figure 11.18 Proboscis length (●) *versus* range of corolla tube depths visited by queens of nine different bumblebee species. (Source: redrawn from Ranta and Lundberg, 1980.)

setting of competing plant species is reduced and the position of the food supplier is strengthened. Thus dandelions in an apple orchard can attract pollinators away from apple trees. As a note of irony it may be mentioned that dandelions are largely apomictic, i.e. seed development occurs without fertilization. The nonetheless bountiful nectar production must therefore have another function than promotion of reproduction. However, hard evidence for this cunning type of antagonism is lacking.

An interesting example of competitive interactions between pollinators relates to two sympatric bumblebee species foraging on two different flower species. Each bee species had an apparent preference for one flower species. However, when all or most of the individuals of either bee species were removed from a local patch, individuals of the remaining bee species would, in addition to their already adopted flower species, start to visit the vacant flower species more frequently [37].

In conclusion, there is strong evidence that competition for pollinators occurs between plants and that the evolutionary outcome of such interactions is resource partitioning

[65] (Fig.11.18) and character displacement. Plants may minimize competition for pollinators by adapting their phenology to the periods during which the chances of fertilization are optimal, as well as evolving structural and physiological characteristics that reduce the spectrum of pollinators they use, but ensure adequate resources to those they use [42, 59].

11.6 EVOLUTION

Angiosperms are by far the largest present-day group of land plants. They are characterized by a spectacular diversity in flower size, shape and colour. This conspicuous variation induced Linnaeus to construct a classification of flowering plants in his *Systema Naturae* based on their sexual organs. The extraordinary evolutionary success of angiosperms results from adaptations of their reproductive organs to pollination by insects. This pollination system, as deduced from paleobotany and systematics, is an ancient mechanism. The first angiosperms and their sister clades were probably already entomophilous, although some of the early angiosperms may have used both insects and the wind as pollen vectors [10, 13].

The dramatic radiation of angiosperms toward the end of the early Cretaceous (between 130 and 90 million years ago), and their takeover of the ancient Mesozoic plant communities of ferns, horsetails and gymnosperms during the late Cretaceous and early Tertiary has often been linked to a simultaneous diversification of pollen- and nectar-collecting insects during these eras (Fig. 11.19). Many of the sophisticated pollination systems that characterize extant angiosperms originated at that time [11].

The advantages of insect pollination as compared to wind pollination are manifold. However, disadvantages exist as well, and **anemophily** (wind pollination) has evolved repeatedly from insect-pollinated ancestors [10, 56] and many families of insect-pollinated plants contain a few members that have

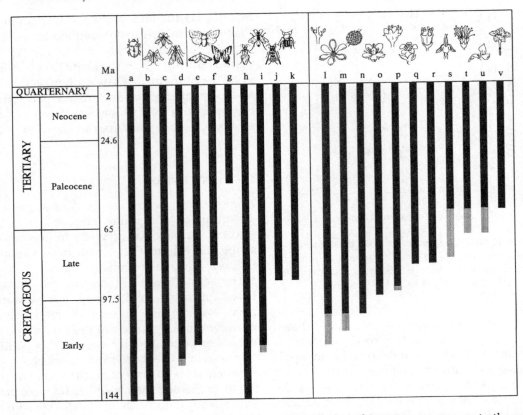

Figure 11.19 Time of appearance on a geological time scale of selected insect taxa germane to the evolution of insect pollination (a–k) as compared to the appearance of major floral types (l–v). Black bars are based on direct fossil evidence; grey bars indicate probable range, inferred rather than based on direct fossil evidence. Ma = million years before present. Insects: (a) Coleoptera; (b–d) Diptera: (b) Tipulidae, (c) Mycethophilidae, (d) Empididae; (e–g) Lepidoptera: (e) Micropterigidae, (f) Noctuidae, (g) Papilionidae; (h–k) Hymenoptera: (h) Symphyta, (i) Sphecidae, (j) Vespidae, (k) Apidae. Plants: (l) small simple flowers with few floral parts; (m) acyclic or hemicyclic flowers with numerous parts; (n) small monochlamydeous flowers; (o) cyclic, heterochlamydeous and actinomorphic flowers; (p) epigynous and heterochlamydeous flowers; (q) sympetalous flowers; (r) epigynous and monochlamydeous flowers; (s) zygomorphic flowers; (t) brush-type flowers; (u) papilionoid flowers; (v) deep funnel-shaped flowers. (Source: redrawn from Crepet and Friis, 1987, Friis and Crepet, 1987 and Grimaldi, 1996.)

become anemophilous, e.g. species of *Fraxinus* (Oleaceae), *Thalictrum* (Ranunculaceae) and *Ambrosia* (Asteraceae). In contrast to anemophily, pollination by insect vectors does not require massive and wasteful pollen production and can operate with smaller pollen grains than the most effective size for wind dispersal. Insect pollination also permits effective outcrossing at lower plant population densities and accurate pollen transfer between widely spaced individuals in multispecies vegetations. In some plant communities, such as moist tropical forests, anemophily is almost completely absent because of lack of sufficient air movement (Table 11.4).

Conversely, wind-pollinated species are prevalent in wind-swept temperate regions such as those of northern latitudes, and in

Table 11.4 Frequencies (percentage of plant species) of different pollination systems in tropical lowland rain forest trees. (Source: reproduced, with permission, from the Annual Review of Entomology, vol. 36, © 1991, by Annual Reviews Inc.)

	Forest stratum	
Pollination type	Canopy	Subcanopy and understorey
Medium-sized to large bee	44.2	21.8
Small bee	7.7	16.8
Beetle	–	15.5
Butterfly	1.9	4.5
Moth	13.5	7.3
Wasp	3.8	1.8
Small diverse insect	23.1	7.7
Bat	3.8	3.6
Hummingbird	1.9	17.7
Wind	–	3.2
Total	100 ($n = 52$)	100 ($n = 220$)

communities of low species diversity. Thus the proportion of anemophilous plants steadily increases with latitude and elevation, reaching 80–100% among the trees of the northernmost regions [68] (Fig. 11.20). Whether reproduction by wind *versus* insect pollination involves higher energy costs overall still needs to be determined.

The evolution of angiosperm flower diversity is commonly interpreted as the result of coevolutionary relationships with pollinating insects. Some insect groups have been more influential than others. Most authors, however, use the term 'coevolution' to mean reciprocal evolutionary change in interacting species [81]. As regards this definition, the evolutionary relationship between plants and pollinating insects is in the great majority of cases very asymmetrical: the pollinators have decisively influenced the evolution of flowering plants, including extensive radiation in many plant taxa, while the plants have hardly affected the evolution (e.g. speciation) of pollinating insects. Coevolution in this sense resulted in the exceptionally tight association

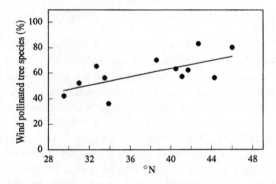

Figure 11.20 Percentage of wind-pollinated tree species in eastern North America as a function of latitude. (Source: redrawn from Regal, 1982.)

between fig and fig wasps and between yuccas and yucca moths [81].

Other insect groups differ greatly in the intensity of selection pressure they have exerted on flowering plants. A pivotal role in floral evolution accrues to the Apoidea. Because bees are completely dependent on floral resources during both the adult and larval stages they have numerous adaptations to a floral diet. Their digestive system can extract nutrients from pollen grains despite the presence of an almost impermeable cuticle [83]. Few other insect groups have succeeded in exploiting this protein-rich plant product. The well-developed learning capacities of Apoidea, together with their advanced flight and navigational abilities, allow for floral constancy and exploitation of widely scattered floral resources [59].

These features promoted flower specialization while blooms, in turn, have evolved structures, such as the floral tube and other corolla characters, that are associated with pollination by bees. Fossil flower remains show that primitive angiosperms had large numbers of stamens, pistils and petals arranged in a spiral, like present-day magnolia and white water lily (*Nymphaea alba*) flowers. In the course of time this developed into a regular radial symmetry, and trends towards flower shapes adapted to relationships with particular groups of insects (Fig. 11.21).

These include a reduction in the number of sepals and petals and the formation of a tubular or spurred corolla with nectaries positioned so that they are accessible to long-tongued insects only. By the late Cretaceous zygomorphic flower types with one plane of symmetry had evolved. Fusion of flower parts, for instance in papilionoid flowers, occurred in the early Tertiary and a proliferation of advanced floral types reflects the beginning of the spectacular evolutionary interaction between hymenopterans and angiosperms. A bee finds two parts of its body difficult to groom: the areas in the middle of the back and beneath the head. Some zygomorphic flowers, such as certain Leguminosae, exploit this limitation of the bee's dexterity and place their pollen loads on these inaccessible sites, thereby preventing transfer to the pollen baskets.

The late Cretaceous–early Tertiary is a time of the greatest rate of appearance of new angiosperm taxa, as well as the apparent period of the appearance of bee pollination,

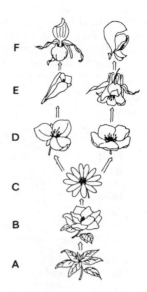

Figure 11.21 Evolutionary trends of flower shapes over 100 million years. **(A)** The earliest flowers had no discernible shape or symmetry. **(B)** Flower of open hemispherical shape, but still without clear symmetry. **(C)** Typical open, radially symmetrical flower, such as that of the yellow adonis. Subsequent divergence often altered shape in monocots (left) and dicots (right). **(D)** Flowers with reduced but fixed number of floral parts (e.g. spiderwort, left, and buttercup, right). **(E)** Flowers of increasingly bilateral symmetry and hidden nectaries, as in the freesia (left) and columbine (right). **(F)** Examples of complex and strongly zygomorphic flower shapes (e.g. the lady's slipper orchid, left, and monkshood, right). (Source: reproduced, with permission, from the *Annual Review of Ecology and Systematics*, volume 13, © 1992, by Annual Reviews Inc.)

suggesting a causal relation of some degree [13, 31, 55]. The eventual transition to increasingly three-dimensional flower types, like those of orchids and monkshood, probably has two significant advantages. First, because of their conspicuous shapes pollinators may easily recognize these flowers from a distance and second, since pollinators will learn how to manoeuvre most efficiently to reach the reward, the position of stamens and pistil can be adapted to the body orientation of their specialized pollinators. The advanced position of hymenopterans in the evolutionary association between insects and flowers is supported by the observation that bee-pollinated plant taxa show a greater diversity than taxa dependent on other groups. Thus bee plants in the southern California flora have an average of 5.9 species per genus whereas only 3.4 species per genus occur in promiscuous insect-pollinated plants. This difference suggests elevated speciation rates in bee plants [29]. Another advantage of pollination by bees is found in their hairy fur. This allows transport of large numbers of pollen grains per visit and the number of ovules in bee-pollinated plant species is accordingly high, resulting in high seed numbers per flower. More than any other group of insects, bees are the driving force of variation in floral design.

The refinement of adaptation to insect pollination culminates in an exceeding multiformity within the Orchidaceae. The monocotyledonous orchids represent the evolutionarily most recent yet the most speciose family of vascular plants, with more than 25 000 species. In orchid flowers the pollen grains cohere to form club-shaped pollen packets, called **pollinia**, usually two to each flower. Each pollinium includes an adhesive tag or clamp, and sticks to the head or another part of the visiting insect. It is then transported to another flower on which, depending on the particular shape of the flower species, the insect lands in such a position that the pollinia are accurately placed on the stigma. One pollinium suffices to fertilize all the ovules from a single pollination, giving rise to many small seeds. The diversity in orchid flower structures represents adaptations to different types of pollinator, about 60% of the orchids being pollinated by bees. Non-social bees especially, such as bumblebees in the northern hemisphere and solitary euglossine bees in the neotropics, are effective in pollinating widely separated plant populations but ensure outcrossing by the extreme precision of pollen transfer and reception. Even when the pollinating bee visits different orchid species reproductive isolation is usually maintained because each flower species snaps its pollinia on a different part of the insect's body. Up to 13 different places where the pollinia can be placed have been recorded.

Orchid diversity is apparent not only in flower shapes but also in floral scent. Flower recognition by pollinating insects is promoted not only through great diversity in flower shapes but also by wide variation in floral scents [41]. A bizarre case of floral deception is found in orchids that lure visitors by faking the female sex pheromone. The flowers of about one-third of all orchid species offer neither nectar nor pollen as reward, but produce a scent that mimics the sex attractant of their pollinators. *Ophrys* flowers, for instance, release volatiles that show chemical similarities with compounds produced in glands of their pollinating insects [6]. Many *Ophrys* flowers, in addition to the odorous lure, have developed visual and tactile stimuli to mimic conspecific female insects (Fig. 11.22).

Patrolling males become sexually excited and upon landing attempt to mate with the flower. Such 'pseudocopulations' rarely lead to the release of sperm but bring the male to touch the pollinia, which become attached to its body. Pollination may occur when the insect is attracted to another flower. This strategy has the advantage that visiting a flower does not extinguish the insect's sex drive and the next flower remains as attractive as the previous one. These cases may be regarded as a kind of 'behavioural parasitism' on the part of the plant, since the insect is exploited without a reward. This tactic is not an unique exception. It has evolved independently at least three times among the orchids and their visitors, and occurs occasionally in other plant taxa as well, involving various insect groups [79].

11.7 NATURE CONSERVATION

Since insect pollination is central to the maintenance of the plant diversity of world ecosystems any significant reduction in natural pollinators may have devastating effects on

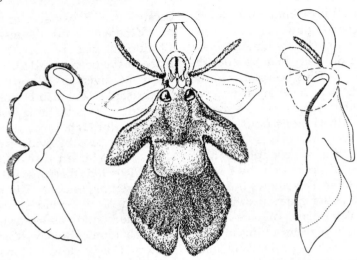

Figure 11.22 The hairy coverings of the fly orchid (*Ophrys insectifera*) and its pollinator, the wasp *Gorytes mystaceus* (left). On the right is the flower contour, with hairs, showing their arrangement and the direction of the nap. The resemblance to a female wasp is increased by the coloration of the flower's labellum. (Source: reproduced from Kullenberg, 1961, with permission.)

the plant world. Bees in particular play a paramount role. As aptly stated by Neff and Simpson:

> there can be little doubt that bees are extremely important, or the most important, group of pollinators in a wide array of plant communities. Indeed it is difficult to envision a world without bees. Other insects (...) might be able to assume the role of bees in some cases but in many communities, large proportions of the flora (...) are obligately dependent on bees as pollinators. Many of these plants (...) would simply disappear if bees were suddenly eliminated from the systems in which they occur.
>
> Neff and Simpson, 1993

Despite this notion conservation ecologists have very little quantitative data on the effects of changes in the pollinator force on plant communities. Two unintentional large-scale 'experiments' may be cited, showing that when native pollinator populations decline seed-set in some plant species in natural habitats or agrosystems is reduced. In the early 1970s large forested areas of New Brunswick (Canada) were sprayed with an insecticide that was highly toxic to bees, to control an outbreak of spruce budworm (*Choristoneura fumiferana*). This severely affected pollination success in blueberry fields. When the use of this insecticide was discontinued a steady recovery could be seen [42]. Misuse of diazinon for aphid control on alfalfa fields in northwestern parts of the USA killed sufficient alkali bees in 1973 to incur an estimated loss in alfalfa seed production worth close to $0.3 million. More than 2 years later alkali bees had regained only 25% of their initial populations [39]. Likewise, large-scale uses of herbicides, which remove alternative food sources for pollinators, may have far-reaching implications for natural vegetations *via* negative effects on wild insect pollinators.

Habitat destruction, including the removal of marginal lands and hedgerows, leads to a reduced diversity of forage plants and nest sites for natural pollinators. It is therefore a major cause of the alarming decline in the diversity and numbers of native bees. This decline, in turn, may feed back on the local flora.

11.8 ECONOMY

An inconspicuous but most important contribution of insects to human food sources is their pollination of crop plants. About 30% of our food is derived from bee-pollinated plants. The role of honeybees as honey producers is only minute in comparison. On a world basis the value of crops pollinated by bees exceeds the value of the annual honey crop by a factor of 50. Although exact figures for crop losses when all honeybees are removed are hard to determine, a figure in the order of US$5000 million annually for 63 crops in the USA seems a realistic estimate [76]. The economic value of honeybee pollination of 177 crops in the European Union amounts to roughly 4000 million ecus (£3000 million) [90]. Poor pollination levels will not only reduce crop yields but, equally importantly, they will reduce the quality of crops, such as apples, melons and other fruits.

Often, approximately 80% of the insect pollination of crops in the Western world is attributed to honeybees, but this figure may be an overestimation and the contribution of honeybees may be considerably lower. There is growing evidence that for many crops native bees are either an important adjunct to honeybees as pollinators or are even superior to them [61]. The decline of wild bee species, which is well documented for several parts of Europe, is therefore a matter of serious concern with regard to future agricultural production. Maintaining some uncultivated land areas as refuge habitats could stop a further decline of unmanaged bees and at the same time serve as havens for insect natural enemies, which are beneficial in the control of pest species. Some recent examples of such measures with positive results are encouraging [61].

To compensate for local shortages in natural pollinators, large numbers of honeybee colonies are often rented and moved, sometimes over great distances [20, 38]. Additional pollination capacity can be obtained by rearing some other bee species. These are nowadays produced and distributed on a commercial scale to enhance pollination success, either in the open field (leafcutter bees) or in greenhouses (bumblebees).

11.9 CONCLUSIONS

The fascinating panorama of partnerships that exist between insects and flowers provides a window on one of the longest relationships in biological history. It shows at the same time a range and a complexity unsurpassed by any other type of interaction between insects and plants. This complexity arises from the interplay of two dynamic systems. Superimposed on mutualism between the plants and their pollinators, the same two partners form competitively interacting systems of (1) plants for pollinators and (2) pollinators for floral resources [42]. The outcome of this complex interplay is often hard to predict but is at the heart of the present-day composition of the earth's biota. The terrestrial ecosystems as we know them would most probably never have reached their present richness in the absence of pollinating insects.

The relationships between flowers and pollinators have been the subject of many books. As a well-written and delightfully illustrated example, H. G. Barth's book [3] *Insects and Flowers*, may be mentioned. Equally informative and superbly written introductions are those by Heinrich [34], Gould and Gould [27], Procter *et al.* [64] and Seeley [72, 73]. Dafni [15] provides a very useful manual of the methods and procedures used in pollination research, with an emphasis on ecological studies. Thompson [81] presents a thorough review of the principles of coevolution and mutualism.

11.10 REFERENCES

1. Alm, J., Ohnmeiss, T. E., Lanza, J. and Vriesenga, L. (1990) Preference of cabbage

white butterflies and honey bees for nectar that contains amino acids. *Oecologia*, **84**, 53–57.
2. Baker, H. G. and Baker, I. (1986) The occurrence and significance of amino acids in floral nectars. *Plant Syst. Evol.*, **151**, 175–186.
3. Barth, H. G. (1985) *Insects and Flowers. The Biology of a Partnership*, Princeton University Press, Princeton, N. J.
4. Bawa, K. S. (1990) Plant-pollinator interactions in tropical rain forests. *Annu. Rev. Ecol. Syst.*, **21**, 399–422.
5. Boch, R. (1956) Die Tänze der Biene bei nahen und fernen Trachtquellen. *Z. Vergl. Physiol.*, **38**, 136–167.
6. Borg-Karlson, A. K. (1990) Chemical and ethological studies of pollination in the genus *Ophrys* (Orchidaceae). *Phytochemistry*, **29**, 1359–1387.
7. Braun, E., MacVicar, R. M., Gibson, D. A., Pankiw, P. and Guppy, J. (1953) Studies in red clover seed production. *Can. J. Agric. Sci.*, **3**, 48–53.
8. Burd, M. (1994) Bateman's principle and plant recognition: the role of pollen limitation in fruit and seed set. *Bot. Rev.*, **60**, 83–139.
9. Chittka, L. and Menzel, R. (1992) The evolutionary adaptation of flower colour and the insect pollinator's colour vision. *J. Comp. Physiol. A*, **171**, 171–181.
10. Cox, P. A. (1991) Abiotic pollination: an evolutionary escape for animal-pollinated angiosperms. *Phil. Trans. Roy. Soc. Lond. B*, **333**, 217–224.
11. Crane, P. R., Friis, E. M. and Pedersen, K. R. (1995) The origin and early diversification of angiosperms. *Nature*, **374**, 27–33.
12. Crepet, W. L. and Friis, E. M. (1987) The evolution of insect pollination in angiosperms, in *The Origins of Angiosperms and Their Biological Consequences*, (eds E. M. Friis, W. G. Chaloner and P. R. Crane), Cambridge University Press, Cambridge, pp. 181–201.
13. Crepet, W. L., Friis, E. M. and K. C. Nixon (1991) Fossil evidence for the evolution of biotic pollination. *Phil. Trans. Roy. Soc. Lond. B*, **333**, 187–195.
14. Dafni, A. (1984) Mimicry and deception in pollination. *Annu. Rev. Ecol. Syst.*, **15**, 259–278.
15. Dafni, A. (1992) *Pollination Ecology. A Practical Approach*, IRL Press, Oxford.
16. Darwin, C. (1859) *The Origin of Species*, John Murray, London.
17. Dobson, H. E. M. (1994) Floral volatiles in insect biology, in *Insect–Plant Interactions*, vol. 5, (ed. E. A. Bernays), CRC Press, Boca Raton, FL, pp. 47–81.
18. Dobson, H. E. M., Bergström, G. and Groth, I. (1990) Differences in fragrance chemistry between flower parts of *Rosa rugosa* Thunb. (Rosaceae). *Israel J. Bot.*, **39**, 143–156.
19. Feinsinger, P. (1983) Coevolution and pollination, in *Coevolution*, (eds D. J. Futuyma and M. Slatkin), Sinauer, Sunderland, MA, 282–310.
20. Free, J. B. (1993) *Insect Pollination of Crops*, 2nd edn, Academic Press, London.
21. Friis, E. M. and Crepet, W. L. (1987) Time of appearance of floral features, in *The Origins of Angiosperms and Their Biological Consequences*, (eds E. M. Friis, W. G. Chaloner and P. R. Crane), Cambridge University Press, Cambridge, pp. 145–179.
22. Frisch, K. von (1967) *The Dance Language and Orientation of Bees*, Harvard University Press, Cambridge, MA.
23. Gegear, R. J. and Laverty, T. M. (1995) Effect of flower complexity on relearning flower-handling skills in bumble bees. *Can. J. Zool.*, **73**, 2052–2058.
24. Gilbert, F. S., Haines, H. and Dickson, K. (1991) Empty flowers. *Funct. Ecol.*, **5**, 29–39.
25. Giurfa, M. and Núñez, J. A. (1992) Honeybees mark with scent and reject recently visited flowers. *Oecologia*, **89**, 113–117.
26. Gould, J. L. (1993) Ethological and comparative perspectives on honey bee learning, in *Insect Learning. Ecological and Evolutionary Perspectives*, (eds D. R. Papaj and A. C. Lewis), Chapman & Hall, New York, pp. 18–50.
27. Gould, J. L. and Gould, C. G. (1988) *The Honey Bee*, W. H. Freeman, New York.
28. Goulson, D. and Cory, J. S. (1993) Flower constancy and learning in foraging preferences of the green-veined white butterfly, *Pieris napi*. *Ecol. Entomol.*, **18**, 315–320.
29. Grant, V. (1949) Pollination systems as isolating mechanisms in angiosperms. *Evolution*, **3**, 82–97.
30. Grant, V. (1950) The flower constancy of bees. *Bot. Rev.*, **16**, 379–398.
31. Grimaldi, D. A. (1996) Captured in amber. *Sci. Am.*, **274**(4), 71–77.
32. Heinrich, B. (1976) The foraging specializations of individual bumblebees. *Ecol. Monogr.*, **46**, 105–128.
33. Heinrich, B. (1979a) 'Majoring' and 'minoring'

by foraging bumblebees, *Bombus vagans*: an experimental analysis. *Ecology*, **60**, 245–255.
34. Heinrich, B. (1979b) *Bumblebee Economics*, Harvard University Press, Cambridge, MA.
35. Heinrich, B. and Raven, P. H. (1972) Energetics and pollination ecology. *Science*, **176**, 597–602.
36. Hess, D. (1983) *Die Blüte*, Ulmer, Stuttgart.
37. Inouye, D. W. (1978) Resource partitioning in bumblebees: experimental studies of foraging behavior. *Ecology*, **59**, 672–678.
38. Jay, S. C. (1986) Spatial management of honey bees on crops. *Annu. Rev. Entomol.*, **31**, 49–65.
39. Johansen, C. A. (1977) Pesticides and pollinators. *Annu. Rev. Entomol.*, **22**, 177–192.
40. Jones, C. E. and Buchmann, S. L. (1974) Ultraviolet floral patterns as functional orientation cues in hymenopterous pollination systems. *Anim. Behav.*, **22**, 481–485.
41. Kaiser, R. (1993) *The Scents of Orchids. Olfactory and Chemical Investigations*, Elsevier, Amsterdam.
42. Kevan, P. G. and Baker, H. G. (1983) Insects as flower visitors and pollinators. *Annu. Rev. Entomol.*, **28**, 407–453.
43. Klinkhamer, P. G. L. and de Jong, T. J. (1993) Attractiveness to pollinators, a plant's dilemma. *Oikos*, **66**, 180–184.
44. Knoll, F. (1956) *Die Biologie der Blüte*, Springer-Verlag, Berlin.
45. Knudsen, J. T., Tollsten, L. and Bergström, G. (1993) Floral scents – a checklist of volatile compounds isolated by head-space techniques. *Phytochemistry*, **33**, 253–280.
46. Kriston, I. (1971) Zum Problem des Lernverhaltens von *Apis mellifica* L. gegenüber verschiedenen Duftstoffen. *Z. Vergl. Physiol.*, **74**, 169–189.
47. Kullenberg, B. (1961) Studies on *Ophrys* L. pollination. *Zool. Bidr. Uppsala*, **34**, 1–340.
48. Laverty, T. M. (1994) Bumble bee learning and flower morphology. *Anim. Behav.*, **47**, 531–545.
49. Laverty, T. M. and Plowright, R. C. (1988) Flower handling by bumblebees, a comparison of specialists and generalists. *Anim. Behav.*, **36**, 733–740.
50. Leppik, E. E. (1971) Origin and evolution of bilateral symmetry in flowers. *Evol. Biol.*, **5**, 49–86.
51. Lewis, A. C. (1993) Learning and the evolution of resources: pollinators and flower morphology, in *Insect Learning. Ecological and Evolutionary Perspectives*, (eds D. R. Papaj and A. C. Lewis), Chapman & Hall, New York, pp. 219–242.
52. Lindauer, M. (1948) Über die Einwirkung von Duft- und Geschmacksstoffen sowie anderer Faktoren auf die Tänze der Bienen. *Z. Vergl. Physiol.*, **31**, 348–412.
53. McKone, M. J., Ostertag, R., Rausher, J. T., Heiser, D. A. and Russell, F. L. (1995) An exception to Darwin's syndrome: floral position, protogyny, and insect visitation in *Besseya bullii* (Scrophulariaceae). *Oecologia*, **101**, 68–74.
54. Maynard Smith, J. (1982) *Evolution and the Theory of Games*, Cambridge University Press, Cambridge.
55. Michener, C. D. and Grimaldi, D. A. (1988) The oldest fossil bee: apoid history, evolutionary stasis, and antiquity of social behavior. *Proc. Natl Acad. Sci. USA*, **85**, 6424–6426.
56. Midgley, J. J. and Bond, W. J. (1991) How important is biotic pollination and dispersal to the success of the angiosperms? *Phil. Trans. Roy. Soc. Lond. B*, **333**, 209–215.
57. Moore, P. D., Webb, J. A. and Collinson, M. A. E. (1991) *Pollen Analysis*, 2nd edn, Blackwell, Oxford.
58. Morse, R. A. (1958) The pollination of bird's-foot trefoil. *Proc. 10th Int. Congr. Ent.*, **4**, 951–953.
59. Neff, J. L. and Simpson, B. B. (1993) Bees, pollination systems and plant diversity, in *Hymenoptera and Biodiversity*, (eds J. LaSalle and I. D. Gauld), CAB International, Wallingford, Oxon, pp. 143–167.
60. Olesen, J. M. and Warncke, E. (1989) Flowering and seasonal changes in flower sex ratio and frequency of flower visitors in a population of *Saxifraga hirculus*. *Holarct. Ecol.*, **12**, 21–30.
61. O'Toole, C. (1993) Diversity of native bees and agrosystems, in *Hymenoptera and Biodiversity*, (eds J. LaSalle and I. D. Gauld), CAB International, Wallingford, Oxon, pp. 169–196.
62. Pham-Delegue, M. H., Etievant, P., Guichard, E., Marilleau, R., Douault, Ph., Chauffaille, J. and Mason, C. (1990) Chemicals involved in honeybee–sunflower relationship. *J. Chem. Ecol.*, **16**, 3053–3065.
63. Pleasants, J. M. (1980) Competition for bumblebee pollinators in Rocky Mountain plant communities. *Ecology*, **61**, 1446–1459.
64. Procter, M., Yeo, P. and Lack, A. (1996) *The Natural History of Pollination*, Harper Collins, London.
65. Ranta, E. and Lundberg, H. (1980) Resource partitioning in bumblebees: the significance of differences in proboscis length. *Oikos*, **35**, 298–302.

66. Reader, R. J. (1975) Competitive relationships of some bog ericads for major insect pollinators. *Can. J. Bot.*, **53**, 1300–1305.
67. Real, L., Ott, J. and Silverline, E. (1982) On the trade-off between the mean and the variance in foraging: effects of spatial distribution and color preference. *Ecology*, **63**, 1617–1623.
68. Regal, P. J. (1982) Pollination by wind and animals: ecology and geographic patterns. *Annu. Rev. Ecol. Syst.*, **13**, 497–524.
69. Sazima, M., Vogel, S., Cocucci, A. and Hausner, G. (1993) The perfume flowers of *Cyphomandra* (Solanaceae): pollination by euglossine bees: bellows mechanism, osmophores, and volatiles. *Plant Syst. Evol.*, **187**, 51–88.
70. Schmitt, U. and Bertsch, A. (1990) Do foraging bumblebees scent-mark food sources and does it matter? *Oecologia*, **82**, 137–144.
71. Schmitt, U., Lübke, G. and Francke, W. (1991) Tarsal secretion marks food sources in bumblebees (Hymenoptera, Apidae). *Chemoecology*, **2**, 35–40.
72. Seeley, T. D. (1985) *Honeybee Ecology*, Princeton University Press, Princeton NJ.
73. Seeley, T. D. (1996) *The Wisdom of the Hive: The Social Physiology of Honeybee Colonies*, Harvard University Press, Cambridge, MA.
74. Snodgrass, R. E. (1956) *Anatomy of the Honey Bee*, Comstock, Ithaca, NY.
75. Snow, A. A., Spira, T. P., Simpson, R. and Klips, R. A. (1996) The ecology of geitonogamous pollination, in *Floral Biology. Studies on Floral Evolution in Animal-Pollinated Plants*, (eds D. G. Lloyd and S. C. H. Barrett), Chapman & Hall, New York, pp. 191–216.
76. Southwick, E. E. and Southwick, L. (1992) Estimating the economic value of honeybees (Hymenoptera, Apidae) as agricultural pollinators in the United States. *J. Econ. Entomol.*, **85**, 621–633.
77. Sprengel, C. K. (1793) *Das entdeckte Geheimniss der Natur im Bau und in der Befruchtung der Blumen*, F. Vieweg, Berlin (reprint published by J. Cramer and H. K. Swann, Lehre, and Weldon and Wesley, Codicote, New York, 1972).
78. Stead, A. D. (1992) Pollination-induced flower senescence, a review. *Plant Growth Regul.*, **11**, 13–20.
79. Stowe, M. K. (1988) Chemical mimicry, in *Chemical Mediation of Coevolution*, (ed. K. S. Spencer), Academic Press, San Diego, CA, pp. 513–580.
80. Surburg, H., Guentert, M. and Harder, H. (1993) Volatile compounds from flowers. Analytical and olfactory aspects, in *Bioactive Volatile Compounds from Plants*, ACS Symposium 525, (eds R. Teranishi, R. G. Buttery and H. Sugisawa), pp. 168–186.
81. Thompson, J. N. (1994) *The Coevolutionary Process*, University of Chicago Press, Chicago, IL.
82. Tollsten, L. and Bergström, G. (1989) Variation and post-pollination changes in floral odours released by *Platanthera bifolia* (Orchidaceae). *Nord. J. Bot.*, **9**, 359–362.
83. Velthuis, H. H. W. (1992) Pollen digestion and the evolution of sociality in bees. *Bee World*, **73**, 77–89.
84. Waddington, K. D. (1985) Cost-intake information used in foraging. *J. Insect Physiol.*, **31**, 891–897.
85. Waddington, K. D. (1990) Foraging profits and thoracic temperature of honey bees (*Apis mellifera*). *J. Comp. Physiol.*, **B160**, 325–329.
86. Waser, N. M. (1986) Flower constancy: definition, cause, and measurement. *Am. Nat.*, **127**, 593–603.
87. Weiss, M. R. (1991) Floral colour changes as cues for pollinators. *Nature*, **354**, 227–229.
88. Weiss, M. R. (1995a) Floral color change, a widespread functional convergence. *Am. J. Bot.*, **82**, 167–185.
89. Weiss, M. R. (1995b) Associative colour learning in a nymphalid butterfly. *Ecol. Entomol.*, **20**, 298–301.
90. Williams, I. H. (1994) The dependence of crop production within the European Union on pollination by honey bees. *Agric. Zool. Rev.*, **6**, 229–257.

INSECTS AND PLANTS: HOW TO APPLY OUR KNOWLEDGE   12

12.1	Which herbivorous insect species become pests and why?	343
12.1.1	Characteristics of herbivorous pest species	343
12.1.2	Consequences of crop plant introductions	344
12.1.3	Agricultural practices promote the occurrence of pest problems	345
12.2	Host-plant resistance	346
12.2.1	Host-plant resistance mechanisms	346
12.2.2	Partial resistance	347
12.2.3	Plant characteristics associated with resistance	348
12.2.4	Methodology of resistance breeding	349
12.3	Polycultures: why fewer pests?	350
12.3.1	The disruptive-crop hypothesis	352
12.3.2	The enemies hypothesis	353
12.3.3	Trap-cropping and crop–weed systems	353
12.3.4	Diversity as a guiding principle	354
12.4	Plant-derived insecticides and antifeedants	355
12.4.1	Antifeedants	355
12.4.2	Neem tree, azadirachtin	356
12.4.3	Outlook for antifeedants as crop protectants	358
12.5	Weed control by herbivorous insects	359
12.5.1	*Opuntia* and *Salvinia*	360
12.5.2	Success rate of biological weed control programmes	360
12.6	Conclusion: diversification holds the clue to the control of pestiferous insects	362
12.7	References	362

In natural ecosystems roughly 10% of a plant's resources are lost to herbivory. Preharvest losses of agricultural crop production when no insecticides are used vary between 10% and 100%. In systems based on the use of insecticides and non-chemical control methods losses to insect herbivory come to an estimated 13% [64]. Alarmingly, losses to animal pests, as viewed over a period of 25 years, are increasing for several major food crops (Fig. 12.1), albeit this is more than compensated by increasing yields per unit of area [59]. Apparently there is a positive correlation between yields of crop plants and their susceptibility to insect pests and other biotic and abiotic constraints.

When attempting to increase agricultural production in order to feed a world population with a present growth rate of 1.6% per year, to reduce the use of synthetic insecticides and to convert current agriculture into more sustainable systems insights gained from insect–plant studies may appear helpful if not indispensable. This chapter discusses aspects of insect–plant interactions that, first, may clarify why some insect species develop the status of a pest species and what measures can be taken to suppress such development and, second, may enable the control of weeds by herbivorous insects.

12.1 WHICH HERBIVOROUS INSECT SPECIES BECOME PESTS AND WHY?

12.1.1 CHARACTERISTICS OF HERBIVOROUS PEST SPECIES

Some insect species are predestined to become pests when a favourable crop plant species becomes available, whereas others, including closely related species, are unable to switch

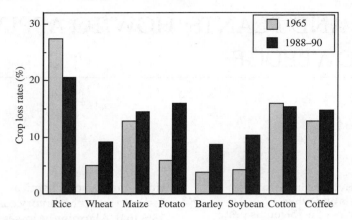

Figure 12.1 Crop losses due to animal pests for eight principal food and cash crops in 1965 and in 1988–90. (Source: data from Oerke et al., 1994.)

easily to the new food resource. Several physiological and behavioural characteristics of a species, such as fecundity, larval diet breadth and voltinism, contribute to the likelihood of an insect attaining pest status when a suitable habitat is made available to it. An analysis of these factors has been made for some insect groups. From the biological characteristics of several *Pieris* species it can be understood why, out of perhaps several dozen species or geographical subspecies of crucifer-feeding pierid butterflies, only two species (*Pieris brassicae* and *P. rapae*) have attained economic pest status on crucifer crops worldwide. Both *Pieris* species exploit a wide range of crucifers relative to other *Pieris* species and this 'euryphagy' seems to be associated with preadaptation to crop hosts. The multivoltinism of both species is another property that enables them to rapidly expand populations and to produce numerous offspring that may colonize new habitats (*Pieris rapae* females may produce more than 800 eggs). A third trait that contributes to the two species reaching pest status is a preference for dense host populations in mesic (i.e. neither extremely dry nor extremely wet) habitats [18]. Thus several factors in combination may render an insect species a potential pest if exposed to a crop plant species that is physiologically and behaviourally an acceptable host.

12.1.2 CONSEQUENCES OF CROP PLANT INTRODUCTIONS

Most insect–plant relationships in natural ecosystems are based on millions of years of evolution (Chapter 10). As a result a balance between plants and herbivores has evolved to the extent that plants are rarely eliminated as a result solely of insect attack. When a plant is confronted with an 'unknown' insect species the situation may be different. As Southwood noted [78] in a discussion on the evolutionary perspectives on insect–plant relationships: 'Even today when a phytophagous species first attacks a new host it often inflicts disproportionately heavy damage.' Many crop plants, especially those in the temperate zones, are introduced species. In most instances their insect pests have moved from feeding on native vegetation to also feeding on the new crop. They invade a new niche in which food is abundant and natural enemies are scarce. Moreover, the resistance of such introduced crop plants is unadapted to local insect species. An example is potato (*Solanum tuberosum*), which originated in South America and was introduced to North America. The Colorado potato beetle (*Leptinotarsa decemlineata*), living on a native *Solanum* species (*S. rostratum*), has since colonized the new food resource very successfully, because potato did not possess constitutive resistance traits against this insect. The beetle then became a serious pest and eventually spread, with potato, to Europe.

In fact, the number of pestiferous insect species in agricultural crops is remarkably small in view of the enormous pool of potential invaders. Because of the high dispersal capacities of insects, plants in natural as well as agricultural communities are normally visited by many herbivorous species. However, only a small fraction of those visitors appear to establish an enduring association with these plants. For example, only about 40 insect species have colonized soybean fields in Illinois (USA), while over a period of 12 years more than 400 herbivorous species were sampled in such fields. And although over 60 species of aphid were trapped in Illinois soybean fields alone, not a single aphid species has been capable to exploit soybean as a permanent host in either North or South America [47]. Adoption of a new food plant, in this case a crop species, even if it is readily accessible in large numbers, is apparently a difficult step for most insects.

Where do insect pests come from? Do they belong to the native fauna or are they immigrants? Out of 148 major insect species that infest crop plants in the USA only 57 (i.e. less than 40%) are foreign-introduced species. Likewise, out of 70 major insect pest species in American forests the majority are native species, with less than 30% originating from Europe or elsewhere [65]. Similarly, in Europe only approximately 20% of insect pests were introduced [64]. Thus, in managed as well as natural ecosystems the majority of insect pest species are native species, although some of the most serious insect pests in forests are introduced species.

12.1.3 AGRICULTURAL PRACTICES PROMOTE THE OCCURRENCE OF PEST PROBLEMS

For reasons of mechanization and efficiency of sowing, planting management, harvesting and processing, agricultural crops are predominantly grown as monocultures, especially so in the Western world. Why are such systems more prone to insect population outbreaks than so-called natural systems [67]? There is no simple answer to this question, because each species or biotype of insect, each species or variety of host plant, each soil type on which they are grown and each microclimate constitutes a specific situation. Because of the multidimensional nature of each of these components of an agroecosystem, factors that cause pest outbreaks can be diverse. Nevertheless, some of the most important factors promoting insect outbreaks in agroecosystems are obvious. They are, on the one hand, the reduced chemical and physical resistance of crop plants compared with those of their ancestral forms or closely related wild species and, on the other, the 'simplification' of the species structure of agroecosystems compared with natural ecosystems. This simplification includes a drastic reduction in plant and animal species, increased genetic uniformity of the crop, the abandonment of crop rotation, and the decrease in landscape diversity by removal of hedges, ditches and other non-crop habitats [67]. Monocultures present favourable habitats for some insect species that thrive once food is unlimited. To quote Nickle [58]: 'insects with reproductive capacities evolved for survival on broadly dispersed wild plants can normally multiply much more rapidly in the "supermarket" utopia of monoculture'.

Some of our knowledge about host-plant resistance and the significance of polycultures for increasing agricultural diversity will be discussed below. Other factors thought to stimulate the development of insect pests are discussed in detail in some recent reviews [11, 12, 46].

12.2 HOST-PLANT RESISTANCE

In nature, host-plant resistance and natural enemies are the two dominant factors controlling herbivorous insect populations. Therefore, modern approaches of pest control consider host-plant resistance breeding as a key method of insect pest regulation in crop plants. Since the beginning of agriculture, probably more than 10 000 years ago, crop plants have been selected for high yields and nutritional value, together with low mammalian toxicity and reasonable resistance against pests and disease. In recent times, however, interest in resistance to insects, plant pathogens and ability to compete with weeds has diminished. As a result very few cultivated species have retained the insect resistance level of their wild progenitors. Concomitantly, the defensive diversity is often reduced as well [64].

Plant resistance breeding is a 20th-century activity that stems from a knowledge of basic genetics and from the methodology of selecting, crossing and hybridizing plants. It was undertaken fervently and became more rigorous in its approach only after the rediscovery of Mendel's laws of heredity in 1900 by H. de Vries. Although development of insect-resistant crop cultivars is a time-consuming and expensive process, the benefits may be enormous in terms of monetary return and reduced burdening of the environment with insecticides. The economic advantage of using pest resistant cultivars is estimated to be a 120-fold greater return on investment and, not less importantly, some new cultivars of cotton, rice and vegetables developed recently contain insect resistance sufficient to eliminate the use of insecticides entirely [76]. There is an abundant literature documenting the genetic control of arthropod resistance and resistance variation in agricultural crops, as evidenced by extensive reviews by Maxwell and Jennings [52], Fritzsche *et al.* [32] and Panda and Khush [61], and a bibliography by Stoner [79]. From these studies it can be concluded that the reason why resistant crop varieties are so rare is not because of a lack of resistant resources. It is the complicated and undesirable large-scale insect bioassays that make breeders reluctant to incorporate resistance to (mobile) insects in their breeding programmes. Therefore new technologies such as genetic modification and molecular-marker-assisted selection (MAS) [51], which avoid insect bioassays, are highly valued for this particular purpose.

12.2.1 HOST-PLANT RESISTANCE MECHANISMS

For a long time researchers in host-plant resistance breeding have been mostly concerned with methods of rapidly identifying resistance genotypes in germplasm banks and monitoring inheritance of resistance in breeding lines, while being less interested in the mechanisms underlying resistance. A scientific basis of the field and a more systematic research approach was introduced by R. H. Painter [60] and in recent overviews [23, 61, 75] plant breeding methodology and analysis of resistance mechanisms are integrated.

Painter [60] recognized three 'causes' of resistance, emphasizing the insect–plant relations that are a feature of insect resistance: (1) non-preference, (2) antibiosis and (3) tolerance. **Non-preference** defines the group of plant characters and insect responses that lead away from the use of a particular plant or variety for oviposition, for food or for shelter, or for combinations of the three. **Antibiosis** denotes reduced fecundity, decreased size, reduced longevity and increased mortality. Antibiosis, in contrast to non-preference, clearly refers to those plant properties that adversely affect the physiology of a herbivore. **Tolerance** is a form of resistance in which the plant shows an ability to grow and reproduce itself or to repair injury to a marked degree in spite of supporting a population approximately equal to that damaging a susceptible host. (When the pest insect is a vector for one or more plant pathogens, tolerance is an undesirable trait, since the insect population may increase on the crop, enhancing the risk of pathogen spreading.) Tolerance is a plant property that expresses itself irrespective of whether an insect (or another agency) is responsible for tissue loss.

Because the term 'non-preference' describes the response of the insect rather than a plant characteristic, it has been replaced by **antixenosis**, defined as plant properties evoking negative (non-preference) responses or total avoidance by insects [48]. Whereas antixenosis and antibiosis lend themselves well to deliberate selection in specific laboratory and standardized field tests, tolerance is a modality of plants that is more difficult to assess because it requires simultaneous observation of insect populations and yield potential of adult plants [23].

So far, tolerance seems to be the least common type of resistance. An examination of more than 200 reports on resistance to arthropod pests in vegetables showed that tolerance was involved in about 10%, whereas the remaining cases were equally attributed to either antixenosis or antibiosis [79]. It is important to realize, however, that although Painter's classification has proved to be a very useful one, resistance is most frequently a combination of two or even all three types of resistance.

12.2.2 PARTIAL RESISTANCE

Although some examples exist, it is often difficult to attain complete resistance to a particular insect species, and only partial resistance can be obtained. An advantage of incomplete resistance is, however, that it poses weaker selection pressure on the insect population and consequently is more durable. In combination with various integrated pest management measures, partial resistance may be sufficient or even preferable because of the reduced risk of the development of new virulent insect biotypes [36, 37]. In this context two more terms need to be introduced.

Horizontal or **polygenic resistance** is based on a mixture of minor resistance genes, which are accumulated in one genotype. **Vertical resistance**, conversely, is resistance governed by one or more genes in the host plant, each of which corresponds to a matching gene for parasitic ability in the pest species (it is therefore sometimes called **gene-for-gene resistance**). Numerous cases of polygenic resistance to insects are known to occur in many crop plant species. Many instances of monogenic resistance have also been reported in the literature. The most extensively studied inheritance of the latter type is that of wheat resistance to the Hessian fly, as a result of which 26 genes for resistance have been identified [61].

There is an important difference between the two resistance types with respect to their stability. Horizontal resistance involves the accumulation of genes from diverse germplasm. To build up a satisfactory level of resistance is a time-consuming process. This is compensated by the fact that this resistance is generally more difficult to overcome by resistance-breaking insect biotypes and thus generally more stable than vertical resistance. Resistance stability is sometimes of short duration, particularly if the resistance level is very high, its inheritance is simple and the resistant cultivars are grown on a large scale. Under these circumstances insects may break the resistance by developing biotypes that possess an inherent genetic capability to overcome host-plant resistance. Cases are known in which as few as three generations were required to select resistance-breaking biotypes, and occasionally the insect's potential to overcome plant resistance is so great that the effect of the resistance is nullified before the resistant cultivar reaches widespread use. This happened, for instance, to cultivars of Brussels sprouts resistant to the cabbage aphid *Brevicoryne brassicae* [28].

Fortunately, adaptation to new cultivars usually takes longer, even under strong selection regimes [30], and several examples of long-lasting resistance are known. The apple variety 'Winter Majetin', for instance, which was reported to be resistant to the woolly aphid (*Eriosoma lanigerum*) as long ago as 1831, still retains this trait. Another often-cited example is the partial resistance of grape vines to the grape phylloxera (*Phylloxera vitifoliae*) in French vineyards, which has been effective since 1890 [61].

Polygenic resistance is probably not *per se* more durable (for instance not when it involves the concentration of a single chemical compound), but when it relates to multiple chemical, physiological or morphological mechanisms the chances that a pest species will break resistance are much lower. When trying to understand why some resistances are easily overcome and others are durable, the insect's adaptability is a critical factor. Conceivably, insects may adapt physiologically to the presence of, for instance, toxic compounds in their food more easily than they can adjust behaviourally to new plant characteristics. In the latter case a series of changes is needed, including adaptation to various cues governing oviposition behaviour and feeding. This view agrees with the fact that plant breeders selecting for insect-resistant genotypes consider the antixenotic type of resistance more valuable than the antibiotic type. This is because in their experience the latter type is generally less durable [23, 79].

12.2.3 PLANT CHARACTERISTICS ASSOCIATED WITH RESISTANCE

Not surprisingly, plant features causing resistance in cultivated plants do not differ from those operative in wild plant species, i.e. physical (Fig. 12.2), chemical or phenological factors.

Many such factors have been identified, and

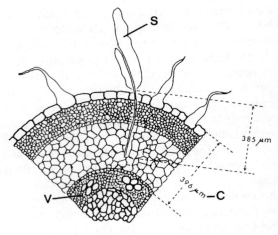

Figure 12.2 Insect resistance caused by plant anatomical characteristics. A thick cortex (C) in stems of *Lycopersicon hirsutum* prevents aphids (*Macrosiphum euphorbiae*) from reaching vascular tissue (V) with their stylet bundles (S). (Source: reproduced from Quiras *et al.*, 1977, with permission.)

numerous examples are given elsewhere in this book. For information on resistance mechanisms identified in specific crop plant species, reviews of the extensive literature [52, 61, 77, 79] should be consulted.

Most forms of plant resistance appear to involve some diversion of resources by the plant to increased production of allelochemicals or extra physical defence structures. Selection in the past of crop species for improved agricultural value has been associated with reduced levels of particular secondary plant substances (Table 12.1) and increased herbivore susceptibility.

Usually, the incorporation of resistance genes into high-yield cultivars requires some sacrifice in yield and therefore can be considered to impose a 'cost'. Thus the best soybean breeding lines resistant to various insect herbivores still yield less than the best available cultivars grown in the absence of pests [47].

When under insect-free conditions two barley cultivars with an isogenic difference in

Table 12.1 Effect of domestication on amounts of secondary metabolites; concentrations of secondary compounds in crop plants are expressed as percentages of those in wild relatives

Compounds	Plant	% of wild species	Reference
Quinolizidine alkaloids	Lupin	0.5	91
Cucurbitacins	Squash	1	40
2-tridecanone	Tomato	1.5	89
Glycoalkaloids	Potato tubers	4	43
Glucosinolates	Cabbage	20	44
Gramine	Barley	20	49

greenbug (*Schizaphis graminum*) resistance were grown in competition, the susceptible cultivar was the better competitor. However, when the cultures were exposed to aphid feeding the outcome was reversed: the resistant cultivar became the better competitor [90]. In other cases such costs are difficult to detect. One of the causes of contradictory results on the costs of resistance is that the relative performance of resistants and susceptibles depends on several environmental conditions, which may easily mask costs spent on resistance traits [13].

Since different insects have different nutritional requirements and show different responses to plant defence factors a particular plant cultivar selected for resistance to one insect species usually remains susceptible to other insect species. Multiple species resistance is often difficult to develop. This is illustrated by the difficulties encountered when developing resistance to three major pest species of cotton. Whereas smooth-leaf cultivars suffer less from the larvae of some *Helicoverpa* species, they are prone to increased feeding damage from tarnished plant bugs (*Lygus lineolaris*). Frago-bract strains, in which the bracts are modified such that the cotton buds are exposed, show reduced infestation by boll weevils (*Anthonomus grandis*) but increased susceptibility to tarnished plant bugs. Despite these obstacles it has been possible to develop cotton cultivars that exhibit resistance to all three insects, as well as to cotton leafhoppers (*Pseudatomoscelis seriatus*) [15]. This shows that satisfactory protection may be attained even to insect pest complexes.

Plant resistance against herbivores has not only a 'direct' but also an 'indirect' component, through an influence on the third trophic level. Different cultivars may differ in the production of entomophage-attracting allelochemicals as, among others, has been observed in bean plants (*Phaseolus vulgaris*) for predatory mites [25]. If in a plant-breeding programme this aspect is neglected such an indirect resistance factor may inadvertently be eliminated. This would result in the selection of cultivars with a reduced net resistance under field conditions, especially where natural enemies play a significant role in herbivore mortality [25].

12.2.4 METHODOLOGY OF RESISTANCE BREEDING

Plant breeding exploits genetic variability within the crop species and its wild relatives and aims to enhance resistance to insect pests and diseases by prudent selection and breeding methods. Nowadays resistance genes may also be transferred by molecular biological techniques. Present-day approaches involve the combination of (1) the use of population growth models for exploring resistance management strategies, (2) developing efficient test procedures, (3) a further exploitation of antixenosis as a resistance modality, and (4) evaluating the potential of molecular biological techniques [23].

As sources of resistance genes there are a number of broad-based germplasm collections, consisting of wild species as well as large numbers of cultivars, located in different parts of the world [38, 74]. They stock numerous gene banks, containing valuable gene pools for crop improvement activities. Wild relatives provide valuable source materials for insect and disease resistance. For instance, high levels of resistance to two planthopper species have been transferred from *Oryza officinalis* to cultivated rice (*O. sativa*) [41].

As mentioned above insect resistance depends sometimes on one locus (monogenic) but more often several independent loci that confer resistance in different ways are involved, or several (oligogenic) or even many loci (polygenic) determine resistance. Traits with simple Mendelian inheritance are relatively easy to work with. Monogenic resistance has frequently been found in crop plants. A classic example is resistance to the brown planthopper (*Nilaparvata lugens*) of rice [61]. In wild plants, however, resistance to an insect is seldom based upon a single resistance mechanism. Several modes of defence (e.g. chemical, physical and imbalance of nutritional factors) are combined and controlled by a complex system with several loci and multiple alleles at one locus. When resistance genes are located in exotic germplasm it requires a lot of work to incorporate them into more agronomically acceptable lines. Depending on the reproductive system of the crop species, i.e. self-pollinating or cross-pollinating, various breeding programs can be used, as described in books on methodology of breeding for insect resistance such as those by Panda and Khush [61] and Smith *et al.* [77].

Traditional selective breeding can now be shortcircuited by ingenious biotechnological methods. Recent advances in molecular biology and tissue culture have made it possible to transfer genes not only from related species but also from unrelated plants and from other still more distantly related organisms, such as animals, bacteria and viruses [34]. Genetic engineering methods permit the introduction of novel genes into crop species that render them resistant to insects. For instance, genes from insect pathogens introduced into the insect's food plant may result in effective insect population control. Thus genes responsible for the production of a toxin derived from the insect pathogen *Bacillus thuringiensis* have been introduced into, among others, tomato, rice, cotton and spruce trees. Transgenic potatoes that, because of the expression of this gene, are resistant to the Colorado potato beetle will be among the first transgenic food crops deployed commercially [7]. Other orally active adverse proteins, such as lectins, amylase inhibitors and proteinase inhibitors, which retard growth and slow down development, have also been produced in transgenically modified plants. For instance, by transferring cDNA that encodes the α-amylase inhibitor occurring in the seeds of *Phaseolus vulgaris* into pea (*Pisum sativum*), resistance to the pea weevil (*Bruchus pisorum*) was conferred. Transgenic pea seeds accumulated the α-amylase inhibitor to a level of 3% of soluble protein. The inhibitory effect on human α-amylase should disappear through cooking [72].

Obviously, genetic engineering opens fascinating avenues for crop improvement and insect resistance and has been assumed to provide the ultimate technique in agricultural production. Valuable though it is, lessons from the past strike a note of caution. As Stoner [79] rightly remarks: 'It is much too soon to abandon traditional approaches to plant resistance to insects. Researchers in the field of plant resistance to insects should take advantage of the opportunities presented by new developments in biotechnology, but should also maintain their unique focus on the behavioural, physiological, ecological, and evolutionary interactions of the insect with its host plant.'

12.3 POLYCULTURES: WHY FEWER PESTS?

From time immemorial farmers have known that growing several crops on one unit of land resulted in increased yields. Pliny the Younger (23–79 AD) wrote in his *Naturalis Historiae* that when rape (*Brassica napus*) and common vetch (*Vicia sativa*) were grown together many insects normally occurring on these crops remained absent. Intercropping is still common in peasant agriculture in the tropics, where the percentage of cropped land devoted to polycultures varies from a low of 17% in India to a high of 94% in Malawi [85]. By contrast, modern intensive agriculture in the Western world has reached a shockingly high degree of biouniformity. Large acreages are planted with monocultures of only one out of a few cultivars, which often possess very low genetic diversity. Increasing vegetational diversity by planting different crops intermingled is one type of cultural control strategy that can make agroecosystems less favourable to the pest insect and/or more favourable to natural enemies.

Terms related to polycultural planting schemes are sometimes used rather loosely and inconsistently. **Intercropping** describes a system whereby more than one crop is grown in an area simultaneously, in such a way that they interact agronomically. Intercrops can be of four types:

1. **mixed cropping**: growing two or more crops simultaneously with no distinct row arrangement;
2. **row intercropping**: one or more of the crops grown simultaneously in rows;
3. **strip intercropping**: two or more crops are grown in strips wide enough to permit independent cultivation, but narrow enough for the crops to interact agronomically;
4. in **trap-cropping** systems one species serves as a trap crop to decoy the pest away from the major crop.

Multiple cropping refers either to intercropping, i.e. crops growing simultaneously, or to sequential cropping, i.e. growing two or more crops in sequence on the same field per year [85].

Interactions between component crops make intercropping systems more complex and at the same time frequently reduce pest attack. Overwhelming evidence suggests that polycultures support a lower herbivore load than monocultures. A survey of 209 published studies on the effects of vegetation diversity in agroecosystems on herbivorous arthropod species showed that 52% of the total herbivore species were found to be less abundant in polycultures than in monocultures, whereas only 15% of the herbivore species exhibited higher population densities in polycultures [4] (Table 12.2).

As might be predicted, the cases of lower abundance in polycultures were predominantly among the food specialists. In contrast, polyphagous species often fared better and exhibited higher densities in polycultures (Table 12.2). Of course not all combinations of

Table 12.2 Relative abundance of arthropod species in polycultures when compared with monocultures. Figures are percentages of total numbers of species. A variable response means that an arthropod species did not consistently have a higher or lower population density in polycultures compared with monocultures when the species response was studied several times. (Source: reproduced, with permission, from the *Annual Review of Entomology*, vol 36, © 1991, by Annual Reviews Inc.)

	% more abundant	% no difference	% less abundant	% variable	Total no. of species
Herbivores					
Monophagous species	8	14	59	19	220
Polyphagous species	40	8	28	24	67
Natural enemies	53	13	9	26	130

crops are equally effective in this respect, and the choice of the partner crop is more important than the simple decision to practice intercropping. Combining wheat and maize, for instance, would actually increase the damage level inflicted by shared pests, such as chinch bugs (*Blissus* spp.) and nematodes, whereas intercropping wheat with potatoes would reduce the damage to wheat.

Although many studies have documented differences in single *versus* multicropping systems in intensity of herbivore attack, precise information is lacking on the mechanisms that generate these effects. Numerous biotic and abiotic factors vary between the two practices, including plant density and structural complexity, microclimatic factors, such as temperature, shadiness and humidity, refuges, alternative food sources for natural enemies (flowers, extrafloral nectaries), masking and repellent odours, and camouflage [10]. Yet the discovery of underlying mechanisms of yield responses to intercropping is vital both for generating predictive theory and for the application of this knowledge in managed systems.

Three theories that attempt to explain reduced pest infestations in polycultures have received much attention: these are (1) the disruptive-crop hypothesis; (2) the enemies hypothesis and (3) the trap-crop hypothesis [2, 85].

12.3.1 THE DISRUPTIVE-CROP HYPOTHESIS

A basic observation in ecology is that consumers tend to concentrate at places where their resources are abundant and easy to find. Root [69] formalized this phenomenon as the 'resource concentration hypothesis' (Chapter 9). The hypothesis predicts that herbivores are more likely to find and remain on host individuals grown in monocultures than host plants grown in spatially diluted systems, i.e. polycultures. Not only may insect populations be influenced directly by the spatial dispersion of their host plants, there can be also a direct effect of associated plant species on the ability of the insect herbivore to find and utilize its host. Volatiles emitted by non-host intercrops may mask the odour of the host plant, thereby disrupting the host-finding behaviour of the pest insect. Such 'olfactory masking' has been shown, for example, in relation to the orientation of Colorado potato beetles to potato odours. In laboratory experiments, starved Colorado potato beetles exhibit strong positive anemotactic responses to air currents blown over potato foliage, whereas responses to air streams with tomato odours do not differ from those to clean air. The attractiveness of host-plant odour, however, is completely masked in an odour blend of the two plant species [80] (see Fig. 5.13). A well-known example of olfactory masking is the old practice of interplanting carrots with onions to prevent attack by carrot flies [84]. Several aromatic herbs, likewise, have been used to repel insects infesting vegetable crops. Brussels sprouts intercropped with the herbs sage (*Salvia officinalis*) and thyme (*Thymus vulgaris*) received fewer eggs from the diamondback moth (*Plutella xylostella*) than pure stands, through an olfactory effect of the labiate herbs [27].

Insects in polycultures also show an increased tendency to leave their host plant, often followed by migration out of the field. In the case of the striped cucumber beetle (*Acalymma vittata*) densities reached in polycultures of cucumber, corn and broccoli were 10–30 times lower than in monocrops of cucumber [8] (Fig. 12.3).

Interestingly, in this case the associated crops also have an indirect effect upon the insect *via* its host plant. When under laboratory conditions the beetles were offered a choice between leaves taken from monocultures and those from cucumber plants intercropped with tomatoes, the insects preferred the foliage from plants in pure stands [9]. This indicates that plant-stand diversity and host-plant quality may interact in a very complex way.

Polycultures: why fewer pests?

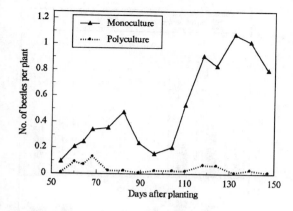

Figure 12.3 Mean number of striped cucumber beetles per plant in high-density plant systems of cucumber plants only (monoculture) and cucumber intercropped with maize and broccoli plants (polyculture). (Source: redrawn from Bach, 1980.)

12.3.2 THE ENEMIES HYPOTHESIS

According to Root's [69] enemies hypothesis, generalist and specialist natural enemies of insect pest species are expected to be more abundant in polycultures than in monocultures because polycultures often offer additional food sources, such as honeydew, nectar and pollen, and more refuges where they can shelter in the shade and encounter higher humidity during hot periods. Also, more alternative prey or herbivore hosts may be available in periods in which the pest species is scarce [19]. Natural enemies show lower emigration rates from diverse plant assemblages, whereas immigration rates are not affected [20]. A literature survey [4] showed that 53% out of a total of 130 natural enemy species did indeed attain higher population densities in polycultures compared to monocultures, whereas in only 9% of the observed cases were lower population densities encountered (Table 12.2). The dramatic yield increase of 100% for maize in a triculture with faba bean (*Vicia faba*) and squash (*Cucurbita moschata*) compared to a maize monoculture was due to reduced population densities of aphids (*Rhopalosiphum maidis*) and spider mites (*Tetranychus urticae*).

Table 12.3 Relative importance of regulating mechanisms in 36 reports on reduced insect pest levels in intercropping systems. (Source: modified from Baliddawa, 1985)

Pest population controlling factor	Occurrence
Lowered resource concentration, trap cropping, microclimate and physical obstruction	9
Reduced colonization	5
Masking and camouflage	5
Repellency	5
Natural enemies	12

In the triculture, aphids experienced higher levels of attack by several species of arthropod predator, including two species of ladybird more constantly associated with aphids in the diversified system [83]. An analysis of the causes of reduced pest insect levels through polycultural practice showed that in 12 out of 36 studies the effect was primarily due to natural enemy action [10] (Table 12.3).

12.3.3 TRAP-CROPPING AND CROP–WEED SYSTEMS

Trap crops are plant stands in the vicinity or in certain parts of a field where the principal crop is grown, which attract pest insects so that the target crop escapes pest infestation. Trap-cropping systems have been found to be particularly useful to subsistence farming in tropical countries. Thus a tomato monoculture in Central America was totally destroyed by *Spodoptera sunia* caterpillars, while an intercrop of tomatoes and beans was effective in reducing the attack to virtually zero. The caterpillars of *S. sunia* were all attracted to the bean plants, which served as a trap crop. To date trap cropping has played a major role in a few crops only: cotton, soybeans, potatoes and cauliflower. Of these, the cotton and soybean trap-cropping systems clearly have the greatest importance worldwide, although plenty of successful examples suggest that this strategy could be used much more than it actually is [39].

Whereas weeds can act as reservoirs of pests, and many pest outbreaks can be traced to locally abundant weeds belonging to the same family as the affected crop plants [81], weeds often harbour a beneficial entomofauna that may affect herbivore populations on adjacent crop plants positively. Because weeds can offer important sources for natural enemies, such as alternative food resources and microhabitats that are not available in weed-free monocultures, certain types of crop pest are less likely to develop in weed-diversified crop systems. Many examples are known of cropping strategies in which the presence of weeds enhanced the biological control of specific crop pests, ranging from fruit crops (e.g. apple), to vegetables (e.g. Brussels sprouts), fibre crops (e.g. cotton), grains (e.g. sorghum) and vineyards [2]. In an experiment to investigate 'green' pest control methods, which abandon the use of conventional insecticides in apple orchards, selected weeds were sown in strips to attract aphidophagous predators. After some time predaceous arthropods were more abundant in the strip-sown area of the orchard than in the control weed-free area. This difference was paralleled by significantly reduced numbers of two detrimental aphid species [92] (Fig. 12.4).

Field experiments involving several crops have also shown that careful diversification of the weedy component of agricultural systems often lowers pest populations significantly. More details of insect manipulation through weed management are given by Altieri [2].

12.3.4 DIVERSITY AS A GUIDING PRINCIPLE

Diversification is probably a key element in future insect control strategies in agriculture [64] and polycultures may provide an important step towards that future. There is an interesting form of polyculture that negates some technical disadvantages of culturing mixtures of two crop species: growing combinations of

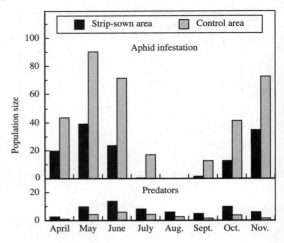

Figure 12.4 Aphid infestation rates and mean number of predaceous arthropods in an apple orchard. In one part of the orchard weed strips were sown in the existing ryegrass; the other part was used as a control area. Aphid population size was recorded in infestation classes and aphidophagous predators were recorded in absolute numbers per tree. During summer the aphids *Dysaphis plantaginea* and *Aphis pomi* live on alternative host plants. (Source: redrawn from Wyss, 1995.)

genetically different crop cultivars. When a cassava cultivar susceptible to whiteflies was grown intercropped with a cultivar that possessed partial resistance to the whitefly *Trialeurodes variabilis*, the overall population density of this insect in the intercropped system was 60% lower than in the monoculture [35]. So far, however, the potential gains of growing mixtures of resistant and susceptible varieties of a crop plant species are still largely unexplored.

Polycultural strategy has often been found to increase yields, sometimes to a considerable extent. An analysis of the mechanism causing the reduction of a pest population is not so easy, especially since several factors are often involved. Table 12.3, although based on a limited number of studies, shows that a variety of mechanisms, including lowered resource concentration, natural enemy action and various diversionary mechanisms, may be operating and be responsible for higher yield under polycultural practices.

12.4 PLANT-DERIVED INSECTICIDES AND ANTIFEEDANTS

In view of the ample evidence that most, if not all, herbivorous insects are inhibited from feeding by secondary compounds in non-host plants, it is a logical step to exploit such substances for protection of our food crops. Indeed since the dawn of civilization man has used plant materials to combat insect pests or alleviate the damage they cause. Although early agricultural writings frequently contain references to the use of plant extracts for pest control, the descriptions of the plants are often so vague as to make identification impossible. Nevertheless, well-documented records show that before 1850 20 plant species belonging to 16 different families were used for control of agricultural and horticultural pests in western Europe and China [57, 74]. The rich knowledge of plants with pesticide properties was not lost in China, as evidenced by a recent report stating that in China different parts or extracts of 276 plant species are used as pesticides [93].

A resurgent interest in plant-derived chemicals to control pest insects stems from the need for pesticide products with less negative environmental and health impacts than the highly effective synthetic insecticides mostly have. Some insecticides of plant origin were used on a large scale before they were outcompeted by synthetic insecticides. Nicotine (**41**), rotenone (**52**) and pyrethrins (**49**) have been extensively used and were effective insecticides which, because they degrade rapidly, do not accumulate in the food chain. Caution is required, however. Although many natural insecticides show lower mammalian toxicity than, for example, most organochlorine compounds, they are not harmless merely because they are natural products, a view that convincing statistics have shown to be a serious misconception [3]. Another reason to remain cautious when searching for new insecticides, whether natural or not, is the risk that target insect species may become resistant to them and, still more importantly, that non-target invertebrates, including natural enemies, are at risk. Compounds that modify the behaviour of target species and have a primarily non-toxic mode of action may in the long term provide the most dependable and environmentally safe method of chemical control. Behaviour-controlling phytochemicals include attractants, repellents and deterrents, several hundreds of which have been discussed in the literature [54]. We will discuss the use of feeding deterrents only as a behavioural method of insect pest management.

12.4.1 ANTIFEEDANTS

Feeding deterrents or 'antifeedants' are chemicals that, when perceived, reduce or prevent insect feeding. When produced by the plant such compounds decrease feeding damage and the risk of being infected with plant pathogens. The insect responds to the sensory detection of antifeedants by reducing food intake, which may either lead to it leaving the plant or to adverse effects on growth, development, survival and reproduction. In contrast to repellents, antifeedants do not cause oriented locomotion away from the source of stimulus [24]. In the presence of an antifeeding compound the insect may starve to death and females do not lay eggs before they find an untreated host. Some antifeedants have been found to be effective at very low doses, i.e. in the order of less than one part per million. Azadirachtin, one of the strongest antifeedants known, inhibits feeding at 0.01 ppm in the polyphagous desert locust, *Schistocerca gregaria*, when applied to palatable foliage [53], and 1 mg of this compound suffices to protect 100 m^2 of leaf surface from this notoriously devastating insect.

Candidate compounds for an antifeedant approach to insect control must possess several essential properties (Table 12.4),

Table 12.4 Criteria for antifeedant compounds as crop protectants

1. No or very low toxicity to vertebrates
2. No or very low phytotoxicity
3. Active at very low concentrations
4. Effective to many pest insect species
5. Harmless to beneficial arthropods (natural enemies, pollinators)
6. Penetration of plant surface and/or uptake by roots and systemic translocation
7. Compatible with other pest management methods
8. Limited persistence in environment
9. Sufficient source material
10. Amenable to commercial development (production costs, etc.)
11. Long shelf life

which, however, are fulfilled by few if any of the compounds assayed so far.

Since less than 1% of all secondary plant substances, estimated to number 400000 or more, have been tested, and then on a limited number of insect species only, several effective compounds may remain to be discovered. Promising chemicals that have attracted attention as potential antifeedants, either as source material for novel analogues or not, are listed in Table 12.5.

To date only azadirachtin-based products have been marketed. Among the drimanes polygodial (**47**), warburganal (**74**) and muzigadial (**38**) are of interest. Polygodial, a sesquiterpenoid extracted from the herb water pepper (*Polygonum hydropiper*) prevents probing behaviour in aphids at very low application rate. It has been found in field trials to reduce barley yellow dwarf virus transmission by the bird-cherry aphid, *Rhopalosiphum padi*, giving a 36% higher grain yield relative to untreated plots. Polygodial can be synthesized, but its action is very dependent on its stereochemistry. The (+) isomer must be removed from racemic mixtures because it is, in contrast to the natural (−) isomer, highly phytotoxic [62].

12.4.2 NEEM TREE, AZADIRACHTIN

Indian farmers, homemakers and folk healers have known for centuries that neem trees have many remarkable properties, including a strong repellency to many insects. More than half a century ago an Algerian agronomist noticed that only neem trees remained unconsumed by a locust plague and showed that leaf extracts were highly unpalatable to desert locusts [86]. With the advent of DDT and a subsequent array of broad-spectrum synthetic insecticides, neem as a potential source of chemicals to manipulate insects remained unnoticed until in the 1970s a German entomologist, H. Schmutterer, stimulated researchers from all over the world to launch studies on the useful properties of neem [71].

The neem tree, *Azadirachta indica* (Meliaceae; mahogany family), probably native to Burma, has been widely cultivated for a long time in tropical Asia and Africa, where it has become extensively naturalized. It is now also widely planted in Central America, because of its rapid growth and fine timber. The tree has proved to be very adaptable and able to with-

Table 12.5 Plant-derived antifeedants with promising properties for application in pest management systems, based on results from field experiments

Chemical class	Botanical source	Insects affected	Reference
Meliacins	Meliaceae	Many species	71
Drimanes	*Polygonum hydropiper*	Aphids	62
Limonoids	*Citrus paradisi*	Colorado potato beetle	55

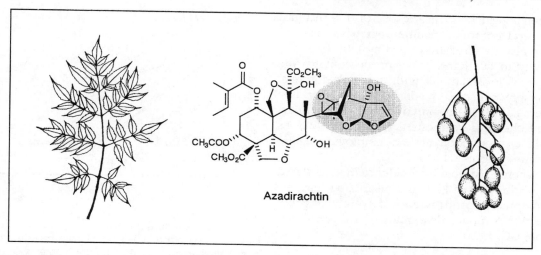

Figure 12.5 Bipinnate leaf and fruiting panicle from neem tree (*Azadirachta indica*) and structural formula of azadirachtin, a potent antifeedant and insect growth regulator. (Source: drawings reproduced from Schmutterer, 1995, with permission.)

stand arid conditions. Its bipinnate leaves are garlic-scented when damaged, and the fruits resemble olives (Fig. 12.5).

The Meliaceae, like most sister families belonging to the order Rutales, produce and accumulate bitter and biologically active nor-triterpenoids called limonoids or meliacins and quassinoids depending on structural features and occurrence. Azadirachtin, only one of more than 70 triterpenes from neem, is a highly oxidized limonoid with many reactive functional groups in close proximity to each other (Fig. 12.5). It occurs predominantly in the seeds of *A. indica* at a concentration of about 3.5 mg/g dry kernel, and is a very potent antifeedant to many insect species, especially lepidopterous larvae and several, but not all, Orthoptera. In addition to the antifeedant action, azadirachtin and related neem-seed derivatives have often pronounced physiological effects as well. After ingestion it causes growth inhibition, malformation and mortality due to interference with the insect's endocrine system [53, 71]. There is evidence suggesting that the left half of the azadirachtin molecule is the antiendocrine part whereas the hydroxyfuranacetal moiety (the right half, i.e. the grey part of the molecule in Fig. 12.5) is particularly important for insect antifeedant activity [16]. As a third mode of action, azadirachtin has been found to negatively affect food utilization through the inhibition of digestive enzymes [82]. To some insects, related compounds, such as salannin (**54**), also present in *A. indica* seeds, and toosendanin (**67**), isolated from the bark of the related *Melia toosendan*, are even more unpalatable than azadirachtin [50]. Whereas small farmers in the Indian continent use neem extracts in various traditional ways, there are now commercial neem products also on the market in some Western countries and several formulations have been patented [31].

Of all plant-derived compounds known to deter insect feeding or oviposition behaviour, azadirachtin currently offers the greatest potential for widespread use [56]. In many respects it fulfils the requirements of an ideal antifeedant, notably its relative safety to beneficial organisms in the environment, its practical non-toxicity to mammals and its systemic transport in crop plants [6], which ensures that piercing–sucking insect species, for instance several notorious planthopper pests on rice, are also deterred from feeding [70]. Its sensitivity to UV light, however, necessitates the use of formulations with sun-screen filters. Freshly collected seeds serve for the time being as the main source of neem compounds, but chemical synthesis is under way [22], as well as production methods employing *in vitro* tissue cultures [1].

There is compelling evidence that, as stated by Schmutterer [71], the neem tree 'has the potential to contribute to "solve global problems" (National Research Council, Washington, DC, 1992)'.

12.4.3 OUTLOOK FOR ANTIFEEDANTS AS CROP PROTECTANTS

As discussed before (Chapters 6 and 7) insects may, after repeated contact, habituate to the presence of a feeding deterrent. This is especially likely in polyphagous insect species [42] and would of course be a serious drawback to its usefulness. Indeed, habituation to low levels of azadirachtin has been observed in several insect species, including the Asian armyworm (*Spodoptera litura*). Interestingly, when a commercial product containing azadirachtin as well as neem oil was tested, no habituation occurred [17] (Fig. 12.6).

Another important issue when considering the development of behaviour modifying natural compounds for pest management is the prospect of resistance development. Long-lasting selection experiments with diamondback moth larvae (*Plutella xylostella*) showed that resistance can be developed to azadirachtin, albeit to a much lower degree than to the insecticide deltamethrin. Resistance to neem seed kernel extracts, containing a spectrum of various molecular agents including azadirachtin, developed still more slowly than to pure azadirachtin [87]. Presumably, the combination of the behavioural and physiological actions of azadirachtin makes it more difficult for the insect to

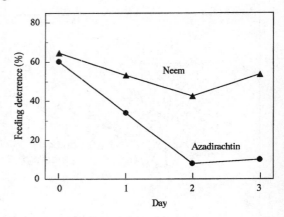

Figure 12.6 Feeding deterrence of cabbage leaf discs treated with 1.3 ng/cm^2 azadirachtin and neem seed extract containing the same absolute amount of azadirachtin in repeated choice assays with fifth instar larvae of *Spodoptera litura*. The values for neem do not change significantly, whereas those for azadirachtin do. (Source: redrawn from Bomford and Isman, 1996.)

develop resistance. Unlike ordinary insecticides based on a single active ingredient, the chemical defence of plants comprises an array of compounds with varying behavioural, physiological and toxicological properties, consequently more difficult for an insect to adapt to. As a matter of fact, it is unlikely that oligophagous insects, for instance, could easily develop general resistance to feeding inhibitory substances, because this would result in rapid changes of their host-plant range, which is primarily determined by the occurrence of such substances in non-host plants. However, such changes are very rare events in nature [42].

A difficulty in identifying antifeedants is that great differences exist between species in their sensitivity to a given antifeedant compound (Tables 6.6; 12.6).

Because most researchers, when testing candidate compounds, employ only a few or even only one insect species, effective antifeedants to a particular insect will easily escape attention. Among seven orthopterans tested for

Table 12.6 Antifeedant concentration (ppm) in wheat flour wafers that reduce food intake 50% in two locust species. (Source: data from Bernays and Chapman, 1978)

	Azadirachtin	Aristolochic acid
Desert locust (*Schistocerca gregaria*)	0.1	0.1
Migratory locust (*Locusta migratoria*)	100	0.01

sensitivity to azadirachtin interspecific differences span six orders of magnitude [53]. Several more caveats regarding searches for natural compounds with antifeedant activity are listed in some recent state-of-the art papers on antifeedants [31, 42].

Is there a realistic future for any large-scale use of antifeedants? They certainly do not constitute the final tool for control of insect pests. However, in view of the environmental strains imposed by present agricultural practices we cannot afford to leave thousands of natural defence substances provided by Nature unexplored. The fact that many plant species rely to a large extent on the presence of such compounds is a strong impetus for continual explorations of the Plant Kingdom. Advances made on the application of neem products seem to support the statement by Frazier and Chyb [31] that 'The practical use of natural product feeding inhibitors in insect control is rapidly becoming a reality'.

12.5 WEED CONTROL BY HERBIVOROUS INSECTS

Many plant species have been either purposely or accidentally transferred by humans to other parts of the world. The alien plants, once outside their natural habitats, have sometimes developed into aggressive invaders, which outcompeted native plant species and caused detrimental effects in natural ecosystems or inflicted significant losses to agricultural production. In several parts of the world 60–97% of the weeds are immigrant species (Table 12.7), demonstrating that plants can become undesirable weeds in foreign habitats.

As a result, crop losses to weeds exceed those attributed to insects, and expenditure on herbicides worldwide is about 30% higher than on insecticides [64]. For obvious reasons biological control has several advantages over other types of weed control [88]. Exotic plant species that have become weeds can sometimes be controlled by introducing host-specific insects from the plant's place of origin. Two outstandingly successful cases of control of invasive weeds by introduction of their herbivores exemplify the principle of biological weed control.

Table 12.7 Origin of weed species in North America and Australia. 'Europe' includes species from Eurasia. (Source: data from Gassmann, 1995 and Pimentel, 1986)

	No. of weeds	Origin of weeds (%)				
		Native	Europe	America	Asia	Africa
Canada	516	40	52	4	3	0
Canada: common weeds	126	2	71	25	2	0
Australia	637	7	39	26	7	18
Australia: state of Victoria	83	4	60	23	4	10
USA: weeds in cultivated crops	80	28	50	8	5	1

12.5.1 OPUNTIA AND SALVINIA

Prickly pears are cactus species native to North and South America. Among the 30 or so species that were in the past introduced to Australia as pot or garden plants, two species, *Opuntia stricta* and *O. inermis*, ran out of control. *Opuntia stricta* was brought to Australia in 1839 in a pot from the southern United States and was planted as a hedge plant in eastern Australia. It gradually developed into a pestilential weed difficult to control by mechanical methods, burning, etc. By 1900 it had occupied 4 million hectares in the coastal regions of Australia, and was by then rapidly spreading inland into immense areas of wheat, rangeland and marginal agricultural land, choking out most other plant life. In 1925 some 25 million hectares were infested in Queensland and New South Wales alone. About one-half of this area was covered with dense growth, over 1 m in height and so dense as to be virtually impenetrable to humans and livestock. Farms were abandoned. In 1920 the Commonwealth Prickly Pear Board was appointed to attempt control of the weed by establishing insects and mites that feed on these cacti. Some species collected from the rich fauna present on American cacti were of some service, but a major breakthrough did not occur until the release between 1927 and 1930 of masses of a small Argentinian moth, *Cactoblastis cactorum*, whose larvae mine in the paddle-like cactus stems. In 2 years time the original stands of prickly pears had collapsed under the onslaught of the moth larvae. Successful accomplishment of the great biological control programme became apparent in 1939, and the Board was disbanded. 'Great tracts of country, utterly useless on account of the dense growth of the weed, have been brought into production. The prickly pear territory has been transformed as though by magic from a wilderness to a scene of prosperous endeavour' as victoriously described by Dodd [26]. At present the moth still maintains prickly pears as a scattered plant at low, stable equilibrium.

Since 1939 salvinia (*Salvinia molesta*), a floating aquatic fern about 2–10 cm long native to south-eastern Brazil, has spread by human agency to many tropical and subtropical parts of the world. Outside its native range its unlimited growth has caused serious problems because it forms mats up to 1 m thick, covering whole lakes and rice paddies and completely blocking all waterways, including slow-moving rivers and irrigation canals. Its growth capacity is evinced, for example, by its proliferation after invading the Sepik River floodplain of Papua New Guinea. A few plants introduced in 1972 grew in 8 years into mats covering 250 km^2 and weighing 2 million tonnes, severely disrupting the normal life of the local population, which was forced to migrate and to abandon whole villages [68]. In Brazil salvinia was found to be attacked by a tiny (2 mm long) weevil species, *Cyrtobagous salviniae*, unknown before then. This insect was distributed throughout Australian salvinia infestations during the early 1980s and turned out to be an extremely effective weed control agent. The weevil population increased in less than 1 year from a few thousands to more than 100 million individuals and destroyed 30 000 tonnes of salvinia. In Africa and India also the insect reduced the sizes of salvinia populations by more than 99%, before new, low-density equilibria were attained [68].

12.5.2 SUCCESS RATE OF BIOLOGICAL WEED CONTROL PROGRAMMES

Certainly, not all attempts of weed control by insects have met with the same spectacular successes as those described above. Even the successful establishment of an imported agent is no guarantee that it makes any impact on the abundance of the weed. Thus, whereas at least 69% of released arthropod species established on alien weed plants (Fig. 12.7(A)), complete control has been achieved on fewer occasions [33, 73] (Fig. 12.7(B)) and the degree of control varies under different circumstances.

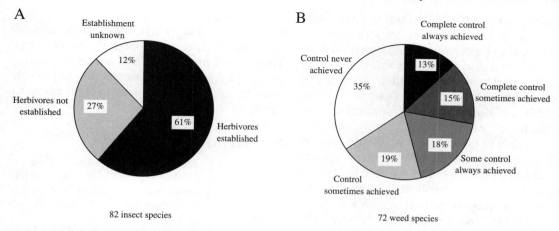

Figure 12.7 Succcess rates of establishment and control of invasive weeds by introduced alien herbivores. **(A)** Proportions of arthropod species established on weeds of European origin **(B)** Degree of weed control by insects which have been introduced and established long enough to permit control assessment. (Source: **(A)** data from Julien, 1992; **(B)** redrawn from Sheppard, 1992, data from Julien, 1992.)

Although control by herbivorous insects is usually considered for cases of introduced weed species, under certain circumstances it is also a potential method for controlling native weeds. Thus, a biocontrol programme is being developed to combat bracken (*Pteridium aquilinum*) in the UK, where this weed is becoming increasingly invasive. The very success of this plant species worldwide has resulted in different assemblages of herbivores attacking it in different parts of the world. A mesophyll-feeding leafhopper (*Eupteryx maigudoi*), native to South Africa, is currently being investigated for its suitability as a biological control agent in the UK [29].

The advantages of weed control by employment of natural enemies need hardly be emphasized (Table 12.8). Its weakest point is the unpredictability of its results. This is caused by the fact that some essential demographic parameters, especially those of the functional and numerical responses, can only be determined after the herbivore has been released, because their values depend so critically on local conditions.

An interesting debate concerns the fundamental question whether or not herbivores

Table 12.8 Advantages and disadvantages of biological control of weeds

Advantages
- Reasonably permanent management of target species
- No harmful side-effects, environmentally safe
- Attack restricted to specific target species (or very small group of closely related species)
- Agents are self-perpetuating, often density dependent, and self-disseminating
- High benefit–cost ratios for successful programmes
- Costs are non-recurrent

Disadvantages
- Relatively slow-acting
- If target weed is related to a crop, the number of usable herbivores is greatly reduced

that are highly adapted to their hosts are likely to be the most effective control agents. Whereas highly specialized insects may flourish once confronted with unlimited food resources, it has also been argued that the plant partner in a less close insect–plant association may be more susceptible to insect attack than plant species with long-standing intimate relationships with their specialist herbivores [29, 64]. *Cactoblastis cactorum*, which turned out to be very successful in controlling *Opuntia stricta* and *O. inermis*, was obtained in South America not from either species but from a different species of *Opuntia*. On the other hand, the weevil that suppressed the invasions of salvinia is a highly adapted species, able to increase its population rapidly at its host's expense. Employing insect species that are strict monophages reduces the risk of them switching to other hosts. A complicating factor is the role of natural enemies in regulating the herbivore's population density in its native region. It is generally thought that uncoupling specialist herbivores from their normal natural enemies is a key part of biological weed control [29].

Unfortunately, weed control programmes still lack a firm theoretical basis. Perhaps an approach based on only one control agent is utterly wrong and a more diversified control system, including plant pathogens, would be more appropriate. *Lantana camara*, which developed into a pestilential weed in Hawaii, may serve as an example. Some control of this species could only be obtained after several insect species had been introduced, that together eventually constituted a large guild of herbivores. A better understanding of the factors regulating plant populations would help to improve weed control methods by natural agents, including insects [21].

12.6 CONCLUSION: DIVERSIFICATION HOLDS THE CLUE TO THE CONTROL OF PESTIFEROUS INSECTS

The common principle for successful insect control based on biological principles is diversification [64]. Resistance breeding depends on the availability of large gene pools, and pest resistance is more effective when a variety of resistance factors are combined. Diversification of crop cultivation practices, such as intercropping and crop rotation, often reduces the risks of serious insect damage. Antifeedant compounds are more effective when applied in mixtures and when they affect various behavioural and physiological mechanisms. The limited numbers of the successful examples of biological weed control possibly result from insufficient attention to the need to diversify.

12.7 REFERENCES

1. Allan, E. J., Eeswara, J. P., Johnson, S., Mordue (Luntz), A. J., Morgan, E. D. and Stuckbury, T. (1994) The production of azadirachtin by in-vitro tissue cultures of neem, *Azadirachta indica*. *Pestic. Sci.*, **42**, 147–152.
2. Altieri, M. A. (1994) *Biodiversity and Pest Management in Agroecosystems*, Food Products Press, New York.
3. Ames, B. N., Profet, M. and Gold, L. S. (1990) Dietary pesticides (99.99% all natural). *Proc. Natl Acad. Sci. USA*, **87**, 7777–7781.
4. Andow, D. A. (1991) Vegetational diversity and arthropod population response. *Annu. Rev. Entomol.*, **36**, 561–586.
5. Anon. (1979) *Lab to Land: Good News to the Tobacco Growers*, Central Tobacco Research Institute, Rajamundry, pp. 1–5.
6. Arpaia, S. and van Loon, J. J. A. (1993) Effects of azadirachtin after systemic uptake into *Brassica oleracea* on larvae of *Pieris brassicae*. *Entomol. Exp. Appl.*, **66**, 39–45.
7. Arpaia, S., Gould, F. and Kennedy, G. (1997) Potential impact of *Coleomegilla maculata* predation on adaptation of *Leptinotarsa decemlineata* to Bt-transgenic potatoes. *Entomol. Exp. Appl.*, **82**, 91–100.
8. Bach, C. E. (1980) Effects of plant density and diversity on the population dynamics of a specialist herbivore, the striped cucumber beetle, *Acalymma vittata* (Fab.). *Ecology*, **61**, 1515–1530.
9. Bach, C. E. (1981) Host plant growth form and diversity: effects on abundance and feeding preference of a specialist herbivore, *Acalymma vittata* (Coleoptera, Chrysomelidae). *Oecologia*, **50**, 370–375.
10. Baliddawa, C. W. (1985) Plant species diversity and crop pest control. An analytical review. *Insect Sci. Appl.*, **6**, 479–487.
11. Barbosa, P. (1993) Lepidopteran foraging on plants in agroecosystems: constraints and consequences, in *Caterpillars. Ecological and Evolu-*

tionary Constraints on Foraging, (eds N. E. Stamp and T. M. Casey), Chapman & Hall, New York, pp. 523–566.
12. Barbosa, P. and Schultz, J. C. (1987) *Insect Outbreaks*, Academic Press, San Diego, CA.
13. Bergelson, J. (1994) The effects of genotype and the environment on costs of resistance in lettuce. *Am. Nat.*, **143**, 349–359.
14. Bernays, E. A. and Chapman, R. F. (1978) Plant chemistry and acridoid feeding behaviour, in *Biochemical Aspects of Plant and Animal Coevolution*, (ed. J. B. Harborne), Academic Press, London, pp. 99–141.
15. Bird, J. V. (1982) Multi-adversity (diseases, insects and stresses) resistance (MAR) in cotton. *Plant Dis.*, **66**, 173–176.
16. Blaney, W. M., Simmonds, M. S. J., Ley, S. V., Anderson, J. C., Smith, S. C. and Wood, A. (1994) Effect of azadirachtin-derived decalin (perhydronaphthalene) and dihydrofuranacetal (furo[2,3-b] pyran) fragments on the feeding behaviour of *Spodoptera littoralis*. *Pestic. Sci.*, **40**, 169–173.
17. Bomford, M. K. and Isman, M. B. (1996) Desensitization of fifth instar *Spodoptera litura* (Lepidoptera, Noctuidae) to azadirachtin and neem. *Entomol. Exp. Appl.*, **81**, 307–313.
18. Chew, F. (1995) From weeds to crops: changing habitats of pierid butterflies (Lepidoptera, Pieridae). *J. Lepidop. Soc.*, **49**, 285–303.
19. Coll, M. (1995) Parasitoid activity and plant species composition in intercropped systems, in *Enhancing Natural Control of Arthropod Pests Through Habitat Management*, (eds C. H. Pickett and R. L. Bugg), Agricultural Access Publishers, David, CA.
20. Coll, M. and Bottrell, D. G. (1996) Movement of an insect parasitoid in simple and diverse plant assemblages. *Ecol. Entomol.*, **21**, 141–149.
21. Crawley, M. J. (1983) *Herbivory. The Dynamics of Animal–Plant Interactions*, Blackwell, Oxford.
22. Denholm, A. A., Jennens, L., Ley, S. V. and Wood, A. (1995) Chemistry of insect antifeedants from *Azadirachta indica*. Part 19: A potential relay route for the synthesis of azadirachtin. *Tetrahedron*, **51**, 6591–6604.
23. De Ponti, O. M. B. and Mollema, C. (1992) Breeding strategies for insect resistance, in *Plant Breeding in the 1990s*, (eds H. T. Stalker and J. P. Murphy), CAB International, Wallingford, Oxon, pp. 323–346.
24. Dethier, V. G., Barton Browne, L. and Smith, C. N. (1960) The designation of chemicals in terms of the responses they elicit from insects. *J. Econ. Entomol.*, **53**, 134–136.
25. Dicke, M., Sabelis, M. A., Takabayashi, J., Bruin, J. and Posthumus, M. A. (1990) Plant strategies of manipulating predator–prey interactions through allelochemicals: prospects for application in pest control. *J. Chem. Ecol.*, **16**, 3091–3118.
26. Dodd, A. P. (1940) *The Biological Campaign Against Prickly Pear*, Government Printer, Brisbane, Queensland.
27. Dover, J. W. (1986) The effect of labiate herbs and white clover on *Plutella xylostella* oviposition. *Entomol. Exp. Appl.*, **42**, 243–247.
28. Dunn, J. A. and Kempton, D. P. H. (1972) Resistance to attack by *Brevicoryne brassicae* among plants of Brussels sprouts. *Ann. Appl. Biol.*, **72**, 1–11.
29. Fowler, S. V. (1993) The potential for control of bracken in the UK using introduced herbivorous insects. *Pestic. Sci.*, **37**, 393–397.
30. França, F. H., Plaisted, R. L., Roush, R. T., Via, S. and Tingey, W. M. (1994) Selection response of the Colorado potato beetle for adaptation to the resistant potato, *Solanum berthaultii*. *Entomol. Exp. Appl.*, **73**, 101–109.
31. Frazier, J. L. and Chyb, S. (1995) Use of feeding inhibitors in insect control, in *Regulatory Mechanisms in Insect Feeding*, (eds R. F. Chapman and G. de Boer), Chapman & Hall, New York, pp. 364–381.
32. Fritzsche, R., Decker, H., Lehmann, W., Karl, E. and Geissler, K. (1987) *Resistenz von Kulturpflanzen gegen tierische Schaderreger*, Springer-Verlag, Berlin.
33. Gassmann, A. (1995) Europe as a source of biological control agents of exotic invasive weeds: status and implications. *Mitt. Schweiz. Entomol. Ges.*, **68**, 313–322.
34. Gatehouse, A. M. R., Shi, Y., Powell, K. S., Brough, C., Hilder, V. A., Hamilton, W. D. O., Newell, C. A., Boulter, D. and Gatehouse, J. A. (1993) Approaches to insect resistance using transgenic plants. *Phil. Trans. Roy. Soc. Biol. Sci.*, **342**, 279–286.
35. Gold, C. S., Altieri, M. A. and Bellotti, A. C. (1989) Effects of cassava varietal mixtures on the whiteflies *Aleurotrachelus socialis* and *Trialeurodes variabilis* in Colombia. *Entomol. Exp. Appl.*, **53**, 195–202.
36. Gould, F. (1986) Simulation models for predicting durability of insect-resistant germ plasm: a deterministic diploid, two-locus model. *Environ. Entomol.*, **15**, 1–10.

37. Gould, F. (1988) Genetics of pairwise and multispecies plant–herbivore coevolution, in *Chemical Mediation of Coevolution*, (ed. K. S. Spencer), Academic Press, San Diego, CA, pp. 13–55.
38. Harlan, J. R. and Starks, K. J. (1980) Germplasm resources and needs, in *Breeding Plants Resistant to Insects*, (eds F. G. Maxwell and P. R. Jennings), John Wiley, New York, pp. 253–273.
39. Hokkanen, H. M. T. (1991) Trap cropping in pest management. *Annu. Rev. Entomol.*, **36**, 119–138.
40. Howe, W. L., Sanborn, J. R. and Rhodes, A. M. (1976) Western corn rootworm adult and spotted cucumber beetle associations with *Cucurbita* and cucurbitacins. *Envir. Entomol.*, **5**, 1043–1048.
41. Jena, K. K. and Khush, G. S. (1990) Introgression of genes from *Oryza officinalis* Well × Watt to cultivated rice, *O. sativa* L. *Theor. Appl. Genet.*, **80**, 737–745.
42. Jermy, T. (1990) Prospects of antifeedant approach to pest control – a critical review. *J. Chem. Ecol.*, **16**, 3151–3166.
43. Johns, T. and Alonso, J. G. (1990) Glycoalkaloid change during domestication of the potato, *Solanum* section *Petota*. *Euphytica*, **50S**, 203–210.
44. Josefsson, E. (1967) Distribution of thioglucosides in different parts of *Brassica* plants. *Phytochemistry*, **6**, 1617–1627.
45. Julien, M. H. (1992) *Biological Control of Weeds*, 3rd edn, CAB International, Wallingford, Oxon.
46. Kim, K. C. and McPheron, B. A. (1993) *Evolution of Insect Pests. Patterns of Variation*, John Wiley, New York.
47. Kogan, M. (1986) Plant defense strategies and host-plant resistance, in *Ecological Theory and Integrated Pest Management Practice*, (ed. M. Kogan), John Wiley, New York, pp. 83–134.
48. Kogan, M. and Ortman, E. F. (1978) Antixenosis – a new term proposed to define Painter's nonpreference modality of resistance. *Bull. Ent. Soc. Am.*, **24**, 175–176.
49. Lovett, J. V. and Hoult, A. H. C. (1992) Gramine: the occurrence of a self defence chemical in barley, *Hordeum vulgare* L., in *Proceedings of the 6th Australian Society of Agronomy Conference, Armidale, NSW*, pp. 426–429.
50. Luo Lin-er, van Loon, J. J. A. and Schoonhoven, L. M. (1995) Behavioural and sensory responses to some neem compounds by *Pieris brassicae* larvae. *Physiol. Entomol.*, **20**, 134–140.
51. Maliepaard, C., Bas, N., van Heusden, S., Kos, J., Pet, G., Verkerk, R., Vrielink, R., Zabel, P. and Lindhout, P. (1995) Mapping of QTLs for glandular trichome densities and *Trialeurodes vaporariorum* (greenhouse whitefly) resistance in an F_2 from *Lycopersicon esculentum* × *Lycopersicon hirsutum* f. *glabratum*. *Heredity*, **75**, 425–433.
52. Maxwell, F. G. and Jennings, P. R. (1980) *Breeding Plants Resistant to Insects*, John Wiley, New York.
53. Mordue (Luntz), A. J. and Blackwell, A. (1993) Azadirachtin: an update. *J. Insect Physiol.*, **39**, 903–924.
54. Morgan, E. D. and Mandava, N. B. (1990) *CRC Handbook of Natural Pesticides, vol. 4. Insect Attractants and Repellents*, CRC Press, Boca Raton, FL.
55. Murray, K. D., Alford, A. R., Groden, E., Drummond, F. A., Storch, R. H., Bentley, M. D. and Sugathapala, P. M. (1993) Interactive effects of an antifeedant used with *Bacillus thuringiensis* var. *san diego* delta endotoxin on Colorado potato beetle (Coleoptera, Chrysomelidae). *J. Econ. Entomol.*, **86**, 1793–1801.
56. National Research Council (1992) *Neem, A Tree for Solving Problems*, National Academy Press, Washington, DC.
57. Needham, J. (1986) *Science and Civilization in China*, vol. 6(1), Cambridge University Press, Cambridge.
58. Nickle, J. L. (1937) Pest situation in changing agricultural systems – a review. *Bull. Entomol. Soc. Am.*, **19**, 136–142.
59. Oerke, E. C., Dehne, H. W., Schönbeck, F. and Weber, A. (1994) *Crop Production and Crop Protection. Estimated Losses in Major Food and Cash Crops*, Elsevier, Amsterdam.
60. Painter, R. H. (1951) *Insect Resistance in Crop Plants*, University Press of Kansas, Lawrence, KS.
61. Panda, N. and Khush, G. S. (1995) *Host Plant Resistance to Insects*, CAB International, Wallingford, Oxon.
62. Pickett, J. A., Dawson, G. W., Griffiths, D. C., Hassanali, A., Merritt, L. A., Mudd, A., Smith, M. C., Wadhams, L. J., Woodcock, C. M. and Zhang, Z. (1987) Development of plant antifeedants for crop protection, in *Pesticide Science and Biotechnology*, (eds R. Greenhalgh and T. R. Roberts), Blackwell, Oxford, pp. 125–128.
63. Pimentel, D. (1986) Biological invasions of plants and animals in agriculture and forestry, in *Ecology of Biological Invasions of North America and Hawaii*, (eds H. A. Mooney and J. A. Drake), Springer-Verlag, New York, pp. 149–162.
64. Pimentel, D. (1991) Diversification of biological

control strategies in agriculture. *Crop Prot.*, **10**, 243–253.
65. Pimentel, D. (1993) Habitat factors in new pest invasions, in *Evolution of Insect Pests. Patterns of Variation*, (eds K. C. Kim and B. A. McPheron), John Wiley, New York, pp. 165–181.
66. Quiras, C. F., Stevens, M. A., Rick, C. M. and Kok-Yokomi, M. K. (1977) Resistance in tomato to the pink form of the potato aphid (*Macrosiphum euphorbiae* Thomas): the role of anatomy, epidermal hairs, and foliage composition. *J. Am. Soc. Hortic. Sci.*, **102**, 166–171.
67. Risch, S. J. (1987) Agricultural ecology and insect outbreaks, in *Insect Outbreaks*, (eds P. Barbosa and J. C. Schultz), Academic Press, San Diego, CA, pp. 217–238.
68. Room, P. M. (1990) Ecology of a simple plant–herbivore system: biological control of Salvinia. *Trends Ecol. Evol.*, **5**, 74–79.
69. Root, R. (1973) Organization of a plant-arthropod association in simple and diverse habitats. The fauna of collards. *Ecol. Monogr.*, **43**, 95–124.
70. Saxena, R. C. (1987) Antifeedants in tropical pest management. *Insect Sci. Applic.*, **8**, 731–736.
71. Schmutterer, H. (1995) *The Neem Tree. Source of Unique Natural Products for Integrated Pest Management, Medicine, Industry and Other Purposes*, VCH Verlagsgesellschaft, Weinheim.
72. Schroeder, H. E., Gollasch, S., Moore, A., Tabe, L. M., Craig, S., Hardie, D. C., Chrispeels, M. J., Spencer, D. and Higgins, T. J. V. (1995) Bean α-amylase inhibitor confers resistance to the pea weevil (*Bruchus pisorum*) in transgenic peas (*Pisum sativum* L.). *Plant Physiol.*, **107**, 1233–1240.
73. Sheppard, A. W. (1992) Predicting biological weed control. *Trends Ecol. Evol.*, **7**, 290–291.
74. Smith, A. E. and Secoy, D. M. (1981) Plants used for agricultural pest control in western Europe before 1850. *Chem. Industry*, **3 Jan.**, 12–17.
75. Smith, C. M. (1989) *Plant Resistance to Insects. A Fundamental Approach*, John Wiley, New York.
76. Smith, C. M. and Quisenberry, S. S. (1994) The value and use of plant resistance to insects in integrated crop management. *J. Agric. Entomol.*, **11**, 189–190.
77. Smith, C. M., Khan, Z. R. and Pathak, M. D. (1994) *Techniques for Evaluating Insect Resistance in Crop Plants*, CRC Press, Boca Raton, FL.
78. Southwood, T. R. E. (1973) The insect/plant relationship – an evolutionary perspective. *Symp. Roy. Entomol. Soc.*, **6**, 3–30.
79. Stoner, K. A. (1992) Bibliography of plant resistance to arthropods in vegetables, 1977–1991. *Phytoparasitica*, **20**, 125–180.
80. Thiéry, D. and Visser, J. H. (1987) Misleading the Colorado potato beetle with an odor blend. *J. Chem. Ecol.*, **13**, 1139–1146.
81. Thresh, J. M. (1981) *Pests, Pathogens and Vegetation: the Role of Weeds and Wild Plants in the Ecology of Crop Pests and Diseases*, Pitman, Boston, MA.
82. Timmins, W. A. and Reynolds, S. E. (1992) Azadirachtin inhibits secretion of trypsin in midgut of *Manduca sexta* caterpillars: reduced growth due to impaired protein digestion. *Entomol. Exp. Appl.*, **63**, 47–54.
83. Trujillo-Arriaga, J. and Altieri, M. A. (1990) A comparison of aphidophagous arthropods on maize polycultures and monocultures, in Central Mexico. *Agric. Ecosystems Environ.*, **31**, 337–349.
84. Uvah, I. I. I. and Coaker, T. H. (1984) Effect of mixed cropping on some insect pests of carrots and onions. *Entomol. Exp. Appl.*, **36**, 159–167.
85. Vandermeer, J. (1989) *The Ecology of Intercropping*, Cambridge University Press, Cambridge.
86. Volkonsky, M. (1937) Sur l'action acridifuge des extraits de feuilles de *Melia azedarach*. *Arch. Inst. Pasteur Algér.*, **15**, 427–432.
87. Völlinger, M. (1995) Studies of the probability of development of resistance of *Plutella xylostella* to neem products, in *The Neem Tree. Source of Unique Natural Products for Integrated Pest Management, Medicine, Industry and Other Purposes*, (ed. H. Schmutterer), VCH Verlagsgesellschaft, Weinheim, pp. 477–483.
88. Wapshere, A. J., Delfosse, E. S. and Cullen, J. M. (1989) Recent developments in biological control of weeds. *Crop Prot.*, **8**, 227–250.
89. Williams, W. G., Kennedy, G. G., Yamamoto, R. T., Thacker, J. D. and Bordner, J. (1980) 2-Tridecanone, a naturally occurring insecticide from the wild tomato *Lycopersicon hirsutum* f. *glabratum*. *Science*, **207**, 888–889.
90. Windle, P. N. and Franz, E. H. (1979) The effects of insect parasitism on plant competition: greenbugs and barley. *Ecology*, **60**, 521–529.
91. Wink, M. (1988) Plant breeding: importance of plant secondary metabolites for protection against pathogens and herbivores. *Theor. Appl. Genet.*, **75**, 225–233.
92. Wyss, E. (1995) The effects of weed strips on aphids and aphidophagous predators in an apple orchard. *Entomol. Exp. Appl.*, **75**, 43–49.
93. Yang, R. Z. and Tang, C. S. (1988) Plants used for pest control in China, a literature review. *Econ. Bot.*, **42**, 376–406.

APPENDIX A: FURTHER READING

BOOKS WHOLLY OR TO A LARGE EXTENT FOCUSED ON INSECT–PLANT INTERACTIONS

Abrahamson, W. G. (ed.) (1989) *Plant-Animal Interactions*, McGraw-Hill, New York, 480 pp.

Ahmad, S. (ed.) (1983) *Herbivorous Insects, Host Seeking Behavior and Mechanisms*, Academic Press, San Diego, CA, 257 pp.

Ananthakrishnan, T. N. (1994) *Functional Dynamics of Phytophagous Insects*, Oxford and IBH Publishing, New Delhi, 304 pp.

Barbosa, P. and Schultz, J. C. (eds) (1987) *Insect Outbreaks*, Academic Press, San Diego, CA, 578 pp.

Barbosa, P. and Letourneau, D. K. (eds) (1988) *Novel Aspects of Insect–Plant Interactions*, John Wiley, New York, 362 pp.

Barbosa, P. and Wagner, M. R. (1989) *Introduction to Forest and Shade Tree Insects*, Academic Press, San Diego, CA, 639 pp.

Barbosa, P., Krischik, V. A. and Jones, C. G. (eds) (1991) *Microbial Mediation of Plant–Herbivore Interactions*, John Wiley, New York, 530 pp.

Bernays, E. A. (ed.) (1989–1994) *Insect–Plant Interactions*, vols 1–5, CRC Press, Boca Raton, FL, 164, 199, 258, 240, 240 pp.

Bernays, E. A. and Chapman, R. F. (1994) *Host-Plant Selection by Phytophagous Insects*, Chapman & Hall, New York, 312 pp.

Boethel, D. J. and Eikenbary, R. D. (eds) (1986) *Interactions of Plant Resistance and Parasitoids and Predators of Insects*, Horwood, Chichester, 224 pp.

Brattsten, L. B. and Ahmad, S. (eds) (1986) *Molecular Aspects of Insect–Plant Interactions*, Plenum Press, New York, 346 pp.

Brues, C. T. (1946) *Insect Dietary*, Harvard University Press, Cambridge, MA (reprinted in 1972 under the title *Insects, Food and Ecology*, Dover Publications, New York), 466 pp.

Crawley, M. J. (1983) *Herbivory. The Dynamics of Animal–Plant Interactions*, Blackwell, Oxford, 437 pp.

Denno, R. F. and McClure, M. S. (eds) (1983) *Variable Plants and Herbivores in Natural and Managed Systems*, Academic Press, New York, 717 pp.

Edwards, P. J. and Wratten, S. D. (1983) *Insect/Plant Relationships*, Edward Arnold, London, 60 pp.

Fritz, R. S. and Simms, E. L. (eds) (1992) *Plant Resistance to Herbivores and Pathogens. Ecology, Evolution and Genetics*, University of Chicago Press, Chicago, IL, 590 pp.

Gilbert, L. E. and Raven, P. H. (eds) (1975) *Coevolution of Animals and Plants*, University of Texas Press, Austin, TX, 246 pp.

Green, M. B. and Hedin, P. A. (eds) (1986) *Natural Resistance of Plants to Pests: Role of Allelochemics*, ACS Symposium 246, Washington, DC.

Harborne, J. B. (ed.) (1978) *Biochemical Aspects of Plant and Animal Coevolution*, Academic Press, New York, 435 pp.

Heinrichs, E. A. (ed.) (1988) *Plant Stress–Insect Interaction*, John Wiley, New York, 492 pp.

Hodkinson, I. D. and Hughes, M. K. (1982) *Insect Herbivory*, Chapman & Hall, London, 77 pp.

Howe, H. F. and Westley, L. C. (1988) *Ecological Relationships of Plants and Animals*, Oxford University Press, Oxford, 273 pp.

Jolivet, P. (1986) *Insects and Plants. Parallel Evolution and Adaptations*, Brill, New York, 197 pp.

Juniper, B. E. and Southwood, T. R. E. (eds) (1986) *Insects and the Plant Surface*, Edward Arnold, London, 360 pp.

Kim, K. C. and McPheron, B. A. (eds) (1993) *Evolution of Insect Pests. Patterns of Variation*, John Wiley, New York, 479 pp.

Mattson, W. J., Levieux, J. and Bernard-Degan, C. (eds) (1988) *Mechanisms of Woody Plant Defenses Against Insects*, Springer-Verlag, Berlin, 416 pp.

Maxwell, F. G. and Jennings, P. R. (eds) (1980) *Breeding *Plants Resistant to Insects*, John Wiley, New York, 683 pp.

Metcalf, R. L. and Metcalf, E. R. (1992) *Plant Kairomones in Insect Ecology and Control*, Chapman & Hall, London, 168 pp.

Miller, J. R. and T. A. Miller (eds) (1986) *Insect–Plant Interactions*, Springer-Verlag, Berlin, 342 pp.

Painter, R. H. (1951) *Insect Resistance in Crop Plants*, University Press of Kansas, Lawrence, KS, 520 pp.

Panda, N. and Khush, G. S (1995) *Host Plant Resistance to Insects*, CAB International, Wallingford, Oxon, 431 pp.

Rosenthal, G. A. and Berenbaum, M. R. (eds) (1991, 1992) *Herbivores. Their Interactions with Secondary Plant Metabolites*, 2nd edn, 2 vols, Academic Press, New York, 468, 493 pp.

Rosenthal, G. A. and Janzen, D. H. (eds) (1979) *Herbivores. Their Interaction with Secondary Plant Metabolites*, Academic Press, New York, 718 pp.

Smith, C. M. (1984) *Plant Resistance to Insects. A Fundamental Approach*, John Wiley, New York, 286 pp.

Spencer, K. V. (ed.) (1988) *Chemical Mediation of Coevolution*, Academic Press, New York, 609 pp.

Stamp, N. E. and Casey, T. M. (eds) (1993) *Caterpillars. Ecological and Evolutionary Constraints on Foraging*, Chapman & Hall, New York, 587 pp.

Strong, D. R., Lawton, J. H. and Southwood, T. R. E. (1984) *Insects on Plants. Community Patterns and Mechanisms*, Blackwell, Oxford, 313 pp.

Tallamy, D. W. and Raupp, M. J. (eds) (1991) *Phytochemical Induction by Herbivores*, John Wiley, New York, 431 pp.

Van Emden, H. F. (ed.) (1973) *Insect/Plant Relationships*, Symposium of the Royal Entomological Society of London 6, Blackwell, Oxford, 213 pp.

Wallace, J. W. and Mansell, R. L. (eds) (1976) *Biochemical Interaction between Plants and Insects*, Plenum Press, New York, 425 pp.

PROCEEDINGS OF INTERNATIONAL SYMPOSIA ON INSECT–PLANT RELATIONSHIPS

Proceedings of the 1st International Symposium, Insect and Foodplant. Wageningen, 1957 (J. de Wilde, ed.). *Entomol. Exp. Appl.*, **1**, 1–118 (1958)

Proceedings of the 2nd International Symposium, Insect and Host Plant. Wageningen, 1969 (J. de Wilde and L. M. Schoonhoven, eds). *Entomol. Exp. Appl.*, **12**, 471–810 (1969)

Proceedings of the 3rd International Symposium, The Host-Plant in Relation to Insect Behaviour and Reproduction, Budapest, 1974 (T. Jermy, ed.). Plenum Press, New York. Also: *Symp. Biol. Hung.*, **16**, 1–322 (1976)

Proceedings of the 4th International Symposium, Insect-Host Plant. Slough, 1978 (R. F. Chapman and E. A. Bernays, eds). *Entomol. Exp. Appl.*, **24**, 201–766 (1978)

Proceedings of the 5th International Symposium, Insect–Plant Relationships. Wageningen, 1982 (J. H. Visser and A. K. Minks, eds). Pudoc, Wageningen (1982)

Proceedings of the 6th International Symposium, Insects–Plants, Pau, 1986 (V. Labeyrie, G. Fabres and D. Lachaise, eds). W. Junk, Dordrecht (1987)

Proceedings of the 7th International Symposium, Insects–Plants, Budapest, 1989 (Á. Szentesi and T. Jermy, eds). *Symp. Biol. Hung.*, **39**, 1–577 (1991)

Proceedings of the 8th International Symposium, Insect–Plant Relationships, Wageningen, 1992 (S. B. J. Menken, J. H. Visser and P. Harrewijn, eds). Kluwer, Dordrecht (1992)

Proceedings of the 9th International Symposium, Insect–Plant Relationships, Gwatt, 1995 (E. Städler, M. Rowell-Rahier and R. Baur, eds). *Entomol. Exp. Appl.*, **80**, 1–324 (1996)

APPENDIX B: STRUCTURAL FORMULAE OF SELECTED SECONDARY PLANT SUBSTANCES

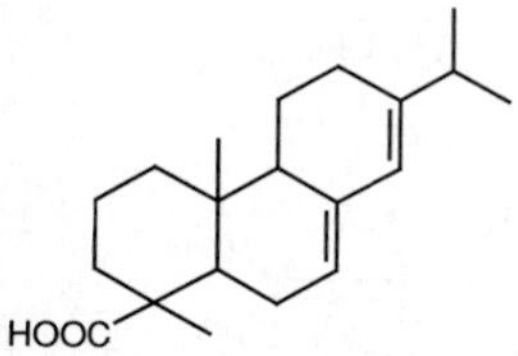

(1) Abietic acid

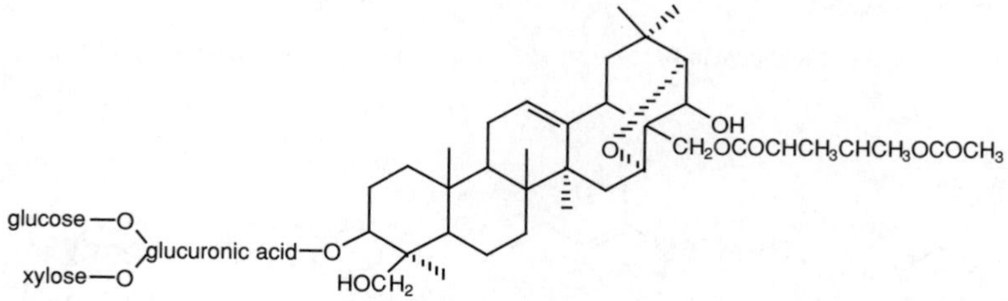

(2) Aescin

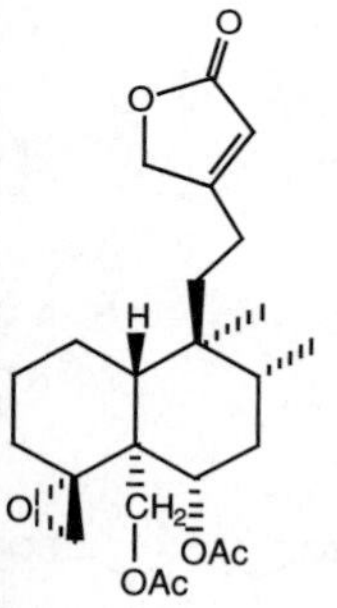

(3) Ajugarin I

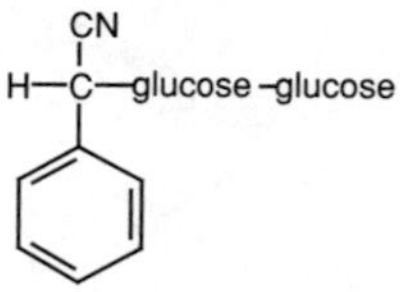

(4) Amygdalin

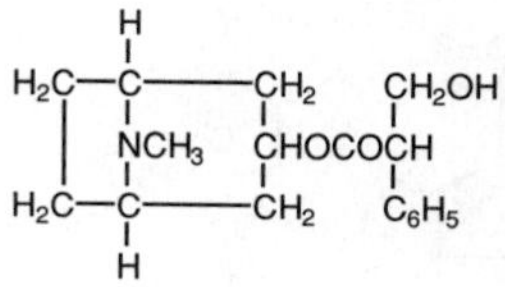

(5) Atropine

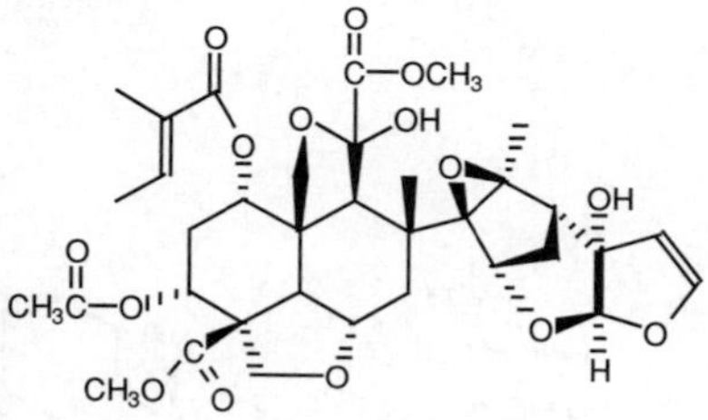

(6) Azadirachtin

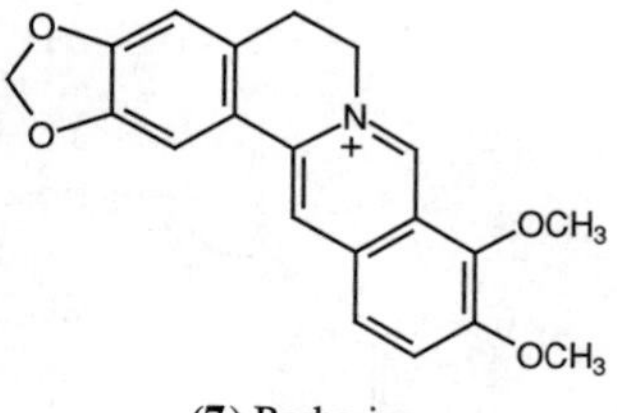

(7) Berberine

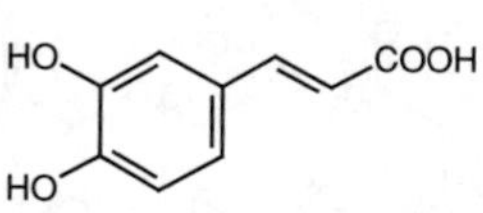

(8) Caffeic acid

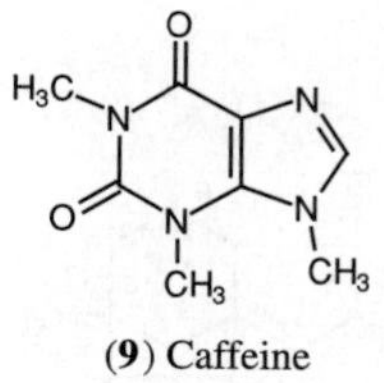

(9) Caffeine

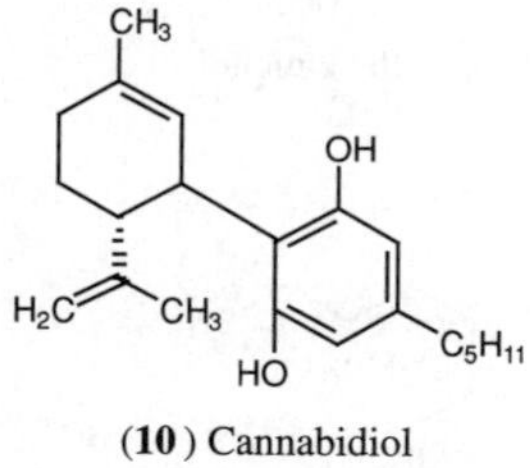

(10) Cannabidiol

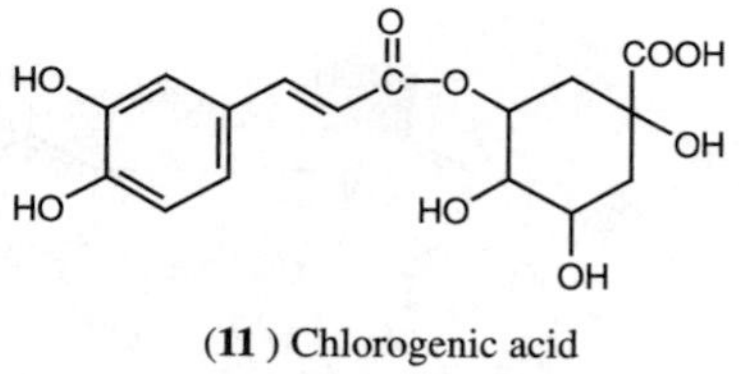

(11) Chlorogenic acid

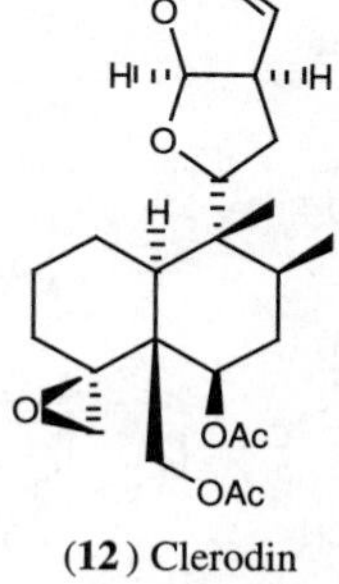

(12) Clerodin

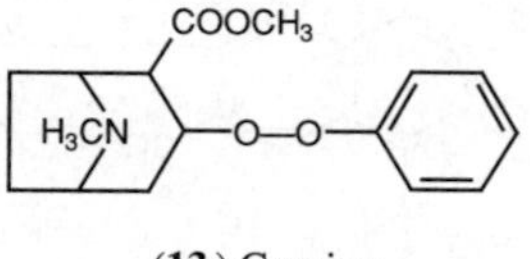

(13) Cocaine

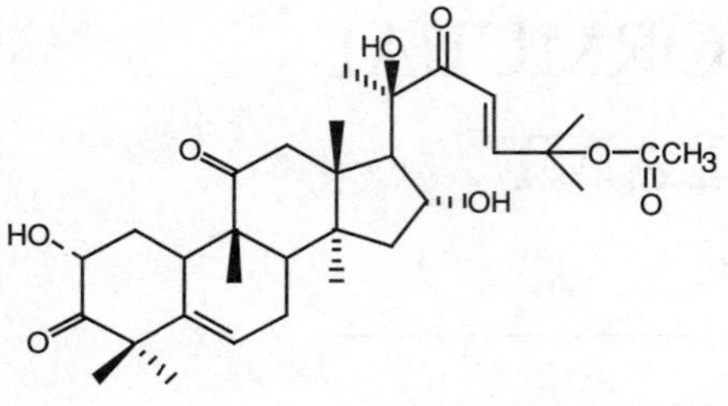

370 *Appendix B: Structural formulae of selected secondary plant substances*

(**14**) Cucurbitacin B

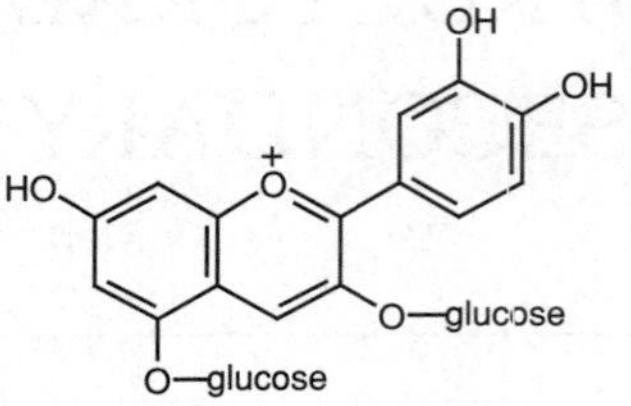

(**15**) Cyanin

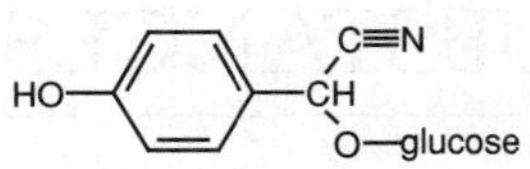

(**16**) Dhurrin

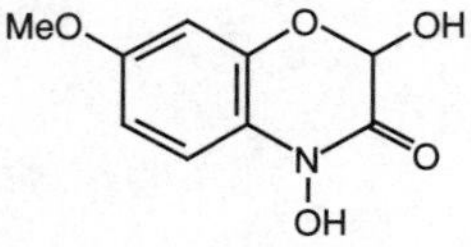

(**17**) DIMBOA

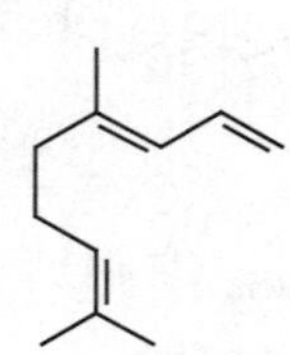

(**18**) 4,8-Dimethyl-1,3(E),7-nonatriene

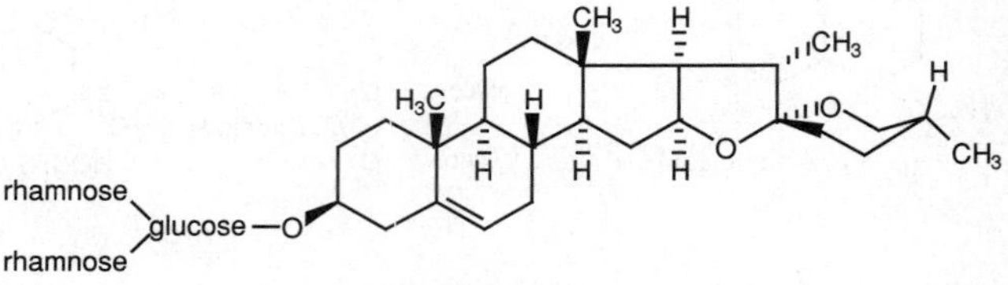

(**19**) Dioscin

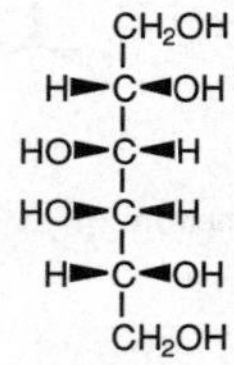

(**20**) Dulcitol

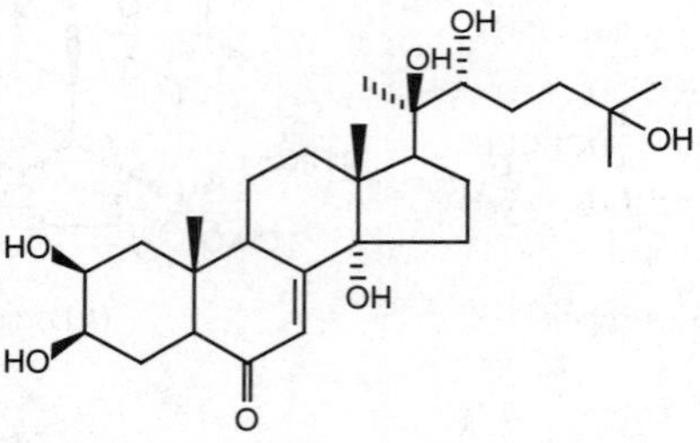

(**21**) β-Ecdyson

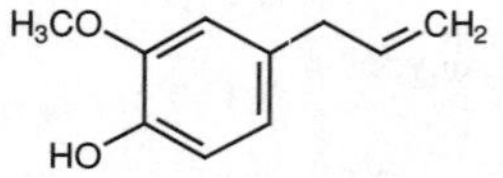

(**22**) Eugenol

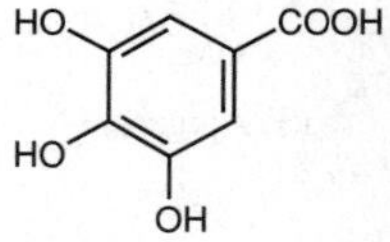

(**23**) Gallic acid

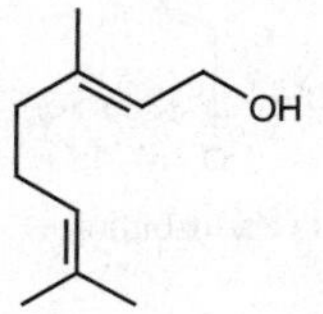

(**24**) Geraniol

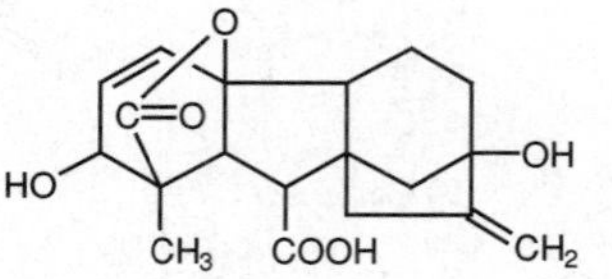

(**25**) Gibberelic acid

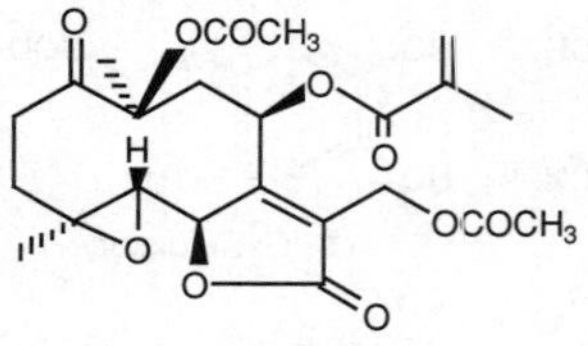

(**26**) Glaucolide-A

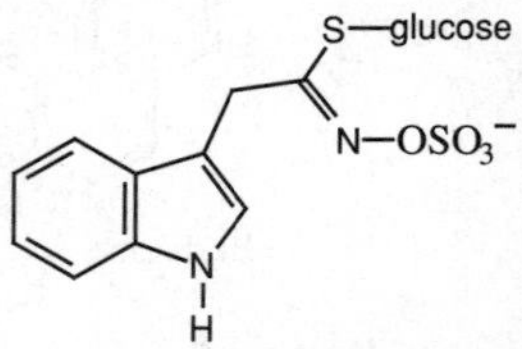

(**27**) Glucobrassicin

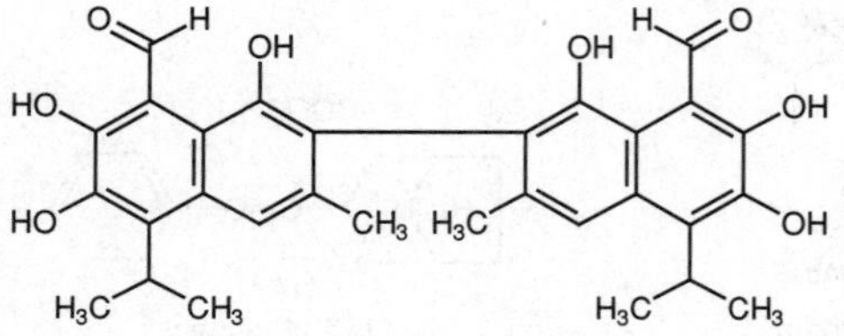

(**28**) Gossypol

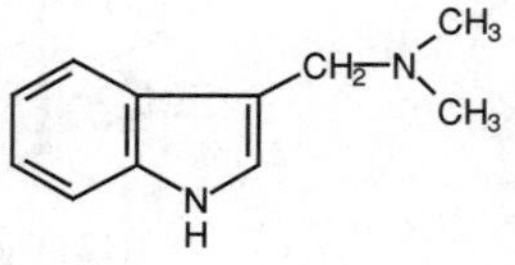

(**29**) Gramine

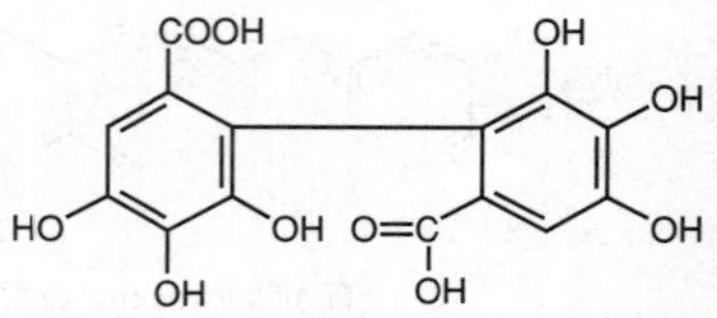

(**30**) Hexahydroxydiphenic acid

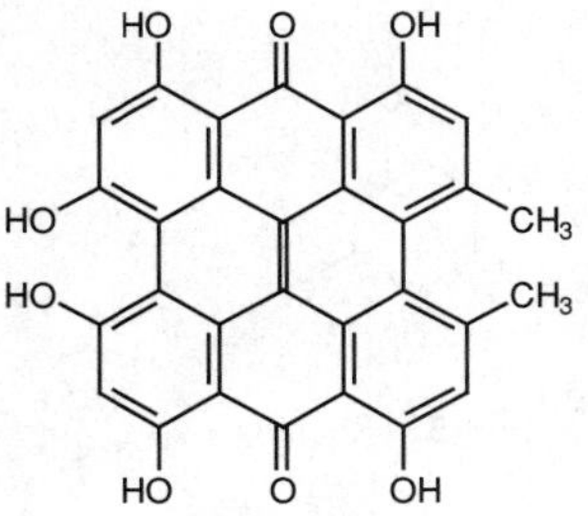

Appendix B: Structural formulae of selected secondary plant substances 371

(**31**) Hypericin

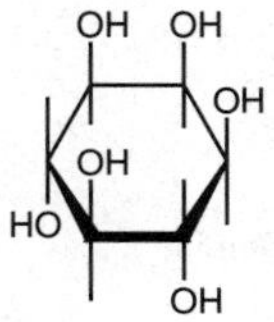

(**32**) Inositol

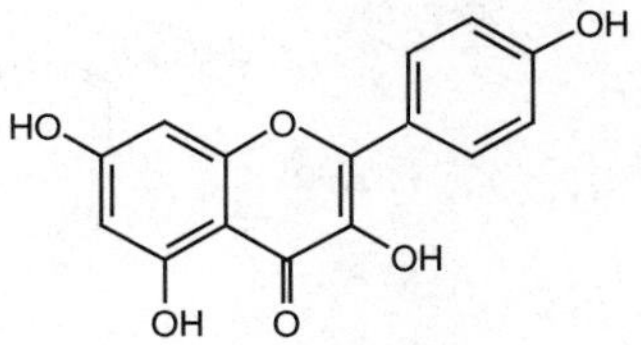

(**33**) Kaempferol

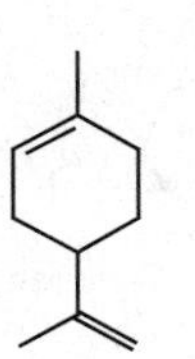

(**34**) Limonene

$CH_3CH_2CH=CHCH_2CH=CHCH_2CH=CH(CH_2)_7COOH$

(**35**) Linolenic acid

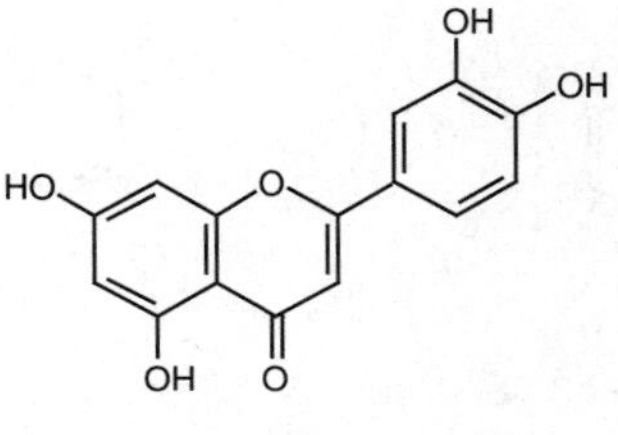

(**36**) Luteolin

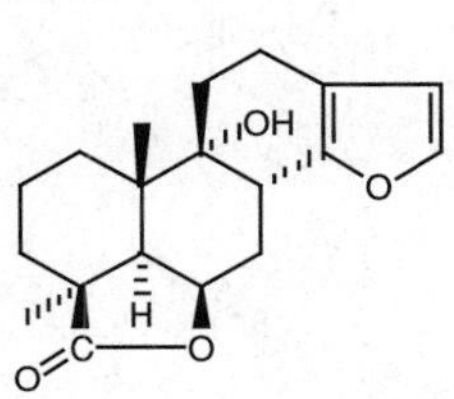

(**37**) Marrubiin

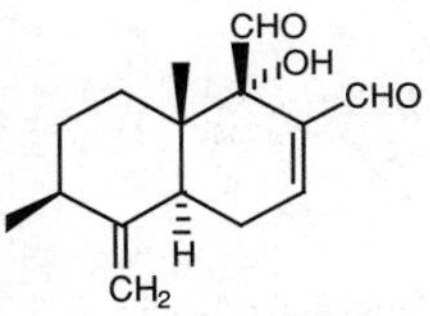

(**38**) Muzigadial

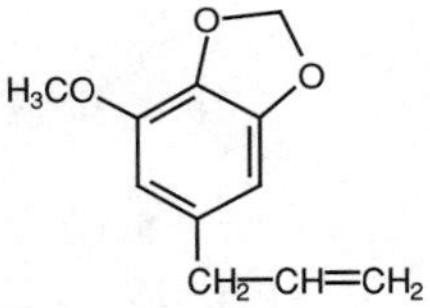

(**39**) Myristicin

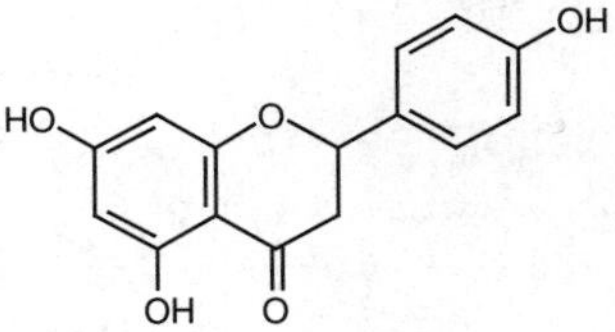

(**40**) Naringenin

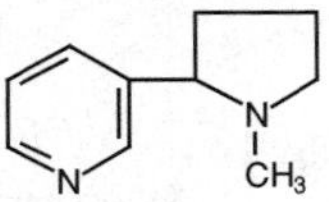

(**41**) Nicotine

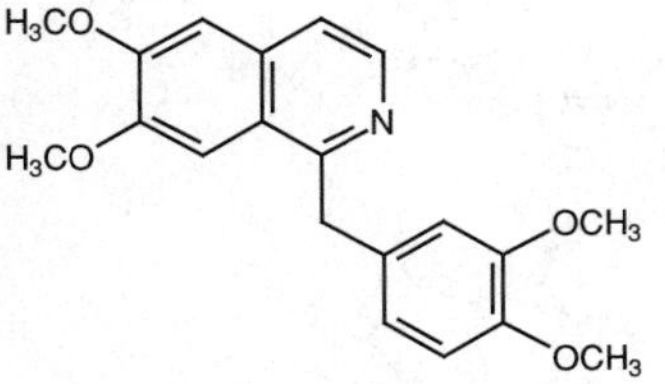

(**42**) Papaverine

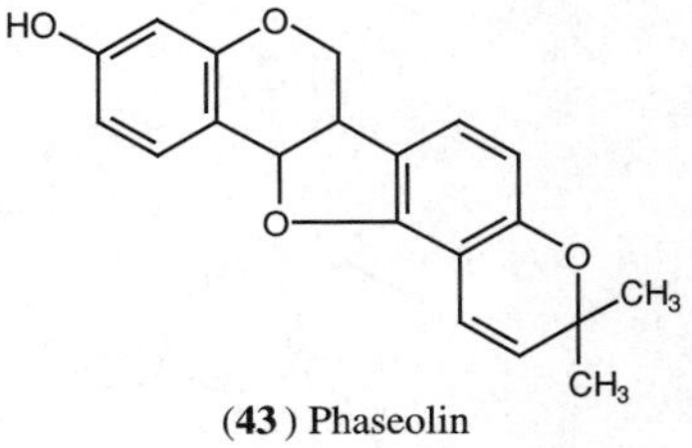

(**43**) Phaseolin

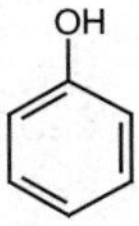

(**44**) Phenol

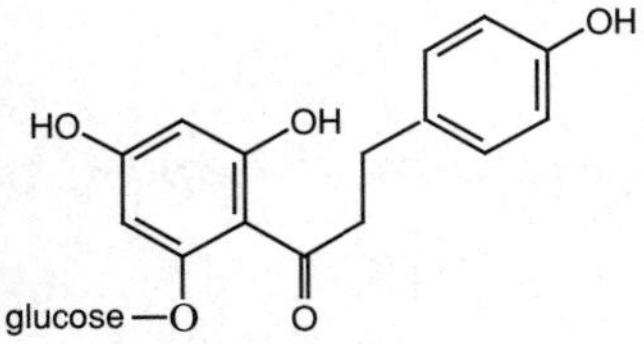

(**45**) Phloridzin

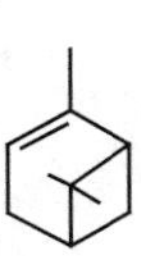

(**46**) α-Pinene

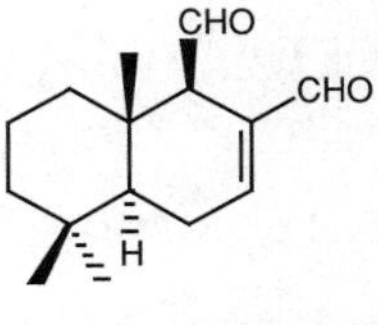

(**47**) Polygodial

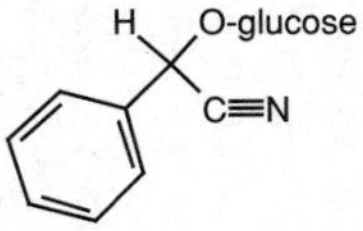

(**48**) Prunasin

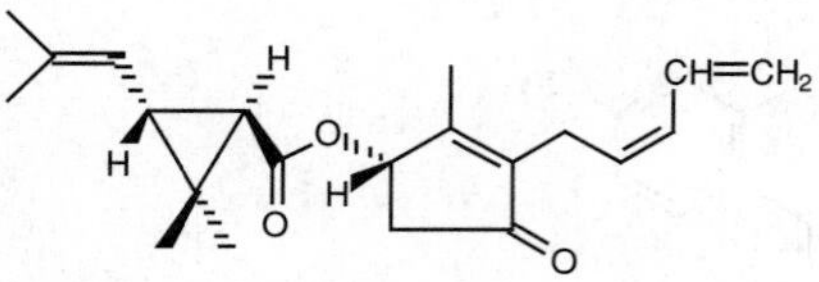

Appendix B: Structural formulae of selected secondary plant substances

(**49**) Pyrethrin I

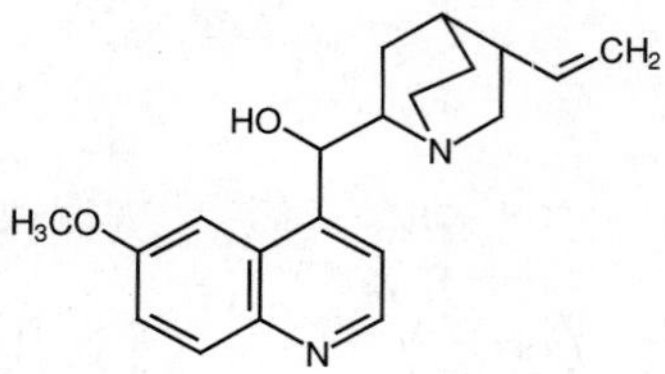

(**50**) Quinine

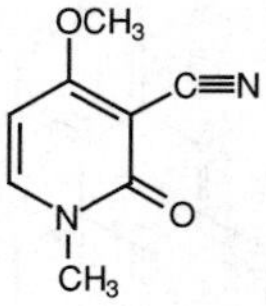

(**51**) Ricinine

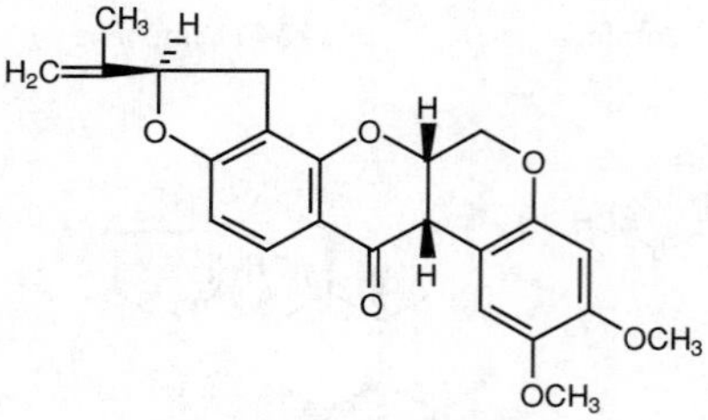

(**52**) Rotenone

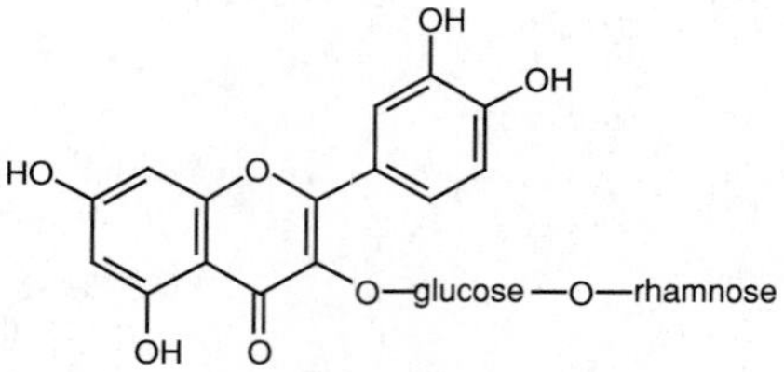

(**53**) Rutin

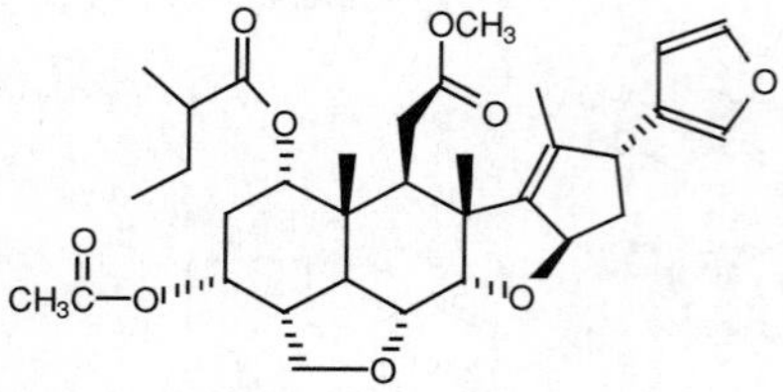

(**54**) Salannin

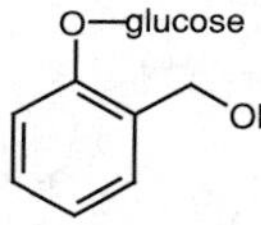

(**55**) Salicin

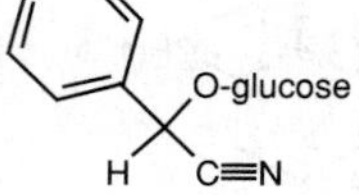

(**56**) Sambunigrin

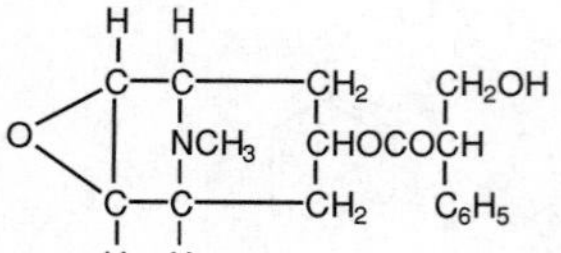

(**57**) Scopolamine

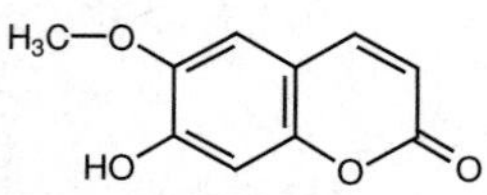

(**58**) Scopoletin

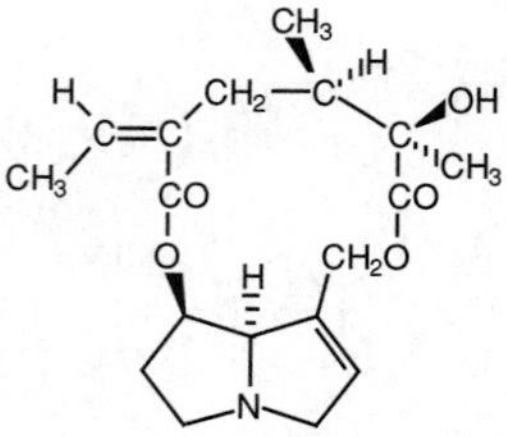

(**59**) Senecionine

(**60**) Sinalbin

(**61**) Sinigrin

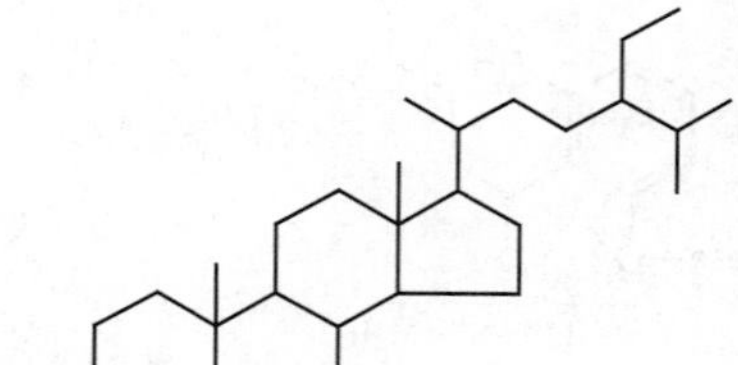

(**62**) Sitosterol

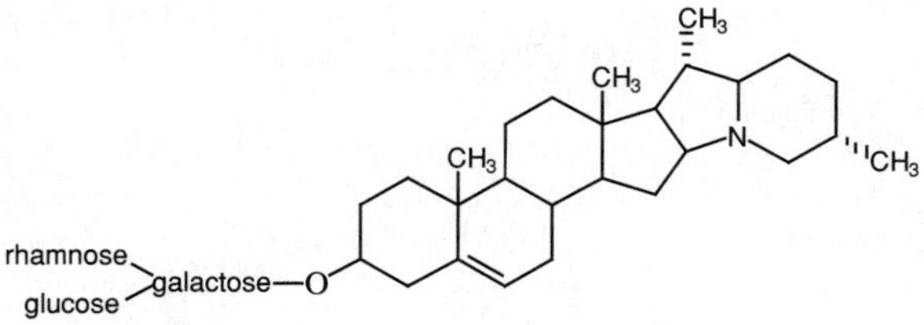

Appendix B: Structural formulae of selected secondary plant substances

(63) α-Solanine

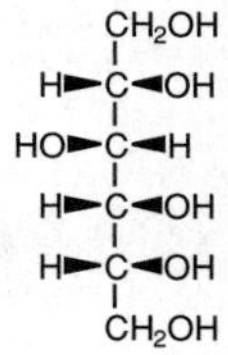

(64) Sorbitol

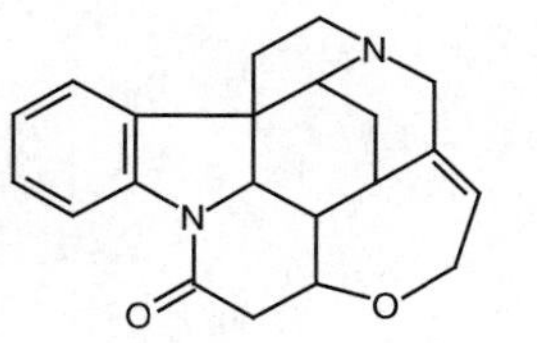

(65) Strychnine

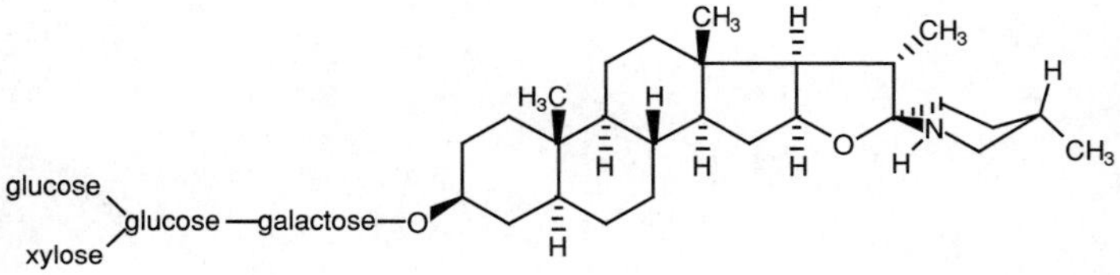

(66) Tomatine

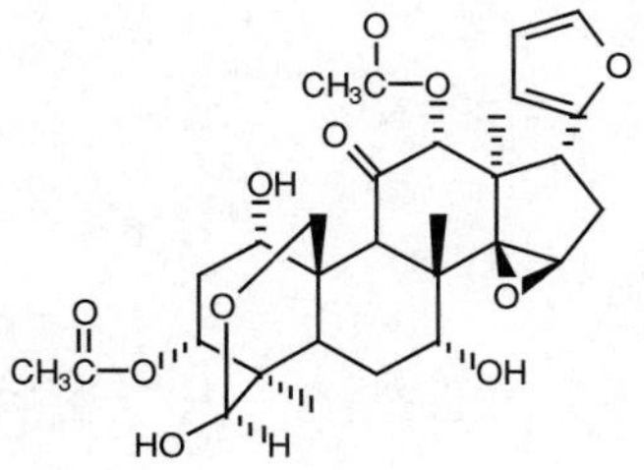

(67) Toosendanin

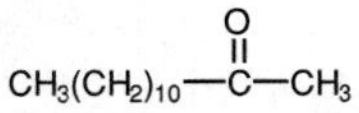

(68) 2-Tridecanone

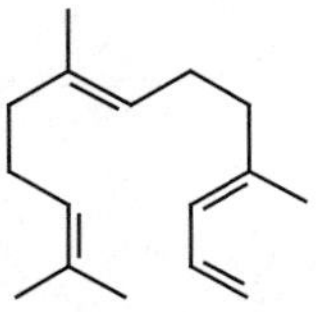

(69) 4,8,12-Trimethyl-1,3(E),7(E),1-tridecatetraene

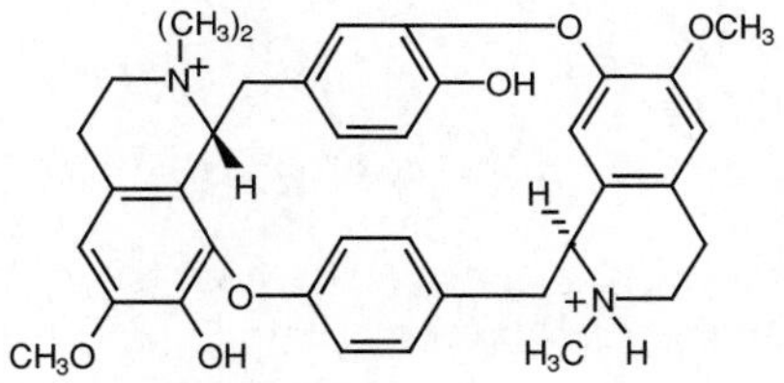

(70) Tubocurarine

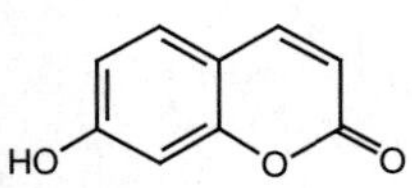

(71) Umbelliferone

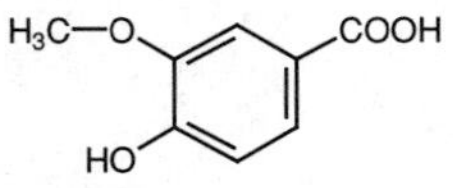

(72) Vanillic acid

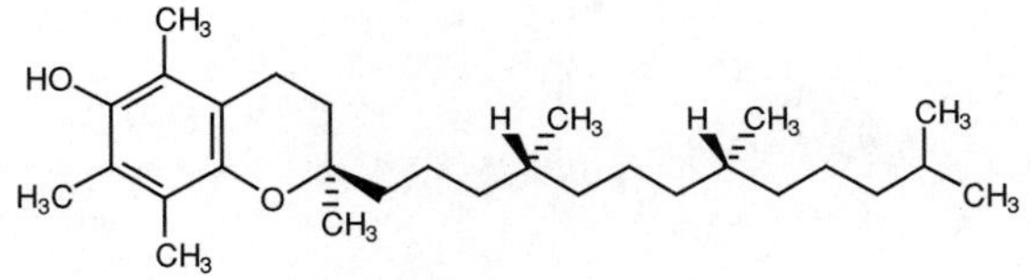

(73) Vitamin E

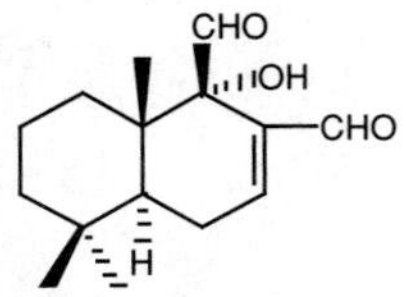

(74) Warburganal

APPENDIX C: METHODOLOGY

C.1	Choice of plants and insects	375
C.1.1	Plants	375
C.1.2	Insects	376
C.2	Behaviour	376
C.2.1	Olfactory orientation	376
C.2.2	Feeding	377
C.2.3	Oviposition	379
C.3	Sensory physiology	379
C.3.1	Ablation	379
C.3.2	Electrophysiology	379
C.4	Plant chemistry	380
C.4.1	Headspace	380
C.4.2	Leaf surface	380
C.4.3	Plant interior	380
C.5	References	380

Because each relationship between an insect species and its host plant has unique aspects, its scientific analysis commonly requires the adaptation of existing standard test procedures. This section lists only briefly a number of procedures that are often used in the study of insect–plant relationships but generally need to be modified to fit a particular case. For more details the reader is referred to the comprehensive reviews by Miller and Miller [47] and Smith et al. [69]. All techniques discussed in this section refer to laboratory studies. Methods for behavioural and experimental studies under field conditions can be found in Miller and Miller [47].

As in other areas of biology the reductionist approach causes a dilemma: often the most sharply defined set of experimental conditions gives the clearest answer, but such an experimental setting is at the same time most distant from the natural situation that one is attempting to understand. This difficulty can be (partly) circumvented by combining the results from different experimental approaches.

The pivotal role of plant chemicals in host choice by herbivorous insects is reflected also in this methodology section, which centres on the identification of factors important in host recognition.

C.1 CHOICE OF PLANTS AND INSECTS

C.1.1 PLANTS

Whole plants growing in their natural environment are the ideal material for studying insect responses. Because this is often impractical, or even impossible, potted plants grown in greenhouses can be good substitutes, although greenhouse plants generally differ substantially from conspecifics grown in the open [26]. Even when plants growing in the open are enclosed in a cage their physiology may change markedly and with it their nutritional value for insects [73]. The responses of small organisms like insects can often be studied conveniently in the laboratory employing plant parts. Although in many cases reliable results will be obtained, the occurrence of wounding effects has to be taken into account (Chapter 3). Instances are known in which an insect, when offered leaf discs from different plant species, showed a preference order reversed in comparison to tests with intact plants [3, 59]. Occasionally insects, when kept on the excised leaves of a normally adequate host plant, show considerably increased mortality rates, indicating fatal changes in chemical composition or moisture content of the food as a result of leaf excision [52].

Sometimes, also, noticeable differences exist in acceptability between leaf discs and excised intact leaves. Young larvae of the leaf beetle *Phyllotreta nemorum* may initiate leaf mining on leaf discs of some non-host plant species, whereas whole leaves of the same plants proved to be fully unacceptable [50]. It is well known that within hours (and probably sooner) of excision leaves undergo biochemical degradation and changes in water relationships [85]. In some cases turgidity can be maintained for long periods by applying water pressure to the cut ends of stems and twigs [35].

C.1.2 INSECTS

The principal sources of test insects are laboratory colonies or field-collected material. Although using laboratory-reared insects is often more convenient, there is a risk that they differ so radically from natural populations in genetic, behavioural and physiological characteristics as to limit their representativeness of the species in the wild [76]. Thus laboratory insects have been reported to lose their ability to grow successfully on their original host plants [25] or have been found to accept plant species totally outside their natural host range [63]. On the other hand, field-collected insects may be infested with pathogens and/or parasitoids that strongly affect behaviour as compared to non-infected individuals.

An insect's feeding history may also markedly influence its behavioural and physiological responses to normal food plants through preference induction [31]. Occasionally, the induction is so rigid that the insect will die of starvation rather than accept one of its other food plant species [27]. Clearly host-plant selection in naive insects may differ markedly from that in experienced insects (Chapter 7).

A generally neglected aspect of test insects is standardization. Since insects collected at different sites or cultured under different conditions may differ greatly in behavioural and physiological characteristics, it is essential for reproducibility of the results and comparison with other studies that the source of the experimental animals is carefully recorded.

C.2 BEHAVIOUR

The techniques used for analysis of insect behaviour are either direct observation or automatic recording and storage followed by retrieval and analysis. Direct observation procedures are facilitated by employing hand-operated event recorders [11, 51]. Automatic techniques for recording insect behaviour include cinematography and video-recording [22, 54, 58], the use of actographs [2, 4] and methods based on electrical registration of feeding activity [10, 77].

Experiments in general, but experiments on behaviour in particular, need a thorough consideration of methodology even at the planning stage. Otherwise, an adequate statistical evaluation of the results may become impossible [46].

C.2.1 OLFACTORY ORIENTATION

Many techniques have been developed for studying olfactory responses to plant volatiles [21]. Methods used to investigate insect orientation to odours vary with insect size and type of locomotion (walking or flight).

(a) Screen test

A very simple test for walking insects, for instance caterpillars, employs a screen between the plant material and the insect [15]. Direct observation of the insect's behaviour or the distribution of insects after some time provides information on the role of olfactory cues [13]. A modified type of screen test has been used to observe caterpillar reactions to attractive and repellent odours released, for instance, by artificial diets [62].

(b) Olfactometer tests

In many cases Y-tube olfactometers prove to be relatively simple yet very useful pieces of apparatus, providing the test insect with a binary choice [12]. Walking as well as flying insects can be tested for their preference for, for instance, an air stream bearing an odour, which passes through one arm of the Y, or for clean air (control), which passes through the other arm. Dimensions and special modifications, for instance a guiding rail for some walking insects, have to be made, according to the size and habits of the insect.

(c) Four-arm olfactometer

A four-arm airflow olfactometer has been used for small walking insects such as hymenopterous parasitoids [79]. It allows testing of more than one odour or different concentrations of one odour at the same time. In a central arena the insect can choose between four different odour fields.

(d) Wind tunnel

Basically, a wind tunnel consists of three parts: (1) an effuser or entrance zone in which the air is accelerated and the flow is smoothed', (2) a working section where the insects are observed and (3) a diffuser or exhaust zone where the air is decelerated [82]. Walking [80] or flying [53] insects are released in the centre or at the downwind end of a tunnel. Various parameters of an insect's response to air streams with or without plant volatiles are recorded. Useful information on planning wind tunnel experiments is given by Finch [21].

(e) Locomotion compensator

The locomotion compensator, or 'Kramer sphere' is a sophisticated instrument [38] allowing accurate measurements of orientation responses to wind-borne volatiles. The test insect stands on a large sphere and every displacement is compensated by a computer-controlled movement of the sphere in the opposite direction. As a result the freely walking insect remains at the same place and its stimulus situation remains constant. It can be operated either in combination with visual stimuli or not, and has been used successfully in combination with a wind tunnel to record various locomotion parameters in different types of walking insects. All recorded movements of the sphere allow automatic data analysis [74].

C.2.2 FEEDING

The fine details of host recognition are undoubtedly in the realm of the insect's contact chemical senses. Taste plays a major role in host-plant choice (Chapter 6) and choice experiments are a simple and indispensable tool in any insect–plant study. Bioassays employing whole plants or plant parts, for instance leaf discs, may be of the no-choice type, or the insects may be offered a choice between two or more alternatives. Choice tests with more than two alternatives, however, should be avoided, because the results may be ambiguous and difficult to analyse. Both no-choice and binary choice designs are suitable for answering different questions. The no-choice situation is generally more representative for the field situation in our agricultural systems, but also for the natural vegetation where a choice, for example, between the leaves of two plant species within a distance of a few millimetres, as in the choice test, is almost never presented. On the other hand, when screening chemicals for behavioural activity binary choice experiments are often much more sensitive. The polyphagous peach aphid, *Myzus persicae*, for instance, readily accepts antifeedant-treated host leaves or artificial foods containing various allelochemicals, but in a choice situation they are often seen to clearly prefer the control lacking the test substance [54, 65]. Statistical evaluation of choice experiments may present special problems, which have been dealt with by several authors [5, 29, 42].

The role of feeding and oviposition stimulants or deterrents can be tested by exposing insects to excised non-host plants with their stems in a stimulant solution [41] or those of host plants in a deterrent solution [8].

(a) Leaf discs

Many test designs are based on the use of leaf discs, mainly to standardize the foliage area exposed to the insect. A commonly used layout is the 'cafeteria test', in which discs punched out of the leaves of two plant species are offered to the insect in a circular array [33]. When pure compounds are to be tested, discs of host plants can be used as a substrate, for instance to test the efficacy of antifeedants. The experimental discs are either dipped into a solution of the test chemical or the compound is applied with a brush or by spraying. The chemical may also be incorporated into an agar or gelatin cover [86] or leaf discs (or whole leaves) may be vacuum-infiltrated [66]. To determine whether a plant is not eaten because of the absence of phagostimulants or because it contains feeding deterrents, tests with 'sandwiches' of leaf discs from the test plant combined with those of a host plant may provide the answer [30].

When leaf discs are used to determine the activity of an antifeedant, the choice of the plant species may affect the insect's sensitivity to the antifeedant compound. Feeding deterrence of azadirachtin, for instance, in *Spodoptera frugiperda* larvae was much higher on cotton leaf discs than on lima bean [55]. Ingestion of leaf discs or neutral substrates can be measured manually (weight, surface area consumed) or automatically [19, 37]. It is important that studies of herbivore consumption report the thickness, density and specific leaf weight of test leaves and in addition provide at least two measures of consumption: leaf area and biomass removed [83], because during the short period (hours) that leaf-disc experiments normally last, the herbivore is mostly using volumetric regulation of meal size [67].

(b) Neutral substrates

Rather than using leaf material as a substrate, which may introduce unwanted sources of variation, neutral substrates may be employed to test responses to particular chemicals. Thus elderberry pith, filter paper, glass fibre discs for locusts [7] and caterpillars [71], styropor lamellae [1] and agar or agar–cellulose blocks [40] can be used. It has been found that chemicals applied to neutral substrates are not always distributed evenly [87]. When testing antifeedant compounds the neutral substrate must be made palatable, usually with sucrose. Several non-nutritional insect phagostimulants may also be useful, and can be obtained commercially [39]. Ingestion in no-choice experiments can also be determined on the basis of dry weight faeces production [9, 34].

(c) Fluid diets

A number of piercing–sucking insect species will feed on artificial diets contained in a Parafilm® sachet [48]. Effects of phagostimulant or antifeedant compounds can be quantified by adding them to the diet [65], or the test chemical can be painted on the Parafilm® membrane. Drinking responses have also been used for a fast assessment of antifeedant effects. In this case the test fluid is administered by a small platinum-wire loop [17] or by a microsyringe to the mouthparts of chewing insects during feeding [64, 66]. Because it is impossible to assess visually when fluid sucking insects, such as aphids [77], thrips [28] and leafhoppers [36], are feeding, an electronic method, EPG, has been developed, which signals various feeding activities once the insect has started to penetrate plant tissues or an artificial diet with its mouthparts [77]. A computer programme for automatic calculation of EPG parameters enables fast processing of abundant data [20].

C.2.3 OVIPOSITION

The females of many herbivorous insect species, when searching for an oviposition site, are guided by a complex of visual, olfactory, contact chemical, form, and/or tactile cues specific to their host plants. For flying insects field cage experiments, as a semi-laboratory method [16], provide conditions that are closest to natural situations. At present, however, a detailed analysis of oviposition behaviour can be carried out only by laboratory experiments. The set-ups for such tests have to be designed specifically almost for each insect species. Here only some references can be given as examples of oviposition assay methods for flying insects, such as lepidopterans [41, 57, 78] and flies [14, 49, 60], as well as for walking insects [32].

C.3 SENSORY PHYSIOLOGY

Since insect feeding behaviour is to a large extent governed by chemosensory information, the analysis of sensory responses to plant chemicals may provide important clues about the role of different chemicals in host-plant recognition. The contribution of the chemical senses to the decision process can be studied by ablation techniques and electrophysiological methods.

C.3.1 ABLATION

The role of specific sensory hairs or organs can be assessed by selective inactivation or by ablating them and observing changes in the insect's behavioural responses to chemical stimuli. Inactivation can be done by applying aggressive chemicals, e.g. hydrochloric acid [70], to the sensillum or by electrical cauterization [9]. Ablation may be effected by microsurgery [40].

C.3.2 ELECTROPHYSIOLOGY

Sensory responses to either pure compounds or mixtures can be recorded from individual olfactory or taste cells by rather specialized electrophysiological techniques [23]. The extracellularly recorded action potentials are small, necessitating the use of special amplifiers [45]. Experiments are usually performed on isolated heads or legs, but recently a method employing intact caterpillars has been described [24]. Since the electrical signals from the sensilla are produced in different neurons, computer programmes were developed that analyse complex spike patterns [43, 68].

Because an insect often possesses many olfactory cells conveniently located on one of the head appendages, i.e. the antennae, electroantennography (EAG) is a useful technique for examining the total response of the olfactory system [23]. EAGs are recorded either from excised antennae or from intact insects. The EAG is thought to reflect the summation of receptor potentials over the whole antenna and the response amplitude is positively correlated with the number of sensilla housing sensitive receptor cells. When classifying plant volatiles for their capacity to evoke olfactory activity, the EAG appears to be a useful technique as it provides a screening of the entire antennal receptor population. An EAG does not, however, allow a conclusion about the specificity of the responding (sub-) populations of antennal olfactory cells. Neither does an EAG allow any conclusion about behavioural attractiveness or repellency of the stimulus. EAGs increase with increasing concentration of the chemical stimulus, until a saturation level is reached (see Fig. 5.15). The EAG technique, which is also applicable to small insects [56, 81], is especially useful in combination with gas chromatography [44, 75]. Direct coupling of both techniques allows the identification of volatiles in complex mixtures and simultaneous determination of the biological activity of individual odour components.

C.4 PLANT CHEMISTRY

C.4.1 HEADSPACE

Plant volatiles to which insects may react are usually produced in very low amounts. Headspace analysis has gradually replaced distillation methods used to extract volatile compounds from plant tissues, because the volatile chemicals present in the plant interior do not necessarily reflect the composition of the mixture of volatiles released into the surrounding air. Volatiles emitted by intact or insect-damaged plants can be collected on to an absorbent material ('odour trap'), from which they can later be readily deabsorbed and, after concentration, be analysed by gas chromatography and mass spectrometry [6].

C.4.2 LEAF SURFACE

Chemicals present on plant surfaces can be extracted by briefly dipping intact plants into organic solvents. Since the waxy surface of most plant species contains a mixture of polar and non-polar compounds, solvents must be chosen that dissolve both. Various methods of extraction and identification of chemicals on plant surfaces are cited in some recent reviews [18, 72].

C.4.3 PLANT INTERIOR

The general problem is that most chemicals that play an important role in host selection by insects originate from within living organisms. Once the plant is prepared in any way for analysis, its metabolic state may change and with it the quantity and quality of its allelochemicals. Most extraction procedures start with homogenizing plant parts in a blender in order to crush all cells, causing the release of their contents into the extracting solvent. All soluble chemicals can then be extracted following any of several methods, including those cited by Smith *et al.* [69]. Often, one is interested in a particular group of chemicals, and chemical analysis is focused on isolation and identification of individual compounds. The numerous and special problems met during the investigation of particular classes of allelochemicals can be found in comprehensive texts, such as Rosenthal and Berenbaum [61] and Waterman and Mole [84].

C.5 REFERENCES

1. Ascher, K. R. S. and Gurevitz, E. (1972) A further use of the styropor method: evaluating the response of the fruit bark beetle, *Scolytus (Ruguloscolytus) mediterraneus* Eggers, to extracts of its host plant. *Z. PflKrankh. PflSchutz*, **79**, 215–222.
2. Ayertey, J. N. (1981) Locomotor activity of *Sitophilus zeamais* and *Sitotroga cerealella* on maize. *Entomol. Exp. Appl.*, **29**, 19–28.
3. Barnes, O. L. (1963) Food plant tests with the differential grasshopper. *J. Econ. Entomol.*, **56**, 396–399.
4. Bernays, E. A. (1979) The use of Doppler actographs to measure locomotor activity in locust nymphs. *Entomol. Exp. Appl.*, **26**, 136–141.
5. Bernays, E. A. and Weiss, M. R. (1996) Induced food preferences in caterpillars: the need to identify mechanisms. *Entomol. Exp. Appl.*, **78**, 1–6.
6. Blaakmeer, A., Geervliet, J. B. F., van Loon, J. J. A., Posthumus, M. A., van Beek, T. A. and de Groot, Æ. (1994) Comparative headspace analysis of cabbage plants damaged by two species of *Pieris* caterpillars: consequences for in-flight host location by *Cotesia* parasitoids. *Entomol. Exp. Appl.*, **73**, 175–182.
7. Blaney, W. M. and Winstanley, C. (1980) Chemosensory mechanism of locusts in relation to feeding: the role of some secondary plant compounds, in *Insect Neurobiology and Pesticide Action*, (Neurotox 79), Chemical Industry, London, pp. 383–389.
8. Blau, P. A., Feeny, P., Contardo, L. and Robson, D. S. (1978) Allylglucosinolate and herbivorous caterpillars, a contrast in toxicity and tolerance. *Science*, **200**, 1296–1298.
9. Blom, F. (1978) Sensory activity and food intake, a study of input-output relationships in two phytophagous insects. *Neth. J. Zool.*, **28**, 277–340.
10. Bowdan, E. (1984) An apparatus for the continuous monitoring of feeding by caterpillars in

choice, or non-choice tests (automated cafetaria test). *Entomol. Exp. Appl.*, **36**, 13–16.
11. Buckley, D. L., Frazer, B. D. and Amour, G. S. (1979) An inexpensive, portable, printing event recorder for behavior studies. *Behav. Res. Meth. Instrum.*, **11**, 561–563.
12. Cannon, W. N. (1990) Olfactory response of eastern spruce budworm larvae to spruce needles exposed to acid rain and elevated levels of ozone. *J. Chem. Ecol.*, **16**, 3255–3261.
13. Chin, C. T. (1950) Studies on the physiological relations between the larvae of *Leptinotarsa decemlineata* Say and some solanaceous plants. *Tijdschr. Pl. ziekt.*, **56**, 1–88.
14. Degen, T. and Städler, E. (1996) Influence of natural leaf shapes on oviposition in three phytophagous flies, a comparative study. *Entomol. Exp. Appl.*, **80**, 97–100.
15. Dethier, V. G. (1947) *Chemical Insect Attractants and Repellents*. Blakiston, Philadelphia, PA.
16. Duan, J. J. and Prokopy, R. J. (1994) Apple maggot fly response to red sphere traps in relation to fly age and experience. *Entomol. Exp. Appl.*, **73**, 279–287.
17. Eger, H. (1937) Über den Geschmackssinn von Schmetterlingsraupen. *Biol. Zbl.*, **57**, 293–308.
18. Eigenbrode, S. D. and Espelie, K. E. (1995) Effects of plant epicuticular lipids on insect herbivores. *Annu. Rev. Entomol.*, **40**, 171–194.
19. Escoubas, P., Lajide, L. and Mitzutani, J. (1993) An improved leaf-disk antifeedant bioassay and its application for the screening of Hokkaido plants. *Entomol. Exp. Appl.*, **66**, 99–107.
20. Febvay, G., Rahbé, Y. and van Helden, M. (1996) MacStylet, software to analyse electrical penetration graph data on the Macintosh. *Entomol. Exp. Appl.*, **80**, 105–108.
21. Finch, S. (1986) Assessing host-plant finding by insects, in *Insect–Plant Interactions*, (eds J. R. Miller and T. A. Miller), Springer-Verlag, New York, pp. 23–63.
22. Frase, B. A. and Willey, R. B. (1981) Slowed motion analysis of stridulation in the grasshopper, *Xanthippus corallipes* (Acrididae, Oedipodinae). *Can. J. Zool.*, **59**, 1005–1013.
23. Frazier, J. L. and Hanson, F. E. (1986) Electrophysiological recording and analysis of insect chemosensory responses, in *Insect–Plant Interactions*, (eds J. R. Miller and T. A. Miller), Springer-Verlag, New York, pp. 285–330.
24. Gothilf, S. and Hanson, F. E. (1994) A technique for electrophysiologically recording from chemosensory organs of intact caterpillars. *Entomol. Exp. Appl.*, **72**, 305–310.
25. Guthrie, W. D. and Carter, S. W. (1972) Backcrossing to increase survival of larvae of a laboratory culture of the European corn borer on field corn. *Ann. Entomol. Soc. Am.*, **65**, 108–109.
26. Hammond, R. B., Pedigo, L. P. and Poston, F. L. (1979) Green cloverworm leaf consumption on greenhouse and field soybean leaves and development of a leaf-consumption model. *J. Econ. Entomol.*, **72**, 714–717.
27. Hanson, F. E. (1976) Comparative studies on induction of food preferences in lepidopterous larvae. *Symp. Biol. Hung.*, **16**, 71–77.
28. Harrewijn, P., Tjallingii, W. F. and Mollema, C. (1996) Electrical recording of plant penetration by western flower thrips. *Entomol. Exp. Appl.*, **79**, 345–354.
29. Horton, D. R. (1995) Statistical considerations in the design and analysis of paired-choice assays. *Environ. Entomol.*, **24**, 179–191.
30. Jermy, T. (1966) Feeding inhibitors and food preference in chewing phytophagous insects. *Entomol. Exp. Appl.*, **9**, 1–12.
31. Jermy, T. (1987) The role of experience in the host selection of phytophagous insects, in *Perspectives in Chemoreception and Behavior*, (eds R. F. Chapman, E. A. Bernays and J. G. Stoffolano), Springer-Verlag, New York, pp. 142–157.
32. Jermy, T. and Szentesi, Á. (1978) The role of inhibitory stimuli in the choice of oviposition site by phytophagous insects. *Entomol. Exp. Appl.*, **24**, 458–471.
33. Jermy, T., Hanson, F. and Dethier, V. G. (1968) Induction of specific food preference in lepidopterous larvae. *Entomol. Exp. Appl.*, **11**, 211–230.
34. Kasting, R. and McGinnis, A. J. (1962) Quantitative relationships between consumption and excretion of dry matter by larvae of the pale western cutworm, *Agrotis orthogonia* Morr. (Lepidoptera, Noctuidae). *Can. Entomol.*, **94**, 441–443.
35. Kendall, R. D. (1957) Simple devices for keeping foodplants fresh. *Lepidop. News*, **11**, 225–226.
36. Kimmins, F. and Bosque-Perez, N. (1996) Electrical penetration graphs from *Cicadulina* spp. and the inoculation of a persistent virus into maize. *Entomol. Ent. Exp.*, **80**, 45–49.
37. Kokko, E. G., de Clerck-Floate, R. A. and Leggett, F. L. (1995) Methods to quantify leaf beetle consumption of leaf disks using image analysis. *Can. Entomol.*, **127**, 519–525.

38. Kramer, E. (1976) The orientation of walking honeybees in odour fields with small concentration gradients. *Physiol. Entomol.*, **1**, 27–37.
39. Lopez, J. D. and Lingren, P. D. (1994) Feeding response of adult *Helicoverpa zea* (Lepidoptera, Noctuidae) to commercial phagostimulants. *J. Econ. Entomol.*, **87**, 1653–1658.
40. Ma, W. C. (1972) Dynamics of feeding responses in *Pieris brassicae* (Linn) as a function of chemosensory input, a behavioural, ultrastructural and electrophysiological study. *Meded. Landbouwhogeschool Wageningen*, **72-11**, 1–162.
41. Ma, W. C. and Schoonhoven, L. M. (1973) Tarsal contact chemosensory hairs of the large white butterfly *Pieris brassicae* and their possible role in oviposition behaviour. *Entomol. Exp. Appl.*, **16**, 343–357.
42. Manly, B. F. J. (1995) Measuring selectivity from multiple choice feeding-preference experiments. *Biometrics*, **51**, 709–715.
43. Marion-Poll, F. (1996) Display and analysis of electrophysiological data under Windows™. *Entomol. Exp. Appl.*, **80**, 116–119.
44. Marion-Poll, F. and Thiéry, D. (1996) Dynamics of EAG responses to host-plant volatiles delivered by a gas chromatograph. *Entomol. Exp. Appl.*, **80**, 120–123.
45. Marion-Poll, F. and van der Pers, J. (1996) Unfiltered recordings from insect taste sensilla. *Entomol. Exp. Appl.*, **80**, 113–115.
46. Martin, P. and Bateson, P. (1986) *Measuring Behaviour. An Introductory Guide*, Cambridge University Press, Cambridge.
47. Miller, J. R. and Miller, T. A. (1986) *Insect–Plant Interactions*, Springer-Verlag, New York.
48. Mittler, T. E. (1988) Application of artificial feeding techniques for aphids, in *Aphids: Their Biology, Natural Enemies, and Control*, vol. 2B, (eds A. K. Minks and P. Harrewijn), Elsevier, Amsterdam, pp. 145–170.
49. Nair, K. S. S., McEwen, F. L. and Snieckus, V. (1976) The relationship between glucosinolate content of cruciferous plants and oviposition preferences of *Hylemya brassicae* (Diptera, Anthomyidae). *Can. Entomol.*, **108**, 1031–1036.
50. Nielsen, J. K. (1989) The effect of glucosinolates on responses of young *Phyllotreta nemorum* larvae to non-host plants. *Entomol. Exp. Appl.*, **51**, 249–259.
51. Noldus, L. P. J. J. (1991) The Observer. A software system for collection and analysis of observational data. *Behav. Res. Meth. Instrum. Comp.*, **23**, 415–429.
52. Olckers, T. and Hulley, P. E. (1994) Host specificity tests on leaf-feeding insects: aberrations from the use of excised leaves. *Afr. Entomol.*, **2**, 68–70.
53. Phelan, P. L., Roelofs, C. J., Youngman, R. R. and Baker, T. C. (1991) Characterization of chemicals mediating ovipositional host-plant finding by *Amyelois transitella* females. *J. Chem. Ecol.*, **17**, 599–613.
54. Powell, G., Hardie, J. and Pickett, J. A. (1995) Responses of *Myzus persicae* to the repellent polygodial in choice and no-choice video assays with young and mature leaf tissue. *Entomol. Exp. Appl.*, **74**, 91–95.
55. Raffa, K. E. (1987) Influence of host plant on deterrence by azadirachtin of feeding by fall armyworm larvae (Lepidoptera, Noctuidae). *J. Econ. Entomol.*, **80**, 384–387.
56. Ramachandran, R. and Norris, D. M. (1991) Volatiles mediating plant–herbivore–natural-enemy interactions: electroantennogram responses of soybean looper, *Pseudoplusia includens* and a parasitoid, *Microplitis demolitor*, to green leaf volatiles. *J. Chem. Ecol.*, **17**, 1665–1690.
57. Reed, D. W., Pivnick, K. A. and Underhill, E. W. (1989) Identification of chemical oviposition stimulants for the diamond-back moth, *Plutella xylostella*, present in three species of Brassicaceae. *Entomol. Exp. Appl.*, **53**, 277–286.
58. Reinecke, P., Buckner, J. S. and Grugel, S. R. (1980) Life cycle of laboratory-reared tobacco hornworms, *Manduca sexta*: a study of development and behavior, using timelapse cinematography. *Biol. Bull.*, **158**, 129–140.
59. Risch, S. J. (1985) Effects of induced chemical changes on interpretation of feeding preference tests. *Entomol. Exp. Appl.*, **39**, 812–884.
60. Roessingh, P. Städler, E., Fenwick, G. R., Lewis, J. A., Nielsen, J. K., Hurter, J. and Ramp, T. (1992) Oviposition and tarsal chemoreceptors of the cabbage root fly are stimulated by glucosinolates and host plant extracts. *Entomol. Exp. Appl.*, **65**, 267–282.
61. Rosenthal, G. A. and Berenbaum, M. R. (1991–1992) *Herbivores. Their Interactions with Secondary Plant Metabolites*, 2nd edn, 2 vols. Academic Press, San Diego, CA.
62. Saxena, K. N. and Schoonhoven, L. M. (1978). Induction of orientational and feeding preferences in *Manduca sexta* larvae for an artificial diet containing citral. *Entomol. Exp. Appl.*, **23**, 72–78.

63. Schoonhoven, L. M. (1967) Loss of hostplant specificity by *Manduca sexta* after rearing on an artificial diet. *Entomol. Exp. Appl.*, **10**, 270–272.
64. Schoonhoven, L. M. and Derksen-Koppers, I. (1973) Effects of secondary plant substances on drinking behaviour in some Heteroptera. *Entomol. Exp. Appl.*, **16**, 141–145.
65. Schoonhoven, L. M. and Derksen-Koppers, I. (1976) Effects of some allelochemics on food uptake and survival of a polyphagous aphid, *Myzus persicae*. *Entomol. Exp. Appl.*, **19**, 52–56.
66. Schoonhoven, L. M. and Jermy, T. (1977) A behavioural and electrophysiological analysis of insect feeding deterrents, in *Crop Protection Agents. Their Biological Evaluation*, (ed. N. R. McFarlane), Academic Press, London, pp. 133–146.
67. Simpson, S. J. (1995) Regulation of a meal: chewing insects, in *Regulatory Mechanisms in Insect Feeding*, (eds R. F. Chapman and G. de Boer), Chapman & Hall, New York, pp. 137–156.
68. Smith, J. J. B., Mitchell, B. K., Rolseth, B. M., Whitehead, A. T. and Albert, P. J. (1990) SAPID Tools: microcomputer programs for analysis of multi-unit nerve recordings. *Chem. Senses*, **15**, 25–270.
69. Smith, C. M., Khan, Z. R. and Pathak, M. D. (1994) *Techniques for Evaluating Insect Resistance in Crop Plants*, CRC, Boca Raton, FL.
70. Städler, E. (1977) Host selection and chemoreception in the carrot rust fly (*Psila rosae* F., Dipt. Psilidae): extraction and isolation of oviposition stimulants and their perception by the female. *Coll. Int. CNRS*, **265**, 357–372.
71. Städler, E. and Hanson, F. E. (1976) Influence of induction of host preference on chemoreception of *Manduca sexta*: behavioral and electrophysiological studies. *Symp. Biol. Hung.*, **16**, 267–273.
72. Städler, E. and Roessingh, P. (1991) Perception of leaf surface chemicals by feeding and ovipositing insects. *Symp. Biol. Hung.*, **39**, 71–86.
73. Stamp, N. E. and Bowers, M. D. (1994) Effects of cages, plant age and mechanical clipping on plant chemistry. *Oecologia*, **99**, 66–71.
74. Thiéry, D. and Visser, J. H. (1986) Masking of host plant odour in the olfactory orientation of the Colorado potato beetle. *Entomol. Exp. Appl.*, **41**, 165–172.
75. Thiéry, D., Bluet, J. M., Pham-Delègue, M. H., Etiévant, P. and Masson, C. (1991) Sunflower aroma detection by the honey bee: study by coupling gas chromatography and electroantennography. *J. Chem. Ecol.*, **16**, 701–711.
76. Tingey, W. M. (1986) Techniques for evaluating plant resistance to insects, in *Insect–Plant Interactions*, (eds J. R. Miller and T. A. Miller), Springer-Verlag, New York, pp. 251–284.
77. Tjallingii, W. F. (1988) Electrical recording of stylet penetration activities, in *Aphids: Their Biology, Natural Enemies and Control*, vol. 2B, (eds A. K. Minks and P. Harrewijn), Elsevier, Amsterdam, pp. 95–107.
78. Traynier, R. M. M. (1979) Long-term changes in the oviposition behaviour of the cabbage butterfly, *Pieris rapae*, induced by contact with plants. *Physiol. Entomol.*, **4**, 87–96.
79. Vet, L. E. M., van Lenteren, J. C., Heymans, M. and Meelis, E. (1983) An airflow olfactometer for measuring olfactory responses of hymenopterous parasitoids and other small insects. *Physiol. Entomol.*, **8**, 97–106.
80. Visser, J. H. (1976) The design of a low-speed wind tunnel as an instrument for the study of olfactory orientation in the Colorado potato beetle (*Leptinotarsa decemlineata*). *Entomol. Exp. Appl.*, **20**, 275–288.
81. Visser, J. H., Piron, P. G. M. and Hardie, J. (1996) The aphids' peripheral perception of plant volatiles. *Entomol. Exp. Appl.*, **80**, 35–38.
82. Vogel, S. (1969) Low speed wind tunnels for biological investigations, in *Experiments in Physiology and Biochemistry*, (ed. G. A. Kerkut), Academic Press, London, pp. 295–325.
83. Waller, D. A. and Jones, C. G. (1989) Measuring herbivory. *Ecol. Entomol.*, **14**, 479–481.
84. Waterman, P. G. and Mole, S. (1994) *Analysis of Phenolic Plant Metabolites*, Blackwell. Oxford.
85. Wolfson, J. L. (1988) Bioassay techniques: an ecological perspective. *J. Chem. Ecol.*, **14**, 1951–1963.
86. Wolfson, J. L. and Murdock, L. L. (1987) Method for applying chemicals to leaf surfaces for bioassay with herbivorous insects. *J. Econ. Entomol.*, **80**, 1334–1336.
87. Woodhead, S. (1983) Distribution of chemicals in glass fibre discs used in insect bioassays. *Entomol. Exp. Appl.*, **34**, 119–120.

TAXONOMIC INDEX

Numbers in *italics* refer to illustrations and tables.

Abies grandis, 66
Acacia farnesiana, 53
  pennata, 61
Acalymma vittata, 299, *352*, *353*
Acanthoscelides obtectus, 232, *233*
Acer rubrum, 265
Aconitum, 323, 324, 326, 336
  henryi, 326
  napellus, 326
Acrolepiopsis assectella, *140*, 167, *233*
Acyrthosiphon pisum, 42, 164, 209, 228, *229*
Adenostyles, 286
Adoxophyes orana, *144*, 179
Adoxus obscurus, 109
Aegopodium podagraria, 286, 287
Aesculus hippocastanum, 38
African cotton bollworm, *see Helicoverpa armigera*
Agapeta zoegana, 246
Agrotis ipsilon, 90
  segetum, 267, 268
Ailanthus, 208
Ajuga remoto, 38
Alder, *see Alnus glutinosa*
Aleyrodes brassicae, 161
Alfalfa, 93, 202, 338
Alfalfa aphid, *see Terioaphis maculata*
Alfalfa weevil, *see Hypera postica*
Alkali bees, 338
Allium, 159, 293
  cepa, 161
Alnus glutinosa, 10
Ambrosia, 334
Anacridium melanorhodon, 103
Andrena, 320
Andromeda glaucophylla, 317
Andropogon scoparius, 60, 61
A. hallii, 61
Androsace lanuginosa, Plate 1
Anigozanthus flavidus, 46
Anoplognatus montanus, 72
Antheraea pernyi, 209
  polyphemus, 209, 232
Anthocaris cardamines, 242
Anthomyiidae, 132
Anthonomus grandis, *140*, 169, 349

Anthophora, 320
Antispila viticordifoliella, 12
Aphanus, 109
Aphis, *124*, 125
  fabae, 71, 88, 135, *140*, 164
  gossypii, 111, *140*
  nerii, 264
  pomi, 354
Aphrophora alni, 15
Apion, 13
Apis mellifera, 320
Apium graveolens, 86
Apocynum androsaemifolium, 326
  sibiricum, 326
Apple, 19, 44, 48, 133, 145, 147, 213, 216, 256, 285, 286, 333, 348, 354
Apple maggot fly, *see Rhagoletis pomonella*
Arctia caja, 203, 211
Arctostaphylos otayensis, 327
Aristolochia, 44, *213*
Armoracia rusticana, 56
Artemisia dracunculus, 217
Asclepias curassavica, 263
Asparagus officinalis, 292, 293
Aspen, *see Populus tremuloides*
Aster tripolium, 19
Atherigona soccata, 161
Athalia rosae, 170
Atropa belladonna, 36
Aulacorthum solani, 286, 287
Azadirachta indica, 21, 356, 357

Bacillus rossius, 209
Bacillus thuringiensis, 109, 265, 350
Bark beetle, 66, 145, 147, 260
Barley, *see Hordeum vulgare*
Bat, 335
Battus philenor, 132, 201, *213*
Bean, *48*, 232
Beet, 131
Beet armyworm, *see Spodoptera exigua*
Beet fly, *see Pegomya betae*
Bellis, 45
Bemisia tabaci, 9, 161

Beta vulgaris, 131, 135, *136*
Betula pendula, 255
  pubescens, *22*, 69, 257
Bifora radians, 10
Birch, 68
Bird cherry, *see Prunus padus*
Birdsfoot trefoil, *see Lotus corniculatus*
Bittercress, *see Cardamine cordifolia*
Black bean aphid, *see Aphis fabae*
Black cherry, 93
Black mustard, *see Brassica nigra*
Black swallowtail, *see Papilio polyxenes*
Black vine weevil, *see Otiorhynchus sulcatus*
Blissus, 352
Blueberry, 145, 286, 338
Boll weevil, *see Anthonomus grandis*
Bombus, 320
Bombyx hesperus, 208
B. mori, 95, 100, 169, 178, 217
Brachys, 12
Bracken fern, *see Pteridium aquilinum*
Brassica, 45, 161
  campestris, 161
  napus, 161, 263, 351
  nigra, 63
  oleracea, 40, 43, 138, 161, 206, 207, 208, 209
Bretschneiderea sinensis, 50
Brevicoryne brassicae, 42, 134, *140*, 161, 348
Broadbean, *see Vicia faba*
Broccoli, *352*, *353*
Brown planthopper, *see Nilaparvata lugens*
Browntail, *see Euproctis chrysorrhoea*
Bruchidius villosus, 263
Bruchus pisorum, 350
Brunsfelsia, 15
Brussels sprouts, 110, 263, 347, 352, 354
Buddleia, 255
Bumblebees, 317–339
Buffalo, 85
Buttercup, 336

Cabbage, 1, 44, 48, 49, 68, 99, 134, 135, 137, 138, 156, 159, 162, 166, 167, 231, 232, 349, 358
  see also *Brassica oleracea*
Cabbage aphid, see *Brevicoryne brassicae*
Cabbage looper, see *Trichoplusia ni*
Cabbage root fly, see *Delia radicum*
Cactus, see *Opuntia*
Cactoblastis cactorum, 360, 362
Caenocoris nerii, 264
Café diable, 208
Callophrys rubi, 105
Callosamia promethea, 209
Calotropis gigantea, 19
Calpodes ethlius, 179
Calystegia sepium, 22
Cameraria, 261
Carausius morosus, 209
Cardamine cordifolia, 62
Cardiaspina densitexta, 62
Carduus nutans, 197
Carex heliophila, 61
Careydon serratus, 233
Carrot, 44, 352
Carrot root fly, see *Psila rosae*
Carya, 12
Cassava, 59, 354
Cassida nebulosa, 170
Castilleja, 198
Catharanthus roseus, 40
Catocala, 11
Cauliflower, 353
Cavariella aegopodii, 140
Cecropia peltata, 54
Celery, see *Apium graveolens*
Centaurea, 293
  maculosa, 246
  scabiosa, 197
Cerinthe major, 328, 329
Ceutorhynchus assimilis, 131, 140
  macula-alba, 260
Chelidonium majus, 55
Chelone glabra, 326
Chenopodium murale, 86
Cherry, 171
Cherry fruit fly, see *Rhagoletis cerasi*
Chickpea, 263
Chilo partellus, 161
Chionaspis, 293, 294
Chlosyne lacinia, 209
Choristoneura occidentalis, 92
Choristoneura fumiferana, 338
Christmas beetle, see *Anoplognatus montanus*
Cidaria albulata, 128
Cigarette beetle, see *Lasioderma serricorne*
Cinnabar moth, see *Tyria jacobaeae*

Cirsium canescens, 244, 245, 246
  vulgare, 13, 271
Citrullus colocynthis, 19
Citrus, 44
  paradisi, 356
  unshiu, 168
Clerodendrum, 294
Clover, 18, 48
Coca, see *Erythroxylon coca*
Coccinella, 264
Cockroach, 89
  see also *Periplaneta americana*
Coffee, 344
Cola nitida, 52
Colias philodice, 215
Collard, 257
Collinsia, 198, 288, 289
Colorado potato beetle, see *Leptinotarsa decemlineata*
Columbine, 336
Commelina tuberosa, 330
Commiphora myrrha, 231
Common vetch, see *Vicia sativa*
Conium maculatum, 297
Convallaria, 293
Convolvulus arvensis, 19
Cordaites, 280
Corn, 44, 50, 67, 90, 93, 95, 108, 157, 234, 352, 353
Corn earworm, see *Helicoverpa zea*
Corylus avellana, 22
Corytes mystaceus, 338
Costelytra zealandica, 39
Cotesia congregata, 262
  glomerata, 263
  marginiventris, 67
Cotton, 45, 67, 70, 71, 102, 107, 108, 156, 157, 217, 266, 263, 344, 349, 350, 353, 354
  see also *Gossypium hirsutum*
Cotton boll weevil, see *Anthonomus grandis*
Cotton whitefly, see *Bemisia tabaci*
Cottonwood, see *Populus*
Cow, 85
Cowpeas, 108
Crataegus, 181, 201, 285
  mollis, 213
  monogyna, 19, 32, 216
Crioceris, 292, 293
Crocidosema plebejana, 11
Cruciferae, 41
Cryptomyzus korschelti, 140
Cucumber, 17, 70, 352, 353
Cucurbita moschata, 299, 353
  pepo, 111
Cutworm, see *Agrotis*
Cyathodes colonsoi, 43
Cymopterus terebinthus, 217

Cynoglossum officinale, 19, 52, 70
Cyrtobagous salviniae, 360

Dactylopius opuntiae, 261
Dacus dorsalis, 140
  oleae, 233
Danaus plexippus, 179, 263
Dandelion, see *Taraxacum officinale*
Dasineura brassicae, 13
  papaveris, 260
Datura stramonium, 300
Deadly nightshade, see *Atropa belladonnna*
Delia, 124, 125, 174, 184
  antiqua, 12, 130, 131, 140, 143, 144, 159, 167
  brassicae, 131, 143, 144
  floralis, 184
  radicum, 12, 123, 131, 133, 134, 140, 157, 160, 167, 168
Dendroctonus, 124, 125, 131
  pseudotsugae, 260
Dendrolimus pini, 178, 183
Desert locust, see *Schistocerca gregaria*
Diabrotica, 130, 262
  adelpha, 70
  balteata, 70
  virgifera virgifera, 204
Diacrisia virginica, 210, 211
Diadegma insulare, 262
Diamondback moth, see *Plutella xylostella*
Digitalis lutea, 331
  purpurea, 40
Dineura virididorsata, 257
Dioscorea, 38
Diplacus aurantiacus, 54
Diploclisia glaucescens, 38
Dock, see *Rumex*
Drosophila, 181, 214, 218, 285
Dysaphis plantaginea, 354
Dysdercus koenigi, 209

Empoasca, 16
  devastans, 131
Encarsia formosa, 16, 17
Enchenopa binotata, 284, 286
Englerina woodfordioides, 40
Entomoscelis americana, 179
Ephestia kuehniella, 260
Epidiaspis leperii, 256
Epilachna niponica, 260
  pustulosa, 209
  tredecimnotata, 262, 299
  varivestis, 112, 113, 157, 161
Epilobium angustifolium, 109
Epirrita autumnata, 65, 68, 69
Equisetales, 86
Erioischia brassicae, 161

Eriosoma lanigerum, 348
   pyricola, 229
Erythroxylon coca, 37
Estigmene acrea, 179
Eucalyptus, 19, 21
Eucalyptus meliodora, 72
Eucosma cana, 272
Euonymus europaeus, 44, 181, 216, 247
Euphydrias, 243
   editha, 198, 199, 217, 288, 289, 295
Euphyllura phillyreae, 233
Euproctis chrysorrhoea, 7, 8, 209
Eupteryx maigudoi, 361
European corn borer, see *Ostrinia nubilalis*
Eurytoma tibialis, 271
   robusta, 271
Euura, 251

Fagus sylvatica, 46
Fall armyworm, see *Spodoptera frugiperda*
Fennel, 321
Fenusa pumila, 11
Fern, see *Polypodium*
Fescue grass, 110
Festuca arundinacea, 43
   pratensis, 246
Ficus, 293
Fig, 316, 317, 335
Fig wasp, 316, 317, 335
Foxglove, see *Digitalis*
Fraxinus, 334
Freesia, 336
Fritillaria, 293

Galerucella lineola, 209
Galium saxatile, 246
Garden tiger moth, see *Arctia caja*
Gastrophysa viridula, 111
Genista tinctoria, 105
Gentiana andrewsii, 326
Geocoris, 266
Geranium, see *Pelargonium hortorum*
Ginkgo biloba, 5, 6
Giraudiella inclusa, 260
Glechoma hederacea, 287
Glycine max, 161
Goat, 85
Goldenrod, see *Solidago*
Gonia ornata, 267, 268
Gossypium, 38
   hirsutum, 11, 70
Grain aphid, see *Sitobion avenae*
Grand fir, see *Abies grandis*
Grape, 348, 354
   see also *Vitis vinifera*
Grape phylloxera, see *Phylloxera vitifoliae*

Graphocephala ennahi, 196
Grapholita glycinivorella, 156
Grass, 60, 61, 89, 111, 211, 246, 255, 268, 323
Grasshopper, 85
Greater celandine, see *Chelidonium majus*
Green peach aphid, see *Myzus persicae*
Green hairstreak, see *Callophrys rubi*
Gypsy moth, see *Lymantria dispar*

Halictus, 320
Haltica lythri, 209
Hamamelis vernalis, 22
Haplopappus, 245
Hawthorn, see *Crataegus monogyna*
Heliconia imbricata, 22
Heliconius, 213
Helicoverpa, 124, 125, 175, 218, 349
   armigera, 209, 262, 263
   subflexa, 140
   virescens, 98, 104, 140, 169, 217, 218
   zea, 89, 91, 103, 131, 157, 179, 206, 207, 209, 234, 235, 236, 266
Hemlock, 16
Hessian fly, see *Mayetiola destructor*
Hickory, see *Carya*
Hieracium, 211
Holcaphis holci, 246
Holcus lanatus, 200
   mollis, 246
Homalodisca coagulata, 213
Homoeosoma electellum, 233, 234, 235
Honeybee, 145, 317–339
Hophornbeam, see *Ostrya*
Hoplasoma, 294
Hordeum vulgare, 54, 57, 161, 344, 348, 349, 356
Horse chestnut, see *Aesculus hippocastaneum*
Horseradish, see *Armoracia rusticana*
Horsetail, see *Equisetales*
Hound's tongue, see *Cynoglossum officinale*
Hummingbird, 335
Hyalophora cecropia, 209
Hyalopterus pruni, 135, 200
Hylesia lineata, 214
Hyoscyamus niger, 295
Hypera postica, 163, 164, 287
Hypericum hirsutum, 40, 51
Hyphantrea cunea, 170, 209

Impatiens, 323, 324
   capensis, 326
Ipomoea purpurea, 247
Ips, 124, 125

   typographus, 140
Isia isabella, 182, 183

Japanese beetle, see *Popillia japonica*
Juglans arizonica, 22
Juncus effusus, 201
Juniperus, 20, 21

Kalmia angustifolia, 327
   polifolia, 317
Knapweed, 196

Laburnum anagyroides, 263
Lamponius portoricensis, 204
Lantana camara, Plate 1, 362
Larch budmoth, see *Zeiraphera diniana*
Large white butterfly, see *Pieris brassicae*
Larinus, 196, 197, 293
Larix decidua, 69
Larrea cuneifolia, 40
Lasioderma serricorne, 109
Lavandula spicata, 46
Leafcutter bees, 339
Ledrum groenlandicum, 317
Leek moth, see *Acrolepiopsis assectella*
Leptinotarsa decemlineata, 6, 47, 50, 126, 131, 137, 138, 139, 140, 142, 143, 145, 146, 147, 157, 158, 163, 164, 169, 170, 176, 179, 196, 198, 202, 209, 230, 295, 345, 350, 352, 356
Leptoterna dolabrata, 91
Lettuce, 70
Lettuce aphid, see *Nasonovia ribisnigri*
Lichens, 5
Ligustrum vulgare, 19
Lilioceris lilii, 293
   merdigera, 293
Lilium, 293
Lima bean, 65, 167
Limenitis archippus, 209
Limenitis hybrid rubidus, 209
Lipaphis erysimi, 140, 161
Lipara lucens, 267
   pullitarsis, 267
Liriomyza sativae, 217
Listroderes obliquus, 140
Lithocollectis ostryarella, 12
Locust, 85
Locust tree, 261
Locusta migratoria, 44, 45, 99, 101, 104, 143, 163, 164, 169, 182, 203, 205, 211, 213, 359
Lodgepole pine, 69
Lolium perenne, 44, 110
Lomatium grayi, 217

Lotus corniculatus, 317
  scoparius, 330
Loxostege sticticalis, 209
Lupin, 299, 331, *349*
Lupinus, 37
  nanus, Plate 1
  polyphyllus, 57, 105
Lycaenidae, 8, 203, 294
Lycopersicon, 57
  esculentum, 40, 48, *156*, 176, *349*, 350, 352, 353
  hirsutum, 137, *139*, 348
Lygus hesperis, 156
  lineolaris, 9, 349
Lymantria dispar, 9, *10*, 11, 60, 62, 94, 109, *110*, 179, 203, 204, *209*, 247, 265
Lymenitis, 209, 218

Macrocentrus ancylivorus, 265
Macrosiphum euphorbiae, 156, 348
Magicicada septendecim, 288
Maize, 24, 48, 241, 263, 299, 344, 352, 353
Malacosoma americana, 179
  castrensis, 211
Malus, 180, *181*, 195
Malva parviflora, 11
Mamestra brassicae, 171, 176, *178*, *179*, 181
  configurata, 13, *171*
Manduca sexta, 95, 105, *124*, *125*, 136, 140, 141, 143, *169*, 175, *199*, 205, 209, 210, 262, 265
Mangrove, *see Rhizophora mangle*
Marrubium vulgare, 57
Mayetiola destructor, 202, 284, 299, 301, 347
Mealy plum aphid, *see Hyalopterus pruni*
Medicago sativa, 215
Megachile, 320
Megoura viciae, 228
Melanoplus sanguinipes, 99, 110
Melia toosendan, 357
Melilotus alba, 56, 215
Mentha piperita, 47, 107
Messa nana, 11
Mexican bean beetle, *see Epilachna varivestis*
Microplitis demolitor, 262, 263
Migratory locust, *see Locusta migratoria*
Millet, 157, 204
Monkshood, *see Aconitum*
Mosses, 5
Mountain birch, *see Betula pubescens*
Mountain pine beetle, 69
Mucuna, 40
Muellerianella fairmairei, 200

Mulberry, 217
Musa acuminata, 19
Muscari, 293
Mustard, 48, 49, 156
Mustard beetle, *see Phaedon cochleariae*
Myosotis, Plate 1
Myzocallis schreiberi, 18
Myzus persicae, 6, 108, 110, *124*, *125*, 134, *161*, *171*, 229, 377

Nasonovia ribisnigri, 70
Nasturtium, 7, 263
Neem tree, *see Azadirachta indica*
Nemoria arizonaria, 229, 230
Nerium oleander, 199, 263, 264
Nettleleaf goose foot, *see Chenopodium murale*
Nicotiana, 262
  attenuata, 65, *66*
  sylvestris, 300
  tabacum, 55, 57
Nilaparvata lugens, 63, 301, 350
Nymphaea alba, 336

Oak, 52, *93*, 229, 230, 232, 245, 247, 260
  see also *Quercus*
Oats, 48
Odontata dorsalis, 261
Oedaleus senegalensis, 204
Oleander, *see Nerium oleander*
Oligolephus tridens, 100
Olive, 233
Oncopeltus fasciatus, 164, 218
Onion, 352
Onion fly, *see Delia antiqua*
Operophtera, 21
  brumata, 103, 240, 241
Ophraella, 295
Ophrys, 317, 337
  insectifera, 338
Oporinia, 21
Opuntia, 249, 250, 261, 360, 362
Orchid, 336, 337, 338
Oreina cacaliae, 286
  globosa, 286
Orthocarpus, 198
Oryza officinalis, 350
  sativa, 350
Oscinella frit, 233
Ostrinia nubilalis, 95, 140, 165, 286
Ostrya, 12
Otiorhynchus sulcatus, 156

Pachylia ficus, 86
Palloptera, 272
Palm, 85
Panonychus ulmi, 50
Papaver somniferum, 57, 60

Papilio, 166, *218*, 291
  aegeus, 209
  glaucus, 196, *209*, 284
  machaon, 10, *209*, 214
  oregonius, 217
  polyxenes, 140, 161, *167*, *171*, 179
  protenor, 168
  zeliacon, 217
Paratrytone melane, 86
Paspalum dilatatum, 111
Pastinaca sativa, 40, *44*, 51, 68, 104
Pea, *see Pisum sativum*
Pea aphid, *see Acyrthosiphon pisum*
Pea weevil, *see Bruchus pisorum*
Peanut, 108
Pear, 229
Pedicularis, 198, 218
Pegomya betae, 131
Pelargonium hortorum, 206, *207*
Pemphigus, 200
  betae, 52, 252
Penstemon, 198
Peppermint, *see Mentha piperita*
Peridroma saucia, 107
Periplaneta americana, 143
Petasites, 286
Petunia hybrida, 210, *211*
Phaedon cochleariae, 156, *159*, *162*
Phaseolus lunatus, 166
  vulgaris, 46, 349
Philophylla heraclei, 233
Phleum, 323
Phorodon humuli, 140
Phragmites australis, 19, 260, 267
  communis, 135, *136*, 200
Phratora vitellinae, 209
Phtorimaea operculella, 233
Phyllobius oblongus, 170
Phyllobrotica, 294
Phyllocolpa, 251
Phyllonorycter blancardella, 195, *196*
Phyllotreta, 140
  albionica, 161
  armoraciae, 167
  cruciferae, 161, 257
  nemorum, 161, 376
  striolata, 257
Phylloxera vitifoliae, 349
Phytodecta fornicata, 170
Phytomyza ilicicola, 52
Picea sitchensis, 43
Pieris, 85, *167*, 174, 180, 219
  brassicae, 1, 5, 7, 41, 67, 68, 99, 132, 133, 146, 160, *163*, *164*, 166, *170*, *171*, 172, 177, *178*, *179*, 181, *182*, *183*, 207, *208*, *209*, 260, 263, 292, 344
  rapae, 58, 62, *63*, 101, 132, *156*, 160, 161, *167*, *171*, 172, *173*, *179*, 184, *209*, 214, 257, 292, 329, 344

napi, 256, 292
  napi napi, 185, 197
  napi oleracea, 167, 184, 185, 197, 256
Pine beetle, 69
Pinus, 265
  sylvestris, 57, 62
Piper, 21
  arieianum, 23
Pipevine swallowtail, *see Battus philenor*
Pisum sativum, 350
Plantago, 198, 211
  lanceolata, 288, 289
Platyprepia virginalis, 16
Plutella, 167
  xylostella, 140, 157, 159, 161, 170, 232, 233, 262, 288, 352, 358
Poison hemlock, *see Conium maculatum*
Polygonatum, 293
Polygonia interrogationis, 209
Polygonum hydropiper, 356
Polypodium feei, 19
  vulgare, 38
Pontania, 251
Popillia japonica, 9, 110, 131, 140, 260
Poplar, *see Populus*
Populus, 72, 265
  angustifolia, 16, 52, 252
  fremontii, 16
  grandidentata, 110
  tremuloides, 72, 110
  trichocarpa, 40, 58, 59
Potato, 24, 47, 48, 71, 147, 158, 230, 231, 295, 344, 349, 350, 352, 353
  see also Solanum tuberosum
Potato aphid, *see Macrosiphum euphorbiae*
Privet hawkmoth, *see Sphinx ligustri*
Procecidochares, 218
Prociphilus, 200
Prunella vulgaris, 326
Prunus, 135, 180, 200, 201, 295
  amygdalus, 40
  avium, 22, 327
  padus, 42
Pseudaletia unipuncta, 99
Pseudatomoscelis seriatus, 349
Psila rosae, 12, 140, 143, 144, 166, 352
Pteridium, 258
  aquilinum, 61, 71, 248, 362
Puccinia, 301
Pulmonaria officinalis, 286, 287

Quadraspidiotus marani, 256
  ostreaeformis, 256
  perniciosus, 256
  pyri, 256

Quercus, 93, 265
  emoryi, 22
  ilex, 18
  nigra, 110
  pubescens, 43, 46
  robur, 21, 40, 58, 240, 241, 255
  rubra, 110

Radish, 134
Ragwort, *see Senecio jacobaea*
Rape, *see Brassica napus*
Raphiolepus umbellata, 331
Raspberry, 156
Red cabbage, 134
Red clover, *see Trifolium pratense*
Red turnip beetle, *see Entomoscelis americana*
Reed, *see Phragmites communis*
Reseda, 7
Rhagoletis, 124, 125
  cerasi, 125, 133, 171, 287, 288
  mendax, 144, 145, 216, 286
  pomonella, 131, 132, 133, 140, 144, 145, 147, 148, 213, 216, 284, 285, 286
Rhamnus cathartica, 19
Rhinanthus, 128
Rhizophora mangle, 248
Rhododendron, 19, 21, 196
  callostrotum, 46
  javanicum, 19
Rhopalosiphum, 200
  maidis, 353
  padi, 18, 140, 356
Ribes odoratum, Plate 1
Rice, 24, 63, 344
  see also Oryza sativa
Ricinus communis, 57
Rickettsia, 287
Rosa, 45, 323
  rugosa, 322
Rosmarinus officinalis, 43
Rothschildia lebeau, 86
Rumex, 111

Saccharum officinarum, 19
Sage, *see Salvia officinalis*
Salix, 70, 180, 181, 251, 332
  capraea, 22
  cinera, 19
Salvia officinalis, 352
Salvinia molesta, 360, 362
Sambucus, 22, 42
Saxifraga hirculus, 317
Scaptomyza nigrita, 62
Schistocerca, 213
  americana, 39, 169, 211, 212
  gregaria, 9, 163, 169, 182, 205, 206, 209, 231, 233, 355, 359
  shoshone, 196

Schizaphis graminum, 161, 171, 209, 301, 349
Scirpophaga incertulas, 87
Scots pine, *see Pinus sylvestris*
Scrobipalpa ocellatella, 233
Scutellaria, 294
Senecio, 201, 286
  jacobaea, 242, 259
  sylvaticus, 21
Silkworm, 85
Sitobion avenae, 22, 161
Sitotroga cerealella, 233
Solanum, 255
  angustifolium, 196
  berthaultii, 157, 158
  carolinense, 196
  demissum, 295
  dulcamara, 22, 202, 231, 323
  elaeagnifolium, 196
  mauritanum, 255
  rostratum, 196, 345
  tuberosum, 137, 138, 139, 157, 176, 202
Solidago, 201
  altissima, 22, 243, 255
  missouriensis, 72
  virgaurea, 255
Sorghum bicolor, 44, 45, 55, 56, 57, 161, 205, 263, 354
Sour cherry, 285
Soybean, 108, 112, 113, 156, 157, 263, 344, 345, 353
Soybeanpod borer, *see Grapholita glycinivorella*
Spartina altiniflora, 22
Spider mite, 50, 65, 66, 67, 70, 349
Spiderwort, 336
Spilothetus pandurus, 264
Spinach, 212
Spindle tree, *see Euonymus europaeus*
Spinx ligustri, 199
Spodoptera, 132, 163, 175
  eridania, 63, 104, 209
  exempta, 171
  exigua, 67, 86, 91, 103
  frugiperda, 107, 108, 111, 161, 171, 378
  littoralis, 102, 104, 140, 171, 209, 212
  litura, 107, 205, 358
  sunia, 353
Spruce budworm, *see Choristoneura fumiferana*
Spruce tree, 350
Squash, 241, 349, 353
  see also Cucurbita
St John's wort, *see Hypericum hirsutum*
Stachys, 294
Stinging nettle, *see Urtica dioica*

Strawberry, 48, 156
Striped cucumber beetle, see
    Acalymma vittata
Subcoccinella, 24-punctata, 209
Sunflower, 44, 235, 263, 322
Sunflower moth, see Homoeosoma
    electellum
Supella longipalpa, 89
Swallowtail butterfly, see Papilio
    machaon
Swedish turnip, 263
Sweet potato, 44
Syntomeida epilais, 264

Taeniopoda eques, 211, 212, 213
Tanymecus dilaticollis, 170
Taraxacum officinale, 199, 206, 207,
    211, 330, 332, 333
Tarnished plantbug, see Lygus
    lineolaris
Taxus baccata, 13
Tea, see Thea sinensis
Tegeticula, 293
Tephritidae, 132
Tephroclystis virgaureata, 201
Terioaphis maculata, 202
Tetranychus urticae, 50, 70, 353
Thalictrum, 334
Thea sinensis, 40, 53
Therioaphis trifolii, 18
Thistle, 126, 127, 196, 244, 271
    see also Cirsium vulgare
Thlaspi arvense, 219
Thrips tabaci, 161
Thymus vulgaris, 47, 352
Thyridia, 15

Tiger swallowtail, see Papilio glaucus
Tilia, 12
Tobacco, 44, 53, 57
Tobacco budworm, 71
Tobacco cutworm, see Spodoptera
    litura
Tobacco hornworm, see Manduca
    sexta
Tomato, see Lycopersicon esculentum
Torymus, 271
Trialeurodes variabilis, 354
Trichogramma evanescens, 267, 268
    pretiosum, 266
Trichoplusia ni, 41, 59, 109, 140, 156
Trifolium pratense, 242, 317
    repens, 299
Triticum aestivum, 161, 270
Tropaeolum majus, 207, 208
Turnip root fly, see Delia floralis
Tyria jacobaeae, 242, 259

Uromyces trifolii, 299
Urophora stylata, 271
Urtica dioica, 13, 14

Vaccinium macrocarpon, 317, 323
Vanduza arquata, 261
Vetch aphid, see Megoura viciae
Vicia cracca, 326
    faba, 71, 88, 229, 265, 353
    narbonensis, 71
    sativa, 351
Vitis vinifera, 12

Water pepper, see Polygonum
    hydropiper

Western corn rootworm, see
    Diabrotica virgifera
Western lygus bug, 156
Western spruce budworm, see
    Choristoneura occidentalis
Wheat, 24, 89, 99, 299, 301, 343, 347,
    352
White melilot, see Melilotus albus
Whitefly, 16
    see also Trialeurodes variabilis
Wild parsnip, see Pastinaca sativa
Wild tomato, 44
Willow, see Salix
Willow herb, 331
Winter moth, see Operophtera
    brumata
Woody nightshade, see Solanum
    dulcamara
Woolly aphid, see Eriosoma lanigerum

Yam, see Dioscorea
Yellow adonis, 336
Yellow stem borer, see Scirpophaga
    incertulas
Yew, see Taxus baccata
Yponomeuta, 143, 144, 179, 180, 181,
    187, 216, 218, 247, 290, 292, 293,
    295, 306
    cagnagellus, 232
    evonymellus, 234
Yucca, 293, 335
Yucca moth, 335

Zeiraphera diniana, 68, 69, 233
Zenilla adamsoni, 263
Zonocerus variegatus, 104

AUTHOR INDEX

Numbers in *italics* refer to reference sections.

Abe, F., 166, *189*
Abe, T., 86, *113*
Abe, Y., 305, *308*
Abisgold, J. D., 101, *113*, *118*, 212, *220*
Abrahamson, W. G., 11, *27*
Adati, T., 44, *74*
Addy, M. N., 270, *275*
Adewusi, S. R. A., 57, *74*
Ågren, G. I., 96, *114*
Ahmad, S., 98, *113*, 140, *148*, 155, *187*
Åhman, I., 13, *25*
Aide, T. M., 21, *26*
Aidley, D. J., 101, *114*
Aikman, D., 126, 131, *148*
Albert, P. J., 379, *383*
Alford, A. R., 356, *364*
Allan, E. J., 358, *362*
Allebone, J. E., 44, *74*
Allen, D. C., 260, *272*
Allen, S. E., 84, *114*
Allen, T. F. H., 244, *272*
Allen-Williams, L. J., 131, 140, *150*
Alm, J., 318, 329, *339*
Alonso, J. G., 71, *77*, 349, *364*
Altieri, M. A., 352, 353, 354, *362*, *363*, *365*
Aluja, M., 131, 147, 148, *148*
Amakawa, T., 181, *191*
Ambrose, H. J., 110, *114*
Ames, B. N., 355, *362*
Amour, G. S., 376, *381*
Anantha Raman, K. V., 95, 98, *114*
Anderbrant, O., 129, *149*
Andersen, J. F., 140, *151*
Andersen, P. C., 89, *115*, 213, *220*
Andersen, R. A., 48, *77*
Anderson, J. C., 174, 181, *192*, 357, *363*
Anderson, T. E., 95, *114*
Andersson, B. Å., 48, *74*
Andow, D. A., 2, *4*, 351, *362*
Andres, L. A., 293, *313*
Appel, H. M., 103, 108, 109, *114*
Appelgren, M., 146, *149*
Applebaum, S. W., 231, *238*
Arbas, E. A., 137, *153*

Archer, T. L., 93, *116*
Arn, H., 166, *190*
Arnone, J. A., 113, *114*
Arora, K., 181, *187*
Arpaia, S., 350, 357, *362*
Arriaga, H. O., 55, *75*
Arteel, G. E., 113, *117*
Asaoka, K., 41, *79*, 217
Ascher, K. R. S., 102, *117*, 132, *151*, 378, *380*
Atkins, M. D., 122, *148*
Auger, J., 140, *152*
Avé, D. A., 45, 47, 49, *81*, 157, 158, *189*
Averill, A. L., 213, 216, *223*
Avison, T. I., 218, *220*
Axelsson, B., 96, *114*
Ayala, F. J., 285, *310*
Ayertey, J. N., 376, *380*
Ayres, M. P., 112, *114*
Ayres, P. G., 111, *116*

Babka, B. A., 44, *78*
Bach, C. E., 352, 353, *362*
Bailey, C. G., 90, *114*
Baker, H. G., 329, 333, 338, 339, *340*, *341*
Baker, I., 329, *340*
Baker, T. C., 137, *148*, 377, *382*
Baldwin, I. T., 64, 66, 68, *74*, 300, *308*
Baliddawa, C. W., 352, 353, *362*
Balogh-Nair, V., 171, *189*
Banteli, R., 174, 181, *192*
Barbehenn, R. V., 86, 103, *114*
Barbercheck, M. E., 109, *114*
Barbosa, P., 9, 19, 23, 25, 108, 109, 111, *114*, 203, *220*, 262, 265, 268, *272*, 274, 291, 296, *308*, 345, *362*, *363*
Barbour, J. D., 64, *74*, 299, *311*
Barkman, J. J., 19, 20, *28*
Barnes, O. L., 211, *220*, 375, *380*
Barth, H. G., 318, 339, *340*
Barton Browne, L., 122, 123, 125, *148*, *149*, 355, *363*
Bas, N., 346, *364*
Bashford, R., 21, *25*

Bateson, P., 376, *382*
Baur, R., 140, *148*, 155, *187*
Bawa, K. S., 335, *340*
Baxter, H., 74, *77*
Baylis, M., 203, *220*
Bazzaz, F. A., 40, 51, *82*
Beck, S. D., 102, *114*
Behan, M., 146, *149*
Bell, E. A., 40, *74*
Bell, W. J., 127, 128, 148, *149*
Bellotti, A. C., 354, *363*
Belsky, A. J., 24, *25*
Benecke, R., 40, 59, *81*
Benedict, J. H., 156, *187*, 266, *277*
Bennett, A. F., 199, *220*
Bentley, M. D., 356, *364*
Benz, G., 69, *74*
Berdegue, M., 86, *114*
Berenbaum, M. R., 19, 29, 32, 70, *74*, 80, 90, 103, 104, *114*, *119*, 290, *308*, *380*, *382*
Berge, M. A., 300, *310*, *311*
Bergelson, J., 349, *363*
Bergstöm, G., 48, *81*, 146, *149*, 322, *340*, *341*, *342*
Berlocher, S. H., 284, *308*
Bernard, R. L., 157, *192*
Bernays, E. A., 6, 7, *25*, *27*, 39, 40, 42, 45, *74*, *75*, *82*, 85, 86, 90, 92, 93, 95, 99, 101, 103, 105, *114*, *115*, *118*, 131, 140, 147, *149*, 156, 160, 162, 163, 164, 165, 169, 170, 172, *187*, *188*, 204, 205, 206, 208, 210, 211, 212, 213, 216, *220*, 221, 222, 225, 289, 290, 291, 305, *308*, 359, *363*, 376, 378, *380*
Berry, R. E., 98, *115*
Berryman, A. A., 66, 69, *79*
Bertsch, A., 330, *342*
Besson, E., 40, *74*
Bierbaum, T. J., 216, *221*
Biggs, D. R., 39, *78*
Bilgener, M., 103, *115*
Bingaman, B. R., 40, *74*
Birch, A. N., 184, *192*
Bird, J. V., 349, *363*
Birgersson, G., 146, *149*

Blaakmeer, A., 44, 48, 68, 74, 81, 87, 129, *149*, 166, 166, 171, *187*, *193*, 355, 357, 359, 364, *380*
Blackwell, A., 38, *79*
Bland, R. G., 173, *187*
Blaney, W. M., 101, *119*, 170, 174, 175, 176, 180, 181, 182, 184, *187*, *188*, *192*, 209, 211, 212, *220*, *224*, 357, *363*, 378, *380*
Blau, P. A., 102, *115*, 171, *188*, 378, *380*
Blight, M. M., 140, 145, *149*
Bloem, K. A., 91, 92, 103, *116*
Blom, F., 171, 177, 181, *188*, *192*, 378, *379*, *380*
Blua, M. J., 112, *115*
Bluet, J. M., 379, *383*
Boch, R., 325, *340*
Boeckh, J., 145, 146, *149*
Bolgar, T. S., 140, *153*
Boller, E. F., 171, *192*, 287, 288, *308*
Bolter, C. J., 48, *74*
Bomford, M. K., 205, *220*, 358, *363*
Bond, W. J., 333, *341*
Bongers, W., 202, *220*, 231, *237*
Booth, C. O., 134, 135, 140, *150*, 161, *189*
Borden, J., 131, 140, *150*
Bordner, J., 349, *365*
Borg-Karlson, A. K., 337, *340*
Bosque-Perez, N., 378, *381*
Bottrell, D. G., 353, *363*
Boufford, D. E., 50, *74*
Boulter, D., 350, *363*
Bouthier, A., 38, *78*
Bouthyette, P. J., 157, 158, *189*
Boutin, J. P., 42, 44, *75*, *80*, 165, *188*
Bouton, C. E., 250, *275*
Bowdan, E., 376, *380*
Bowers, M. D., 58, 59, 70, *80*, 166, *188*, *191*, 375, *383*
Boys, H. A., 204, *220*
Bracken, G. K., 13, *25*
Braimah, H., 262, *272*
Brattsten, L. B., 104, 106, 108, *115*
Braun, E., 317, *340*
Breden, F., 204, *225*
Breen, J. P., 110, *115*
Breer, H., 142, 146, *151*, *152*
Brewer, J. W., 92, *115*
Bright, K. L., 211, 212, 213, *220*
Broadway, R. M., 59, *74*
Brodbeck, B. V., 89, 90, *115*, 213, *220*
Brody, A. K., 16, *26*
Brooks, D. R., 305, *311*
Brooks, G., 113, *119*
Brooks, T. M., 5, *28*
Brough, C., 350, *363*
Brown, B. J., 244, *272*

Brown, I. L., 243, *273*
Brown, V. C., 112, *115*
Brown, V. K., 249, 252, 269, 272, 296, 300, *308*
Brown, W. V., 72, *76*
Bruin, J., 67, 74, 349, *363*
Bryant, J. P., 19, *26*, 59, *75*, 296, 300, *309*
Bryce, T. A., 44, *74*
Buchmann, S. L., 330, *341*
Buckingham, J., 35, *74*
Buckley, D. L., 376, *381*
Buckley, R., 261, *272*
Buckner, J. S., 376, *382*
Bugg, R. L., 265, *272*
Buonaccorsi, J. P., 147, *148*
Burd, M., 317, 331, 337, *340*
Burdon, J. J., 111, *115*, 302, *312*
Buser, H. R., 44, *80*, 155, *192*
Bush, G. L., 140, 145, *150*, 216, 217, 218, *221*, 222, 284, 285, 286, 288, 301, *308*, *309*
Buttery, R. G., 47, 48, *75*, 234, *237*
Byers, J. A., 129, 146, 148, *149*

Calderon, M., 266, *276*
Caldwell, M. M., 300, *309*
Callaghan, C. J., 5, *26*
Cambiazo, V., 215, *225*
Campbell, B. C., 91, 92, 103, 108, *115*, *116*
Campbell, C. A. M., 140, *149*
Cannon, W. N., 377, *381*
Cantelo, W. W., 140, *149*
Cappuccino, N., 148, *149*, 243, *276*
Cardé, R. T., 142, 147, *148*, *151*
Carlisle, D. B., 231, 232, *237*
Carlson, R. W., 265, *272*
Carroll, M. R., 260, *272*
Carson, H. L., 283, 287, *309*
Carter, M., 44, *75*, *76*, 166, *188*, *191*
Carter, S. W., 376, *381*
Castro, A. M., 55, *75*
Cates, R. G., 8, *26*, 52, *75*, 196, *220*
Cavalier-Smith, T., 53, *75*
Chamberlain, J. D., 156, *188*
Chameides, W. L., 50, *75*
Champagne, D., 39, 40, *74*, 169, *187*
Chapin, F. S., 19, *26*
Chapin, T., 59, *75*, 296, 300, *309*
Chaplin, J. F., 77, *150*
Chapman, R. B., 19, *26*
Chapman, R. F., 9, *26*, 45, *75*, 82, 84, 85, *115*, 129, 131, 140, 141, 143, 147, 148, *149*, *153*, 156, 157, 160, 162, 165, 170, 172, 173, 176, *187*, *188*, 196, 211, *220*, *225*, 290, 291, *308*, 359, *363*
Chapya, A., 171, *189*

Charles, P. J., 9, *27*
Chatterjee, S. K., 40, *76*
Chauffaille, J., 50, *79*, 322, *341*
Cherrett, J. M., 19, *26*
Chew, F. S., 41, 42, 58, *75*, *78*, 170, 184, *188*, 219, 220, 256, 260, *272*, *276*, 344, *363*
Chiang, H. S., 157, *188*
Chilcote, C. A., 285, 286, *309*
Chin, C. T., 376, *381*
Chippendale, G. M., 204, *225*
Chittka, L., 323, *340*
Choo, G. M., 40, *81*
Chopin, J., 40, *74*
Chrispeels, M. J., 350, *365*
Chyb, S., 166, 170, 179, 183, *189*, *190*, 203, 204, *220*, *224*, 357, 359, *363*
Cincotta, R. P., 210, *223*
Claridge, M. F., *26*, 286, *309*
Clay, F. K., 111, *115*
Coaker, T. H., 140, *150*, 352, *365*
Cobinah, J. R., 11, *28*
Cock, M. J. W., 9, *26*
Cocucci, A., 331, *342*
Cohen, A. C., 95, *115*
Cole, R. A., 40, 64, *75*
Coleman, J. S., 68, 70, *75*, 240, *274*
Coley, P. D., 19, *26*, 54, 59, *75*, 296, 300, *309*
Coll, M., 353, *363*
Collier, R., 134, *152*
Collinge, S. K., 62, *75*, 245, 246, *275*
Collinson, M. A. E., 320, *341*
Collum, D. H., 71, *77*
Colvin, A. A., 59, *74*
Compton, S. G., 248, 258, *275*
Conn, E. E., 53, *80*
Connell, J. H., 290, *309*
Connor, E. F., 21, *28*
Contardo, L., 102, *115*, 171, *188*, 378, *380*
Cooley, S. S., 213, 216, *223*
Coombe, P. E., 134, 136, *149*
Cooper-Driver, G. A. N., 14, *26*, 61, *75*, 103, *115*
Cooper, R., *78*
Corbet, S. A., 214, *220*, 260, *272*
Cornelius, M., 291, *308*
Cory, J. S., 320, *340*
Cottrell, C. B., 203, *221*
Courtney, S., 242, *272*
Courtney, S. P., 5, *26*, 219, 221, 242, *272*
Cox, C. B., 20, *26*
Cox, D. F., 95, *116*
Cox, P. A., 333, *340*
Craig, S., 350, *365*
Crane, P. R., 279, *309*, 333, *340*

Crawley, M. J., 7, 22, *26*, 245, 247, 269, *273*, 360, *363*
Crepet, W. L., 333, 334, 336, *340*
Crnjar, R. M., 175, *188*
Croteau, R., 47, 57, 58, *76*
Cullen, J. M., 359, *365*

Dafni, A., 320, 328, 329, 339, *340*
Dahlman, D. L., 110, *115*, 276
Dale, D., 64, *75*
Damman, H., 19, 21, 22, *26*, 101, *117*, 256, 261, 262, *273*
Daniel, T. L., 88, *117*
Danilevski, A. S., 230, *236*
Danks, H., 231, *236*
Darwin, C., 208, *221*, 315, *340*
Data, E. S., 57, *75*
Datta, R. K., 95, 98, *114*
Davidson, J., 71, *75*
Davies, R. G., 87, *115*
Davis, B. N. K., 13, 14, *26*
Davis, N. B., 260, *274*
Davis-Hernandez, K. M., 232, *237*
Dawson, G. W., 140, 145, *151*, 356, *364*
De Boer, G., 176, *188*, 210, *221*
De Bruyn, L., 267, *273*
De Candolle, A. P., 160, *188*
De Clerck-Floate, R. A., 378, *381*
De Groot, Æ., 38, 44, 48, 68, 74, *81*, 129, *149*, 166, 168, 171, *187*, *193*, 380
De Jong, R., 140, 142, *149*, 153
De Jong, T. J., 51, *81*, 319, 328, 331, *341*
De Ponti, O. B. M., 346, 347, 348, 349, *363*
De Vries, H., 346
De Wilde, J., 131, 147, *149*, 231, *237*
Dearn, J. M., 72, *76*
Decker, H., 346, *363*
DeClerck, R. A., 11, *26*
Degen, T., 379, *381*
Dehne, H. W., 343, 344, *364*
Delfosse, E. S., 359, *365*
Delisle, J., 232, 233, 234, 235, *237*
Dellamonica, G., 40, *74*
Delobel, A., 233, *237*
Den Boer, P. J., 243, *273*
Den Otter, C. J., 146, *149*, *153*, 234, *238*
Denholm, A. A., 174, 181, *192* , 358, *363*
Denno, R. F., 32, 57, 70, *75*, *78*, 219, *221*, 257, *273*
Derksen Koppers, I., 171, *192*, 378, *383*
Derridj, S., 42, 44, *75*, *80*, 165, *188*
Derrien, A., *75*, 165, *188*

Deshon, R. E., 92, *115*
Dethier, V. G., 93, *115*, 122, 123, 125, 126, 131, 148, *149*, 161, 172, 174, 175, 182, 183, *188*, *192*, 206, 207, 208, 210, 211, *221*, 222, 250, *273*, 290, 291, 305, *309*, 355, *363*, 376, 378, *381*
Dettner, K., 230, *237*
Dicke, M., 48, 50, 64, 65, 66, 67, 68, 74, *75*, *78*, *81*, 129, *152*, 264, 265, 268, *273*, 277, 349, *363*
Dickens, J. C., 140, 146, *149*, *150*, 234, *237*
Dickinson, J. L., 204, *221*
Dickson, K., 328, 330, *340*
Diehl, S. R., 284, 301, *309*
Dieleman, F. L., 70, *75*
Dijkman, H., 65, *75*
Dingle, H., 218, *222*
Diniz, I. R., 22, *28*
Dixon, A. F. G., 200, 201, *221*, *222*, *223*
Dobson, H. E. M., 322, *340*
Dobzhansky, Th., 279, *309*
Docherty, M., 113, *119*
Dodd, A. P., 360, *363*
Donnell, D. J., 39, *78*
Doskotch, R. W., 40, *76*
Doss, R. P., 156, *188*
Douault, P., 50, *79*, 322, *341*
Douglas, A. E., 108, *115*
Douwes, P., 128, *150*
Dover, J. W., 352, *363*
Dowd, P. F., 108, 109, *115*, *116*, *118*
Dreyer, D. L., 171, *188*
Drosopoulos, S., 201, *221*
Drummond, F. A., 356, *364*
Du, Y. J., 166, 174, 184, 185, *188*, 197, *221*
Duan, J. J., 379, *381*
Dudt, J. F., 61, *76*
Dueben, B. D., 232, *237*
Duffey, S. S., 64, 65, 68, *80*, 91, 92, 103, *116*, 291, *309*
Duncan, A., 100, *117*
Dunn, I., 89, *120*
Dunn, J. A., 348, *363*
Dussourd, D. E., 44, *76*, 302, *309*

Eastop, V., 16, *26*
Edwards, P. B., 72, *76*, 89, *120*
Eeswara, J. P., 358, *362*
Eger, H., 378, *381*
Ehrlich, P. R., 1, *4*, 219, *223*, 243, *273*, 289, 294, 302, 307, *309*
Eichenseer, H., 110, *115*, 166, 179, 183, *190* , 204, 220
Eigenbrode, S. D., 16, *26*, 42, 44, *76*, 156, 159, 161, 162, 165, 168, *188*, 380, *381*
Eisenbach, J., 263, *273*
Elkinton, J. S., 60, *78*, 142, *151*
Ellis, P. E., 231, 232, 233, *237*
Ellis, R. T., 265, 272
Elmore, J. S., 49, *80*
English-Loeb, G. M., 16, *26*
Epstein, E., 64, *76*, 86, *116*
Ernst, B., 171, *192*
Ernst, K. D., 145, 146, *149*
Ernst, W. H. O., 22, *26*
Errera, L., 1, *4*
Escoubas, P., 378, *381*
Espelie, K. E., 16, *26*, 42, 44,*76*, 156, 159, 161, 162, 165, 168, *188*, 380, *381*
Estesen, B. J., 39, 40, *74*, 169, *187*
Etiévant, P., 50, *79*, 322, *341* , 379, *383*
Evans, H. E., 11, *26*
Evans, K. A., 131, 140, *150*
Everaarts, T. C., 132, *153*

Fabre, J. H., 15, *26*, 160, *188*
Faeth, S. H., 69, *76*, 261, *273*
Fairbairn, J. B., 60, *76*
Fajer, E. D., 112, *117*
Farrell, B. D., 293, 294, 295, 302, 306, 307, *309*
Febvay, G., 379, *381*
Feder, J. L., 284, 285, 286, *309*
Federici, B. A., 98, 109, *118*
Feeny, P., 40, 44, 52, 58, *75*, *76*, 97, 101, 102, 103, 103, *115*, *116*, *119*, 140, *148*, 155, 166, 171, *187*, *188*, 191, 241, *273*, 290, 291, 296, 297, *309*, 313, 378, *380*
Fein, B. L., 146, *150*
Feinsinger, P., 319, *340*
Fellows, L. E., 37, *76*
Fenwick, G. R., 167, 174, *191*, 379, *382*
Fenwick, R., 184, *192*
Ferguson, J. E., 140, *151*
Fernandes, G. W., 266, *276*
Ferro, D. N., 19, *26*
Fiala, V., 44, *75*, 165, *188*
Fiedler, K., 8, *26*, 105, *116*, 294, *309*
Field, M. D., 90, *118*
Finch, S., 45, 61, *75*, *76*, 131, 134, 136, *150*, *152*, 376, 377, *381*
Firn, R. D., 68, *76*, 171, *189*, 281, 305, *310*
Fitter, A. H., 18, *26*, 33, *76*, 269, *273*
Fitting, H., 6, *26*
Flath, R. A., 234, *237*
Flint, W. P., 288, *311*
Floate, K. D., 16, *26*

Author index

Fontaine, A. R., 85, *116*
Ford, E. B., 284, *309*
Foster, S. P., 148, *150*
Fowler, S. V., 361, 362, *363*
Fox, L. R., 6, *26*, 263, *273*, 290, 303, *309*
Fraenkel, G. S., 3, *4*, 33, 51, *76*, 161, 166, 169, *189*, 281, *309*
França, F. H., 348, *363*
Francke, W., 330, *342*
Franz, E. H., 349, *365*
Frase, B. A., 376, *381*
Fraser, J., 173, *188*
Frazer, B. D., 376, *381*
Frazier, J. L., 166, 170, 179, 182, 183, 184, *189*, *190*, 204, *220*, 357, 359, 379, *381*
Free, J. B., 317, 339, *340*
French, R. A., 93, *117*
Frentz, W. H., 143, *153*
Frey, J. E., 140, 145, *150*, 216, 217, *221*, 286, *309*
Frick, K. E., 293, *313*
Friedman, S., 90, 100, *119*
Friis, E. M., 279, *309*, 333, 334, 336, *340*
Frisch, K. von 327, *340*
Fritz, R. S., 199, *221*, 257, 261, *273*, 297, *309*
Fritzsche, R., 346, *363*
Frohne, D., 40, 74, *76*
Frost, S. W., 12, *26*, 85, *116*
Fukami, H., 40, 44, *74*, 79
Fung, S. Y., 32, *76*
Funk, D. J., 295, *310*
Futuyma, D. J., 8, *26*, 195, 219, *221*, 225, 290, 291, 295, 299, *310*

Galeffi, C., 44, *82*
Galina, A., 285, *310*
Gall, L. F., 12, 27, 52, *76*
Gallun, R. L., 284, 299, *310*
Games, S. B., 292, *310*
Gange, A. C., 269, *272*
Ganzle, H., 146, *151*
Gara, R. I., 131, *153*
Garrec, J. P., 42, 75, 165, *188*
Garth, J. K. L., 156, *188*
Gassmann, A., 246, *273*, 359, 360, *363*
Gaston, K. J., 243, *273*
Gatehouse, A. M. R., 350, *363*
Gatehouse, J. A., 350, *363*
Gaugler, R., 111, *116*
Gause, G. F., 257, *273*
Geervliet, J. B. F., 48, *74*, 129, *149*, 380
Gegear, R. J., 320, *340*
Geiger, R., 18, *27*
Geissler, K., 346, *363*

Gershenzon, J., 47, 53, 54, 55, 57, 58, 62, *76*, 299, *310*
Getz, W. M., 146, *152*
Gibson, D. A., 317, *340*
Giebink, B. L., 196, *224*
Gilbert, F. S., 328, 330, *340*
Gilbert, L. E., 214, *221*, 290, *310*
Gillot, C., 140, *151*
Gittleman, J. L., 5, *28*
Giurfa, M., 330, *340*
Glendinning, J. I., 180, *189*
Godfray, H. C. J., 252, 254, *273*
Goeden, R. D., 256, *273*
Gold, C. S., 354, *363*
Gold, L. S., 355, *362*
Gollasch, S., 350, *365*
Gols, G. J. Z., 181, *190*
Gothilf, S., 379, *381*
Gould, C. G., 339, *340*
Gould, F., 103, *116*, 195, 218, *221*, 224, 225, 290, 299, *310*, 347, 350, 362, *363*
Gould, J. L., 321, 339, *340*
Gould, S. J., 217, *221*
Goulson, D., 320, *340*
Grabstein, E. M., 215, *221*
Graham, M., 7, *25*, 289, 290, *308*
Grant, V., 320, 337, *340*
Grayer, R. J., 40, *77*
Greathead, D. J., 261, *273*
Green, P. C. W., 174, *192*
Green, T. R., 64, *76*
Green, W., 181, *192*
Greene, E., 230, *237*
Greenslade, P. J. M., 297, *310*
Gref, R., 62, *76*
Gregoire, V., 165, *188*
Gregory, P., 44, *80*, 157, 158, 179, *189*, *190*
Grevillius, A. Y., 8, *27*
Grewal, P. S., 111, *116*
Grewal, S. K., 111, *116*
Griepink, F. C., 44, *81*, 166, 168, *193*
Griffiths, D. C., 356, *364*
Grimaldi, D. A., 334, 336, *340*, *341*
Grime, J. P., 296, 297, *310*
Grimshaw, H. M., 84, *114*
Groden, E., 356, *364*
Gross, P., 250, *275*
Grossman, R. B., 174, 181, *192*
Groth, I., 322, *340*
Grugel, S. R., 376, *382*
Gueldner, R. C., 45, *77*
Guentert, M., 322, *342*
Guenther, E., 53, *76*
Guerin, P. M., 140, 144, *150*
Guichard, E., 50, *79*, 322, *341*
Guppy, J., 317, *340*

Gupta, P. D., 159, *189*
Gurevitz, E., 40, *78*, 378, *380*
Guthrie, F. E., 105, *118*
Guthrie, W. D., 95, *116*, 284, 299, *310*, 376, *381*
Gutteridge, C., 174, 181, *192*

Haack, R. A., 9, *27*, 93, *117*
Hadley, H. H., 157, *192*
Hagen, K. S., 265, 266, *273*
Hagenbeek, D., 68, 74, 171, *187*
Haines, H., 328, 330, *340*
Hairston, N. G., 243, *273*
Hajek, A. E., 109, *116*
Haley Sperling, J. L., 176, *189*
Hamamura, Y., 40, *76*
Hamann, B., 132, *151*
Hamilton, J. G., 11, *27*
Hamilton, R. J., 44, *74*
Hamilton, W. D. O., 350, *363*
Hamilton-Kemp, T. R., 48, *77*, 260, *275*
Hammond, A. M., 111, *115*, *116*, 263, *273*
Hammond, R. B., 375, *381*
Han, K., 54, *77*
Hanhimäki, S., 8, *28*, 257, *275*
Hansen, R. M., 204, *225*
Hanson, F. E., 138, *152*, 166, 181, *191*, 206, 207, 208, 210, 215, 218, *221*, 222, 376, 378, 379, *381*, *383*
Hansson, B. S., 146, *150*
Harborne, J. B., 31, 36, 38, 39, 40, 74, *77*, 111, *116*
Harder, H., 322, *342*
Harder, R., 6, *26*
Hardie, D. C., 350, *365*
Hardie, J., 140, 145, *151*, 229, *237*, 376, 377, 379, *382*, *383*
Hardy, T. N., 111, *115*, *116*, 263, *273*
Hare, J. D., 98, 109, 110, *117*, *118*, 196, *221*, 231, *237*
Harlan, J. R., 350, *363*
Harper, J. L., 21, *27*, 244, 246, *273*
Harrewijn, P., 228, 229, *237*, 378, *381*
Harris, K. F., 116
Harris, K. F., 88, *116*
Harris, M. O., 148, *150*, 159, *189*
Harrison, G. D., 179, *190*
Hart, E. R., 40, *74*
Hartley, S. E., 67, *77*
Hartmann, T., 32, 54, 57, 73, *77*, *81*
Hassanali, A., 356, *364*
Hatanaka, A., 47, *77*
Hatcher, P. E., 111, *116*, 268, *273*
Hatt, H., 142, 146, *152*
Haukioja, E., 24, *27*, 65, 68, 69, *77*, 79, 257, *275*

Hausner, G., 331, *342*
Hawkes, C., 140, *150*
Hawkins, B. A., 266, 267, *273*
Hawkins, C. P., 257, *274*
Hay, R. K. M., 18, *26*, 33, *76*
Hayashi, N., 166, *189*
Hayashiya, K., 40, *76*
Healey, P. L., 44, *80*
Heard, T., 69, *77*
Heath, R. R., *152*, 232, *237*
Hedin, P. A., 45, 71, *77*
Hegnauer, R., 74, *77*
Heidorn, T., 33, *77*
Heinrich, B., 318, 323, 324, 326, 327, 328, 331, 332, 339, *340*, *341*
Heinrichs, E. A., 63, *77*
Heiser, D. A., 331, *341*
Helander, M., 110, *118*
Heliövaara, K., 112, *116*
Hendrikse, A., 234, *237*
Hendrix, S. D., 5, *27*, 305, *310*
Henstra, S., 87, *118*
Hering, M., 15, *27*
Herms, D. A., 9, *27*
Herrebout, W. M., 32, *76*, 170, 180, *191*, 234, *237*, 292, 293, *311*
Herzog, G. A., 140, *150*
Hesler, L. S., 131, *150*
Hess, D., 330, *341*
Hewitt, G., 126, 131, *148*
Heymans, M., 377, *383*
Hick, A. J., 145, *151*
Higashi, M., 86, *113*
Higgins, T. J. V., 350, *365*
Hildebrand, J. G., 142, *150*, 174, *189*
Hilder, V. A., 350, *363*
Hille Ris Lambers, D., 16, *27*
Hillyer, R. J., 232, 233, *237*
Hinks, C. F., 110, *116*
Hiratsuka, E., 100, *116*
Hirsh, I. S., 109, *114*
Ho, T. H., 288, *310*
Hochuli, D. F., 86, *116*
Hodgson, E., 105, *118*
Hoffman, L. R., 5, *28*
Hogen Esch, T., 88, *119*
Hokkanen, H. M. T., 354, *364*
Holden, C., 1, *4*
Hollister, B., 166, 179, 183, *190*, 204, *220*
Holman, R. T., 48, *74*
Holt, R. D., 260, 261, *274*
Holtzer, T. O., 93, *116*
Honda, I., 140, *150*
Honda, K., 44, *77*, 166, 168, *189*
Hopkins, A. D., 214, *221*
Horn, J. M., 70, *80*
Horner, J. D., 11, *27*
Horton, D. R., 378, *381*

Horváth, J., 126, 128, 131, 147, 148, *150*
Hoult, A. H. C., 349, *364*
Howard, G., 208, *221*
Howard, J. J., 6, *27*, 39, 40, *74*, 169, 187, 211, 212, 213, *220*
Howe, W. L., 349, *364*
Hruska, A. J., 42, *77*
Hsiao, C., *310*
Hsiao, T. H., 163, 165, 169, 170, *189*, 196, 198, 222, 287, 305, *310*
Huang, X., 167, *189*, 205, 208, 222, *223*
Huettel, M. D., 218, *222*
Hughes, C. G., 300, *312*
Hughes, P. R., 112, 113, *116*
Huh, S., 300, *308*
Hulley, P. E., 70, *79*, 255, *275*, 376, *382*
Hunt, T. A., 284, *309*
Hurter, J., 166, 167, 168, 171, 174, *191*, *192*, 379, *382*
Hyer, A. H., 156, *187*

Ibenthal, W. D., 52, *79*
Inouye, D. W., 333, *341*
Ishaaya, I., 40, *78*, 102, *117*
Ishikawa, I., 140, *150*
Isman, M. B., 205, *220*, 358, *363*
Itagaki, H., 174, *190*
Ives, P. M., 216, *222*
Iwasa, Y., 51, *81*

Jackson, D. M., 44, *77*, 140, *150*
Jacob, F., 305, *310*
Jacobson, M., 140, *149*
Jaenike, J., 214, 217, 218, 219, *222*, 291, 297, 301, *310*
Jang, E. B., 234, *237*
Janzen, D. H., 40, *74*, 77, *80*, 86, *115*, 303, *310*
Jay, S. C., 339, *341*
Jeffree, C. E., 42, 43, 44, 46, *77*, 159, 162, *189*
Jena, K. K., 350, *364*
Jenkins, J. N., 71, *77*
Jennens, L., 174, 181, *192*, 358, *363*
Jennings, P. R., 346, 348, *364*
Jensen, U., 40, *74*, *76*
Jermy, T., 68, *77*, 126, 128, 131, 147, 148, *150*, 161, 169, 170, *189*, 197, 198, 204, 205, 206, 207, 208, 209, 214, 216, 222, 225, 259, 268, 270, 271, *274*, 289, 290, 291, 292, 305, *310*, 358, 359, *364*, 376, 377, 378, 379, *381*, *383*
Jetter, R., 42, 44, *77*
Jobe, R. T., 42, 58, *78*
Jobin, A., 255, *274*

Joern, A., 33, 60, 61, *77*, *79*, 255, *274*, 292, *310*
Johansen, C. A., 338, *341*
Johns, T., 71, *77*, 349, *364*
Johnson, A. W., 44, *77*, 140, *150*
Johnson, R. H., 112, *117*
Johnson, S., 358, *362*
Jones, C. E., 330, *341*
Jones, C. G., 68, 70, *75*, *76*, 171, *189*, 240, 250, 268, 272, *274*, 281, 305, *310*, 378, *383*
Jones, G. A., 157, *191*
Jones, K. C., 171, *188*
Jones, O. G., 108, 109, *114*
Jones, R. E., 216, *222*
Jördens-Röttger, D., 44, *77*
Josefsson, E., 349, *364*
Joshi, A., 291, *310*
Joshi, S., 181, *187*
Judd, J. G. R., 131, 140, *150*
Julien, M. H., 361, *364*
Juniper, B. E., 44, 45, *77*
Justus, K. A., 41, *79*
Juvik, J. A., 44, *78*

Kaila, L., 8, *27*
Kaiser, R., 337, *341*
Kaitaniemi, P., 62, *80*
Kalinova, B., 140, *153*
Kallet, C., 216, *223*
Kamerman, J. W., 16, *29*
Kamm, J. A., 48, *75*
Karban, R., 6, 16, *26*, *27*, 53, 64, *78*
Kareiva, P. M., 126, 128, 147, 148, 149, *151*, 242, 257, *274*
Karl, E., 346, *363*
Karowe, D. N., 60, *78*, 103, *116*, 215, 222, 264, *274*
Karsten, G., 6, *26*
Kasting, R., 378, *381*
Kaszab, Z., 243, *274*, 293, *310*
Katz, R. B., 170, *188*
Kaufman, T., 211, *222*
Kawada, K., 229, *237*
Kays, S. J., 57, *75*, *79*
Kearby, W. H., 260, *272*
Keating, S. T., 109, 110, *116*, *117*, *118*, 265, *276*
Keena, M. A., 94, *117*
Keese, M. C., 295, *310*
Kelly, W., 44, *74*
Kempton, D. P. H., 348, *363*
Kendall, R. D., 376, *381*
Kennedy, C. E. J., 13, 15, 18, *27*, 89, *117*, 255, 277, 291, *311*
Kennedy, G. G., 64, *74*, 196, 221, 299, *311*, 349, 350, 362, *365*
Kennedy, J. S., 16, *27*, 126, 134, 135, 135, 140, *150*, 161, *189*

Kent, K. S., 174, *189*
Kerr, A., 301, *311*
Kershaw, W. J. S., 134, 135, 140, *150*
Kester, K. M., 263, 273
Kevan, P. G., 333, 338, 339, *341*
Khan, Z. R., 86, 87, 117, 156, *189*, 348, 350, *365*, 375, 380, *383*
Khattar, P., 132, *152*
Khush, G. S., 157, *191*, 202, *223*, 346, 347, 348, 349, 350, *364*
Kiang, C. S., 50, *75*
Kibota, T. T., 5, *26*
Kim, K. C., 219, 222, 265, 274, 345, *364*
Kim, M., 40, *74*
Kimmerer, T. W., 11, *27*, 52, *78*
Kimmins, F., 378, *381*
Kindler, S. D., 202, *223*
Kingan, T. G., 232, 235, *237*
Kingsolver, J. G., 88, *117*
Kingston, J. D., 32, *79*
Kinney, K. K., 113, *117*
Kirby, W., 1, *4*, 206, *222*
Kite, G. C., 37, *76*
Kjaer, A., 50, *74*
Klausnitzer, B., 7, *27*, 255, *274*
Klekowski, R. Z., 100, *117*
Klingauf, F., 40, 44, *78*, 162, *189*
Klingler, J., 138, *151*
Klinkhamer, P. G. L., 319, 328, 331, *341*
Klips, R. A., 331, *342*
Klos, R., 201, *222*
Klun, J. A., *311*
Knoll, A. H., 279, *311*
Knoll, F., 324, *341*
Knudsen, J. T., 322, *341*
Kogan, M., 13, 18, *27*, 95, 97, *117*, 210, *222*, 345, 347, 348, *364*
Koh, H. S., 40, *74*
Kok-Yokomi, M. K., 156, *191*, 348, *365*
Kokko, E. G., 378, *381*
Kolb, G., 133, *152*
Kolodny-Hirsch, D. M., 25, *29*
Kooi, R. E., 170, 180, *191*, 259, *277*
Koptur, S., 265, *274*
Körner, C., 113, *114*
Kos, J., 346, *364*
Kosztarab, M., 243, *274*, 294, *311*
Kovalev, B. G., *153*
Koveos, D. S., 233, *237*
Kozár, F., 243, 256, *274*, 294, 304, *311*
Kramer, E., 377, *381*
Krebs, J. R., 260, *274*
Krieger, J., 146, *151*
Krischik, V. A., 57, 70, *78*, 108, 109, *114*, *117*, 265, 268, 272, *274*, 296, *308*

Kriston, I., 321, *341*
Kruess, A., 242, *274*
Krug, E., 105, *116*
Kruse, L., 215, *225*
Kubo, I., 171, *189*
Kubo, T., 51, *81*
Kuch, J. H., 175, 182, 183, *188*
Kullenberg, B., 338, *341*
Kundu, R., 200, *222*
Kutchan, T. M., 32, *78*

Labandeira, C. C., 281, 282, 283, *311*
Labeyrie, V., 291, *311*
Lábos, E., 291, 305, *310*
Lack, A., 339, *341*
Lafont, R., 38, *78*
Lajide, L., 378, *381*
Lamb, R. J., 140, *151*
Lampman, R., 140, *151*
Lance, D. R., 9, *27*, 60, *78*, 131, *150*
Landolt, P. J., 140, *151*, 232, *237*
Lane, G. A., 39, *78*
Langenheim, J. H., 53, *78*
Langer, H., 132, *151*
Lanza, J., 318, 329, *339*
Larsen, J. R., 18, *27*
Larsen, L. M., 40, *79*, 166, *190*
Larson, K. C., 25, *29*
Larsson, S., 58, *78*
Laverty, T. M., 320, 323, 324, 326, 340, *341*
Lavie, D., 40, *78*
Lawrence, R. K., 9, *27*
Lawrey, J. D., 5, *27*
Lawton, J. H., 6, *27*, 28, 67, *77*, 203, *223*, 241, 242, 248, 249, 250, 251, 252, 255, 257, 258, 260, 261, 266, 267, 273, 274, 275, 277, 280, 300, 308, *312*
Lecomte, C., 140, *152*
Lederhouse, R. C., 200, 218, *224*
Lee, J. C., 212, *222*
Lee, Y. L., 18, *27*
Lee, Y. W., 171, *189*
Lees, A. D., 228, 229, *237*
Leggett, F. L., 378, *381*
Lehmann, W., 346, *363*
Leigh, T. F., 156, *187*
Leppik, E. E., 336, *341*
Leppla, N. C., 95, *114*
Leslie, J. F., 218, *222*
Levin, D. A., 285, 300, *311*
Levins, R. A., 285, *311*
Levy, E. C., 40, *78*
Lewinsohn, T. M., 248, 258, *275*
Lewis, A. C., 92, 93, *115*, *118*, 320, *341*
Lewis, J. A., 167, 174, *191*, 379, *382*

Lewis, W. J., 49, 67, *81*, 265, 277
Lewontin, R. C., 217, *221*
Ley, S. V., 170, 181, *188*, *192*, 357, 358, *363*
Li Zhao Hua 16, *29*
Libbey, L. M., 260, *276*
Light, D. M., 234, *237*
Lincoln, D. E., 54, 77, 112, *117*
Lindauer, M., 332, *341*
Lindhout, P., 346, *364*
Lindroth, R. L., 108, 113, *117*
Lindsay, E., 113, *119*
Lindsay, R. W., 50, *75*
Lindström, J., 8, *27*
Ling, L. C., 47, 48, *75*
Lingren, P. D., 378, *382*
Linhart, Y. B., 299, *311*
Lipke, H., 3, *4*
Llewellyn, M., 202, *222*
Loader, C., 101, *117*
Löfqvist, J., 129, 146, *149*, *150*
Logan, P., 196, *222*
Lopez, J. D., 378, *382*
Louda, S. M., 41, 62, *75*, *78*, 244, 245, 246, 256, *273*, *275*
Loughrin, J. H., 48, *77*, 260, *275*
Lovett, G. M., 247, *275*
Lovett, J. V., 349, *364*
Lu, W., 196, *222*
Lübke, G., 330, *342*
Luckner, M., 35, 57, *78*
Lundberg, H., 332, 333, *341*
Lundgren, L., 48, *74*
Luo, Lin-er 171, 181, 183, *190*, *192*, 357, *364*
Lupton, F. G. H., 156, *190*
Luthy, B., 57, *78*

Ma, W. C., 140, 143, 146, *151*, 163, 164, 170, 171, 179, *190*, 207, 208, *222*, 378, 379, *382*
MacArthur, R. N., 252, 257, 259, *275*
MacFayden, A., 96, *118*
Machinek, R., 52, *79*
Mackenzie, A., 200, 214, *223*
MacMahon, J. A., 257, *274*
MacVicar, R. M., 317, *340*
Madore, M. A., 112, *115*
Madsen, J. O., 44, 50, *74*
Maes, F. W., 141, 146, *149*
Magadum, S. B., 95, 98, *114*
Mägdefrau, K., 280, *311*
Magurran, W. E., 243, *275*
Maier, C. T., 196, *223*
Maindonald, J. M., 39, *78*
Maini, S., 286, *311*
Maliepaard, C., 346, *364*
Mandava, N. B., 170, *190*, 355, *364*
Mangel, M., 260, *276*

Author index

Manglitz, G. R., 202, *223*
Manly, B. F. J., 378, *382*
Mannila, R., 8, *28*
Marilleau, R., 44, 50, *79*, 322, *341*
Marini Betollo, G. B., *82*
Marino, B. D., 32, *79*
Marion-Poll, F., 379, *382*
Markham, K. R., 40, *74*
Marques, E. S. A., 22, *28*
Marquis, R. J., 21, 23, *27*
Marshall, D. B., 42, 58, *78*
Martin, M. M., 97, 103, *114*, *119*
Martin, P., 376, *382*
Martinat, J., *25*
Martinet, P., 203, *220*
Maschinski, J., 25, *29*
Masson, C., 50, *79*, 322, *341*, 379, *383*
Masters, R. A., 245, 246, *275*
Matile, P., 56, 57, *78*
Matsuda, K., 44, *74*
Matsumoto, Y., 140, *150*, *151*, 166, *190*
Matsuura, K., 40, *77*
Matthews, J. R., 5, *28*
Matthews, R. W., 5, *28*
Mattiacci, L., 67, *78*
Mattoo, A. K., 232, 235, *237*
Mattson, W. J., 9, 27, 84, 91, 93, *117*, 270, *275*
Matzinger, D. F., 55, *81*
Maxwell, F. G., 346, 348, *364*
Maxwell, H. D., 157, *193*
May, R. M., 1, 4, 5, 27, 283, *311*
Maynard Smith, J., 328, *341*
Mayoral, A., 88, *119*
Mayr, E., 239, *275*, 284, 285, 287, *311*
Mazza, M., 143, *153*
McCashin, B. G., 176, *190*
McClure, M. S., 32, *75*, 219, 221, 257, *273*
McEvoy, P. B., 101, *117*
McEwen, F. L., 166, *190*, 379, *382*
McGinnis, A. J., 378, *381*
McIver, S. B., 157, *190*
McKone, M. J., 331, *341*
McNaughton, S. J., 64, *78*
McNeil, J. N., 232, 233, 234, 235, *237*
McNeill, S., 91, *117*
McPheron, B. A., 219, 222, 265, *274*, *275*, 345, *364*
Meade, T., 110, *117*
Mechaber, R. A., 42, 58, *78*
Mechaber, W. L., 42, 58, *78*
Medrano, F. G., 63, *77*
Meelis, E., 377, *383*
Meerman, J., 99, *118*, 215, *224*
Mehta, I., 44, *80*
Meinecke, C. C., 132, *151*

Meisner, J., 102, *117*, 132, *151*
Mellanby, K., 93, *117*
Menken, S. B. J., 234, *237*, 289, 291, 292, 293, 295, 305, *311*
Menninger, M. S., 219, *223*
Menzel, R., 323, *340*
Merritt, L. A., 356, *364*
Merz, E., 203, 211, *223*
Messchendorp, L., 181, *190*
Metcalf, C. L., 288, *311*
Metcalf, E. R., 38, 47, *78*, 131, 146, *151*, *190*
Metcalf, R. L., 38, 47, *78*, 131, 140, 146, *151*, *190*
Meyer, G. A., 23, *28*, 89, *117*
Michael, R. R., 260, *276*
Michener, C. D., 336, *341*
Midgley, J. J., 333, *341*
Mihályi, F., 243, *275*
Mikhail, A. A., 40, *76*
Miles, P. W., 89, *118*
Millar, J. G., 232, *237*
Miller, J. R., 122, 132, *151*, *152*, 159, 172, *189*, *190*, 375, *382*
Miller, T. A., 375, *382*
Mitchell, B. K., 41, *79*, 163, 171, 174, 176, 179, 182, *189*, *190*, 192, *193*, 379, *383*
Mitchell, E. R., 140, *152*
Mithen, R., 174, 184, *192*
Mitter, C., 293, 294, 295, 302, 305, 306, 307, *311*
Mittler, T. E., 108, *117*, 229, *238*, 378, *382*
Mitzutani, J., 378, *381*
Mizell, R. F., 89, *115*, 213, *220*
Moeckh, H. A., 140, *151*
Moericke, V., 134, 135, 136, *151*
Mohamad, B. M., 110, *117*
Mole, S., 38, 41, 60, 61, 62, *78*, *79*, *81*, 380, *383*
Mollema, C., 346, 347, 348, 349, *363*, 378, *381*
Molnár, I., 291, 305, *310*
Montague, R. A., 142, *150*
Monteith, L. G., 267, *275*
Montgomery, M. E., 166, *190*
Mooney, H. A., 21, *28*
Moore, A., 350, *365*
Moore, P. D., 20, *26*, 320, *341*
Moore, R. F., 94, *119*
Mopper, S., 286, *311*
Morais, H. C., 22, *28*
Moran, N., 291, *311*
Moran, N. A., 200, *223*
Moran, V. C., 249, 250, *255*, *275*, 277, 300, *311*
Mordue (Luntz), A. J., 38, *79*, 355, 357, 358, 359, *362*, *364*

Morgan, E. D., 170, *190*, 355, 358, *362*, *364*
Morgan, F. D., 11, *28*
Morgan, M. E., 32, *79*, 260, *276*
Morris, W. F., 126, 128, 147, *151*
Morrow, P. A., 6, *26*, 290, *309*
Morse, R. A., 317, *341*
Morton, T., 295, *310*
Mothes, K., 33, *79*
Moya, A., 285, *310*
Mudd, A., 356, *364*
Mulkern, G. B., 126, *151*
Müller, F. P., 286, 287, *311*
Müller-Schärer, H., 246, *275*
Mullin, C. A., 89, *117*, 166, 179, 183, *190*, 204, *220*
Murdock, L. L., 378, *383*
Murlis, J., 130, 142, *151*
Murphy, D. D., 219, *223*, 243, *273*
Murray, D. A. H., 262, 263, *275*
Murray, K. D., 356, *364*
Murtagh, G. J., 57, *76*
Mustaparta, H., 138, 140, *151*, *153*
Muthukrishnan, J., 63, *79*
Myers, J. H., 6, 27, 62, 64, *78*, *79*

Nair, K. S. S., 166, *190*, 379, *382*
Naito, K., 40, *76*
Nakanishi, K., 171, *189*
Nash, R. J., 37, *76*
Nault, L. R., 42, *79*
Navon, A., 98, 109, *118*
Neal, J. J., 11, *28*, 99, 104, *114*, *118*
Needham, J., 355, *364*
Neff, J. L., 333, 336, 338, *341*
Nentwig, W., 255, *274*
Neuvonen, S., 68, 69, *79*, 110, *118*, 257, *275*
New, T. R., 21, *28*
Newell, C. A., 350, *363*
Ng, D., 218, *223*
Nickle, J. L., 345, *364*
Nielsen, J. K., 40, *79*, 166, 167, 174, *190*, *191*, 376, 379, *382*
Niemelä, P., 8, 27, 65, *77*
Niklas, K. J., *29*, *311*
Nilson, A., 62, *80*
Nishida, J., 40, *77*
Nishida, R., 44, *79*, 166, *190*
Nishijima, Y., 156, *190*
Nixon, K. C., 333, 336, *340*
Nobel, P. S., 18, *28*
Nöcker-Wenzel, K., 162, *189*
Noldus, L. P. J. J., 376, *382*
Nordenhem, H., 138, *151*
Nordlander, G., 138, 151
Norman, J. M., 93, *116*
Norris, D. M., 156, 157, *188*, *189*, 379, *382*

Nottingham, S. F., 44, 57, *75*, *79*, 140, 145, *151*
Nuckols, M. S., 21, *28*
Numata, H., 231, *237*
Nuñez, J. A., 330, *340*
Nuorteva, P., 15, *28*, 69, *79*

Obasi, B. N., 59, *79*
Odell, T. M., 94, *117*
Oerke, E. C., 343, 344, *364*
Ogutuga, D. B. A., 52, *79*
Ohgushi, T., 260, *275*
Ohmart, C. P., 58, *78*
Ohnmeis, T. E., 66, 68, *74*, 300, *308*, 318, 329, *339*
Ohsaki, N., 16, *28*
Okolie, P. N., 59, *79*
Olckers, T., 70, *79*, 255, *275*, 376, *382*
Olesen, J. M., 317, *341*
Olsen, N., 85, *116*
Ortman, E. F., 347, *364*
Osborne, D. J., 231, 232, *237*
Osman, S. F., 40, *80*
Ostertag, R., 331, *341*
Ott, J., 325, *342*
Ott, J. R., 257, *273*
Owen, D. F., 247, *275*, 296, *311*
Owens, E. D., 131, *152*
Ozaki, K., 181, *191*
Ozaki, M., 181, *191*
Ozinkas, A. J., 300, *311*
O'Neill, K. M., 92, *115*
O'Toole, C., 339, *341*

Packham, J., 246, *275*
Paige, K. N., 25, *29*
Painter, R. H., 346, *364*
Paliniswamy, P., 140, *151*
Panda, N., 157, *191*, 202, *223*, 346, 347, 348, 350, *364*
Pankiw, P., 317, *340*
Papaj, D., 132, *151*, 200, 213, 216, *223*
Parkinson, J. A., 84, *114*
Parmesan, C., 199, *224*, 288, 289, *312*
Parrott, W. L., 71, *77*
Parsons, J. A., 263, *276*
Pashley, D. P., 6, *28*
Pasteels, J. M., 263, *276*, 286, *312*
Patana, R., 95, *115*
Pathak, M. D., 348, 350, *365*, 375, 380, *383*
Patterson, C. G., 48, *77*
Paul, N. D., 111, *116*
Pedersen, K. R., 279, *309*, 333, *340*
Pedigo, L. P., 375, *381*
Pellmyr, O., 217, *225*
Pelosi, P., 143, *153*
Peng, Z., 89, *118*
Penman, D. R., 19, *26*

Pereyra, P. C., 166, *191*
Perring, T. M., 112, *115*
Pescador, A. R., 214, 215, *223*
Pet, G., 346, *364*
Petersen, C. E., 97, *120*
Peterson, S. C., 166, 170, 180, 181, *191*, 195, *221*, 290, 299, *310*
Petrusewicz, K., 96, *118*
Pettersson, J., 140, *149*, *151*
Pham-Delègue, M. H., 50, *79*, 322, *341*, 379, *383*
Phelan, P. L., 108, *118*, 377, *382*
Phillipson, J., 100
Pianka, E. R., 271, *275*
Pickett, J. A., 140, 145, *149*, *151*, 356, *364*, 376, 377, *382*
Pierce, N., 203, *220*
Pimentel, D., 2, 3, 4, 23, *28*, 343, 345, 346, 354, 359, 362, *364*
Pimm, S. L., 5, *28*
Piron, P. G. M., 379, *383*
Pittendrigh, B. R., 233, *237*
Pivnick, K. A., 140, *151*, 233, *237*, 379, *382*
Plaisted, R. L., 348, *363*
Pleasants, J. M., 332, *341*
Plowright, R. C., 324, *341*
Pluthero, F. G., 290, *311*
Podoler, H., 231, *238*
Podolsky, R., 218, *225*
Pospisil, J., 140, *151*
Posthumus, M. A., 48, 66, 67, 68, *74*, *78*, *81*, 129, *149*, *152*, 349, *363*, 380
Poston, F. L., 375, *381*
Potter, D. A., 11, 27, 52, *78*, 260, *275*
Potvin, M. A., 245, 246, *275*
Poulton, J. E., 42, 56, *79*
Pourmohseni, H., 52, *79*
Pouzat, J., 232, 233, *237*
Powell, G., 376, 377, *382*
Powell, J. A., 293, *311*
Powell, K. S., 350, *363*
Powell, R. J., 19, *26*
Price, P. W., 22, 27, *28*, 239, 247, 248, 250, 251, 257, 259, 266, 273, *275*, *276*, 297, *309*, *311*
Prins, A. H., 25, *28*
Procter, M., 339, *341*
Profet, M., 355, *362*
Prokopy, R. J., 131, 132, 133, 134, 147, *148*, *151*, *152*, 171, *191*, 213, 216, *223*, 379, *381*
Proksch, P., 105, *116*
Prophetou-Athanasiadou, D. A., 233, *237*
Provenza, F. D., 210, *223*
Pschorn-Walcher, H., 266, *275*
Punzo, F., 215, *223*

Quarmby, C., 84, *114*
Quiras, C. F., 156, *191*, 348, *364*
Quisenberry, S. S., 346, *365*

Radke, C. D., 44, *79*, 166, 174, 184, *191*, *193*
Radtkey, R. R., 289, *311*
Rafaeli, A., 235, *237*
Raffa, K. F., 66, 69, *79*, 171, *191*, 378, *382*
Rahbé, Y., 379, *381*
Raina, A. K., 232, 234, 235, 236, *237*
Ramachandran, R., 86, 87, *117*, 379, *382*
Raming, K., 146, *151*
Ramp, T., 166, 167, 168, 174, *191*, 379, *382*
Rank, N. E., 70, *80*
Ranta, E., 332, 333, *341*
Ranta, H., 110, *118*
Rathore, V. S., 95, *116*
Raubenheimer, D., 6, 27, 101, *118*, 211, 212, 213, *220*
Raupp, M. J., 64, 68, *81*
Rausher, J. T., 331, *341*
Rausher, M. D., *79*, 99, *118*, 202, 214, 216, *223*, 296, 298, *311*
Raven, J. A., 89, *118*
Raven, P. H., 1, 4, 289, 294, 302, 307, 309, 327, 328, *341*
Reader, P. M., 296, *312*
Reader, R. J., 317, *341*
Real, L., 325, *342*
Reavy, D., 203, *223*
Redfern, M., 13, *28*
Reed, D. W., 140, *151*, 379, *382*
Reed, G. L., 95, *116*
Rees, C. J. C., 40, 51, *79*, 166, *191*
Reese, J. C., 90, 96, 102, 114, *118*
Regal, P. J., 335, *342*
Regupathy, A., 110, *114*
Reichart, G., 252, 253, *276*
Reichelderfer, C. F., 265, 274, *276*
Reichstein, T., 263, 264, *276*
Reinecke, P., 376, *382*
Reissig, W. H., 146, *150*
Remberg, G., 52, *79*
Renwick, J. A. A., 41, 44, *75*, *79*, 166, 167, 170, 171, 174, 184, 185, *188*, *189*, *191*, *193*, 197, 205, 208, *221*, *222*, *223*, 260, *276*
Reveal, J. L., 305, *312*
Reynolds, S. E., 357, *365*
Rhoades, D. F., 40, *80*
Rhodes, A. M., 349, *364*
Rice, R. E., 234, *237*
Richardson, J., 50, *75*
Rick, C. M., 156, *191*, 348, *364*
Riddiford, L. M., 232, *238*

Riederer, M., 44, 77
Ring, R. A., 85, *116*
Risch, S., 241, *276*
Risch, S. J., 70, *80*, 345, *365*, 375, *382*
Rizk, A. F., 140, *152*
Robberecht, R., 300, *309*
Robert, P. C., 232, 233, *238*
Roberts, F. M., 86, *116*
Robinson, A. G., 5, *28*
Robinson, S. J., 157, *191*
Robinson, T., 57, 59, 60, *80*
Robson, D. S., 378, *380*
Roden, D. B., 132, *152*
Rodrigues, V., 181, *187*
Rodriguez, E., 44, *80*
Rodriguez, J. G., 48, *77*, 96, 113, *119*
Roelofs, C. J., 377, *382*
Roelofs, W. L., 146, *150*
Roessingh, P., 42, 45, *80*, 93, *118*, 134, *152*, 156, 159, 160, 166, 167, 168, 174, *191*, *192*, 379, 380, *382*, *383*
Rohfritsch, O., 12, *28*
Roitberg, B. D., 171, *191*, 260, *276*
Roitberg, C. A., 213, 216, *223*
Rolseth, B. M., 176, *190*, 379, *383*
Room, P. M., 360, *365*
Root, R., 352, 353, *365*
Root, R. B., 23, *28*, 241, 242, 243, 256, *276*
Rose, H. A., 106, 107, *118*
Rosenberry, L., 44, *76*, 166, *188*
Rosenthal, G. A., 74, *80*, 300, *311*, 380, *382*
Roskam, J. C., 305, *312*
Ross, H. H., 85, *118*
Ross, J. A. M., 61, *79*
Rothschild, M., 171, *191*, 260, 263, 264, *276*
Röttger, U., 131, *152*, 162, *189*
Roush, R. T., 217, *223*, 348, *363*
Rowe, W. J., 260, *275*
Rowell-Rahier, M., 263, *276*, 286, *312*
Rudinsky, J. A., 260, *276*
Ruehlmann, T. E., 5, *28*
Ruesink, A. E., 247, *275*
Rumi, C. P., 55, *75*
Russ, K., 260, *276*, 288, *308*
Russell, F. L., 331, 341
Russell, G. J., 5, *28*
Rutledge, C. E., 68, *82*
Ryan, C. A., 64, *76*
Ryan, J. D., 44, *80*
Ryker, L. C., 260, *276*
Rynne, K. P., 262, 263, *275*

Sabelis, M. A., 67, *74*, 349, *363*
Sacchi, C. F., 257, *273*
Sachdev-Gupta, K., 44, *75*, *76*, *79*, 166, 174, 184, *188*, *191*, *193*

Saikkonen, N., 110, *118*
Salama, H. S., 140, *152*
Salim, M., 86, *118*
Salt, D. T., 113, *119*
Sanborn, J. R., 349, *364*
Sanderson, E. D., *223*
Sanderson, G. W., 53, *80*
Sandlin, E. A., 204, *223*
Sato, Y., 16, *28*
Sattaur, O., 23, 24, *28*
Saxena, K. N., 131, 132, *152*, 210, *223*, 376, *382*
Saxena, R. C., 86, *118*, 131, *152*, 358, *365*
Sazima, M., 331, *342*
Schaffner, U., 255, *274*
Schalk, J. M., 40, *80*, 202, *223*
Scherer, C., 133, *152*
Schlee, D., 32, *80*
Schmidt, D. J., 96, 102, *118*
Schmitt, U., 330, *342*
Schmutterer, H., 356, 357, 358, *365*
Schneider, D., 105, *120*
Schneider, J. C., 217, *223*
Schofield, A. M., 37, *76*
Schönbeck, F., 343, 344, *364*
Schöne, H., 126, 128, *152*
Schöni, R., 166, *191*
Schooneveld, H., 231, *237*
Schoonhoven, L. M., 10, 15, *28*, 33, 44, 68, *74*, *80*, *81*, 87, 93, 99, *115*, *118* , 146, *152*, 160, 161, 162, 166, 168, 169, 170, 171, 174, 175, 176, 177, 178, 179, 180, 181, 182, 183, 184, 186, *187*, *190*, *191*, *192*, *193*, 199, 206, 207, 208, 209, 210, 215, *223*, *224*, 260, 264, 274, *276*, 295, *312*, 357, *364*, 376, 377, 378, 379, *382*, *383*
Schreiber, K., 295, *312*
Schröder, D., 248, 249, *274*
Schroeder, H. E., 350, *365*
Schroeder, L. A., 96, 97, *118*
Schultz, J. C., 9, 19, *25*, 31, 52, *80*, 109, 110, *116*, *117*, *118*, 265, *276*, 281, 291, *312*, 345, *363*
Schuster, M. F., 266, *276*
Schwalbe, C. P., 60, *78*
Schwerdtfeger, F., 260, *276*
Scriber, J. M., 3, 8, *28*, 63, *80*, 91, 92, 93, 95, 96, 97, 98, 99, 100, 101, 113, *118*, *119*, 196, 200, 208, 215, 218, 221, 224 , 284, *312*
Secoy, D. M., 355, *365*
Seeley, T. D., 318, 325, 339, *342*
Segers, J. C., 266, *277*
Seigler, D., 32, *74*
Seigler, D. S., 5, *28*, 53, *80*, 140, *152*
Self, L. S., 105, *118*

Selvan, S., 63, *79*
Sepkoski, J. J., 281, 282, 283, *311*
Sergel, R., 196, *224*
Sethi, S. L., 229, *238*
Severson, R. F., 44, *77*, *79*, 140, *150*
Shah, M. A., 159, *192*
Shanbhag, S., 181, *187*
Shanks, C. H., 156, *188*
Shaposhnikov, G. Ch., 200, *224*
Sharaby, A., 140, *152*
Sharkey, T. D., 37, *80*
Shear, W. A., 280, *312*
Sheck, A. L., 218, *224*
Sheehan, W., 267, *276*
Shen, S. K., 109, *118*
Shepard, M., 265, *276*
Sheppard, A. W., 246, *276*, 360, 361, *365*
Sherwood, C. B., 243, *273*
Shi, Y., 350, *363*
Shields, O., 305, *312*
Shields, V. D. C., 138, *153*, 171, 179, 182, *192*
Shorthouse, J. D., 11, 12, *26*, *28*
Shure, D. J., 61, *76*
Siddiqi, O., 181, *187*
Siegel, M. R., 110, *115*
Sierp, H., 6, *26*
Silverline, E., 325, *342*
Silverstein, R. M., 40, *82*
Simberloff, D. S., 248, *276*, 303, *312*
Simmonds, M. S. J., 37, *76*, 101, *119*, 170, 174, 175, 176, 180, 181, 182, 182, 184, *187*, *188*, *192*, 209, 211, 212, 220, *224*, 357, *363*
Simmons, G. A., 132, *152*
Simms, E. L., 55, *80*, 199, 221, 247, *276*, 298, 299, *312*
Simons, J. N., 40, *82*
Simpson, B. B., 333, 336, 338, *341*
Simpson, C. L., 98, 100, 101, *118*, 203, *224*
Simpson, R., 331, *342*
Simpson, S. J., 98, 100, 101, 113, *118*, *119*, 140, *149*, 163, 164, *187*, 203, 212, 220, *224*, 378, *383*
Singer, M. C., 198, 199, 219, *224*, 243, *273* , 288, 289, 290, 310, *311*, *312*
Singh, B. B., 157, *192*
Singh, P., 94, *119*
Singh, R. J., 290, *311*
Singla, C. L., 85, *116*
Singleton, T. A., 5, *26*
Singsaas, E. L., 37, *80*
Skinner, G., 131, *150*
Skrydstrup, T., 50, *74*
Slansky, F., 8, *28*, 92, 93, 95, 96, 97, 98, 99, 100, 101, 102, 107, 113, *118*, *119*

Slobodkin, L. B., 243, 273
Smallegange, R. C., 132, 153
Smiley, J., 290, 312
Smiley, J. T., 70, 80
Smith, A. E., 355, 365
Smith, B. D., 166, 192
Smith, B. H., 146, 152
Smith, C. M., 157, 192, 346, 348, 350, 365, 380, 383
Smith, C. N., 122, 123, 149, 355, 363
Smith, F. E., 243, 273
Smith, J. J. B., 375, 379, 383
Smith, M. C., 356, 364
Smith, P. M., 74, 80
Smith, S. C., 174, 181, 192, 357, 363
Snider, D., 196, 224
Snieckus, V., 379, 382
Snodgrass, R. E., 318, 319, 342
Snow, A. A., 331, 342
Soldaat, L. L., 42, 80
Son, K. C., 44, 79
Sørensen, H., 40, 79, 166, 190
Southwick, E. E., 339, 342
Southwick, L., 339, 342
Southwood, T. R. E., 6, 13, 15, 16, 27, 28, 45, 77, 89, 91, 92, 117, 119, 157, 192, 241, 242, 248, 249, 252, 255, 257, 272, 276, 277, 280, 296, 297, 300, 311, 312, 344, 365
Spence, W., 1, 4, 206, 222
Spencer, D., 350, 365
Spencer, K. C., 5, 28
Spencer, W. A., 204, 225
Spira, T. P., 331, 342
Sprengel, C. K., 315, 316, 342
Spurr, D. T., 110, 116
St. Leger, R. J., 109, 116
Städler, E., 42, 44, 45, 79, 80, 134, 138, 140, 144, 148, 150, 152, 155, 156, 159, 160, 165, 166, 167, 168, 171, 173, 174, 184, 187, 191, 192, 193, 378, 379, 380, 381, 382, 383
Stammitti, L., 42, 75, 165, 188
Stamp, N. E., 58, 70, 80, 375, 383
Starks, K. J., 284, 299, 310, 350, 363
Stead, A. D., 330, 331, 342
Steffan, A. W., 293, 312
Steinberg, S., 231, 238
Steinly, B. A., 103, 119
Stengl, M., 142, 146, 152
Stenhagen, G., 48, 74
Stephenson, M. G., 44, 77, 140, 150
Stevens, M. A., 156, 191, 348, 364
Stewart, A. J. A., 257, 258, 277
Stinner, B. R., 108, 118
Stockoff, B. A., 203, 204, 224
Stoner, K. A., 156, 193, 346, 347, 348, 350, 365
Storch, R. H., 356, 364
Storer, J. R., 49, 80

Stork, A., 156, 171, 187
Stork, N. E., 159, 162, 193
Stout, M. J., 64, 65, 68, 80
Stoutjesdijk, P., 19, 20, 28
Stowe, M. K., 337, 342
Straatman, R., 218, 224
Stradling, D. J., 19, 26
Strickler, K. L., 122, 151, 172, 190
Stride, G. O., 218, 224
Strong, D., 90, 115
Strong, D. R., 6, 28, 241, 242, 252, 255, 257, 275, 277, 280, 312
Stuckbury, T., 358, 362
Styer, W. E., 42, 79
Sugathapala, P. M., 356, 364
Suomela, J., 62, 80, 110, 118, 257, 275
Surburg, H., 322, 342
Sutcliffe, J. F., 179, 193
Sutherland, O. R. W., 39, 78, 229, 238
Sutter, G. R., 131, 150
Swain, T., 41, 61, 75, 80
Swenson, K. G., 229, 238
Sword, G. A., 196, 225
Sword, S., 211, 220
Szelegiewicz, H., 200
Szelényi, G., 250, 257, 260, 270, 271, 274, 277
Szentesi, Á., 126, 128, 131, 147, 148, 150, 161, 189, 197, 198, 204, 205, 206, 208, 209, 214, 216, 222, 225, 263, 277, 379, 381

Taanman, J. W., 146, 153
Tabashnik, B. E., 101, 119, 218, 219, 225
Tabe, L. M., 350, 365
Tada, R., 166, 189
Takabayashi, J., 48, 66, 67, 68, 81, 129, 152, 349, 363
Takahashi, S., 48, 67, 68, 81
Tallamy, D. W., 64, 68, 81
Tang, C. S., 355, 365
Tanner, J. A., 94, 117
Tanton, M. T., 156, 193
Tarpley, M. D., 204, 225
Tarrants, J. L., 64, 78
Tempère, G., 15, 28
Tenow, O., 21, 28, 62, 76
Terofal, F., 292, 312
Terriere, L. C., 98, 115
Thacker, J. D., 349, 365
Theuring, C., 54, 57, 81
Thibout, E., 140, 152
Thieme, H., 59, 81
Thiéry, D., 131, 137, 138, 139, 140, 147, 152, 352, 365, 377, 379, 382, 383
Thomas, C. D., 288, 289, 312
Thomas, G., 146, 153, 234, 238
Thompson, A. C., 45, 77

Thompson, J. N., 11, 29, 217, 218, 225, 275, 287, 289, 290, 291, 297, 301, 302, 303, 304, 307, 310, 312, 317, 335, 339, 342
Thompson, R. F., 204, 225
Thompson, V., 89, 119
Thorpe, W. H., 214, 225
Thorsteinson, A. J., 159, 166, 189, 190, 232, 233, 237
Thresh, J. M., 354, 365
Thurston, R., 157, 191
Tichenor, L. H., 140, 152
Tiffney, B. H., 279, 311
Timmermann, E. A., 44, 78
Timmins, W. A., 357, 365
Tinbergen, L., 202, 225
Ting, I. P., 25, 29
Tingey, W. M., 44, 80, 157, 158, 189, 193, 348, 363, 376, 383
Tingle, F. C., 140, 152
Tjallingii, W. F., 84, 88, 119, 376, 378, 381, 383
Tokunaga, F., 181, 191
Tollsten, L., 48, 81, 322, 331, 341, 342
Tommerås, B. A., 140, 153
Toogood, P. L., 174, 181, 192
Topazzini, A., 143, 153
Traynier, R. M. M., 132, 153, 214, 225, 379, 383
Treacy, M. F., 266, 277
Trujillo-Arriaga, J., 353, 365
Trumble, J. T., 25, 29, 86, 114
Tscharntke, T., 242, 260, 274, 277
Tully, T., 215, 225
Tumlinson, J. H., 49, 50, 67, 81, 265, 277
Turlings, T. C. J., 49, 50, 67, 81, 264, 265, 277
Turnipseed, S. G., 157, 193
Tzanakakis, M. E., 233, 237

Ueckert, D. N., 204, 225
Underhill, E. W., 379, 382
Uvah, I. I. I., 352, 365
Uvarov, B., 9, 29

Väisänen, R., 112, 116
Vallo, V., 288, 308
Valterova, I., 140, 153
Van Baarlen, P., 65, 75
Van Beek, T. A., 38, 44, 48, 68, 74, 81, 129, 149, 166, 168, 171, 187, 193, 380
Van Dam, N. M., 51, 52, 54, 57, 68, 70, 81
Van der Ent, L. J., 131, 153
Van der Meijden, E., 52, 81, 242, 247, 259, 277, 299, 312
Van der Pers, J. N. C., 143, 144, 146, 150, 153, 234, 238, 379, 382

Author index

Van Drongelen, W., 142, 180, 181, 193, 216, 225
Van Duyn, J. W., 157, 193
Van Eeuwijk, F. A., 162, 179, 193
Van Emden, H. F., 49, 80, 110, 117, 262, 272
Van Helden, M., 379, 381
Van Heusden, S., 346, 364
Van Lenteren, J. C., 16, 17, 29, 377, 383
Van Loon, J. J. A., 44, 48, 68, 74, 81, 96, 97, 99, 119, 129, 132, 142, 143, 149, 153, 162, 166, 168, 171, 174, 177, 179, 180, 181, 182, 183, 184, 185, 187, 188, 190, 193, 197, 216, 221, 225, 357, 362, 364, 380
Van Steenis, C. G. G. J., 283, 313
Van Valen, L., 247, 277, 303, 313
Van Veldhuizen, A., 171, 187
Van Wijk, C. A. M., 54, 81, 259, 277
Vandenberg, P., 55, 81
Vandermeer, J., 351, 352, 365
Van't Hof, H. M., 97, 119
Vedder, A. L., 40, 81
Velthuis, H. H. W., 336, 342
Vereijken, B. H., 22, 29
Verkaar, H. J., 25, 28, 247, 277
Verkerk, R., 346, 364
Verpoorte, R., 52, 81
Verschaffelt, E., 1, 4, 41, 81, 160, 165, 166, 170, 193
Vet, L. E. M., 67, 81, 265, 277, 377, 383
Via, S., 11, 29, 217, 219, 225, 348, 363
Vickery, B., 35, 73, 81
Vickery, M. L., 35, 73, 81
Virtanen, T., 110, 118
Visser, J. H., 45, 47, 48, 49, 74, 81, 128, 129, 131, 137, 138, 139, 140, 142, 143, 145, 146, 147, 149, 150, 151, 152, 153, 352, 365, 377, 379, 383
Vité, J. P., 131, 153
Vogel, S., 331, 342, 377, 383
Voland, M. L., 112, 113, 116
Volkonsky, M., 356, 365
Völlinger, M., 358, 365
Von Euw, J., 263, 264, 276
Von Frisch, K., 327, 340
Vos-Bünnemeyer, E., 234, 237
Vray, V., 52, 79
Vrba, E. S., 291, 313
Vrieling, K., 54, 68, 70, 81
Vrielink, R., 346, 364
Vriesenga, L., 318, 339
Vrkoc, J., 153
Vuorinen, P., 110, 118

Wäckers, F. L., 265, 277

Waddington, K. D., 326, 328, 342
Wadhams, L. J., 140, 145, 149, 151, 356, 364
Waldbauer, G. P., 90, 96, 100, 119
Waldvogel, M., 9, 25, 203, 218, 220, 225
Waller, D. A., 378, 383
Walsh, J. S., 295, 310
Wang, J., 109, 114
Wanjura, W. J., 72, 76
Wapshere, A. J., 359, 365
Ward, J. T., 156, 189
Ward, K. E., 232, 237
Waring, G. L., 266, 276
Warncke, E., 317, 341
Warrington, S., 112, 119
Warthen, J. D., 166, 181, 191
Waser, N. M., 320, 342
Wassel, G., 60, 76
Wasserman, S. S., 219, 225
Waterhouse, D. F., 16, 29
Waterman, P. G., 38, 40, 61, 62, 79, 81, 380, 383
Watson, M. A., 71, 81
Watt, A. D., 113, 119
Watts, D., 40, 81
Webb, J. A., 320, 341
Weber, A., 343, 344, 364
Wehling, W., 218, 225
Weiss, A. E., 19, 29, 250, 275
Weiss, M. R., 208, 216, 220, 320, 330, 331, 342, 378, 380
Wellso, S. G., 75
Wensler, R. J. D., 166, 193
Wessels, R., 65, 75
Wheeler, A. G., 5, 29
Wheeler, G. S., 101, 119
White, P., 212, 224
White, P. R., 143, 153
White, R. R., 243, 273
White, T. C. R., 62, 82, 91, 92, 119
White, W. H., 71, 77
Whitehead, A. T., 379, 383
Whitham, T. G., 16, 25, 26, 29, 52, 72, 73, 82, 252, 277
Whitman, D. W., 262, 277
Whittaker, J. B., 111, 112, 113, 116, 119, 290, 313
Whittaker, R. H., 34, 82, 283, 313
Wiebes, J. T., 234, 237, 292, 293, 311, 313
Wiegert, R. G., 97, 120
Wightman, J. A., 96, 120
Wijn, M., 247, 277
Wiklund, C., 10, 29, 214, 225
Willey, R. B., 376, 381
Williams, C. B., 248, 277
Williams, C. M., 232, 238
Williams, E. M., 39, 78

Williams, I. H., 339, 342
Williams, M. A. J., 12, 29
Williams, W. G., 349, 365
Willig, M. R., 204, 223
Willis, M. A., 137, 153
Willmer, P., 18, 29
Wilson, D. D., 44, 79
Wilson, I. D., 38, 78
Windle, P. N., 349, 365
Wink, M., 55, 82, 105, 120, 263, 277, 349, 365
Winstanley, C., 211, 220, 378, 380
Witte, L., 54, 57, 81
Wolfson, J. L., 63, 82, 376, 378, 383
Wood, A., 174, 181, 192, 357, 358, 363
Wood, T. K., 284, 286, 313
Woodcock, C. M., 140, 145, 149, 151, 356, 364
Woodhead, S., 42, 44, 45, 82, 378, 383
Wooster, M. T., 260, 272
Workman, J., 68, 80
Workman, K. V., 64, 80
Wratten, S. D., 89, 120, 243, 277
Wu, B. R., 42, 75, 165, 188
Wyss, E., 354, 365

Xu Rumei, 16, 29

Yamamoto, K., 231, 237
Yamamoto, R. T., 349, 365
Yamauchi, T., 166, 189
Yang, R. Z., 355, 365
Yencho, C. G., 157, 158, 193
Yendol, W. G., 109, 110, 116, 117
Yeo, P., 339, 341
Yost, M. T., 210, 221
Young, O. P., 9, 29
Youngman, R. R., 377, 382
Yu, S. J., 98, 108, 115, 120

Zabel, P., 346, 364
Zacharuk, R. Y., 138, 153
Zaller, J. G., 113, 114
Zalom, F. G., 234, 237
Zalucki, M. P., 11, 27
Zandee, M., 305, 312
Zandt, H., 113, 114
Zangerl, A. R., 40, 51, 68, 70, 74, 82
Zenk, M. H., 32, 78
Zhang, Z., 356, 364
Ziegler, C., 113, 114
Zielske, A. F., 40, 82
Zohren, E., 123, 153
Zucker, W. V., 52, 82
Zur, M., 102, 117
Zwölfer, H., 196, 197, 225, 271, 277, 293, 313

SUBJECT INDEX

Numbers in *italics* refer to illustrations

Abietic acid, 38
Ablation of sensilla, 379
Abundance, 242, 243, *248*, 303
  of herbivores, 22, 250, *259*, 298
  of plants, 295
Acceptance, *see* Host plant
Acid rain, 112
Across-fibre patterns, *see* Sensory coding
Active space
  definition, 129
  odour, 130, 131
Aescin, 38
Age, 58, 133, 135
  see also Leaf age
Agriculture, 2, 343–362
Air pollution, 112–113
Ajugarin, 38, 171
Alkaloids, 36–37, 44, 55, 105, *300*
  toxicity, *105*
Allelochemics
  definition, 34
  and food utilization, 102–104
Allelopathy, 33
Allomone, 262
Allopatric speciation, 199, 285
Amino acid receptor, 179
Amino acids, 44, 90, 162, 164, 165
Anemotaxis, *125*, 127, 136, 137, 147
Angiosperms radiation, *see* Evolution
Anthocyanins, 39
Antiaggregation pheromone, 260
Antibiosis definition, 347
Antifeedant index, *181*
Antifeedants, 355–359, 378
Antixenosis definition, 347
Ants, 19, 261, 266
Aphids
  host selection behaviour, *125*
  polymorphism, *228*
  sexual forms, 229
  wing development, *228, 229*
Apparency hypothesis, 296
Apple maggot fly (*Rhagoletis pomonella*)
  colour vision, *133*

EAG, *145*
  host selection, *148*
  searching, 147
  vision, 147
Approximate digestibility, 96
Arrestant definition, 123
Arrestment, 126, 155
Artficial diet, 60, 89, 93–95, 102, 107, 165, 181, 203, 206, 378
  composition, 94
  for aphids, 88, 109, 229, 378
Associative learning, *see* Learning
Attractant definition, 123
Atropine, 36
Aversion learning, *see* Learning
Azadirachtin, 38, 171, *182*, 355–358, 378

Bark beetles
  host selection behaviour, *125*, 147
Bees as pollinators, 317–339
Benzyl isoquinoline, 36
Berberine, 36
Bioassay, 70, 377
  guided fractionation, 165
Biological communities, 239
Biological control, *see* Weed control
Biomass, 2, 271
  of insects, 1
  loss of plant biomass, 19
  of plants, 5
Biosynthesis, 32, 35
Biotype, 196, 284, 308
Botanical instinct, 15, 16
Boundary layer, *18*, 128, 155
Bracken fern biological control, 361
Bud break, 240, *241*
Budget equation, 96
Bumblebees
  energy requirements, 327

C3 and C4 plants, 33
Cabbage root fly
  colour vision, *134*
  oviposition, *134, 160*
Caffeic acid, 38
Caffeine, 37, 52

Caloric value, 83
Canavanine, 299
Cannabidiol, 38
Carbohydrates, 90
  see also Sugars
Carbon dioxide (CO_2) 112
Cardenolides, 180, *264*
Casting, 136
Catechin, 39
Chalcones, 39
Character displacement, 333
Chemical legacy hypothesis, 214
Chemoreception, 138
Chemoreceptors
  generalists, 144, 145
  internal, 174
  specialists, 144
Chemotaxis, 127, 138
Chemotaxonomy, 73
Cherry fruit fly (*Rhagoletis cerasi*)
  host marking, 171
  host selection, *125*
  olfaction, 144, 147
  vision, *125*
Chlorogenic acid, 169, 171
Choice experiments, 377
Cholesterol, 38
Cibarium, 88
Cladograms, *292, 294*
Clerodin, 38
Clines, 300
Clone, 72
Cocaine, 37
Coevolution, 289, 302–303, 306, 335
  diffuse, 303, 306
  diversifying, 289
Colonization, 13
  of introduced plants, 252
Colorado potato beetle
  anemotaxis, 147, *138, 139*
  EAG, *142*
  green leaf volatiles, *143*
  and leaf age, *230*
  orientation, 137–140, 147
  taste hair responses, 176
Colour vision, 130, 131, 132, 134, 147
Community definition, 247

Subject index

Compartmentation, *see* Secondary plant substances
Compensation, 23–25, 246, 247, 269
 overcompensation, 24, 25, 246, 298
Compensatory feeding, 98, 100–101
Competition, 244, 283, 331
 for pollinators, 331, 333
 among insects, 256, 257
 interspecific, 247, 252, 258, 283, 289
 intraspecific, 259
Competitive exclusion principle, 257
Conditioning, *see* Learning
Constitutive resistance, 64
Constraints, 7, 101
Consumption rate, 95, 97, 101
Contact chemoreception, 377
Contact chemoreceptors, 163, 172
Contact testing, 124
 evaluation, 159
Convergence neural, 142
Cost of growth, 99
 of detoxification, 113
Costs-benefits, 298
 flower visitation, 324
Costs of production, *see* Secondary plant substances
Coumarins, 38
Crops, 267
 losses to herbivory, 3, 343, 344
 pollination, 339
Cross-habituation, 205
Cucurbitacin, 38, 70, 262, 299
Cultivars, 16
 concentration plant compounds, 70–71
 glandular trichomes, 157
 volatiles, 49
Cyanin, 39
Cyanogenics, 34, 41–42, 53, 61
Cyclic outbreaks, 68
Cytochrome P450, 106, *108*

Damage, 18–23, *50*, 246
 in agroecosystems, 23
 caused by insects, 22, 267, 297
 artificial, 64, 66, 67
 volatiles, 50
 see also Compensation
Defence, 31, 296
 constitutive, 64
 direct, 64
 indirect, 64
Defoliation, 19, 21, 65, 68, 69, 259, 269
 artificial, 23
Deme, 286
Demography, 239
Desert locust reproduction, 231

Desiccation, 92, 141
Deterrents, 123, 169, 170, 171, 172, 262
 definition, 123
 and evolution, 290
 receptors, 177, 180, *181*, *182*, 217
Detoxification, 89, 105
 enzymes, 94, 104
 enzyme induction, 107
 of plant allelochemicals, 104–108
Development, 227
Dhurrin, 55, 57
Diapause and leaf age, 230
Diet
 breadth, 7, 8
 self-selection, 211
 see also Artificial diet
Digestibility, 86, 98
Digestion, 102
 adaptation, 255
Dioscin, 38
Directed movement, 126
Diseases plant viruses, 111
Dispersal, 126, 260, 286
Disruptive-crop hypothesis, 352
Disturbance, 296
Diterpenoids, 38
Diversification, 89, 279, 283, 362
Diversity
 flower types, 335
 insects, *254*, 283
 and pest insect damage, 354
 plant structure and chemistry, 295
 plants, 267
Drinking, 92
 tests, 378
Drought stress, 93
Drumming, *133*, 155
Dual discrimination theory, 161
Dulcitol, 179, 295

EAG, *see* Electroantennogram
Ecdyson, 38
Ecosystem, 18, 247, 337
 agroecosystem, 23, 345
Efficiency
 of conversion of digested food, 96
 of conversion of ingested food, 96
 metabolic, 96–100
Egg load, 156
Electrical penetration graph (EPG), 378, 379
Electroantennogram (EAG), 141, 143
 technique, 379
 of sibling species, *145*
Electrophysiology, 144, 379
Empodia, 18
Encounter-frequency hypothesis, 248
Endophyte, 110, *111*

Enemy escape hypothesis, 250, 290
Enemy hypothesis, 242, 353
Enemy-free space, 16, 252, 260
Energy
 budget, *100*
 flow, 270, *271*
 optimization, 147
Entomopathogens susceptibility, 109
EPG, *see* Electrical penetration graph
Epideictic pheromone, *see* Pheromone
Epidermis, 11, 44, 52, 89
 vacuole contents, 55
Epipharyngeal sensilla, 172, 176
Essential oils, 37, 44
Ethylene, 234, *235*
Euryphagy, 7, 305
Evolution, 219, 279–308
 Angiosperms, 280, *281*, 333
 chemical, 281
 insect taxa, *281*, *282*, 283
 nervous system, 291
 plant defence, 282
 sequential, 305–307
Excretion, 105
Exotic plants, *see* Neophytes
Experience, 123, 204–214
Extrafloral nectaries, *see* Nectaries

Facilitation among insects, 260
Feeding
 deterrents, *see* Deterrents
 periods, 101
 rhythms, 59
Feeding stimulants, *see* Phagostimulants
Ferns, 14
Fertilizer, 60, 62, *63*, 91
Fitness
 insect, 32, 68, 123, 215, 259
 parasitoid, 263
 see also Plant
Flavonoids, 38, 61
Flavonols, 34, 39
Flower
 age, 330, *331*
 automimicry, 328
 colour, 322
 colour changes, 330
 constancy, 319–324
 diversity, 335
 evolution of shape, 336
 handling, 323, *324*
 odour trail, 322
 recognition, 320–323
 scents, 320
 spacing, 327, 331
 types, 326
Flowering time, 332

Food plant
  effects on entomopathogens, 109
  preference test, *207*
  quality, 83, 89–95
Food web, 270
Forest
  fertilization effects, 63
  insects, 21, 247, 269
  insect feeding strategy, 52
  pest outbreaks, 93, 112, 259
Fossil insects, *282*
Furanocoumarins, 44, 51, 104

Gallic acid, 41
Galls induced by insects, 12, 52, 266–267, 280
  distribution, 12, 72, 251
Gene banks, 351
Gene-for-gene interactions, 301
Generalists, 9, 232, 255, 291, 299, 351
  definition, 6
Genetic changes, 307
Genetic covariance, 217, 219, 297
Genetic modification, 346
Genetic variation, 196, 199, 216
Geographic mosaic theory of coevolution, 304
Geographical race, 284
Geographical variation of host range, 195
Geographical range of plants, *249*
Germplasm collections, 350
Gestalt, 210
Gibberellin, 38, *231*
Gibberellic acid, 232
Glabrous leaves, 16
Glaucolide-A, 38
Glaucous plant surface, *162*
Glomeruli, 143
Glossy plant surface, 159, *162*
Glucobrassicin, 44
Glucosinolate receptor, 177, 184, *168*
Glucosinolates, *34*, 41, 44, 167
  as oviposition stimulants, *167*, 184
Glutathione, 112, *113*
Gossypol, 38, 70, *71*, 102
Grass fungal endophytes, 110
Grasshopper polyphagy, *292*
Gravimetric method, 96
Green leaf volatiles, 45, 47, *49*, 143, 232
Greenhouse
  effects on plant growth, 64, 375
  pollination, 339
Growth rate
  and air pollution, 112
  differences between species, 99
  herbivores and carnivores, 100

insects, 97
  locusts, 99
  plants and phenolics, 54
Guild, 98, 256, 257, 362
Gustation, 138, 161, 174
Gustatory coding, 174–176
Gypsy moth (*Lymantria dispar*)
  change in food preference, 203
  change in nutritional requirements, 203
  feeding rhythm, 60
  frass effects, 247
  susceptibility to virus, 109, 265

Habitat diversity, 248
Habitat templet model, 296, *297*
Habitat-heterogeneity hypothesis, 248
Habituation to deterrents, 204, 205, 206, 215, 358
Hairs, see Trichomes
Half-life, 57
Haustellate mouthparts, 86, *87*
Headspace, 45, 47, 48, 49, 380
Helicoverpa
  host selection behaviour, *125*
  taste receptors, 175
Hemicellulose, 33
Honey production, 339
Honeybee foraging distance, 325
Honeydew, 246
  contents, 266
  food for ants and predators, 261, 265, 266
  food for herbivores, 147
Hopkins host-selection principle, 214
Hormones, see Plant
see also Neuroendocrine system
Host alternation, 200, 201
Host finding, 291
Host marking, 171
Host-plant
  acceptability, 172
  acceptance, 156
  acceptance definition, 122
  effects on diapause, *230*
  effects on mating, 232
  effects on morphism, 228, *230*
  effects on reproduction, 231–235
  more than food plant, 16–17
  morphology, 16–18, 156
  preference definition, 122
  quality and natural enemies, 262
  range, 7, 9–11, 195, 376
  recognition definition, 122
  resistance, 90
  selection, 122–126, 155
  specialization, 5

Host preference
  age effects, 203
  change, 289
  developmental stage, 203, 204
  genetic changes, 291
  genetic variation, 216
  induction, 206–210, 215, *210*
  seasonal changes, 200
  sex, 204
  temperature effects, 202
Host race, 284, *285*, 292
Host range expansion, 196
Host shift, 180, 187, 213, 289, 292, 294, 295
Host specialization adaptive value, 289
Hybrid incompatibility, 287
Hybrids, 16, 180
  host preference, 216, 217
Hydrogen cyanide (HCN), 56, 104
Hygroreceptors, 141
Hypericin, 51

Imprinting, see Learning
Indole alkaloids, *34*, 37
Induced preference, see Learning
Induced resistance, 64–70
  delayed responses, 68
  heritability, 70
  and natural enemies, 264
  resin, 66
Induction
  see Detoxification
  see Learning
  see Resistance
Inositol receptor, 179
Insect demography and host quality, 240
Insect phenology, 251
Insecticide susceptibility, 107, 110
Insecticide treatment, 246, 268
  seed production, 245
Insecticides, *107*, 355
Intercropping definition, 351
Introduced species
  insects, 255
  plants, see Neophytes
  weeds, 361
Isoprene, *34*, 37, 50
Isoquercitrin, 169
Isothiocyanates, 41, 49

Juvenile hormone, 229

Kaempferol, 39
Kairomone, 262
Key-lock model, 186
Kineses, 126
Klinokinesis, 126

Labelled line, see Sensory coding
Latex, 57, 60
Laticifer, 57
Leaf
    age, 58, 133, 135, *231*, 241
    age and diapause, 230
    age and reproduction, 231–232
    boundary layer, 155
    discs, 70, 375, 377, 378
    shape, 131, 202, 213, 300
    size, 18, 134, 300, 301
    surface, *43*, *46*, *47*, 130
    surface chemistry, 42, 44, 380
    surface waxes, 44
Leafminers, 7, 11, *12*, 280
Learning, 204–211
    adaptive value, 215–216
    associative, 204, 210, 213
    aversion learning, 210, 212, 215, *211*
    conditioning, 214
    flower, *321*
    flower handling, 323, *324*
    food imprinting, 206
    induction, 206–210, 215, 376
    oviposition behaviour, 213
    peripheral, 206, 208
    see also Habituation
Light intensity, 60
Lignan, 104
Lignocellulose, 32
Limonene, 37, *66*
Limonoids, 38, 357
Locomotion compensator, 137, 377
Locust food-plant range, 9
Losses
    to insects, 19, 343
    to sucking insects, 19, 22
Lupine alkaloids, 37
Lycaenids
    change in food choice, 203

Macapine, 32
Maintenance, 99
Mandibles, 84
Mandibulate mouthparts, 84
Marking pheromone, *see* Pheromone
Mating and host plants, 232–235
Maxilla, 85
Maxillary taste hairs, 174, *175*
Meal size, 169
Mechanoreceptors, 85, 127, 141, 159, 173
Meristem, 24
Mesophyll, 11, 52, 55, 88
Metabolism, 33, 92, 96, 99
Metabolic load hypothesis, 99
Metapopulations, 285
Microclimate, 18, *20*, 241

Microorganisms, 247, 268, 269
Mineral flow, 270
Mixed cropping, 351
Mixed diet, 211
Mixed-function oxydases (MFOs), 106
Mixtures, 146, *168*, 181, 183, 186
Modular structure of plants, 24, 301
Monoculture, 241, *242*, 345, 351
Monophagy, 5, 7
Monoterpenoids, 37, *66*, 106, 322
Morphine, 36, 57, *60*
Mosaic resistance, 72, 73
Motivation definition, 126
Mouthparts, 84
    morphology, 85–87
Multiple cropping definition, 351
Mustard oil glucosides, 41
Mutations, 306
Mutualism, 315, 317–319
Mycetome, 108
Myristicin, 104

Naringenin, 39
Natural enemies, 242, 246, 252, *256*, 260, 262, 267, *268*, 352, 353
Nectar
    amino acid contents, 328
    flow, 319, 332
    guide, 321
    as insect food, 219, 315, 317
    status, 330
    sugar contents, 315, 318, 332
Nectaries
    extrafloral, 265–266
    extrafloral and natural enemies, 353
    location, 324, 332, 336
Neophilia, 212
Neophytes, 252, 255
Neuroendocrine system
    affected by host plant, 227, 231, 232, *236*
Niche, 252, 256, 288, 303
    saturation hypothesis, 252
    segregation, 257, 259
    'vacant', 258
Nicotine, 53, *55*, 57, *66*, 105, *106*, 262, 265, 300, 355
Nitric oxide, 112
Nitriles, 41
Nitrogen, 63, 90–92, *93*, 300
    availability, 55
    in plant tissues, *84*
Non-preference definition, 346
Number of species
    crop plants, 23, *24*
    herbivorous insects, 5, 6
    insect pests, 23

    insect species per plant, 13, *15*, 249
    plants, 5, *6*
Nutritional feedback, 100
Nutritional indices, 96
Nutritional quality, 228
    and air pollution, 112
Nutritional requirements, 89, 204, 219
    age effects, 203

Oak (*Quercus* spp.)
    catkins herbivory, 229
    number of insects, 13, 89
    phenology and herbivory, 240
    seasonal effects on chemistry, 58, 103
    seed production, 245
    tannins in palisade tissue, 52
    volatiles, 232, 261
Odour
    distance attraction, 130
    gradient, 137, 142
    masking, 137, 147, 352
    plume, *130*, 136
    trap, 380
Olfaction, 138
    central processes, 142
Olfactometer, 377
Olfactory chemoreceptors, 138
    chemoreceptor sensitivity, 142
    coding, 143, *145*
    coding across-fibre patterns, 146
    coding labelled lines, 146
Olfactory orientation, 136–140, 376–377
Oligophagy, 5
Oogenesis, 232
Optimal foraging theory, 324
Orchids
    pollination, 337
Orientation, 129–138
    Colorado potato beetle, 137
    methods, 377
    to odour, 129, 136, 376
    to visual cues, 132
Orthokinesis, 126
Overcompensation, *see* Compensation
Oviposition, 297, 379
    cabbage root fly, *123*
    deterrents, 68, 170–172, 259
    induced preference, 213–214, 216
    mistakes, 200
    on non-host plants, 219
    preference and larval performance, 10, 217, 219, 297
    stimulant definition, 123
    stimulants, *167*, *168*, 184
Ozone, 112

Subject index

Palpation, *45*, 155, 211
Papaverine, 36
Papilio oviposition, 10, 166, *168*
Parasitization, 16, *263*
Parasitoids, *67*, 100, *267*
Parenchyma, 89
Pathogens of insects, 265
    see also Plant pathogens
PBAN, 234, *236*
Performance, *10*, *11*, 97, 291, 297
Periferal interactions, 181
Pest insects, 343
    number of species, 345
Pest outbreak factors, 345
Phagostimulant, 123, 162, 170, 378
    definition, 123
Phaseollin, 39
Phenolics, 38–41, 45, 52, 53, 61, 68, 294
Phenology, 13, 62, 219, 227, 240, 333
Pheromone(s)
    antiaggregation, 260
    epideictic, 171, 259, 260
    and evolution, 286
    and flower visitation, 330, 337
    marking, 171, 259, 260
    production, 232–234, *235*
    sex, 123
Phloem, 88, 89
Phloridzin, 44
Phospholipids, 146
Photomenotaxis, 127
Photoperiodism, 200
Photoreceptor, 131
Phytoalexins, 111
Phytochemistry, 16, 31–41
Phytoecdysteroids, *34*, 38
Piercing-sucking insects, 86–89, 98, 280, 378
Pieris
    colour vision, 132, *133*
    glucosinolate receptor, 177, 184, *168*, *185*
    oviposition, 166
    oviposition deterrent, 171
    tarsal taste hairs, *173*
Pinene, 37, 50, *66*
Plant
    abundance, 295
    chemistry, 160, 250
    communites, 268
    competitiveness, 246
    demography, 244
    diseases and natural enemies, *263*
    effects on hormone production, 227–232, *236*
    effects on pheromone production, 232, *235*, *236*
    epicuticle, 168

fitness, 21, 24, 51, 54, 297–299
heterogeneity, 250
pathogens, 110
phenology, 240
size, 15, 131, 266
structural diversity, 249
succession, 252, *254*, *269*
surface, 42–44, 155–157, 159, 161, 162
texture, 157
volatiles, 45–50
Plant architecture, 14, 249
    and natural enemies, 266, *267*
    and number of insects, *248*, *250*
Plasmalemma, 57
Pollen
    basket, *319*, 336
    digestion, 336
    as insect food, 317
Pollination, 315–339
    energetics, 324–325, 328, *329*
    evolution, *334*
    wind, 318, 333, *335*
Polycultures, 345, 350
Polygodial, 37, 356
Polyphagy, 6, 9
Polyphenism, 227
Polysubstrate monooxygenases (PSMOs), 99, 106, *108*
Pool exhaustion hypothesis, 255
Population dynamics, 239
Potato
    odour, 137
Predators, 100, *354*
Preference
    and developmental stage, 203
    induction, *see* Learning
    order, 375
    performance relationship, 10
    ranking, 195, 196
    tests, *207*
    see also Host plant
Primary host, 200
Primary plant metabolism, 32
    effect of sun and shade, 61, 62
Primary plant metabolites
    and food selection, 162–165
    production costs, 53
Probing, *123*, 124, 156, 165
Production costs, *see* Secondary plant substances
Protease, 102
    inhibitors, 52, 64
Protein
    and air pollution, 112
    amounts in plants, 91
    different tissues, 52
    digestibility, 86
    effects of sun and shade, 62

and gossypol, 102
and tannins, 103
Protein-carbohydrate ratio, 89, 203
Proximate factors, 3
Prunasin, 42
Purine alkaloids, 37
Pyrethrins, *106*, 355
Pyrrolizidine alkaloids (PAs), *34*, 37, 52

Quassinoids, 357
Quinine, 37
Quinolizidine alkaloids, 37, 57

Radius
    of detection, 147
    of effective attraction, 129
Rainy season, 231
Reaction chains, 122
Receptor
    potential, 141
    sensitivity, 145
    sensitivity change, 203, 206, *209*, 212
    specificity, 145
Recognition, *see* Host plant recognition
Recording techniques, 376
Reflectance, 133, *134*
Rejection, *45*
Relative consumption rate, 97
Relative growth rate, 97
Relative humidity, 18
Repellents, 123, 355
    definition, 123
Reproduction, 83, 95, 231
    host effects on oogenesis, 232, 233
Reproductive isolation, *237*, 285
Resin ducts, 57
Resins, 38, 42, 66
Resistance
    constitutive, 64–66
    definition, 31, 296
    horizontal, 347
    induced, 64–66
    mechanisms, 346
    and molecular biology, 346
    monogenic, 347
    mosaic, 72
    partial, 347
    polygenic, 347, 348, 350
    and secondary plant substances, 348
    stability, 347
    to antifeedant treatment, 358
    to insect herbivory, 295, 296, 344
    to insect pests, 346
    vertical, 347
Resistance breeding, 157, 346

Subject index

and biotechnology, 350
methods, 349
and natural enemies, 349
Resource availability hypothesis, 59, 296
Resource concentration hypothesis, 240, 352
Resource partitioning, 333
Respiration, 271
Respirometry, 99
Rhythm feeding, 59
Roots, see Secondary plant substances, 53
Rotenone, 39, *106*, 355
Ruderals, 296, 297
Rutin, 39, 103, 109

Salannin, 357
Salicin, 70, 205
Saliva, 68
  aphids, 87, 89
Sandwich test, 169, 378
Saponins, 38, 52
Scopoletin, 38
Screen test, 376
Search, *128*, 147
  image, 202, 214
  random, 126
Searching
  definition, 121
  mechanisms, 126–128
  patterns, *126*
Seasonal effects on insects, 251, 253
Secondary host, 200
Secondary plant substances, 32, 33–36, 165–169
  age, 51
  autotoxicity, 55
  compartmentation, 55–57
  concentrations, 40, 51, 70
  concentrations in crop plants, 348, 349
  day/night effects, 59–60
  different plant parts, 51
  effect of fertilizers, 62
  function, 299
  interyear variation, 60
  number, 35
  production costs, 50, 53–57, 104
  in roots, 53
  seasonal variation, 57–59
  sequestration, 263, *264*
  storage, 55
  toxicity, 55
  turnover, 57
Seed
  dispersal, *244*
  predation, *245*
  production, 244

production in crops, 317
Selection, see Host plant
Semiochemicals, 148
  definition, 122
Senecionine, 37
Senescence, 57
Sensilla basiconica, 138
Sensilla styloconica, see Maxillary taste hairs
Sensory coding, 174–186
Sensory coding
  across-fibre patterns, 174, 178, 185, *176*
  deterrents, 184
  labelled line, 174, 178, 185
Sequential evolution, 305–307
Sequestration, 263, *264*, 291
Sesquiterpenoids, 37, 322
Sex ratio parasitoids, 262
Sexupara, 200
Shade, 60, 61, 62, 251
Shelter, 262
Sibling species, 284
  host preference, *145*, 166
Sign stimuli, 161, 165, 169, 177
Silicium, 64, 86, 90, *87*, 156, 296
Sinalbin, 41
Single-cell-recording, 144
Sitosterol, 38
Size
  body size, 1, 8, 31, 86, 99, 121, 203
  and feeding strategy, 89
  food particles, 86
  fruit, 147, 213
  leaf, 18, 134, 300, 301
  meal, 169
  see also Plant
Soil factors, 62
Solar radiation, see Sun exposure
Somatic mutation, 72
Sorbitol, 32, 179, 372
Specialists, 8, 232, 255, 291, 299, 351
  definition, 6
Specialization
  and body size, 8
  and colonization of neophytes, 255, 344
  and insensitivity to toxicants, 104
  on plant parts, 11–13
Speciation, 293, 295
  allopatric, 199, 285
  insects, 284
  rate, 287
  reciprocal, 289
  sympatric, 285
Species rarity, 242–243
Species-area relationships, 248
Specific hunger, 212

Stenophagy, 7, 305
Steroids, 37–38
Sterols, 89
Stomata, *43*, 45, *49*
Structural formulas, 74, 371–372
Strychnine, 37
Stylet pathway, 88
Stylets, 87, *348*
Styropor, 378
Suboesophageal ganglion, 174, 186
Sugars, 44, 162, 165
  oviposition, 165
  phagostimulants, 163, *164*
  receptors, 178–179
Sulphur dioxide (SO_2), 112
Sun, 60
  effect on herbivory, 9, 62
  exposure, 9, 18
Symbionts, 108–109, 268, 287, 289
Synchronization of life cycle, 227
Synergism, 103, *104*, 146, 159, 182, *183*
Synomone, 262

Tannins, 8, 39–41, 53, *54*, *58*, 59, 61, 103, 230, *241*, 265, 296
  condensed, *34*, 41
  and food utilization, 103
  hydrolysable, 35, 41
  non-hydrolysable, 41
Target-site insensitivity, 104
Tarsal taste hairs, *173*
  neural responses, *168*, *185*
Taste, 138, 377
  hairs, see Maxillary taste hairs
  receptors, see Contact chemoreceptors
Taxis, 127
Temperature, *19*, 20
  at leaf surface, 18, 19
  and pollination, 326
Terpenoids, 37–38, 45, 49, 53, 66
Test biting, 124, 156, 165
Thermoreceptors, 141
Tobacco hornworm (*Manduca sexta*)
  host selection behaviour, *125*
  taste receptors, 175
Tocopherol, 38
Token stimuli, see Sign stimuli
Tolerance, 246, 299
  definition, 346
Tonoplast, 156
Toosendanin, 183, 357
Toughness, 58, 85, 95, 112
Toxicants and evolution, 290
Toxicity, see Secondary plant substances
Transduction, 141
Trap cropping, 353

Trichomes, 16, *17*, 44, *46*, *47*, 157, *158*, 262, 266
  glandular, 157
Triterpenoids, 38, 357
Tropane alkaloids, 36
Trophic levels, 262–267, 271
Tropotaxis, 127
Turnover, *see* Secondary plant substances

Ultimate factors, 4
Ultraviolet
  perception, 131, *133*, 330
  and secondary compounds, 299
Umbelliferone, 38
Utilization, 95–99
  efficiency, 215, 290

Vacuoles, 39, 42, 55, *56*
Vanillic acid, 38
Variability in host-plant preference, 195
Variation
  individual, 199–200
  interpopulational, 199
  interspecific, 216
  intraspecific, 198, 217
  seasonal, 200
Virulence of insect pathogens, 265
Virus, 265, 356
Vision
  colour, 130, 131, *135*
  silhouette, 133
  see also Leaf shape
Volatiles, 45–50, 66

Warburganal, 37, 356
Water, 92–93
  content of leaf, 58, 92
  receptor, 174
  requirement, 92, 96
  stress, 92
Waxes, 42, *43*, *46*, 159, *162*
Weed control, 210, 246, 256, 261, 359–362
Wind speed, 134
Windtunnel, 377
Wounding effects, 375

Xenobiotics, 109
Xylem, 11, 88, 89

Yellow attractivity, 135, 136, 147, *135*
Yew number of insects, 13
Yponomeuta
  EAG, *144*
  host switch, 180, 295, 306
  hybrids, 180, 216
  phyletic relation with hosts, 290, 292
  plant volatiles and mating, 232, *234*
  taste receptors, 179, 306

Zigzag flight, 136